Tsunamis in the Pacific Ocean: 2011-2012

Edited by
Alexander B. Rabinovich
Jose C. Borrero
Hermann M. Fritz

Previously published in *Pure and Applied Geophysics* (PAGEOPH), Volume 171, No. 12, 2014

Editors
Alexander B. Rabinovich
Russian Academy of Sciences
P.P. Shirshov Institute of Oceanology
Nakhimovsky Pr.36
117997 Moscow
Russia
and
Institute of Ocean Sciences
9860 West Saanich Road
Sidney, BC, V8L 4B2
Canada

Hermann M. Fritz
Georgia Institute of Technology
School of Civil & Environmental Engineering
790 Atlantic Drive
Atlanta, GA 30332
USA

Jose C. Borrero
University of Southern California
Tsunami Research Center
Los Angeles, CA 90089
USA
and
eCoast Marine Consulting and Research
P.O. Box 151
Raglan
New Zealand

ISBN 978-3-0348-0864-4   ISBN 978-3-0348-0865-1 (eBook)
DOI 10.1007/978-3-0348-0865-1

Library of Congress Control Number: 2014958671

Springer Basel Heidelberg New York Dordrecht London
© Springer Basel 2015

Cover illustration: The sightseeing catamaran Hamayuri (weight: 109 ton, length: 27 meters), overtopped the seawall and washed ashore by the March 11, 2011 Tohoku tsunami, rests atop a two-story hostel in Otsuchi, Iwate prefecture, Japan on April 12, 2011 (photo credit: Hermann M. Fritz, Georgia Institute of Technology).

Cover design: deblik, Berlin.

Printed on acid-free paper

Springer Basel AG is part of Springer Science+Business Media

www.springer.com

Contents

Pure Appl. Geophys. 171 (2014), 3175–3182
© 2014 Springer Basel
DOI 10.1007/s00024-014-0894-8

Introduction to "Tsunamis in the Pacific Ocean: 2011–2012"

ALEXANDER B. RABINOVICH,[1,2] JOSE C. BORRERO,[3,4] and HERMANN M. FRITZ[5]

Abstract—With this volume of the Pure and Applied Geophysics (PAGEOPH) topical issue "Tsunamis in the Pacific Ocean: 2011–2012", we are pleased to present 21 new papers discussing tsunami events occurring in this two-year span. Owing to the profound impact resulting from the unique crossover of a natural and nuclear disaster, research into the 11 March 2011 Tohoku, Japan earthquake and tsunami continues; here we present 12 papers related to this event. Three papers report on detailed field survey results and updated analyses of the wave dynamics based on these surveys. Two papers explore the effects of the Tohoku tsunami on the coast of Russia. Three papers discuss the tsunami source mechanism, and four papers deal with tsunami hydrodynamics in the far field or over the wider Pacific basin. In addition, a series of five papers presents studies of four new tsunami and earthquake events occurring over this time period. This includes tsunamis in El Salvador, the Philippines, Japan and the west coast of British Columbia, Canada. Finally, we present four new papers on tsunami science, including discussions on tsunami event duration, tsunami wave amplitude, tsunami energy and tsunami recurrence.

Key words: Tsunami, 2011 Tohoku earthquake, 2012 El Salvador, 2012 Philippines, 2012 Haida Gwaii, December 2012 Japan earthquakes and tsunamis, source parameters, tsunami field survey, PACIFIC Ocean, DART, tsunami records, seiches, tsunami numerical modelling, spectral analysis.

1. Introduction

The Tsunami Commission was established within the International Union of Geodesy and Geophysics (IUGG) following the 1960 Chile tsunami, generated by the largest (M_w 9.5) instrumentally recorded earthquake. The 1960 tsunami propagated throughout the entire Pacific Ocean, affecting areas located far from the source, and demonstrated the necessity of international cooperation. As organized, the Tsunami Commission is supported by two International Union of Geodesy and Geophysics (IUGG) associations: the International Association of Seismology and Physics of the Earth's Interior (IASPEI) and the International Association for the Physical Sciences of the Oceans (IAPSO). Since its foundation, the Tsunami Commission has held biannual International Tsunami Symposia and published special volumes of selected tsunami papers (SATAKE *et al.* 2007, 2011a, b; CUMMINS *et al.* 2008; 2009).

Two recent volumes (SATAKE *et al.* 2013a, b) were associated with the 25th International Tsunami Symposium, held in Melbourne, Australia from 1 to 4 July, 2011, just 4 months after the catastrophic Tohoku tsunami of 11 March 2011. This tsunami was generated by the largest instrumentally recorded earthquake (M_w 9.0) in the history of Japan, causing nearly 20,000 tsunami casualties and devastating tsunami damage on the Pacific coast of Honshu and Hokkaido islands. Numerous papers have already been published on the Tohoku earthquake and tsunami, including special issues in *Earth, Planets and Space* (Vol. 63, No.7, Editors: KANAMORI and YOMOGIDA 2011), *Geophysical Research Letters* (Vol. 39, No. 7, 2012), *Coastal Engineering Journal* (Vol. 54, No.1, Editor: SATO 2012), *Earthquake Spectra* (Vol. 29, No. S1, Editors: MORI and EISNER 2013), among others. Thus, it was natural that the entire Volume I of the 2013 *PAGEOPH* topical tsunami issue was on the 2011 Tohoku event (Vol. 170, Nos. 6–8, Editor: SATAKE *et al.* 2013a). However, results from new investigations into the 2011 Tohoku

[1] P.P. Shirshov Institute of Oceanology, Russian Academy of Sciences, 36 Nakhimovsky Pr., Moscow 117997, Russia. E-mail: A.B.Rabinovich@gmail.com

[2] Department of Fisheries and Oceans, Institute of Ocean Sciences, 9860 West Saanich Rd., Sidney, BC V8L 4B2, Canada.

[3] eCoast Ltd., Box 151, Raglan 3225, New Zealand. E-mail: jose@ecoast.co.nz

[4] Department of Civil and Environmental Engineering, University of Southern California, Los Angeles, CA, USA. E-mail: jborrero@usc.edu

[5] School of Civil and Environmental Engineering, Georgia Institute of Technology, Atlanta, GA 30332, USA. E-mail: fritz@gatech.edu

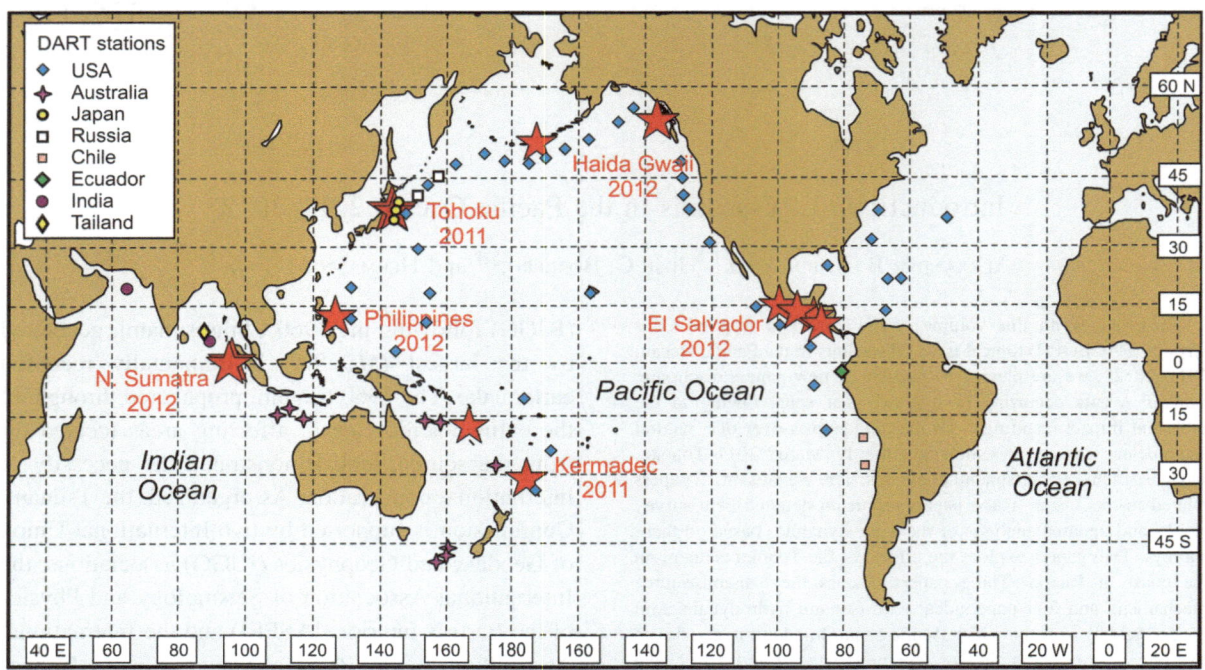

Figure 1
Red stars show the epicenter locations of tsunamigenic earthquakes occurring in 2011 and 2012; green diamonds show the locations of open
ocean DART tsunameters

followed, requiring additional publication outlets. Additionally, in the remainder of 2011 through 2012 several other tsunami events occurred (Fig. 1). While these tsunamis were not as strong or devastating as the 2011 Tohoku event, they were nevertheless scientifically interesting and societally important.

These days, we are living in a new, "instrumental" era of tsunami science with ever more detailed (and sometimes overwhelming!) amounts of data available after each event. The types of data include precise seismological data from broadband IRIS and open-ocean bottom seismographs, GPS measurements of ground deformations on land, hydroacoustic and satellite altimetry data, not to mention the results of tsunami field surveys and high-resolution coastal tide gauge records. Most importantly, however, deep-ocean "tsunameters" now routinely provide direct measurements of tsunami waves in the open ocean, significantly improving our understanding of tsunami physics, the quality of tsunami modelling and the reliability and accuracy of tsunami warnings (MOF-JELD 2009; TITOV 2009; RABINOVICH 2014). When a tsunami in the open ocean was first recorded (FILLOUX

1982; DYKHAN et al. 1983), it was a scientific sensation. Nowadays, DART[1] stations located throughout the Pacific, Indian and Atlantic Oceans (Fig. 1) record every tsunami no matter how small. The low level of background noise in the deep ocean and the high precision of DART instruments enable us to reliably identify and examine tsunamis with wave heights of only a few millimeters. Moreover, in recent years, Japan has developed a dense and effective network of cable tsunameters comprised of scores of stations (KAWAGUCHI et al. 2012; RABINOVICH 2014). These stations provide invaluable information about tsunamis generated by earthquakes in this region (e.g., MATSUMOTO and KANEDA 2013; SATAKE et al. 2013c). Similarly, in Canada, the University of Victoria, BC, established in 2009 the NEPTUNE-Canada[2] geophysical observatory on the southwestern shelf of Vancouver Island, which has already

---

[1] DART = Deep-ocean Assessment and Reporting of Tsunami.

[2] NEPTUNE-Canada = Canadian North-East Pacific Underwater Networked Experiments.

recorded more than ten tsunami events (THOMSON *et al.* 2011; RABINOVICH *et al.* 2013).

In general, we can say that the tools available for tsunami research have dramatically improved over the last 35 years. While the 1980s were rife with scientific ideas and theories about tsunami waves, there was very little empirical data; now we have an abundance of observational data, but sometimes not enough ideas to explain the new findings! In any case, the ongoing publication of high-quality scientific results is very important, and for that reason, the IUGG Tsunami Commission decided to prepare the additional, "inter-symposium" topical tsunami issue presented here. Altogether, 30 papers were submitted with 21 selected for publication.

2. The 2011 Tohoku Tsunami

This volume presents 12 new papers discussing the 2011 Tohoku tsunami: three papers deal with field data including previously unpublished data from surveys in the Fukushima exclusion zone, three papers investigate the seismic source, two papers deal with tsunami data recorded off the coast of Russia and four papers look at the tsunami waves recorded in the far-field (Marshall Islands, Hawaii, California and basin-wide).

Starting the group of field survey papers, TSUJI *et al.* (2014) compare the results from the Tohoku post-tsunami field surveys to runup and inundation data from past great earthquakes along the Honshu coast. They show that along the Sanriku coast, tsunami heights from the 2011 event were positively correlated with the previous Sanriku tsunamis, indicating that for near-field tsunamis, local variations in tsunami height resulting from the irregular coastline may be more dominant than the earthquake location, type, or magnitude. Along that coast, correlations to heights from the far-field Chilean tsunamis (1960, 2010) were less significant due to differences between the local and trans-Pacific tsunamis. However, along the Ibaraki and Chiba coasts, wave heights from the 2011 Tohoku and the Chilean tsunamis are positively correlated, showing a general decrease toward the south with small local variations such as large heights near peninsulas.

SHIMONZONO *et al.* (2014) next describe in detail the maximum runup height of nearly 40 m measured in a funneling coastal valley of the Aneyoshi district north of the entrance to Yamada Bay. Wave records from offshore GPS-buoys are introduced in a numerical model to analyze the measured localized runup amplification. The results indicated that a spectral component with a relatively short dominant period of 4–5 min in the leading wave plays a key role in the tsunami runup amplification by local bathymetry and topography. A final field survey paper by SATO *et al.* (2014) reports on detailed field data collected from inside the former 20-km exclusion zone around the Fukushima Dai-ichi nuclear power plant (NPP). The field surveys in the exclusion zone were delayed for more than a year due to elevated radiation levels caused by the multiple meltdowns at the NPP. Detailed distributions of the measured tsunami heights are presented in combination with observed infrastructure damage. Large tsunami heights exceeding 10 m were limited to areas within 500 m from the shoreline, while onshore profiles of the maximum inundation levels were dependent on inland topography. Tsunami flood levels in the coastal plains were affected by the extent of seawall damage, and remnant seawalls that survived the tsunami overflow provide valuable design lessons.

Next, we present two papers dealing with the seismological and wave generation aspects of the 2011 earthquake. First RODKIN and TIKHONOV (2014) use data from the Japan Meteorological Agency (JMA) to examine the recent historical seismicity of the Tohoku earthquake source region. They find that this earthquake occurred in a region where over the previous 7 years (2005–2011), there was a considerable decrease in the number of earthquakes and associated *b*-values. The authors then describe some short-term (20 days) precursor anomalies of the Tohoku earthquake, in particular, an increase in the number of earthquakes and a decrease in hypocentral depths in a 120 km radius around the epicenter of the 2011 mega event. The unexpectedly high tsunami waves associated with a rather small rupture area of the Tohoku earthquake demonstrate that similar events could occur in some other subduction zones where such cases had previously been considered impossible.

This is followed by a review paper by PARARAS-CARAYANNIS (2014), who revisits the earthquake and tsunami source mechanism of the great Tohoku-Oki earthquake of 11 March 2011. An examination of the source mechanism determined that the great tsunami resulted from a combination of crustal deformations of the ocean floor due to up-thrust tectonic motions augmented by additional uplift due to the quake's slow and long rupturing process, as well as to large coseismic lateral movements that compressed and deformed the compacted sediments along the accretionary prism of the overriding plate. The author suggests that as with the 1896 Sanriku tsunami, additional ocean floor deformation and uplift of the sediments was responsible for the higher waves generated by the 2011 earthquake.

One of the most critical issues relevant to tsunami forecasting, and tsunami modeling in general, is the characterization of the tsunami source. A tsunami source can be modeled based on data from seismometers, GPS measurements of crustal deformation, deep-ocean tsunameters, and other advanced instruments that provide near real-time information. Using data from the 2011 Tohoku earthquake, WEI et al. (2014) applied several different source models, based on different data sources, to simulate the generation, propagation and inundation of tsunami waves in the near field. They compare their results to measured data, including inundation extents and water level records from nearby tsunameters. Their results highlight the critical role these deep-ocean tsunami measurements play in the rapid determination of an approximate tsunami source suitable for issuing reliable and accurate forecasts. They also show that results from tsunameter-derived source models are compatible with results based on other real-time geodetic data, and may also provide a more nuanced understanding of tsunami generation from both tectonic and, non-seismic processes such as submarine landslides.

Our last cluster of papers on the 2011 Tohoku tsunami highlights the findings of researchers who studied tsunami records from various locations around the Pacific. Two papers from the Russian coast give a regional perspective, while four other papers examine tsunami waves at far-field locations or over the Pacific basin as a whole. This group of

papers starts with the work of RAZJIGAEVA et al. (2014), who examined tsunami sedimentation in the bays of the southernmost group of the Kuril Islands (the Lesser Kuril Islands). They report on the unusual situation where the sea during the tsunami was covered with ice, and describe changes in the erosional capacity of the tsunami due to ice fragments carried along with the waves. Their paper is a continuation of their previous work (RAZJIGAEVA et al. 2013), which was based on the direct post-event studies. Their newer studies showed well-preserved tsunami deposits, evident one-and-a-half years after the event. A comparative analysis between tsunami deposits in the Kuril Islands relative to near-field (Honshu Island) sites indicated that differences in runup heights contributed to considerable differences in erosion, sedimentation, distribution of tsunami deposits, the formation of sedimentary structures, grain-size composition, and diatom and foraminifera assemblages. On the coasts of the Lesser Kuril Islands, the mud layers in sections of coastal lowlands in closed bays contained the grain-size composition of the 2011 mud, and preserved detailed geological records of paleotsunamis.

This is followed by the contribution of SHEVCHENKO et al. (2014), who examine deep-water and coastal observations and data of the 2011 tsunami along the Far East coast of Russia. Key findings from their study show that tsunami waves reached up to 2.5 m along the Kuril Islands and that tsunami waves affecting the region northeast of the source exhibited much longer periods than waves propagating directly to the east. The authors estimated the major characteristics of the recorded waves and compared the observed tsunami series to the results from numerical models. In contrast to the open-ocean stations where the first waves were the highest, at Russian coastal sites, the highest waves occurred several hours after the arrival of the first tsunami wave.

HEIDARZADEH and SATAKE (2014a) present evidence that extremely long modes of the Pacific Ocean were excited by the 11 March 2011 Tohoku tsunami. A numerical approach was employed to calculate the basin-wide modes of the Pacific Ocean, resulting in 49 modes in the range of 2–48 h. Spectral analysis of tide-gauge records showed that some of the calculated basin-wide modes were indeed excited by the

Tohoku tsunami. The tide gauge signals were classified into three groups: (1) basin-wide modes (> 1.5 h), (2) the tsunami source periods (20–90 min), and (3) local bathymetric effects (<20 min). The average contributions to the total tsunami energy were 6.4 % for the basin-wide mode, 64.1 % for the tsunami source, and 29.5 % for the local bathymetry. Simulations suggest that the amount of contribution of basin effects to the total tsunami energy depends on the location of the tsunami source.

Next, FORD et al. (2014) report on an interesting mid-basin data set of tsunami waves from the 2011 Tohoku event. In their study, tide gauge data and numerical simulations supplement a serendipitous recording of the tsunami on an array of bottom-mounted pressure sensors at Majuro and Kwajalein Atolls in the Republic of the Marshall Islands. They show that tsunami oscillations in the lagoon were substantially more energetic and longer lasting than observed on the reefs or modeled in the deep ocean, and that the tsunami excited the normal modes of the atoll lagoons. They also showed that the propagation of the tsunami across the reef flats is tidally dependent, with amplitudes increasing/decreasing shoreward at high/low tide. Most importantly, from a hazard management viewpoint, they showed that while the peak wave heights in the Marshall Islands coincided with low tide, the observed amplitudes could have caused inundation had they coincided with a higher tide stage.

In the far field, ZHOU et al. (2014) and ADMIRE et al. (2014) investigate both tsunami wave heights and currents. Firstly, ZHOU et al. (2014), examine the effects of frequency dispersion on tsunami waves and currents from the 2011 Tohoku event. In a series of numerical experiments, they compare the trans-oceanic propagation and tsunami dynamics in the vicinity of the Hawaiian Islands using both a weakly dispersive Boussinesq model and a shallow-water model that neglects dispersion effects. The model results indicate that dispersion effects generally result in reduced amplitudes of the leading tsunami waves. They also show that a model neglecting dispersion effects could underestimate wave heights and current speeds of the trailing waves developing in coastal waters.

This is followed by ADMIRE et al. (2014), who present some relatively rare tsunami current speed data from northern California, including instrumental data from the 2010 Chile and 2011 Tohoku tsunamis in Humboldt Bay and current speeds derived from video camera footage at the entrance to Crescent City Harbor during the 2011 Tohoku tsunami. During the 2011 event, the tsunami signal was evident for more than 40 h in Humboldt Bay, with a peak current speed of 0.8 m/s occurring approximately 1 h after arrival. At Crescent City, within the first 3 h of tsunami activity, peak surface currents were estimated to have exceeded 4.5 m/s on one wave cycle and 3 m/s on six others, and were the cause of most of the damage experienced there. Their study also presents numerical model results in an effort to calibrate to measured data.

3. New Tsunami Events

Four new tsunami events occurred in 2012, and in this volume we present papers describing aspects of each. In August 2012, tsunamis occurred in El Salvador and Nicaragua on August 27 and in the Philippines on August 31. BORRERO et al. (2014) conducted a post-tsunami reconnaissance survey of the El Salvador event and measured tsunami heights of up to 6 m with a maximum inundation distance of over 300 m. This is an important finding in that the causative earthquake for this event was relatively small, with a moment magnitude of only 7.3. Seismological analyses of the earthquake showed that it was a 'slow' or 'tsunami' earthquake (KANAMORI 1972), making it the second such event in this region in 20 years—an important consideration when assessing tsunami hazards for this region. HEIDARZADEH and SATAKE (2014b) also investigated aspects of the El Salvador tsunami as well as the August 31 Philippines tsunami. In their study, they focused on the far-field signature of these events as recorded on coastal tide gauges and deep-water tsunameters. They performed a detailed spectral analysis (using Fourier and wavelet techniques), and showed that the spectral content of the tsunami signal is related to the location of the recording station relative to the orientation of the tsunami source.

On 27 October 2012, a relatively large (M_w 7.7) earthquake occurred off the west coast of Haida Gwaii in the Canadian province of British Columbia. CAS-SIDY *et al.* (2014) provide an overview of the seismo-tectonics of the fault region and put this event in context. They explain that although the earthquake source region is predominantly a strike-slip transform fault boundary, there is a component of oblique convergence between the Pacific and North America plates off Haida Gwaii. The October earthquake, the second largest instrumentally recorded earthquake in that region, had a primarily thrust mechanism and was responsible for generating a tsunami that caused significant local runup. LEONARD and BEDNARSKI (2014) specifically examined the near-field effects of the 2012 Haida Gwaii tsunami. Despite a lack of evidence suggesting damaging waves along the coast of British Columbia (largest amplitudes of ∼0.5–1 m were measured in Hawaii), field surveys conducted in the weeks and months following the earthquake revealed that much of the unpopulated and un-instrumented coastline of western Haida Gwaii was impacted by significant tsunami waves reaching a maximum height of 13 m with runup exceeding 3 m along 200 km of the coastline. The greatest impacts were evident at the heads of narrow inlets and bays on western Moresby Island, where natural and anthropogenic debris was found on the forest floor and caught in tree branches, suggesting flow depths up to 2.5 m. The authors indicate that lessons learned from their study of the impacts of the Haida Gwaii tsunami may prove useful to future post-tsunami and paleotsunami surveys.

Finally, BERNARD *et al.* (2014), report on a small tsunami produced by a M_W 7.3 earthquake offshore of Japan, adjacent to the source region for the 2011 Tohoku event. They present the deep-water tsunameter data from this event, recorded on instruments that were deployed just 2 weeks before the causative earthquake. They show that the data recorded by these two instruments helped to improve the speed and accuracy of tsunami forecasts for the coast of Japan and the Pacific Ocean in general.

4. Tsunami Science

Four papers in this issue relate to general topics in tsunami science, warning and hazard mitigation, and use examples from recent events to provide new insights or answers. In particular, KIM and WHITMORE (2014) examined the relationship between tsunami maximum amplitude and signal duration based on 89 historical data sets from 13 tsunamis occurring between 1952 and 2011. The problem of event duration is quite important for effective tsunami warnings, hazard assessment and emergency response. They evaluated the tsunami sea level time series and used a linear least squares fit to get a quantitative estimate of amplitude decay function for their study region. The confidence interval was found to be roughly 20 h over the range of maximum tsunami amplitudes; this relatively large interval likely resulted from local resonance, late-arriving reflections, and other effects.

NYLAND and HUANG (2014) discussed specific problems related to tsunami warnings and advisories based on the West Coast and Alaska Tsunami Warning Center (WCATWC) experience during the 2011 Tohoku tsunami event. Specifically, they looked at the problem of how long a tsunami warning or advisory should be in place. To address this issue, the WCATWC developed a technique to estimate the amplitude decay of tsunami waves recorded at tide stations within the WCATWC Area of Responsibility (AOR). Based on an analysis of the real-time tide gauge records, they estimated exponential decay curves, which were then combined with an average West Coast decay function to provide an initial tsunami amplitude-duration forecast.

NOSOV *et al.* (2014) used detailed slip distribution data from recent earthquakes and the OKADA (1985) formulae to calculate the vector fields of coseismic sea-floor deformations, the displaced water volume and the gravitational potential energy of the tsunami source. They suggest that in many cases, the horizontal components of the bottom deformation provide an additional contribution to the displaced water volume without reducing the contribution of the vertical component. This increase varies from 0.07 to 55 % with an average increase of 14 %. The authors go on to examine measures of a tsunami's initial energy as functions of the moment magnitude, and show that a tsunami captures from 0.001 to 0.34 % of the earthquake energy, and on average 0.04 %.

And finally, closing this volume, KAISTRENKO (2014) describes a Poissonian probability model for tsunami runup heights, with emphasis on a tsunami recurrence function. He shows that a general tsunami recurrence function contains at least two important scaling parameters: (1) the asymptotic frequency of big tsunamis f at a study site, and (2) a characteristic tsunami height H^* for that site. These parameters and their variations are statistically evaluated using observational data from tsunami catalogues. The author also considers theoretical and applied problems related to tsunami recurrence, shows an example of a two-parameter tsunami hazard map, and discusses issues related to probabilistic tsunami hazard estimation.

Acknowledgments

We, the guest editors, would like to thank Dr. Renata Dmovska, Editor-in-Chief of PAGEOPH for arranging and supporting the continued production of these topical volumes on tsunamis. We feel very strongly that they provide an important outlet for many high-quality research articles related to all aspects of tsunami science. We would also like to thank Ms. Priyanka Ganesh in the Springer Journals Editorial Office for her timely editorial assistance. Finally, we thank each of the authors for their excellent contributions and hard work in making this volume a success.

REFERENCES

ADMIRE, A., DENGLER, L., CRAWFORD, G., USLU, B., BORRERO, J., GREER, D., MONTOYA, J., and WILSON, R. (2014), *Observed and modeled currents from the Tohoku-oki, Japan and other recent tsunamis in northern California*, Pure Appl. Geophys., 171 (this issue); doi:10.1007/s00024-014-0797-8.

BERNARD, E., TANG, L., WEI, Y., and TITOV, V. (2014), *Impact of near-field, deep-ocean tsunami observations on forecasting the December 7, 2012 Japanese tsunami*, Pure Appl. Geophys., 171 (this issue); doi:10.1007/s00024-013-0720-8.

BORRERO, J., KALLIGERIS, N., LYNETT, P.J., FRITZ, H.M., NEWMAN, A.V., and CONVERS, J.A. (2014), *Observations and modeling of the August 27, 2012 earthquake and tsunami affecting El Salvador and Nicaragua*, Pure Appl. Geophys. (this issue); doi:10.1007/s00024-014-0782-2.

CASSIDY, J.F., ROGERS, G.C., and HYNDMAN, R.D. (2014), *An overview of the October 28, 2012 Mw 7.7 earthquake in Haida Gwaii, Canada: A tsunamigenic thrust event along a predominantly strike-slip margin*, Pure Appl. Geophys. (this issue); doi:10.1007/s00024-014-0775-1.

CUMMINS, P.R., KONG, L.S.L., and SATAKE, K. (2008), *Tsunami Science Four Years after the 2004 Indian Ocean Tsunami. Part I: Modelling and Hazard Assessment*, Pure Appl. Geophys. 165, Topical Issue.

CUMMINS, P.R., KONG, L.S.L., and SATAKE, K. (2009), *Tsunami Science Four Years after the 2004 Indian Ocean Tsunami. Part II: Observation and Data Analysis*, Pure Appl. Geophys. 166, Topical Issue.

DYKHAN, B.D., JAQUE, V.M., KULIKOV, E.A. et al. (1983), *Registration of tsunamis in the open ocean*, Marine Geodesy, 6, 303–310.

FILLOUX, J.H. (1982), *Tsunami recorded on the open ocean floor*, Geophys. Res. Lett., 9 (1), 25–28.

FORD, M., BECKER, J.M., MERRIFIELD, M.A. and SONG, Y.T. (2014), *Marshall Islands fringing reef and atoll lagoon observations of the Tohoku tsunami*, Pure Appl. Geophys. (this issue); doi:10.1007/s00024-013-0757-8.

HEIDARZADEH, M., and SATAKE, K. (2014a), *Excitation of basin-wide modes of the Pacific Ocean following the 11 March 2011 Tohoku tsunami*, Pure Appl. Geophys. (this issue); doi:10.1007/s00024-013-0731-5.

HEIDARZADEH, M., and SATAKE, K. (2014b), *The El Salvador and Philippines tsunamis of August 2012: insights from sea level data analysis and numerical modeling*, Pure Appl. Geophys. (this issue); doi:10.1007/s00024-014-0790-2.

KAISTRENKO, V. (2014), *Tsunami recurrence function: structure, methods of creation, and application for tsunami hazard estimates*, Pure Appl. Geophys. (this issue); doi:10.1007/s00024-014-0791-1.

KANAMORI, H. (1972), *Mechanism of tsunami earthquakes*, Phys. Earth Planet. Inter., 6, 346–359.

KANAMORI, H. and YOMOGIDA, K. (2011), *First results of the 2011 Off the Pacific Coast of Tohoku earthquake*, Earth, Planets Space, 63, 511–902.

KAWAGUCHI, K., KANEDA, Y., ARAKI, E. et al. (2012), *Reinforcement of seafloor surveillance infrastructure for earthquake and tsunami monitoring in western Japan*, Proceed. Oceans2012 Mts/Ieee Yeosu, May 21–24, 2012. Yeosu, Republic of Korea.

KIM, Y.T. and WHITMORE P. (2014), *Relationship between maximum tsunami amplitude and duration of signal*, Pure Appl. Geophys. (this issue); doi:10.1007/s00024-013-0674-x.

LEONARD, L. and BEDNARSKI, J. (2014), *Field survey following the 28 October 2012 Haida Gwaii tsunami*, Pure Appl. Geophys. (this issue); doi:10.1007/s00024-014-0792-0.

MATSUMOTO, H., and KANEDA, Y. (2013), *Some features of bottom pressure records at the 2011 Tohoku earthquake—Interpretation of the far-field DONET data*. Proceed. 11th SEGJ Intern. Symp.

MOFJELD, H.O. (2009), *Tsunami measurements*. In: *The Sea, Vol. 15, Tsunamis*, A.Robinson, E.Bernard (Eds.). USA, Harvard University Press, Cambridge, p.201–235.

MORI, J. and EISNER, R. (2013), *2011 Tohoku-Oki Earthquake and Tsunamis*, Earthquake Spectra, 29(S1), vii-S499, Special issue.

NOSOV, M., BOLSHAKOVA, A., and KOLESOV, S. (2014), *Displaced water volume, potential energy of initial elevation and tsunami intensity: analysis of recent tsunami events*, Pure Appl. Geophys. (this issue); doi:10.1007/s00024-013-0730-6.

NYLAND, D. and HUANG, P. (2014), *Forecasting wave amplitudes after the arrival of a tsunami*, Pure Appl. Geophys. (this issue); doi:10.1007/s00024-013-0703-9.

OKADA (1985), *Surface deformation due to shear and tensile faults in a half-space*. Bull. Seism. Soc. Amer., 75(4), 1135–1154.

PARARAS-CARAYANNIS, G. (2014), *The Great Tohoku-Oki earthquake and tsunami of March 11, 2011 in Japan: A critical review and evaluation of the tsunami source mechanism*, Pure Appl. Geophys. (this issue); doi:10.1007/s00024-013-0677-7.

RABINOVICH, A.B. (2014), *Tsunami observations in the open ocean*, Atmosph. Oceanic Physics, *54* (5) (in press).

RABINOVICH, A.B., THOMSON, R.E., and FINE, I.V. (2013), *The 2010 Chilean tsunami off the west coast of Canada and the northwest coast of the United States*. Pure Appl. Geophys. *170*, 1529–1565; doi:10.1007/s00024-012-0541-1.

RAZJIGAEVA, N.G., GANZEY, L.A., GREBENNIKOVA, T.A., IVANOVA, E.D., KHARLAMOV A.A., KAISTRENKO, V.M., and SHISHKIN, A.A. (2013), *Coastal sedimentation associated with the Tohoku tsunami of 11 March 2011 in South Kuril Islands, NW Pacific Ocean,* Pure and Applied Geophysics, *170*, 1081–1102.

RAZJIGAEVA, N.G., GANZEY, L.A., GREBENNIKOVA, T.A., IVANOVA, E.D., KHARLAMOV, A.A., KAISTRENKO, V.M., ARSLANOV, KH.A., and CHERNOV, S.B. (2014), *The Tohoku tsunami of 11 March 2011: The key event to understand tsunami sedimentation on the coasts of closed bays of the Lesser Kuril Islands*, Pure Appl. Geophys. (this issue); doi:10.1007/s00024-014-0794-y.

RODKIN, M.V., and TIKHONOV, I.N. (2014), *Seismic regime in the vicinity of the 2011 Tohoku mega earthquake (Japan, Mw = 9)*, Pure Appl. Geophys. (this issue); doi:10.1007/s00024-013-0768-5.

SATAKE, K., OKAL, E.A. and BORRERO, J.C. (2007), *Tsunami and its Hazards in the Indian and Pacific Oceans*, Pure Appl. Geophys., *164*, Topical Issue.

SATAKE, K., RABINOVICH, A.B., KÂNOĞLU, U. and TINTI, S. (2011a), *Tsunamis in the World Ocean: Past, Present, and Future. Volume I*, Pure Appl. Geophys., *168* (6-7), Topical Issue.

SATAKE, K., RABINOVICH, A.B., KÂNOĞLU, U. and TINTI, S. (2011b), *Tsunamis in the World Ocean: Past, Present, and Future. Volume II*, Pure Appl. Geophys., *168* (11), Topical Issue.

SATAKE, K., RABINOVICH, A.B., DOMINEY-HOWES, D., and BORRERO, J.C. (2013a), *Historical and Recent Catastrophic Tsunamis in the World: Volume 1. The 2011 Tohoku tsunami*, Pure Appl. Geophys., *170* (6–8), Topical Issue.

SATAKE, K., RABINOVICH, A.B., DOMINEY-HOWES, D., and BORRERO, J.C. (2013b), *Historical and Recent Catastrophic Tsunamis in the World: Volume 2. Tsunamis from 1755 to 2010*, Pure Appl. Geophys., *170* (9–10), Topical Issue.

SATAKE, K., FUJII, Y., HARADA, T., and NAMEGAYA, Y. (2013c), *Time and space distribution of coseismic slip of the 2011 Tohoku earthquake as inferred from tsunami waveform data*, Bull. Seism. Soc. Amer., *103*, 1473–1492.

SATO, S. (2012), *Special Anniversary Issue on the 2011 Tohoku earthquake tsunami*. Coastal Engineering Journal, 54 (1).

SATO, S., OKAYASU, A., YEH, H., FRITZ, H.M., TAJIMA, Y., and SHIMOZONO, T. (2014), *Delayed survey of the 2011 Tohoku Tsunami in the former exclusion zone in Minami-Soma, Fukushima Prefecture*, Pure Appl. Geophys. (this issue); doi:10.1007/s00024-014-0809-8.

SHEVCHENKO, G., IVELSKAYA, T., and LOSKUTOV, A. (2014), *Characteristics of the 2011 Great Tohoku tsunami on the Russian Far East coast: Deep-water and coastal observations*, Pure Appl. Geophys. (this issue); doi:10.1007/s00024-014-0727-1.

SHIMOZONO, T., CUI, H., PIETRZAK, J.D., FRITZ, H.M., OKAYASU, A., and HOOPER. A.J. (2014), *Short wave amplification and extreme runup by the 2011 Tohoku tsunami*, Pure Appl. Geophys. (this issue); doi:10.1007/s00024-014-0803-1.

THOMSON, R., FINE, I., RABINOVICH, A., MIHÁLY, S., DAVIS, E., HEESEMANN, M., and KRASSOVSKI, M. (2011), *Observation of the 2009 Samoa tsunami by the NEPTUNE-Canada cabled observatory: Test data for an operational regional tsunami forecast model*, Geophys. Res. Lett., *38*, L11701, doi:10.1029/2011 GL046728.

TITOV, V.V. (2009), *Tsunami forecasting*. In: *The Sea, Vol.15, Tsunamis*, Eds.: A. Robinson, E. Bernard, Harvard Univ. Press, Cambridge, p. 371–400.

TSUJI, Y., SATAKE, K., ISHIBE, T., HARADA, T., NISHIYAMA, A., and KUSUMOTO, S. (2014), *Tsunami heights along the Pacific coast of northern Honshu from the 2011 Tohoku and past great earthquakes*, Pure Appl. Geophys. (this issue); doi:10.1007/s00024-014-0779-x.

WEI, Y., TITOV, V.V., NEWMAN, A.V., HAYES, G.P., and TANG, L. (2014), *Seismic regime in the vicinity of the 2011 Tohoku mega earthquake (Japan, Mw = 9)*, Pure Appl. Geophys. (this issue); doi:10.1007/s00024-014-0xxxxx.

ZHOU, H., WEI, Y., WRIGHT, L., and TITOV, V.V. (2014), *Waves and currents in Hawaii waters induced by the dispersive 2011 Tohoku tsunami*, Pure Appl. Geophys. (this issue); doi:10.1007/s00024-014-0781-3.

(Received June 26, 2014, accepted June 27, 2014)

Pure Appl. Geophys. 171 (2014), 3183–3215
© 2014 The Author(s)
This article is published with open access at Springerlink.com
DOI 10.1007/s00024-014-0779-x

Pure and Applied Geophysics

Tsunami Heights along the Pacific Coast of Northern Honshu Recorded from the 2011 Tohoku and Previous Great Earthquakes

Yoshinobu Tsuji,[1] Kenji Satake,[1] Takeo Ishibe,[1] Tomoya Harada,[1,2] Akihito Nishiyama,[1] and Satoshi Kusumoto[1]

Abstract—The 2011 Tohoku earthquake generated a huge, destructive tsunami with coastal heights of up to 40 m recorded along northern Honshu. The Sanriku coast experienced similar tsunamis and damage from the 1896 and 1933 Sanriku earthquakes, whereas the only damaging tsunamis on both the Ibaraki and Chiba coasts in the previous century were from the 1960 and 2010 Chile earthquakes. We summarized 12 field surveys in which the height of the 2011 tsunami was recorded at 296 points, along with descriptions of the survey method, reliability, and accuracy. We then compared them with the above-mentioned tsunamis at locations for which specific measurements were given in previous reports. On the Sanriku coast, the 2011 tsunami heights are positively correlated with the previous Sanriku tsunamis, indicating that local variations resulting from the irregular coastline were more dominant factors than the earthquake location, type, or magnitude for near-field tsunamis. The correlations with the Chilean tsunami heights are less significant due to the differences between the local and trans-Pacific tsunamis. On the Ibaraki and Chiba coasts, the 2011 Tohoku and the two Chilean tsunami heights are positively correlated, showing the general decrease toward the south with small local variations such as large heights near peninsulas.

Key words: 2011 Tohoku tsunami, 1896 Sanriku tsunami, 1933 Sanriku tsunami, 1960 Chilean tsunami, 2010 Chilean tsunami, Sanriku coast.

1. Introduction

On March 11, 2011, a giant earthquake, officially named the "2011 off the Pacific coast of Tohoku earthquake" by the Japan Meteorological Agency (JMA), occurred along the Japan Trench where the Pacific plate subducts beneath the Okhotsk plate (Fig. 1). This earthquake, which we refer to as the 2011 Tohoku earthquake in this paper, was the largest earthquake in Japan since the beginning of modern instrumental observations. It had a moment magnitude, M_w, of 9.0. It caused 15,883 fatalities and 2,656 people were reported missing (National Police Agency as of August 9, 2013); more than 90 % of the casualties were caused by the tsunami. The maximum tsunami height was nearly 40 m based on ∼5,900 measurements compiled by the 2011 Tohoku Earthquake Tsunami Joint Survey Group (MORI *et al.* 2011, 2012). Such a gigantic earthquake (∼M9) was unexpected in Japan, but was the huge tsunami also a surprise?

The Pacific coast of the Tohoku region has suffered from many large tsunamis generated by both near-field and far-field earthquakes. The 1896 Sanriku tsunami caused ∼ 22,000 casualties, which is more than the 2011 Tohoku earthquake. The 1896 Sanriku earthquake was a "tsunami earthquake" (KANAMORI 1972), that is, one that produces a tsunami that is much larger than that expected from the earthquake magnitude (surface wave magnitude M_s 7.2; ABE 1994). The 1933 Sanriku earthquake (M_s 8.5) also generated significant tsunami damage with ∼3,000 fatalities. The 1960 Chile earthquake was the largest earthquake (M_w 9.5) of the last century and a transoceanic tsunami struck the Pacific coasts of Japan ∼23 h after the earthquake, causing 142 fatalities. The 2010 Chile earthquake (M_w 8.8) also generated a trans-Pacific tsunami, which caused property damage to aquaculture facilities such as fishery rafts in Japan.

[1] Earthquake Research Institute, The University of Tokyo, 1-1-1 Yayoi, Bunkyo-ku, Tokyo 113-0032, Japan. E-mail: satake@eri.u-tokyo.ac.jp
[2] Center for Integrated Disaster Information Research (CIDIR), Interfaculty Initiative in Information Studies, The University of Tokyo, 7-3-1 Hongo, Bunkyo-ku, Tokyo 113-0033, Japan.

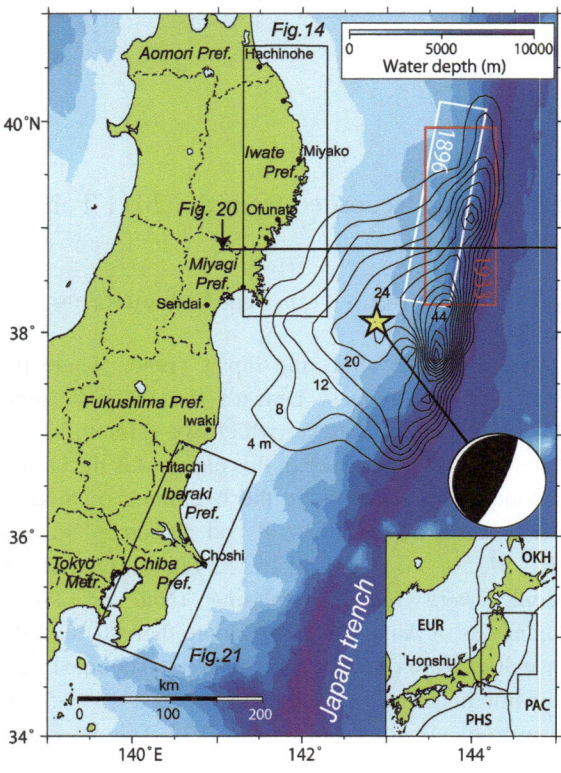

Map of northern Honshu where the 2011 Tohoku earthquake caused tsunami damage. Slip distribution of the 2011 Tohoku earthquake estimated from tsunami waveform inversion (SATAKE *et al.* 2013) is shown by contours with 4-m intervals. The *yellow star* indicates the epicenter of the mainshock, as determined by the JMA. The focal mechanism is provided by the Global Centroid Moment Tensor project. The *black rectangles* indicate regions shown in Figs. 14 and 21. The *white* and *red rectangles* indicate fault models of the 1896 (TANIOKA and SENO 2001) and 1933 Sanriku (AIDA 1977) earthquakes, respectively. *Dashed lines* indicate prefectural boundaries. The location of the profile in Fig. 20 is also shown. In the *inset*, OKH, EUR, PAC, and PHS indicate the Okhotsk, Eurasia, Pacific, and Philippine Sea Plates, respectively

This paper first summarizes our field surveys (TSUJI *et al.* 2011) in which the tsunami heights of the 2011 Tohoku earthquake were recorded, along with descriptions of the survey method, reliability, and accuracy. Then, we compare the tsunami heights with those from previous earthquakes both near Japan and across the Pacific Ocean. For the Sanriku coast, we compare the 2011 tsunami heights and inundation areas with two local tsunamis (i.e., the 1896 and 1933 Sanriku tsunamis) and two trans-Pacific tsunamis (i.e., 1960 and 2010 Chilean tsunamis) at selected

sites where direct comparisons can be made. For the Pacific coasts of Ibaraki and Chiba prefectures, we similarly compare the 2011 tsunami heights with the 1960 and 2010 Chilean tsunami heights because no damaging tsunamis were recorded from near-field earthquakes in the last century. On the central Sanriku coast, the sawtooth coastal topography is a major factor that strongly affects tsunami heights more so than the earthquake location, type, or magnitude. On the other areas of the Sanriku coast, and the Ibaraki and Chiba coasts, local variations are smaller and the general pattern of tsunami heights reflects the location, slip distribution, and magnitude of the parent earthquake.

2. Field Surveys of Tsunami Heights from the 2011 Tohoku Earthquake

We conducted 12 field surveys between March 16 and October 24, 2011. Locations and tsunami heights above sea level were generally measured with handheld GPS receivers and auto-levels, laser rangefinders, or total stations. In the surveys, we first sought traces that indicate the tsunami heights, which were classified as inundation heights, runup heights, and tsunami heights in ports. For tsunami inundation, we measured flow depths above ground based on watermarks or other physical evidence. The highest inundation on a slope where the flow velocity is considered to have been zero is classified as the runup height. Runup heights were measured from debris carried by the tsunami, the absence of leaves, or similar evidence on a slope. However, as most physical evidence had disappeared by June 2011, later surveys mainly relied on eyewitness accounts. In ports where the tsunami did not inundate above the wharfs, we measured tsunami heights based on eyewitness accounts and classified them as tsunami heights in ports, as proposed by TSUJI *et al.* (2010).

The 2011 tsunami heights in this paper are heights that were above sea level at the time of the maximum tsunami arrival. The measured heights were corrected for differences in tide levels between the measurement time and the arrival time of the maximum tsunami. The arrival times of the maximum tsunami at Hachinohe, Miyako, Kamaishi, Ofunato, Ayukawa,

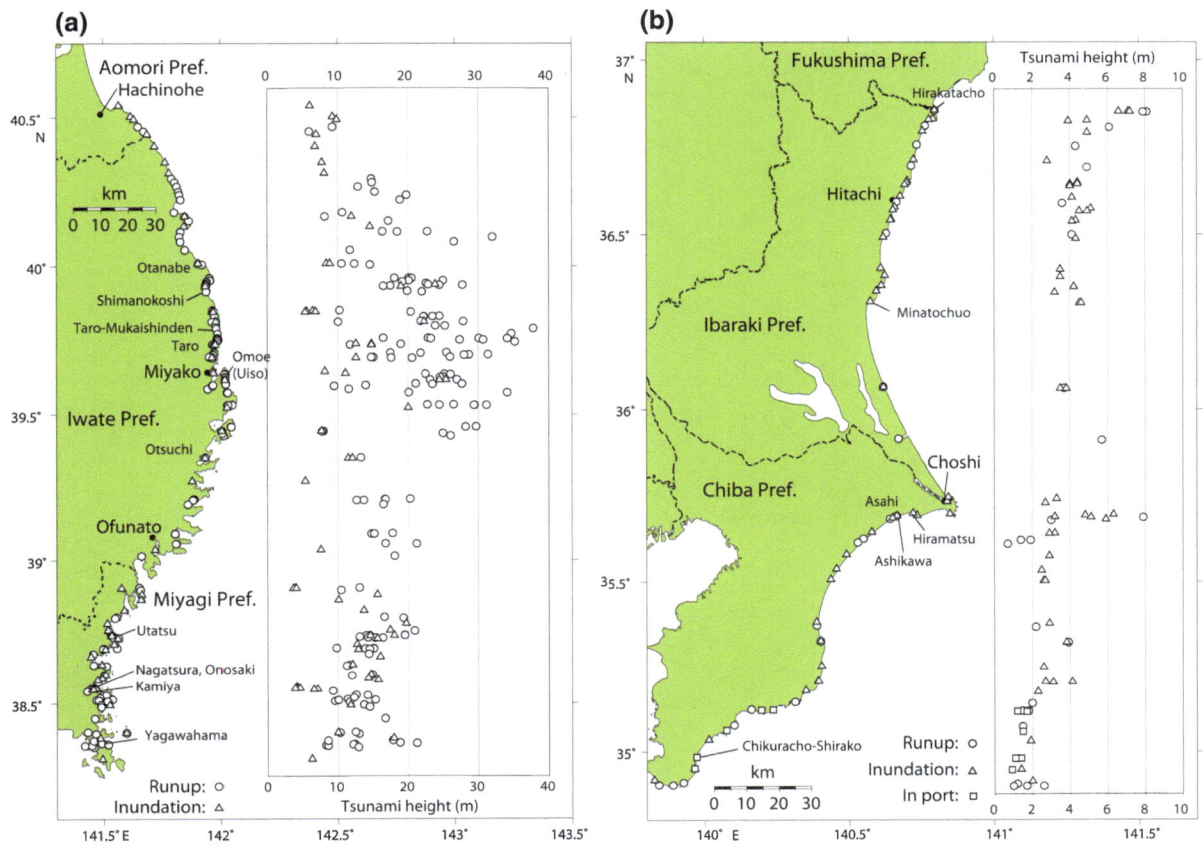

Figure 2

a Distribution of tsunami heights in Aomori, Iwate, and Miyagi prefectures (Tsuji *et al.* 2011). *Circles* and *triangles* indicate runup and inundation heights, respectively. The *dashed lines* indicate prefectural boundaries. **b** Distribution of tsunami heights in Ibaraki and Chiba prefectures. *Circles*, *triangles*, and *squares* indicate runup heights, inundation heights, and tsunami heights in ports, respectively

Onahama, Oarai, Choshi Fishing Port, and Mera (JMA 2011) were used for tide correction. At some locations where the measurement points were far from the sea, we measured altitudes above Tokyo Peil (TP), the leveling datum of Japan. In such cases, we considered land subsidence due to the mainshock, which was as large as 1 m, as recorded by continuous GPS data (OZAWA *et al.* 2011). The altitudes were converted to heights above mean sea level, then to heights above tide level at the time of the maximum tsunami arrival.

The reliability of the evidence was categorized into three classes (A, B, and C), as some tsunami heights were obtained from clear watermarks, whereas others were based on less objective eyewitness accounts (e.g., SHUTO and UNOHANA 1984). Class A indicates the most reliable data, which are based on clear physical evidence or objective eyewitness accounts. Class B indicates evidence based mostly on natural traces such as leaves, grass roots, or debris; while class C indicates the least reliable data that are based on equivocal evidence such as fishing floats in trees or broken windowpanes. Other catalogs such as the NOAA/WDC Tsunami runup Database introduce another type of validity (doubtful), without classifying the reliability.

The measurement accuracy was also categorized into three classes (a, b, and c). Class a means measurement errors are considered to be <0.2 m. The error for class b ranges 0.2–0.5 m because repeated measurements were performed using an auto-level or the sea was rough at the time of measurement. Class c means the errors are > 0.5 m because the sea level could not be directly measured or laser measurements were performed without a reflector.

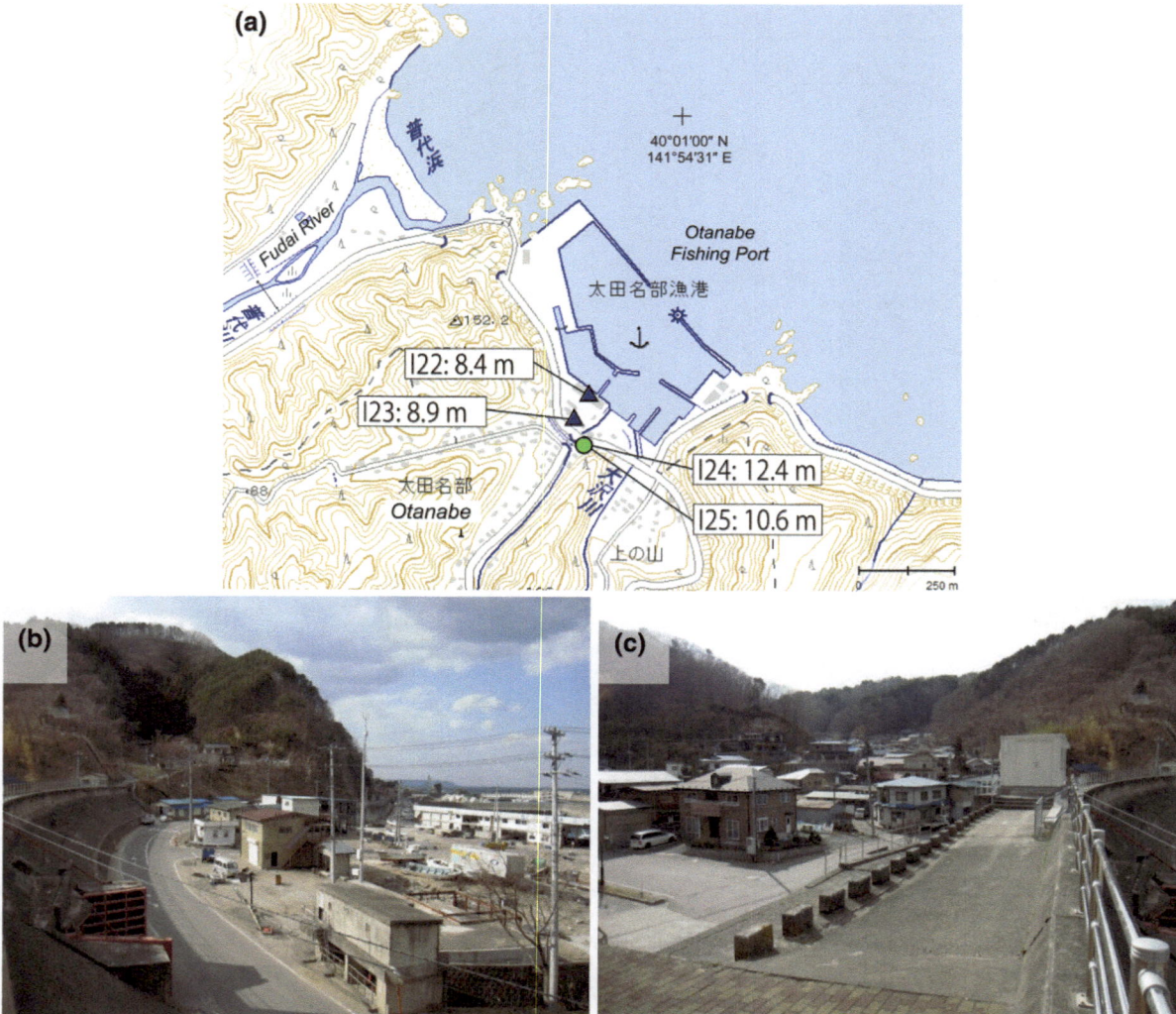

Figure 3

a Tsunami heights at Otanabe (Otanabe Fishing Port), Fudai Village, Iwate Prefecture (with 1:25,000 digital topographic map added from the Geospatial Information Authority of Japan). *Circles* and *triangles* indicate runup and inundation heights, respectively. *Closed circles* and *triangles* are *color* coded according to tsunami heights (i.e., 0–10 m *blue*; 10–20 m *green*, 20–30 m *yellow*, 30 m or above *red*). **b** Destructive damage outside the coastal levee (Otanabe Fishing Port). **c** Very minor damage inside the coastal levee (residential district)

3. Tsunami Heights from 2011 Tohoku Earthquake

Measurements of the 2011 tsunami were made at 296 points on the Sanriku coasts of Aomori (Northern Sanriku), Iwate (Central Sanriku), and Miyagi (Southern Sanriku) prefectures, and the Pacific coasts of Ibaraki and Chiba prefectures (Fig. 2a, b). While the details were reported by TSUJI *et al.* (2011) with the survey points shown on 1:25,000 maps from the Geospatial Information Authority of Japan and photographs of measured tsunami traces, we summarize the tsunami heights at some typical locations. In this paper, we employ the measurement numbers given by TSUJI *et al.* (2011), which start with a letter indicating the prefecture (i.e., A, I, M, B, and C for Aomori, Iwate, Miyagi, Ibaraki, and Chiba prefectures, respectively) and are followed by the sequence number in each prefecture.

Figure 4
a Tsunami heights at Shimanokoshi, Tanohata Village, Iwate Prefecture. The *symbols* and their meanings are the same as in Fig. 3. **b, c** View of tsunami damage at Shimanokoshi

3.1. Sanriku Coast

Along the northern Sanriku coast of Aomori Prefecture, five tsunami heights ranging 6.0–9.9 were measured. Along the central Sanriku coast of Iwate Prefecture, 136 tsunami heights were measured. Tsunami heights were mostly over 10 m, while they were above 30 m at 10 measurement points. Along the southern Sanriku coast of Miyagi Prefecture, 76 tsunami heights were measured. The tsunami heights were mostly 10–20 m, slightly lower than those in Iwate Prefecture, indicating that the highest tsunami height was not recorded directly landward of the largest slip region near the Japan Trench (Figs. 1, 2a).

In the Otanabe district of Fudai Village, a 15-m-high coastal levee had been constructed between the fishing port area and residential area (Fig. 3). The inundation heights were measured as 8.4 and 8.9 m in the fishing port area, and two runup heights of 10.6 m

Figure 5
a Tsunami height at Taro-Mukaishinden, Miyako City, Iwate Prefecture. The *symbol* and its meaning are the same as in Fig. 3. **b** View of tsunami damage at Taro-Mukaishinden. The *white* pole in the *lower-right* of the picture was bent by the tsunami. **c** The survey point where the largest tsunami runup height of 37.8 m in our field surveys was measured

(class B) and 12.4 m (class C) were measured on the coastal levee (I22–I25). The floodgate, which was closed before the arrival of tsunami, completely protected the residential area from the devastating tsunami, while the outside fishing port area was severely damaged.

At Shimanokoshi in Tanohata Village, a bridge and Shimanokoshi Station of the Sanriku Railway were completely destroyed by the tsunami, and all the houses were swept away or leveled except for two that were located on a hill (Fig. 4). A runup height of 19.9 m was measured at the northern slope near the dai-ni (second) Shimanokoshi Tunnel (I41). A large

amount of wood building materials and debris was deposited throughout the tunnel. The runup height of 22.0 m was measured at the front yard of one of the surviving houses (I42).

At Taro-Mukaishinden near Koborinai Fishing Port in Miyako City, a runup height of 37.8 m (I59), the maximum height in our field surveys, was measured on the basis of the upper limit of debris and absence of leaves (Fig. 5). Three firefighters, who were advising fishermen to evacuate, were killed at an elevation of about 30 m.

At Taro in Miyako City, coastal levees with a height of 10 m and a total length of ~2.4 km had

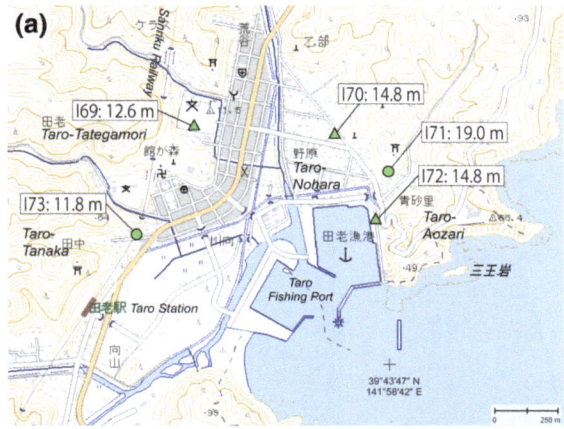

Figure 6

a Tsunami heights at Taro-Tategamori (I69), Taro-Nohara (I70), Taro-Aozari (I71 and I72), and Taro-Tanaka (I73), Miyako City, Iwate Prefecture. The *symbols* and their meanings are the same as in Fig. 3. **b** Part of the breakwater destroyed by the 2011 tsunami. The building in the background is a hotel that the tsunami damaged up to the third floor (14.8 m; I70)

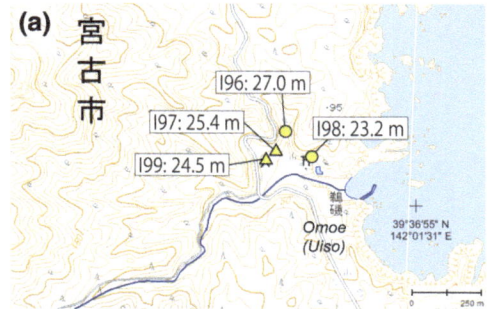

Figure 7

a Tsunami heights at Omoe (Uiso), Miyako City, Iwate Prefecture. The *symbols* and their meanings are the same as in Fig. 3. **b** View of tsunami damage at Uiso Elementary School

been constructed; however, the 2011 tsunami destroyed a portion of these coastal levees, and transported blocks and other debris from the structure more than 100 m inland (Fig. 6). Almost all houses and fishing facilities were swept away or were completely leveled. The first three floors of a hotel were severely damaged, indicating an inundation height of 14.8 m (I70). Tsunami traces on the wall of a Japan Fisheries Cooperatives icehouse indicate the same inundation height (I72). Runup heights of 11.8 m (I73) and 19.0 m (I71) were measured at the western and eastern part of the residential area, respectively. Tsunami trace at a middle school indicates an inundation height of 12.6 m (I69).

At Omoe (Uiso), which is located on the east coast of Miyako City on the Omoe Peninsula, the tsunami arrived at Uiso Elementary School, which is sited on a hill having an altitude of >20 m (Fig. 7). Panes of glass on the first floor were broken and a large amount of debris was scattered over the school playground. The measured inundation heights were 25.4 and 24.5 m, while the runup heights were 27.0 and 23.2 m (I96–I99).

At Utatsu-Namiita and Utatsu-Minato in Minamisanriku Town, most houses in the lowlands were leveled or swept away (Fig. 8). It was found from eyewitness accounts that the second tsunami arrival was the largest. At the northern settlement, one inundation height of 18.0 m and two runup heights of 19.4 and 19.5 m were measured (M15–M17). Although the southern settlement is located nearby, a lower inundation height of 14.6 m and runup heights of 14.5 and 14.0 m were measured (M18–M20).

15

Figure 8
a Tsunami heights at Utatsu-Namiita (M15, M16, and M17) and Utatsu-Minato (M18, M19, and M20), Minamisanriku Town, Miyagi Prefecture. The *symbols* and their meanings are the same as in Fig. 3. **b, c** Tsunami damage at Utatsu-Namiita (*left*) and Utatsu-Minato (*right*)

Around Nagazuraura, which is a brackish lake connected to Oppa Bay in Ishinomaki City, almost all the houses were inundated by the 2011 tsunami (Fig. 9). Japanese black pines and houses at the mouth of the Kitakami River were swept away. Many houses and rice paddies were submerged due to significant ground subsidence, and sand was deposited all over the residential district. In Onosaki district, which is located eastward of Nagazuraura,

inundation heights of 4.2 and 3.9 m were measured (M43, M46). In Nagatsura district to the west of Nagazuraura, four inundation heights ranging 4.1–7.1 m were obtained (M44, M45, M47 and M48). Some residents drowned at the temple to which they had evacuated, erroneously assuming it was at a safe elevation.

At Okawa Elementary School in Kamaya, Ishinomaki City, 10 teachers and 74 of the 108 students

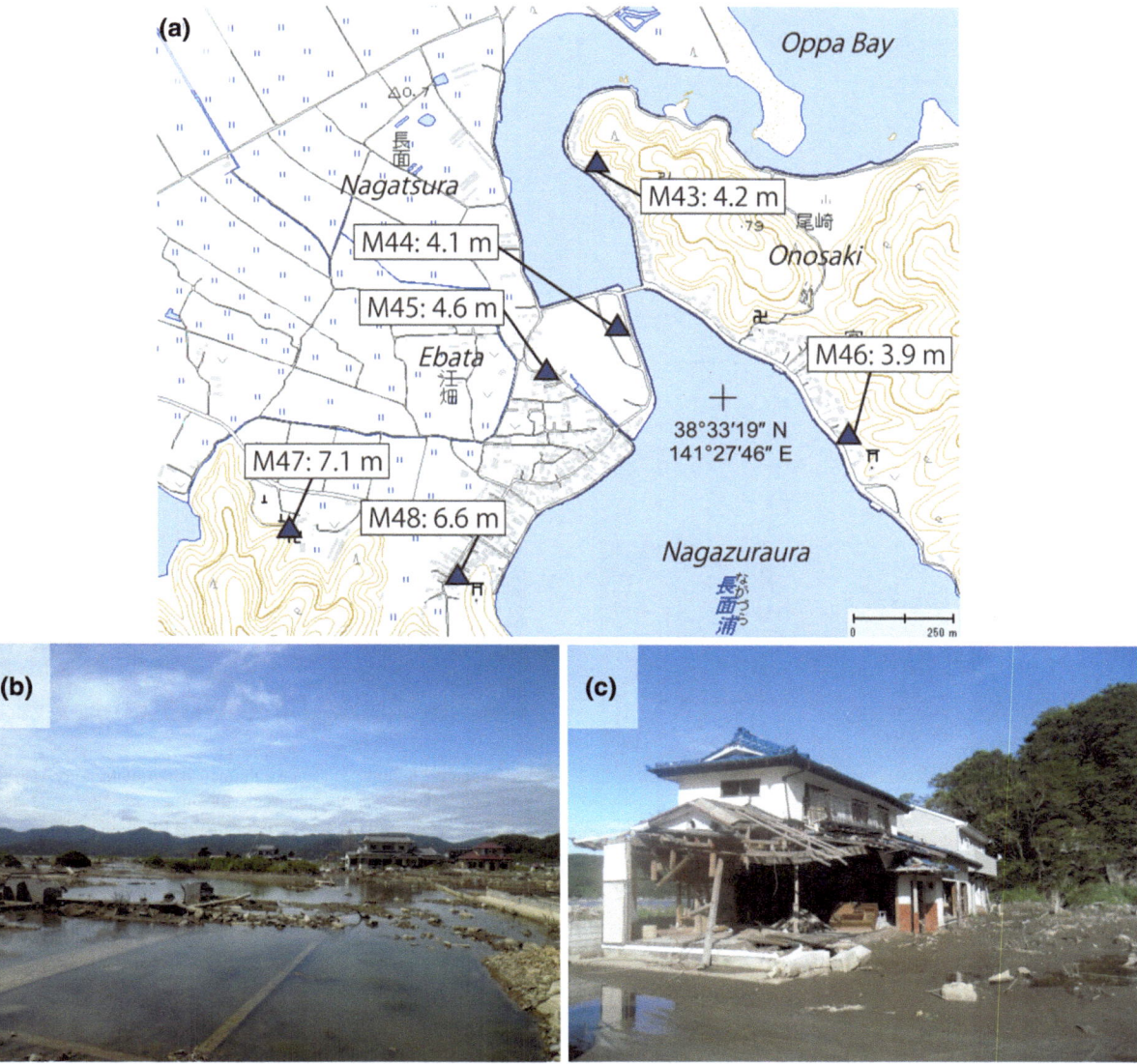

Figure 9
a Tsunami heights at Onosaki (M43, M46) and Nagatsura (M44, M45, M47, and M48), Ishinomaki City, Miyagi Prefecture. The *symbols* and their meanings are the same as in Fig. 3. **b**, **c** View of tsunami damage and ground subsidence at Onosaki and Nagatsura

died on the way to an evacuation site. It was found that the tsunami, which had run up the Kitakami River, inundated to a height above the ceiling of the second floor of the school (Fig. 10). A runup height of 9.3 m was measured at a slope behind the school (M49).

At Yagawahama in Ishinomaki City, almost all the houses were swept away by the tsunami, and farmland was covered by tsunami sediment deposits that included sand, shell fragments and gravel (Fig. 11). A large amount of debris, which was

caught in trees on the side of a road, indicate an inundation height of 18.7 m (M67). The maximum runup height of 21.2 m in the Miyagi Prefecture was measured at the cemetery behind Tofuku-ji Temple (M68).

3.2. Ibaraki and Chiba Coasts

On the Ibaraki coast (between 35.7° and 36.9°N), 36 measurements of the 2011 Tohoku tsunami ranged 2.8–8.1 m, and the heights generally decreased

Figure 10
a Tsunami height at Kamaya in Ishinomaki City, Miyagi Prefecture. The *symbol* and its meaning are the same as in Fig. 3. **b** View of tsunami damage at the Okawa Elementary School

Figure 11
a Tsunami heights at Yagawahama, Ishinomaki City, Miyagi Prefecture. The *symbols* and their meanings are the same as in Fig. 3. **b** View from the survey point with a runup height of 21.2 m (M68)

toward the south (Fig. 2b). The typical tsunami height was 4 m and the local variations were significantly smaller than those along the Sanriku coast. Along the coast of Chiba Prefecture (between 34.9° and 35.7°N), 43 measurements of the 2011 tsunami heights were generally <4.0 m. They gradually decreased toward the south, and most tsunami heights were <2 m along the coast of southern Chiba Prefecture. However, tsunami heights were locally high (5.1–7.9 m) around Asahi City.

At Hirakatacho in Kitaibaraki City, the fishing port and residential area were severely damaged (Fig. 12). The first floors of many houses were destroyed by the tsunami and some houses collapsed completely. Watermarks on buildings indicated inundation heights ranging 6.6–7.2 m, while debris indicated runup heights of 7.9 and 8.1 m (B1–B5).

At Minatochuo in Oarai Town, which is located in the central part of Ibaraki Prefecture, inundation heights of 4.5 and 4.6 m were measured from the watermarks on the ferry terminal building (B30, B31).

At Asahi City, the tsunami heights were locally high. Residential areas along the Pacific coast were badly damaged and 13 people were killed (Fig. 13). In the Hiramatsu district where buildings were densely distributed, a clear watermark on the windows of a store indicated the inundation height as 6.3 m (C4). In the Ashikawa district, an inundation height of 5.1 m and a runup height of 7.9 m were measured (C5, C6). At the Shirako Fishing Port in Chikuracho-Shirako in Minamiboso City, according to eyewitness accounts, the first tsunami arrived at 15:15–15:30 and the seawater rose to near the top of the quay. The measured tsunami heights in this port ranged 1.0–1.4 m (C35, C36). The eyewitnesses also

Figure 12
a Tsunami heights at Hirakatacho, Kitaibaraki City, Ibaraki Prefecture. The *symbols* and their meanings are the same as in Fig. 3. **b** View of tsunami damage at Hirakatacho. **c** Watermarks on windowpanes of entrance doors of the guesthouse Yanagiya (0.89 m above ground level, B3)

reported that the second tsunami arrival was one hour after the first arrival and that the height was similar.

4. Tsunami Height Data from Past Earthquakes

4.1. 1896 and 1933 Sanriku Earthquakes

A number of field surveys and investigations have been conducted for the tsunamis caused by the June 15, 1896 and March 3, 1933 Sanriku earthquakes to determine the heights and inundation areas (e.g., IMAMURA 1934; SHUTO and GOTO 1985a, b; IMAMURA and WATANABE 1990; HATORI 1995). In this study, we examined the reports of the original surveys published soon after the earthquakes (i.e., in the 1890s and 1930s), rather than secondary or recent papers. We briefly describe these original reports below.

YAMANA (1896; reproduced by UNOHANA and OTA 1988) reported that his survey was conducted between July 28 and September 9, 1896, in all the

Figure 13
a Tsunami height at Hiramatsu, Asahi City, Chiba Prefecture. The *symbol* and its meanings are the same as in Fig. 3. **b** A damaged house in Hiramatsu district

villages along the Sanriku coast. He reported not only tsunami heights, but also casualty numbers and property damage for each village, along with providing 168 maps illustrating the disaster. The largest heights of 180 shaku, or 55 m (1 shaku = 0.303 m) were reported at Kosode (I12, I13) and Ryori-Shirahama (I133). However, the accuracy of Yamana's measurements varies at different locations. For example, the reported runup height is 132 shaku (40.0 m) at Shimanokoshi (I41, I42), whereas the height is 130–180 shaku (39–55 m) at Ryori–Shirahama (I133).

Iki (1897) surveyed the Sanriku coast from June 20 to July 21, 1896, and his report contains list of locations and tsunami heights. However, it contains very few figures, making it difficult to identify the exact measurement locations. The reported tsunami heights in feet (1 foot = 0.305 m) were from sea level at the time of measurement. He measured tsunami heights from changes in vegetation color or watermarks due to tsunami inundation, which are considered to be highly reliable, or from debris or scratches on trees. He also distinguished heights measured by himself and those based on eyewitness accounts.

Matsuo (1933) reported tsunami heights from the 1896 and 1933 Sanriku tsunami based on surveys he conducted during March 3–10 and May 19–June 4, 1933, with the support of the Iwate and Miyagi prefectural governments. The tsunami heights from the 1896 Sanriku tsunami were measured 37 years after the event, based on eyewitness accounts. The often quoted maximum height of 38 m at Ryori–Shirahama (I133) from the 1896 Sanriku tsunami was based on this report. Matsuo (1934) also reported the 1933 tsunami heights and inundation areas of the 1896 and 1933 tsunamis along the coasts of Hokkaido and Aomori prefectures.

Kunitomi (1933) reported tsunami heights together with arrival times from both eyewitness accounts and tide gauge records, the number of tsunami waves, and periods from the 1933 Sanriku tsunami in Hokkaido, along the Sanriku coasts of Aomori, Iwate, Miyagi prefectures, and along the Pacific coast of Fukushima Prefecture, based on field surveys from meteorological observatories. Tsunami heights were measured on the basis of traces left at the coast and/or eyewitness accounts. Appendix figures indicating tsunami heights and inundation areas on maps with a scale of 1:200,000, and photographs recording severe damage are also shown. Kunitomi (1933) also compared the 1933 heights with the 1896 heights from Iki (1897) and surveys from the Civil Engineering Division of the Iwate prefectural government.

The Earthquake Research Institute (1934) reported the survey results for the Pacific coasts from Hokkaido to Ibaraki Prefecture. It reported tsunami heights with maps showing estimated inundation limits, damage to houses and other structures, relations between the severity of tsunami effects and topographic conditions, photographs, and tide gauge records. In addition, Otuka (1934) discussed the relationship between tsunami heights and the topography of the Sanriku coast.

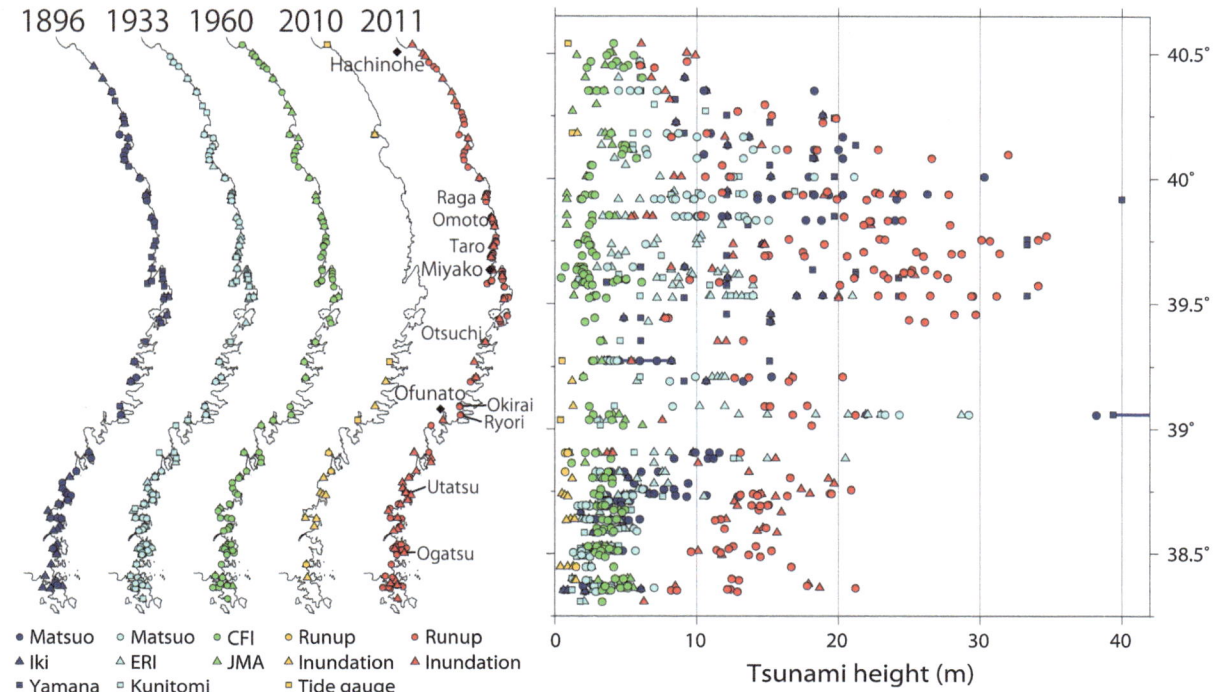

Figure 14

Comparison of 1896 and 1933 Sanriku tsunami heights, 1960 and 2010 Chilean tsunami heights with 2011 tsunami heights. *Bars* indicate the range of measured heights. The measurement locations are indicated on the maps on the *left*. Only data measured at the same locations for two or more tsunamis are shown. *Blue circles*, *triangles*, and *squares*, respectively, indicate the 1896 tsunami heights given by MATSUO (1933), IKI (1897), and YAMANA (1896), which were reproduced by UNOHANA and OTA (1988). *Light blue circles*, *triangles*, and *squares*, respectively indicate the 1933 heights from MATSUO (1933, 1934), the EARTHQUAKE RESEARCH INSTITUTE (1934), and KUNITOMI (1933). *Green circles* and *triangles* indicate the 1960 Chilean tsunami heights from CFI (1961), and JMA (1961), respectively. *Orange circles*, *triangles*, and *squares*, respectively, indicate the 2010 runup, inundation heights, and heights from tide gauges by TSUJI *et al.* (2010). *Red circles* and *triangles*, respectively, indicate the 2011 runup and inundation heights from TSUJI *et al.* (2011)

4.2. 1960 and 2010 Chile Earthquakes

The COMMITTEE for FIELD INVESTIGATION of the CHILEAN TSUNAMI of 1960 (CFI, 1961) consisted of 100 investigators who conducted surveys along the entire Pacific coast of Japan from Hokkaido to Kyushu. They compiled measured tsunami heights above TP (nearly equal to mean sea level), reliability (1–5), and arrival time of the maximum tsunami. The report contains detailed maps showing the exact measurement locations.

Japan Meteorological Agency (JMA, 1961) contains field survey results made by the agency. The reported tsunami heights were either measured on tide gauges or were based on eyewitness accounts. The definition of tsunami heights and datum varies according to location; and detailed maps indicate measurement locations.

IMAI *et al.* (2010) reported tsunami heights from the 2010 Chile earthquake along the coasts of the Kanto and Tokai districts measured from field surveys or tide gauges. TSUJI *et al.* (2010) reported the 2010 tsunami heights along the Sanriku coast measured from field surveys or tide gauges. The tsunami heights were above sea level at the time of maximum tsunami and were classified as inundation heights, runup heights, and tsunami heights in ports. The survey locations were measured using handheld GPS devices showing latitudes and longitudes.

4.3. Selection of Locations for Comparison

We selected locations for which measurements for the 2011 tsunami and at least one of the previous tsunamis were available in order to enable a direct comparison. While the precise measurement points of the 1896 and 1933 Sanriku tsunamis are unknown, comparisons were made at locations (village or smallest bay). We attempted to reduce the local variability in tsunami heights as much as possible. For the 1960 and 2010 Chilean tsunamis, we selected tsunami height data measured in the same ports or similar coastal locations by examining maps of the measurement points. For the comparison of tsunami heights between the 2011 and previous tsunamis, we use the median height of multiple measurements in each report. For an example, three tsunami heights (by the Earthquake Research Institute, median of five heights by Matsuo, and by Kunitomi) were compared with the 2011 height (7.8 m) at I3 (Taneichi, Hirono Town). For another example, the median height of three measurements of the 2011 tsunami (I16–I18) was compared with the previous tsunamis at the Noda Fishing Port in Noda Village.

5. Comparison of Tsunami Heights on the Sanriku Coast

Along the Sanriku coasts of Aomori, Iwate, and Miyagi prefectures, Tsuji et al. (2011) measured 217 tsunami heights at 120 locations. Of these, the 1896 and 1933 heights were reported at 80 and 94 locations, respectively (Fig. 14; Table 1). The 1896 tsunami heights are smaller than the 2011 tsunami heights (median ratio is 0.69; Table 3), and the 1933 heights are even smaller (median ratio 0.33). The 1960 and 2010 Chilean tsunami heights can be compared with the 2011 tsunami heights at 98 and 19 locations, respectively (Table 2). The 1960 tsunami heights are also smaller than the 2011 tsunami heights (median ratio 0.25), and the 2010 heights are much smaller (median ratio 0.07).

5.1. Central Sanriku Coast

Along the central Sanriku coast with latitudes between 39.0° and 40.2°N, the 2011 tsunami heights ranged from 5 to 40 m (Fig. 14). In this region, the 1896 heights were approximately similar (median ratio is 0.85; Table 3) to the 2011 heights, whereas the 1933 heights were smaller (median ratio 0.47). The 1960 and 2010 Chilean tsunami heights were

much smaller than the 2011 heights (median ratio is 0.16 and 0.09, respectively).

At Raga in Tanohata Village, the 1896 and 2011 tsunami heights were much larger than the 1933 tsunami height. The measured 2011 tsunami heights ranged from 23 to 28 m (I34–I37), while the reported 1896 tsunami heights varied from 60 shaku (18 m; Yamana 1896), 75 feet (23 m; Iki 1897), and 24–26 m (Matsuo 1933). During our survey, a local resident testified that the 1896 Sanriku tsunami washed away their former family house, which was at the same location as the current one, whereas the 2011 tsunami did not. Furthermore, the 2011 tsunami inundation limit is located just below the tsunami stone transported by the 1896 tsunami (Fig. 15). Takeda (1987) measured the level of the tsunami stone as 28.2 ± 1.2 m above TP and concluded that the maximum runup height of the 1896 Sanriku tsunami at Raga may have been higher than 30 m. The 1933 tsunami height was reported to be 13 m, while the 1960 tsunami height was 1 m.

At Omoto in Iwaizumi Town, the 1896, 1933, and 2011 tsunamis had similar inundation areas (Fig. 16). The 1960 Chilean tsunami reportedly inundated about 1 km from the river mouth and raised the water level by 1 m (CFI 1961). The 2011 tsunami topped the coastal levee and floodgate, which was constructed in 1990. The measured inundation height was 5 m in the residential area east of the Omoto River (I48). West of the Omoto River, the tsunami inundated farmland and the tsunami heights ranged from 6 to 10 m (I43–I46). Seaward of the Omoto floodgate, the measured runup height was 20 m (I47). The 1896 tsunami heights were between 10 and 20 m. The reported 1933 tsunami heights were between 3 and 13 m. The 1960 tsunami heights were reported as 4 m near I47, 3–4 m at the river mouth by CFI (1961), and 1 m by JMA (1961).

At Taro in Miyako City, the 2011 tsunami was larger than the 1933 tsunami, but it is unclear whether it was larger than the 1896 tsunami. Tsunami heights from the 2011 event ranged from 12 to 19 m (I69–I73). The 1896 heights were reported as being 110 shaku (33 m; Yamana 1896) or 48 feet (15 m; Iki 1897), while the heights from the 1933 tsunami ranged from 4 to 10 m. The damage to an icehouse of the Japan Fisheries Cooperatives indicates that the

Table 1

Tsunami heights for Sanriku coasts of Aomori, Iwate, and Miyagi prefectures from the 1896, 1933, and 2011 earthquakes

No.	Location Name	Latitude		Longitude		2011 height[a]	rel.[b]	acc.[c]	type	1896 height[d]			1933 height[e]		
		deg	min	deg	min.	(m)				Iki (feet)	Yamana	Matsuo (m)	ERI (m)	Matsuo (m)	Kunitomi (m)
A3	Daisakutai, Samemachi, Hachinohe City (Okuki Fishing Port)	40	30	141	38	9.9	A	a	I					3.7	
A4	Oja, Dobutsu, Hashikami Town (Oja Fishing Port)	40	28	141	39	9.3	A	a	R				6.0	6.0	6.0
A5	Kominato, Dobutsu, Hashikami Town (Kominato Fishing Port)	40	27	141	40	6.0	A	a	R	20				4.5	4.5
I2	Taneichi, Hirono Town (Taneichi Fishing Port)	40	24	141	43	6.8	B	a	I	30			4.5		6.0
I3	Taneichi, Hirono Town (Yagi Port)	40	21	141	46	7.8	B	b	I	Δ35	20	10.5 18.3	3.5	2.5 4.5 5.3 5.9 7.2	6.0
I4	Uge, Hirono Town (Uge Fishing Port)	40	19	141	47	8.1	A	a	I		28				
I5	Nakano, Hirono Town (Koge Fishing Port, Koge District)	40	18	141	48	14.8	B	a	R						7.0
I7	Samuraihamacho-Mukaicho, Kuji City (Kawatsunai Fishing Port, Kawatsunai District)	40	16	141	49	12.9	B	a	R						10.6
I8	Samuraihamacho-Shiromae, Kuji City (Shiromae Fishing Port, Shiromae District)	40	15	141	49	15.3	A	a	R	62	40				
I9	Samuraihamacho-Honnami, Kuji City (Shiromae Fishing Port, Honnami District)	40	14	141	50	19.8	B	a	R		65				
I10	Samuraihamacho-Mugyo, Kuji City (Mugyo Fishing Port)	40	13	141	50	18.9	B	a	R	28	50				
I11	Osanaicho (Tamanowaki), Kuji City (Tamanowaki Fishing Port)	40	11	141	48	10.7	A	a	R		30	11.0	3.3 4.0	6.5 7.9 8.7	5.5
I12	Ubecho (Kosode), Kuji City (Kosode Fishing Port, Kosode District)	40	10	141	51	12.0	B	a	I	45	180	20.3 20.3		10.0 13.3	8.2
I13	Ubecho (Kosode), Kuji City	40	10	141	51	8.2	A	a	R						
I15	Ubecho (Kuki), Kuji City (Kuki Fishing Port)	40	8	141	51	14.6	A	b	I	40	70		3.3		5.5
I16	Noda, Noda Village (Noda Fishing Port)	40	7	141	50	22.8	C	a	R			18.3	5.5	6.4	5.5
I17		40	7	141	50	16.4	B	a	R			20.0		14.0	
I18		40	7	141	50	18.5	B	a	R					15.6	
I19	Noda (Maita), Noda Village	40	6	141	50	32.0	B	b	R			10.5		6.5	
I20	Tamagawa, Noda Village (Noda Tamagawa Station)	40	5	141	50	26.6	B	a	R	60	60	20.3 20.3	4.2	10.0 13.3	5.8
I21	Tamagawa (Shimoakka), Noda Village	40	3	141	51	11.8	B	a	R		50		5.0		
I22	Otanabe, Fudai Village (Otanabe Fishing Port)	40	1	141	54	8.4	A	c	I	50	65	17.9 30.3	7.4	12.3	13.0
I23		40	1	141	54	8.9	A	a	I			30.3		12.6	
I24		40	1	141	54	12.4	C	a	R					18.3	
I25		40	1	141	54	10.6	B	a	R					21.1	
I32	Aketo, Tanohata Village	39	57	141	57	19.2	B	a	R	Δ40				8.4	16.9
I33		39	57	141	57	22.6	B	a	R					8.4	
I34	Raga, Tanohata Village	39	56	141	56	23.9	A	a	I	75	60	24.3		13.3	13.0
I35		39	56	141	56	22.9	A	a	R			24.3		13.3	
I36		39	56	141	56	27.8	B	a	R			26.3		13.3	
I37		39	56	141	56	24.5	B	a	R						
I38	Raga (Hiraiga), Tanohata Village	39	56	141	56	17.5	B	a	R		40	14.3 16.3	8.2 8.3	9.3 10.3	8.2
I39		39	56	141	56	16.5	B	a	R			18.3		10.3	
I40	Wano, Tanohata Village	39	56	141	56	19.0	B	a	I			19.3 20.3		10.3	
I41	Shimanokoshi, Tanohata Village	39	55	141	56	19.9	B	a	R		132	14.3 14.3 15.3	4.5 4.5 6.0	7.0 7.5 8.3	9.7
I42		39	55	141	56	22.0	A	a	R			19.9 24.1		8.5 9.8 11.1	
I43	Omoto, Iwaizumi Town	39	51	141	58	10.3	B	a	R	40	65	10.0	3.4–4.4	8.5	
I44		39	51	141	58	6.5	A	a	I			11.8	5.0	9.1	
I45		39	51	141	58	5.6	A	a	I					9.4	
I46		39	51	141	58	6.9	A	a	I					10.1 10.4	
I47		39	51	141	58	20.4	B	a	R					10.6	13.0
I48		39	51	141	58	5.4	A	a	I					13.0 13.4	
I49	Moshi, Omoto, Iwaizumi Town (Moshi Fishing Port)	39	50	141	59	24.5	B	a	R			17.7		13.6	
I50		39	50	141	59	23.5	B	a	R			18.8		14.2	
I51	Moshi, Omoto, Iwaizumi Town	39	50	141	58	22.3	B	a	R			24.0		15.1	
I52		39	50	141	58	22.2	B	a	R					15.1	

Table 1

continued

No.	Location Name	Latitude deg	min	Longitude deg	min.	2011 height[a] (m)	rel.[b]	acc.[c]	type	1896 height[d] Iki (feet)	Yamana	Matsuo (m)	1933 height[e] ERI (m)	Matsuo (m)	Kunitomi (m)
I53	Taro-Shimosettai, Miyako City	39	49	141	59	21.8	B	a	R		45		11.8		
I54		39	49	141	59	27.9	B	a	R						
I55		39	49	141	59	22.3	A	a	I						
I60	Taro-Aonotakiminami, Miyako City (Aonotaki Fishing Port)	39	46	141	59	34.7	B	b	R				10.3		
I61	Taro-Otobeno, Miyako City	39	45	141	59	22.9	B	a	R		110		7.2		
I62		39	45	141	59	16.5	B	a	R				10		
I63		39	45	141	59	23.3	B	a	R				10.2		
I64		39	45	141	59	30.1	B	a	R						
I65		39	45	141	59	34.1	B	a	R						
I69	Taro-Tategamori, Miyako City	39	44	141	58	12.6	A	a	I	48	110		4	6.4	
I70	Taro-Nohara, Miyako City	39	44	141	59	14.8	A	c	I				7		
I71	Taro-Aozari, Miyako City	39	44	141	59	19.0	B	b	R				10		
I72		39	44	141	59	14.8	A	c	I						
I73	Taro-Tanaka, Miyako City	39	44	141	58	11.8	B	a	R						10.1
I74	Taro-Nishimukaiyama, Miyako City (Kashinai Fishing Port)	39	43	141	58	25.5	B	a	R				5		
I75		39	42	141	59	17.5	B	a	R						
I76	Taro-Kashinai, Miyako City (Kashinai Fishing Port)	39	42	141	58	21.8	B	a	R						
I77	Sakiyama (Mattsuki), Miyako City	39	42	141	58	31.4	B	a	R				4		
I78		39	42	141	58	28.7	B	a	R						
I79		39	42	141	58	28.0	B	a	R						
I80	Sakiyama (Onappe), Miyako City	39	42	141	58	15.1	A	a	R		40		4.5		7.5
I81		39	42	141	58	14.8	B	a	R						
I82		39	41	141	58	12.6	B	a	I						
I83		39	41	141	58	17.6	B	a	R						
I84		39	41	141	58	20.6	B	a	R						
I85		39	41	141	58	26.0	B	a	R						
I86	Kuwagasaki (Nakamachi), Miyako City	39	39	141	58	8.2	A	a	I	Δ30	20		3 4	3.7 5.5	
I88	Tsugaruishi (Norinowaki), Miyako City	39	35	141	57	11.6	B	a	R		12		1.5		1.6
I89	Akamae, Miyako City	39	36	141	58	9.5	B	a	R		15				1.7
I90		39	36	141	58	14.0	B	a	R						
I91	Omoe (Tatehama), Miyako City	39	38	142	1	22.5	B	a	R		60		8.2		
I92		39	38	142	1	25.2	B	a	R				12		
I93		39	38	142	1	26.1	B	a	R						
I94	Omoe (Shukuhama), Miyako City	39	38	142	1	25.1	B	a	R		70		9.6-11.5		
I95		39	38	142	1	24.6	B	a	R						
I96	Omoe (Uiso), Miyako City	39	37	142	1	27.0	B	a	R		70		9.7		4.5
I97		39	37	142	1	25.4	A	a	I				12.8		
I98		39	37	142	1	23.2	B	a	R				13		
I99		39	37	142	1	24.5	A	a	I						
I100	Omoe (Aramaki), Miyako City	39	36	142	1	21.1	B	a	R	Δ40	80		7.2		7.6
I101		39	36	142	1	27.7	B	a	R				8.7		
I102		39	36	142	1	23.5	B	a	R						
I103	Omoe (Sato), Miyako City (Omoe Fishing Port)	39	34	142	2	20.1	B	a	R		40		8.3		10.9
I104		39	34	142	2	34.1	C	a	R				8.8 14		
I105	Omoe (Aneyoshi), Miyako City	39	32	142	3	26.5	B	a	R	62	110		12.8	14.0	12.4
I106		39	32	142	3	22.8	B	a	R				14.0		
I107		39	32	142	3	24.5	B	c	R				20 21.0		
I108	Omoe (Chikei), Miyako City	39	32	142	2	29.5	A	a	R	56	80		10.9		6.0
I109		39	32	142	2	31.2	A	a	R				11.0		13.6
I110		39	32	142	2	29.4	A	a	R				12.1 13.2		
I111	Omoe (Ishihama), Miyako City	39	31	142	2	20.0	A	a	I		30		7.2 8.7		12.0
I112	Funakoshi (Ohura), Yamada Town	39	27	142	0	8.0	B	a	R	16	20				
I113		39	27	142	0	7.7	A	a	I						
I114		39	27	142	0	8.0	B	a	R						
I115		39	27	142	0	7.8	B	a	R						
I116	Funakoshi (Sukuiso), Yamada Town	39	27	142	3	29.7	B	a	R	50	40				
I117		39	27	142	3	28.2	B	a	R						
I118	Funakoshi (Koyadori), Yamada Town	39	26	142	1	26.1	A	a	R	50	50		6.6		
I119		39	26	142	1	25.0	B	a	R						
I120	Akahama, Otsuchi Town	39	21	141	56	13.3	B	a	R		20				4.6
I121	Akahama, Otsuchi Town	39	21	141	56	11.5	A	a	I						
I122	(International Coastal Research Center, Atmosphere and Ocean Research Institute, the University of Tokyo)	39	21	141	56	12.1	A	a	I						

Table 1

continued

No.	Location Name	Latitude deg	min	Longitude deg	min.	2011 height[a] (m)	rel.[b]	acc.[c]	type	1896 height[d] Iki (feet)	Yamana	Matsuo (m)	1933 height[e] ERI (m)	Matsuo (m)	Kunitomi (m)
I123	Omachi, Kamaishi City	39	16	141	53	5.4	A	b	I	15–27	50	6.0 6.0 7.2	3.0 3.5 3.7 3.8 3.9	2.7 3.6 3.9 4.2 4.2 4.4 4.4	5.4
I124	Osone, Tonicho,	39	13	141	53	20.3	A	a	R			15.3	11.1	9.9	6.0
I125	Kamaishi City	39	13	141	53	16.7	C	a	I			15.3	11.5	9.9	
I126		39	13	141	53	16.8	B	a	R				11.8		
I127	Sakuratoge, Tonicho,	39	12	141	53	13.7	B	a	R				12.0		
I128	Kamaishi City	39	12	141	53	12.7	B	a	R						
I129	Arakawa, Tonicho, Kamaishi City	39	12	141	52	16.5	B	a	R	Δ35	30	13.3 13.3	5.8	8.1 8.1	
I130	Horei, Sanrikucho-Okirai, Ofunato City	39	6	141	48	17.8	B	a	R		50		12 10.2	8.3	4.2
I131		39	6	141	48	14.8	B	a	R						
I132		39	6	141	48	15.2	B	a	R						
I133	Shirahama, Sanrikucho-Ryori, Ofunato City	39	4	141	49	16.8	B	a	R	72	130–180	38.2	13 15	23.3 23.3	23.0
I134	Okubo, Sanrikucho-Ryori, Ofunato City	39	3	141	49	21.2	B	a	R				18.4 20.7 22.1 23.0 28.7	24.3 29.2	
I135	Miyanomae, Ofunatocho, Ofunato City	39	2	141	43	7.6	A	a	I				3.2 3.7 4.1		3.0
I136	Yonesakicho, Rikuzentakata City	39	1	141	40	18.1	B	b	R						3.2
M1	Karakuwacho-Baba, Kesennuma City	38	54	141	39	13.1	A	a	R	32		9.9 10.9 11.6	7 8.1		5.6
M3	Karakuwacho-Kakehama, Kesennuma City	38	53	141	40	15.6	A	a	I			8.4 10.8 11.2	13 15 20.5		12.6
M4	Karakuwacho-Tsumoto, Kesennuma City	38	52	141	40	10.1	A	a	I				7.0		
M5	Minamimachikaigan, Kesennuma City	38	54	141	34	4.0	A	a	I						1.0
M6	(Kesennuma Port)	38	54	141	34	3.7	A	a	I						
M7		38	54	141	34	4.1	A	a	I						
M8		38	54	141	35	4.1	A	a	I						
M9	Hajikamisuginoshita, Kesennuma City	38	50	141	35	13.7	A	b	I			4.0 4.9 6.1 7.3	2.9 4.5 4.6 7.6 7.8	2.9 3.2 3.8	
M10	Motoyoshicho-Amagasawa, Kesennuma City (Hikado Fishing Port)	38	48	141	33	16.6	A	a	R	17			3.3 4.0 4.0	3.3	3.0
M11	Motoyoshicho-Maehama, Kesennuma City	38	48	141	33	19.3	B	a	I			9.3 9.5	5.8		
M12	Motoyoshicho-Toyomazawa, Kesennuma City	38	47	141	31	19.7	A	a	I			5.6 9.4	7-8		
M13	Motoyoshicho-Nijuichihama, Kesennuma City	38	46	141	31	17.4	A	a	I	20		6.3 7.7 8.4	5.9	6.2	
M14		38	45	141	31	20.9	B	a	R						
M15	Utatsu-Namiita, Minamisanriku Town	38	45	141	32	19.4	A	a	R	22		5.4 6.7 7.0	3.5	3.2	
M16		38	45	141	32	18.0	A	a	I						
M17		38	45	141	32	19.5	A	a	R						
M18	Utatsu-Minato, Minamisanriku Town	38	44	141	32	14.5	A	a	R						
M19		38	44	141	32	14.0	A	a	R						
M20		38	44	141	32	14.6	A	a	I						
M21	Utatsu-Tanoura, Minamisanriku Town	38	44	141	33	13.1	B	a	R	16		3.7 8.5	4.5 5.1	5.0 5.5	
M22	Utatsu-Kaminoyama, Minamisanriku Town	38	44	141	33	14.3	A	a	R			10.7			
M23		38	44	141	33	16.4	B	a	R						
M25	Utatsu-Ishihama, Minamisanriku Town	38	44	141	34	15.6	B	a	I			9.3 12.9 13.0	10.5		7.6
M27	Utatsu-Tatehama, Minamisanriku Town	38	43	141	33	12.7	A	a	I			3.9 4.4 4.7 5.3	3.9 5.2	3.3 3.6 3.7 3.8	

Table 1

continued

No.	Location Name	Latitude deg	min	Longitude deg	min.	2011 height[a] (m)	rel.[b]	acc.[c]	type	1896 height[d] Iki (feet)	Yamana	Matsuo (m)	1933 height[e] ERI (m)	Matsuo (m)	Kunitomi (m)
M28	Utatsu-Niranohama, Minamisanriku Town	38	42	141	30	12.9	A	a	I			3.5 3.6 3.7	2.9	1.6 2.4 2.5	
M29	Shizugawa-Nishida, Minamisanriku Town	38	42	141	30	15.1	B	a	R	12		3.8	2.5	1.9	3.6
M30		38	42	141	30	15.0	B	a	R			4.7		1.9	
M31	Shizugawa-Hosoura, Minamisanriku Town	38	42	141	30	13.9	B	a	R			4.7		2.2	
M32		38	42	141	30	14.5	A	a	R						
M33	Shizugawa-Omori, Minamisanriku Town	38	41	141	27	14.4	C	a	R						5.4
M34	Shizugawa-Hayashi, Minamisanriku Town	38	40	141	27	16.0	A	a	I	6			3.2		
M35	Mitobe, Tokura, Minamisanriku Town	38	38	141	27	11.7	B	a	R	Δ8		2.7 2.9	1.9 2.1	2.2 2.2	
M36	Takihama, Tokura, Minamisanriku Town	38	38	141	30	12.1	B	a	I	13		4.7 6.0	2.4	2.5 2.5	2.4
M37	Nagashizu, Tokura, Minamisanriku Town	38	38	141	31	11.3	B	a	R	16		4.9 5.2	4.6	4.0 4.2	2.4
M38	Kozashi, Kitakamicho-Jusanhama, Ishinomaki City	38	36	141	30	14.7	A	a	I	15			4.6		4.8
M39	Aikawa, Kitakamicho-Jusanhama, Ishinomaki City	38	36	141	30	14.9	A	a	R	15			5.5		4.8
M40	Kodomari, Kitakamicho-Jusanhama, Ishinomaki City	38	36	141	30	12.0	A	a	R				5.0	5.6	4.5
M41	Omuro, Kitakamicho-Jusanhama, Ishinomaki City	38	36	141	30	14.3	C	a	I	13			3.5 3.7	5.8 5.8	3.0
M42	Shirahama, Kitakamicho-Jusanhama, Ishinomaki City	38	35	141	28	15.7	A	a	I	9			3.2	4.1	2.1
M50	Ogatsucho-Naburi, Ishinomaki City	38	32	141	30	14.2	B	a	R	11			2.7 3.3	4.0	4.2
M51	Ogatsucho-Funakoshi, Ishinomaki City	38	32	141	31	12.6	B	a	R				3.7 4	4.0	4.5
M52	Ogatsucho-Osu, Ishinomaki City	38	31	141	32	11.4	B	a	R				3.8	4.2	
M53	Ogatsucho-Kuwanohama, Ishinomaki City	38	30	141	32	11.8	A	a	I				2.3	1.9 1.9 1.9	1.5
M54	Ogatsucho-Tachihama, Ishinomaki City	38	30	141	31	11.7	B	a	R				2.0	1.3 1.8	1.8
M55		38	31	141	31	9.6	B	a	R					1.9 2.2 2.2	
M56	Ogatsucho-Myojin, Ishinomaki City	38	31	141	29	12.3	B	a	R	8			2.0	2.2 2.2	1.8
M57	Funatoshinmei, Ogatsucho-Ogatsu, Ishinomaki City	38	31	141	28	15.3	A	a	R	10		2.9 3.0 3.1 3.2 3.6 3.8 4.8	3.5 3.65 3.90 3.98	3.3 3.4 3.4 3.7 3.8 4.2 4.3 4.5 5.7	
M58	Karakuwa, Ogatsucho-Ogatsu, Ishinomaki City	38	31	141	29	10.1	C	a	I	6			2.1	1.9 1.9	1.8
M59	Wakehama, Ogatsucho-Wakehama, Ishinomaki City	38	30	141	29	13.7	B	a	R	7			1.8	1.9 1.9	1.5
M60	Namiita, Ogatsucho-Wakehama, Ishinomaki City	38	29	141	29	14.5	A	a	R	8			2.1	2.5 2.5	1.5
M61	Ishihama, Onagawa Town	38	27	141	28	16.7	A	a	R	8			2.6	2.2 2.6	2.4
M62	Oishiharahama, Onagawa Town	38	24	141	28	13.0	B	a	R					2.2 2.2	2.4
M65	Samenoura, Ishinomaki City	38	23	141	29	17.9	C	a	I	10		2.6	5.0	3.2	4.8
M66	Oyagawahama, Ishinomaki City	38	22	141	29	17.8	B	a	R			2.0		2.8 4.0	5.2
M67	Yagawahama, Ishinomaki City	38	22	141	29	18.7	C	a	I	11		3.0	5.2	4.0	4.8
M68		38	22	141	29	21.2	B	a	R					7.0	

Table 1

continued

No.	Location Name	Latitude deg	min	Longitude deg	min.	2011 height[a] (m)	rel.[b]	acc.[c]	type	1896 height[d] Iki (feet)	Yamana a	Matsuo (m)	1933 height[e] ERI (m)	Matsuo (m)	Kunitomi (m)
M69	Tomarihama, Ishinomaki City	38	22	141	31	12.4	A	a	R	x20			3.7		
M70		38	22	141	31	12.2	A	a	R						
M71	Kugunarihama, Ishinomaki City	38	19	141	30	6.3	A	a	I				2.1		1.8
M72	Koamikurahama, Ishinomaki City	38	21	141	27	12.9	B	a	R	7			3.0	2.9	3.0
M73	Fukkiura, Ishinomaki City	38	21	141	27	8.2	B	a	R				2.7		1.2
M74	Kitsunezakihama, Ishinomaki City	38	21	141	25	8.6	B	a	R	x2–3					
M75	Kozumihama, Ishinomaki City	38	22	141	27	8.5	B	a	I				2.9	3.5	2.7
M76	Momonoura, Ishinomaki City	38	24	141	26	12.5	B	a	R	4			1.2		1.2

R runup height; *I* inundation height; *P* tsunami height in ports

[a] 2011 heights above sea level at time of maximum tsunami

[b] Rel.: reliability, A: most reliable based on clear physical evidence or eyewitness account; B: mostly based on natural traces; C: least reliable based on equivocal evidence

[c] Acc.: accuracy, a: measurement error <0.2 m; b 0.2 ≤ error ≤ 0.5 m; c error >0.5 m

[d] Iki: runup heights taken from IKI (1897). 1 foot = 0.305 m. Δ: visual measurements; *x* eyewitness accounts; Yamana: measured by YAMANA (1896) reproduced by UNOHANA and OTA (1988). 1 shaku = 0.303 m, Matsuo: taken from MATSUO (1933, 1934) measured with 1933 heights based on eyewitness accounts

[e] ERI: taken from EARTHQUAKE RESEARCH INSTITUTE (1934), Matsuo: taken from MATSUO (1933, 1934), Kunitomi: taken from KUNITOMI (1933)

2011 inundation tsunami height was 15 m (I72). On the back cliff behind the icehouse, two white markers indicate the heights of the 1896 (15 m; IKI 1897) and 1933 (10 m; EARTHQUAKE RESEARCH INSTITUTE 1934) Sanriku tsunamis (Fig. 17). These show that the 2011 tsunami was larger than the 1933 tsunami, but similar to the 1896 tsunami. According to YAMASHITA (2003), 1,867 of 2,248 residents in the affected area were killed by the 1896 tsunami (fatality rate 83 %), and 911 among 2,773 residents died during the 1933 tsunami (fatality rate 32 %). The 2011 tsunami killed ~200 of 4,434 (fatality rate ~5 %). The 1960 tsunami heights were 2–3 m as reported by CFI (1961), and 2 m as reported by JMA (1961), and the coastal levee completely protected the residential area from the 1960 and 2010 Chilean tsunamis.

At Akahama in Otsuchi Town, the 2011 tsunami was larger than the other historical tsunamis. Most of the houses were swept away by the 2011 tsunami. The International Coastal Research Center, Atmosphere and Ocean Research Institute of the University of Tokyo was severely damaged up to the third floor, indicating inundation heights of ~12 m (I121, I122). The runup height behind of the building was measured as 13 m (I120). During the 1896 tsunami, a tsunami height of 6 m, reportedly caused two houses to collapse, 16 to be washed away, and 26 fatalities. The 1933 tsunami height was 5 m, while the 1960 tsunami height was 3 m.

At Sanrikucho-Okirai in Ofunato City, the 2011, 1896, and 1933 tsunamis caused similar inundations and runup heights; however, 1960 tsunami height was much smaller. The 2011 tsunami inundated areas up to ~150 m from the coast near Horei Station of the Sanriku Railway. Tsunami inundation would have been larger if the railway track had not been raised to a height of ~15 m. The 2011 runup heights ranged from 15 to 18 m (I130–I132). The 1896 tsunami completely destroyed a coastal levee with a height of 15 shaku (~5 m), inundating areas up to 500 ken (~900 m; 1 ken = 1.818 m) from the coast with a reported tsunami height of 15 m. At the time of the 1933 tsunami, the inundation distance was 300–400 m with reported tsunami heights of 4–12 m. The 1960 tsunami height was 2 m (CFI 1961).

At Sanrikucho-Ryori in Ofunato City, the 2011 tsunami height was lower than the 1896 and 1933 Sanriku tsunami heights, but higher than the 1960 tsunami. The 1896 tsunami height was reported as 130–180 shaku (approximately 39–55 m; YAMANA 1896), 72 feet (22 m; IKI 1897), or 38 m (MATSUO 1933) (Fig. 18), and it killed 204 of 240 residents. According to MATSUO (1933), the maximum heights of the 1896 Sanriku tsunami (38 m) and 1933 Sanriku tsunami (29 m) were recorded here. The 2011 runup

Table 2

Tsunami heights for Sanriku coasts of Aomori, Iwate, and Miyagi prefectures and Pacific coasts of Ibaraki and Chiba prefectures from the 1960, 2010, and 2011 earthquakes

No.	Location Name	Latitude		Longitude		2011 height[a] (m)	rel.[b]	acc.[c]	type	1960 height						2010 height	
		deg	min	deg	min					CFI[d] (m)	CFI Corrected[e] (m)	CFI acc.[f]	JMA[g] (m)	JMA datum	JMA Corrected[e] (m)	Tsuji/Imai[h] (m)	type
A1	Shimomekurakubo, Samemachi, Hachinohe City (Hachinohe Fishing Port, Ebisuhama District)	40	32	141	34	6.1	A	a	I	4.2	4.2	4				0.9	T
A2	Yoboishi, Samemachi, Hachinohe City (Tanesashi Fishing Port)	40	30	141	37	9.3	A	a	I	3.9	3.9	3	4.1	tp	4.1		
A3	Daisakutai, Samemachi, Hachinohe City (Okuki Fishing Port)	40	30	141	38	9.9	A	a	I	3.7 5.6	3.7 5.6	4 4	1.6	tp	1.6		
A4	Oja, Dobutsu, Hashikami Town (Oja Fishing Port)	40	28	141	39	9.3	A	a	R	4.4	4.4	4	2.6	tp	2.6		
A5	Kominato, Dobutsu, Hashikami Town (Kominato Fishing Port)	40	27	141	40	6.0	A	a	R	4.1–5.2	4.1 –5.2	2	3.6	tp	3.6		
I1	Taneichi (Kadonohama), Hirono Town (Kadonohama Fishing Port)	40	27	141	41	7.0	B	a	R	4.9	4.9	2	2.3	tp	2.3		
I2	Taneichi, Hirono Town (Taneichi Fishing Port)	40	24	141	43	6.8	B	a	I	6.2	6.2	2	2.2	tp	2.2		
I3	Taneichi, Hirono Town (Yagi Port)	40	21	141	46	7.8	B	b	I	3.0 3.5	3.0 3.5	4 4	2.4	tp	2.4		
I5	Nakano, Hirono Town (Koge Fishing Port, Koge District)	40	18	141	48	14.8	B	a	R				3.0	tp	3.0		
I7	Samuraihamacho-Mukaicho, Kuji City (Kawatsunai Fishing Port, Kawatsunai District)	40	16	141	49	12.9	B	a	R				1.3	tp	1.3		
I11	Osanaicho (Tamanowaki), Kuji City (Tamanowaki Fishing Port)	40	11	141	48	10.7	A	a	R	4.0	4.0	4	3.6	tp	3.6	1.2 1.2	T I
I12	Ubecho (Kosode), Kuji City (Kosode Fishing Port, Kosode District)	40	10	141	51	12.0	B	a	I				3.8	tp	3.8		
I13	Ubecho (Kosode), Kuji City	40	10	141	51	8.2	A	a	R								
I15	Ubecho (Kuki), Kuji City (Kuki Fishing Port)	40	8	141	51	14.6	A	b	I	4.8 5.0	4.8 5.0	3 3	4.1	tp	4.1		
I16	Noda, Noda Village (Noda Fishing Port)	40	7	141	50	22.8	C	a	R	4.9	4.9	1	5.2	tp	5.2		
I17		40	7	141	50	16.4	B	a	R								
I18		40	7	141	50	18.5	B	a	R								
I19	Noda (Maita), Noda Village	40	6	141	50	32.0	B	b	R	4.9	4.7	4					
I20	Tamagawa, Noda Village (Noda Tamagawa Station)	40	5	141	50	26.6	B	a	R	5.9 5.9	5.7 5.7	4 4	8.1	tp	7.9		
I21	Tamagawa (Shimoakka), Noda Village	40	3	141	51	11.8	B	a	R	2.7 2.7 2.8	2.5 2.5 2.6	4 4 4					
I22	Otanabe, Fudai Village (Otanabe Fishing Port)	40	1	141	54	8.4	A	c	I	2.5	2.3	4	2.4	tp	2.2		
I23		40	1	141	54	8.9	A	a	I	2.6	2.4	3					
I24		40	1	141	54	12.4	C	a	R								
I25		40	1	141	54	10.6	B	a	R								
I34	Raga, Tanohata Village	39	56	141	56	23.9	A	a	I				x 1.0	tp	0.8		
I35		39	56	141	56	22.9	A	a	R								
I36		39	56	141	56	27.8	B	a	R								
I37		39	56	141	56	24.5	B	a	R								
I38	Raga (Hiraiga), Tanohata Village	39	56	141	56	17.5	B	a	R	2.3	2.1	3	x 1.0	tp	0.8		
I39		39	56	141	56	16.5	B	a	R	2.3	2.1	4					
I40	Wano, Tanohata Village	39	56	141	56	19.0	B	a	I	2.8 3.0	2.6 2.8	4 3					
I41	Shimanokoshi, Tanohata Village	39	55	141	56	19.9	B	a	R	2.0	1.8	4	x 1.0	tp	0.8		
I42		39	55	141	56	22.0	A	a	R	2.1	1.9	4					
I43	Omoto, Iwaizumi Town	39	51	141	58	10.3	B	a	R	2.8 4.0	2.6 3.8	4 4	x 1.0	tp	0.8		
I47		39	51	141	58	20.4	B	a	R	4.4	4.2	4					
I51	Moshi, Omoto, Iwaizumi Town	39	50	141	58	22.3	B	a	R	2.8	2.6	4					
I52		39	50	141	58	22.2	B	a	R								
I53	Taro-Shimosettai, Miyako City	39	49	141	59	21.8	B	a	R	2.6	2.4	4	x 1.0	tp	0.8		
I54		39	49	141	59	27.9	B	a	R								
I55		39	49	141	59	22.3	A	a	I								
I60	Taro-Aonotakiminami, Miyako City (Aonotaki Fishing Port)	39	46	141	59	34.7	B	b	R	2.4	2.2	4					
I61	Taro-Otobeno, Miyako City	39	45	141	59	22.9	B	a	R	2.3	2.1	4					
I62		39	45	141	59	16.5	B	a	R								
I63		39	45	141	59	23.3	B	a	R								
I64		39	45	141	59	30.1	B	a	R								
I65		39	45	141	59	34.1	B	a	R								
I66	Taro-Wano, Miyako City	39	45	142	0	30.7	B	a	R	2.5	2.3	4					
I67		39	45	141	59	27.4	B	a	R								
I70	Taro-Nohara, Miyako City	39	44	141	59	14.8	A	c	I	1.8	1.6	4					

Table 2

continued

No.	Location Name	Latitude deg	min	Longitude deg	min	2011 height[a] (m)	rel.[b]	acc.[c]	type	1960 height CFI[d] (m)	CFI Corrected[e] (m)	CFI acc.[f]	JMA[g] (m)	JMA datum	JMA Corrected[e] (m)	2010 height Tsuji/Imai[h] (m)	type
I71	Taro-Aozari, Miyako City	39	44	141	59	19.0	B	b	R	1.8	1.6	4					
I72		39	44	141	59	14.8	A	c	I	2.6 2.7	2.4 2.5	4 4	2.6	tp	2.4		
I74	Taro-Nishimukaiyama, Miyako City	39	43	141	58	25.5	B	a	R	2.8	2.6	3					
I75	(Kashinai Fishing Port)	39	42	141	59	17.5	B	a	R								
I76	Taro-Kashinai, Miyako City (Kashinai Fishing Port)	39	42	141	58	21.8	B	a	R								
I80	Sakiyama (Onappe), Miyako City	39	42	141	58	15.1	A	a	R	2.1	1.9	4					
I81		39	42	141	58	14.8	B	a	R	3.1	2.9	4					
I82		39	41	141	58	12.6	B	a	I								
I83		39	41	141	58	17.6	B	a	R								
I84		39	41	141	58	20.6	B	a	R								
I85		39	41	141	58	26.0	B	a	R								
I86	Kuwagasaki (Nakamachi), Miyako City	39	39	141	58	8.2	A	a	I	1.8	1.6	4	2.0 2.2	tp	1.8 2.0	0.7	T
I87	Rinkodori, Miyako City	39	38	141	58	11.1	C	a	I	2.1	1.9	4	2.2	tp	2.0		
I88	Tsugaruishi (Norinowaki), Miyako City	39	35	141	57	11.6	B	a	R	6.3	6.1	4	4.3	tp	4.1	2.0 2.1	I I
I89	Akamae, Miyako City	39	36	141	58	9.5	B	a	R	4.6	4.4	4	5.2	tp	5.0		
I90		39	36	141	58	14.0	B	a	R	5.2	5.0	2					
I91	Omoe (Tatehama), Miyako City	39	38	142	1	22.5	B	a	R				1.7	tp	1.5		
I92		39	38	142	1	25.2	B	a	R								
I93		39	38	142	1	26.1	B	a	R								
I94	Omoe (Shukuhama), Miyako City	39	38	142	1	25.1	B	a	R	2.1	1.9	3					
I95		39	38	142	1	24.6	B	a	R								
I96	Omoe (Uiso), Miyako City	39	37	142	1	27.0	B	a	R	2.5	2.3	3	2.2	tp	2.0		
I97		39	37	142	1	25.4	A	a	I								
I98		39	37	142	1	23.2	B	a	R								
I99		39	37	142	1	24.5	A	a	I								
I100	Omoe (Aramaki), Miyako City	39	36	142	1	21.1	B	a	R	2.6	2.4	4	0.6	tp	0.4		
I101		39	36	142	1	27.7	B	a	R								
I102		39	36	142	1	23.5	B	a	R								
I103	Omoe (Sato), Miyako City (Omoe Fishing Port)	39	34	142	2	20.1	B	a	R	2.5	2.3	3	1.6	tp	1.4		
I104		39	34	142	2	34.1	C	a	R	2.8	2.6	4					
I105	Omoe (Aneyoshi), Miyako City	39	32	142	3	26.5	B	a	R	3.0	2.8	4	1.5	tp	1.3		
I106		39	32	142	3	22.8	B	a	R	3.1	2.9	4					
I107		39	32	142	3	24.5	B	c	R								
I108	Omoe (Chikei), Miyako City	39	32	142	2	29.5	A	a	R	3.1	2.9	4	3.0	tp	2.8		
I109		39	32	142	2	31.2	A	a	R								
I110		39	32	142	2	29.4	A	a	R								
I111	Omoe (Ishihama), Miyako City	39	31	142	2	20.0	A	a	I	2.6 3.8	2.4 3.6	4 4	2.7	tp	2.5		
I112	Funakoshi (Ohura), Yamada Town	39	27	142	0	8.0	B	a	R	3.0	2.8	4					
I113		39	27	142	0	7.7	A	a	I								
I114		39	27	142	0	8.0	B	a	R								
I115		39	27	142	0	7.8	B	a	R								
I118	Funakoshi (Koyadori), Yamada Town	39	26	142	1	26.1	A	a	R	2.3	2.1	4					
I119		39	26	142	1	25.0	B	a	R								
I120	Akahama, Otsuchi Town	39	21	141	56	13.3	B	a	R	3.4	3.2	4	3.6	tp	3.4		
I121	Akahama, Otsuchi Town	39	21	141	56	11.5	A	a	I								
I122	(International Coastal Research Center, Atmosphere and Ocean Research Institute, the University of Tokyo)	39	21	141	56	12.1	A	a	I								
I123	Omachi, Kamaishi City	39	16	141	53	5.4	A	b	I	2.8 2.9 2.9	2.6 2.7 2.7	4 4 4	3.5	tp	3.3	0.5	T
I124	Osone, Tonicho,	39	13	141	53	20.3	A	a	R	2.2	2.0	4	2.2	tp	2.0		
I125	Kamaishi City	39	13	141	53	16.7	C	a	I								
I126		39	13	141	53	16.8	B	a	R								
I127	Sakuratoge, Tonicho,	39	12	141	53	13.7	B	a	R								
I128	Kamaishi City	39	12	141	53	12.7	B	a	R								
I129	Arakawa, Tonicho, Kamaishi City	39	12	141	52	16.5	B	a	R							1.2	I
I130	Horei, Sanrikucho-Okirai, Ofunato City	39	6	141	48	17.8	B	a	R	2.6	2.4	4					
I131		39	6	141	48	14.8	B	a	R								
I132		39	6	141	48	15.2	B	a	R								
I133	Shirahama, Sanrikucho-Ryori, Ofunato City	39	4	141	49	16.8	B	a	R	2.9 4.0	2.7 3.8	3 4	4.3 4.7	tp tp	4.1 4.5		
I134	Okubo, Sanrikucho-Ryori, Ofunato City	39	3	141	49	21.2	B	a	R								
I135	Miyanomae, Ofunatocho, Ofunato City	39	2	141	43	7.6	A	a	I	2.6 4.2	2.4 4.0	4 4	5.13	tp	5.0	0.4	T
I136	Yonesakicho, Rikuzentakata City	39	1	141	40	18.1	B	b	R	5.0	4.8	4	6.36	tp	6.2		
M1	Karakuwacho-Baba, Kesennuma City	38	54	141	39	13.1	A	a	R	2.0	2.2	2					
M3	Karakuwacho-Kakehama, Kesennuma City	38	53	141	40	15.6	A	a	I	3.8	4.0	2					
M4	Karakuwacho-Tsumoto, Kesennuma City	38	52	141	40	10.1	A	a	I	1.0	1.2	3	2.0	tp	2.2		
M5	Minamimachikaigan, Kesennuma City	38	54	141	34	4.0	A	a	I	2.7	2.9	4	2.9	tp	3.1	0.7	R
M6	(Kesennuma Port)	38	54	141	34	3.7	A	a	I							0.9	R
M7		38	54	141	34	4.1	A	a	I							1.0	I
M8		38	54	141	35	4.1	A	a	I							1.2 1.2	I I
M9	Hajikamisuginoshita, Kesennuma City	38	50	141	35	13.7	A	b	I	3.6	3.8	4				0.3 0.7	I R

Table 2

continued

No.	Location Name	Latitude		Longitude		2011 height[a] (m)	rel.[b]	acc.[c]	type	1960 height						2010 height	
		deg	min	deg	min					CFI[d] (m)	CFI Corrected[e] (m)	CFI acc.[f]	JMA[g] (m)	JMA datum	JMA Corrected[e] (m)	Tsuji/Imai[h] (m)	type
M10	Motoyoshicho-Amagasawa, Kesennuma City (Hikado Fishing Port)	38	48	141	33	16.6	A	a	R	2.5 3.4	2.7 3.6	4 4	3.2	tp	3.4		
M11	Motoyoshicho-Maehama, Kesennuma City	38	48	141	33	19.3	B	a	I	3.9	4.1	4				1.2	I
M12	Motoyoshicho-Toyomazawa, Kesennuma City	38	47	141	31	19.7	A	a	I	3.9	4.1	4					
M13	Motoyoshicho-Nijuichihama, Kesennuma City	38	46	141	31	17.4	A	a	I	3.1	3.3	4					
M14		38	45	141	31	20.9	B	a	R								
M15	Utatsu-Namiita, Minamisanriku Town	38	45	141	32	19.4	A	a	R	2.6	2.8	4				0.5	I
M16		38	45	141	32	18.0	A	a	I								
M17		38	45	141	32	19.5	A	a	R								
M18	Utatsu-Minato, Minamisanriku Town	38	44	141	32	14.5	A	a	R								
M19		38	44	141	32	14.0	A	a	R								
M20		38	44	141	32	14.6	A	a	I								
M21	Utatsu-Tanoura, Minamisanriku Town	38	44	141	33	13.1	B	a	R	3.0	3.2	4				0.7	I
M22	Utatsu-Kaminoyama, Minamisanriku Town	38	44	141	33	14.3	A	a	R								
M23		38	44	141	33	16.4	B	a	R								
M28	Utatsu-Niranohama, Minamisanriku Town	38	42	141	30	12.9	A	a	I	2.8 3.1	3.0 3.3	4 4					
M29	Shizugawa-Nishida, Minamisanriku Town	38	42	141	30	15.1	B	a	R	3.2	3.4	4					
M30		38	42	141	30	15.0	B	a	R	3.4	3.6	4					
M31	Shizugawa-Hosoura, Minamisanriku Town	38	42	141	30	13.9	B	a	R								
M32		38	42	141	30	14.5	A	a	R								
M33	Shizugawa-Omori, Minamisanriku Town	38	41	141	27	14.4	C	a	R	4.0 4.8	4.2 5.0	4 4	x 4.4	tp	4.6		
M34	Shizugawa-Hayashi, Minamisanriku Town	38	40	141	27	16.0	A	a	I	4.6	4.8	4					
M35	Mitobe, Tokura, Minamisanriku Town	38	38	141	27	11.7	B	a	R	4.2	4.4	4				1.1 1.4	I I
M36	Takihama, Tokura, Minamisanriku Town	38	38	141	30	12.1	B	a	I	3.2	3.4	4					
M37	Nagashizu, Tokura, Minamisanriku Town	38	38	141	31	11.3	B	a	R	3.6	3.8	4				0.8	I
M38	Kozashi, Kitakamicho-Jusanhama, Ishinomaki City	38	36	141	30	14.7	A	a	I	2.8 2.8	3.0 3.0	4 4					
M39	Aikawa, Kitakamicho-Jusanhama, Ishinomaki City	38	36	141	30	14.9	A	a	R	3.8	4.0	4				0.7	I
M40	Kodomari, Kitakamicho-Jusanhama, Ishinomaki City	38	36	141	30	12.0	A	a	R	3.3	3.5	4					
M41	Omuro, Kitakamicho-Jusanhama, Ishinomaki City	38	36	141	30	14.3	C	a	I	3.2 4.0	3.4 4.2	4 4	3.2	tp	3.4		
M42	Shirahama, Kitakamicho-Jusanhama, Ishinomaki City	38	35	141	28	15.7	A	a	I	2.9	3.1	4					
M50	Ogatsucho-Naburi, Ishinomaki City	38	32	141	30	14.2	B	a	R	2.6 2.8	2.8 3.0	4 4	3.0	tp	3.2		
M51	Ogatsucho-Funakoshi, Ishinomaki City	38	32	141	31	12.6	B	a	R	3.4	3.6	4				0.7	P
M53	Ogatsucho-Kuwanohama, Ishinomaki City	38	30	141	32	11.8	A	a	I	3.2	3.4	4					
M54	Ogatsucho-Tachihama, Ishinomaki City	38	30	141	31	11.7	B	a	R	2.7	2.9	4					
M55		38	31	141	31	9.6	B	a	R								
M56	Ogatsucho-Myojin, Ishinomaki City	38	31	141	29	12.3	B	a	R	3.3 3.7	3.5 3.9	4 4	3.6	tp	3.8		
M57	Funatoshinmei, Ogatsucho-Ogatsu, Ishinomaki City	38	31	141	28	15.3	A	a	R	4.3	4.5	4	4.0	tp	4.2		
M58	Karakuwa, Ogatsucho-Ogatsu, Ishinomaki City	38	31	141	29	10.1	C	a	I	3.1	3.3	4					
M59	Wakehama, Ogatsucho-Wakehama, Ishinomaki City	38	30	141	29	13.7	B	a	R	3.0	3.2	4					
M60	Namiita, Ogatsucho-Wakehama, Ishinomaki City	38	29	141	29	14.5	A	a	R	3.4	3.6	4					
M61	Ishihama, Onagawa Town	38	27	141	28	16.7	A	a	R	4.0	4.2	4				0.3 0.4 0.9 1.3 1.3 1.5	I I I I I R
M62	Oishiharahama, Onagawa Town	38	24	141	28	13.0	B	a	R	4.4	4.6	4	2.9	tp	3.1	1.4 1.4 1.5	I I I
M65	Samenoura, Ishinomaki City	38	23	141	29	17.9	C	a	I	4.5	4.7	4					

Table 2

continued

No.	Location Name	Latitude		Longitude		2011 height[a] (m)	rel.[b]	acc.[c]	type	1960 height						2010 height	
		deg	min	deg	min					CFI[d] (m)	CFI Corrected[e] (m)	CFI acc.[f]	JMA[g] (m)	JMA datum	JMA Corrected[e] (m)	Tsuji/Imai[h] (m)	type
M66	Oyagawahama, Ishinomaki City	38	22	141	29	17.8	B	a	R	5.0	5.2	2	5.4	tp	5.6		
M67	Yagawahama, Ishinomaki City	38	22	141	29	18.7	C	a	I	4.0	4.2	4					
M68		38	22	141	29	21.2	B	a	R	4.6 4.7	4.8 4.9	4 4					
M69	Tomarihama, Ishinomaki City	38	22	141	31	12.4	A	a	R	3.2	3.4	4				0.8	T
M70		38	22	141	31	12.2	A	a	R								
M71	Kugunarihama, Ishinomaki City	38	19	141	30	6.3	A	a	I	3.6 3.8	3.8 4.0	4 2					
M72	Koamikurahama, Ishinomaki City	38	21	141	27	12.9	B	a	R	3.2 3.8	3.4 4.0	4 4					
M73	Fukkiura, Ishinomaki City	38	21	141	27	8.2	B	a	R	3.4	3.6	4					
M75	Kozumihama, Ishinomaki City	38	22	141	27	8.5	B	a	I	5.0	5.2	4	4.4 x 5.0	tp tp	4.6 5.2		
M76	Momonoura, Ishinomaki City	38	24	141	26	12.5	B	a	R	4.1 4.7	4.3 4.9	4	5.0 x 5.2	tp tp	5.2 5.4		
B1	Hirakatacho,	36	51	140	48	6.6	A	a	I				5	da	2.5		
B2	Kitaibaraki City	36	51	140	48	7.1	A	a	I								
B3		36	51	140	48	7.2	A	a	I								
B4		36	51	140	48	8.1	B	b	R								
B5		36	51	140	48	7.9	B	b	R								
B6	Otsucho, Kitaibaraki City (Otsu Fishing Port)	36	50	140	47	4.9	A	a	I	1.5	1.4	3	5.7	da	2.9	1.0	I
B19	Osecho, Hitachi City (Ose Fishing Port)	36	35	140	39	5.1	A	a	I				3	msl	3.0	0.8 1.3 1.5 1.5 1.5 1.5 1.6 1.8 1.8	P R R R R R R R R
B24	Kujicho, Hitachi City (Kuji Fishing Port)	36	30	140	38	4.1	B	b	R	2.3	2.2	3	3	un			
B27	Isozakicho, Hitachinaka City (Isozaki Fishing Port)	36	23	140	37	3.5	A	a	I				3	da	1.5		
B28	Hiraisocho, Hitachinaka City (Hiraiso Fishing Port)	36	21	140	37	4.2	A	a	I				2.5	da	1.3		
B29	Kaimoncho, Hitachinaka City (Nakaminato Fishing Port)	36	20	140	36	3.2	A	a	I	1.25 2.1	1.41 2.2	1 5	2	da	1.0		
B30	Minatochuo, Oarai Town (Ibaraki Port, Oarai Port District)	36	19	140	34	4.5	A	a	I							1.2 1.2	P P
B31		36	19	140	34	4.6	A	a	I								
B32	Takeigama, Kashima City	36	4	140	37	3.5	A	b	I				2	un		1.5	R
B33		36	4	140	37	3.7	A	b	R								
B34	Hamatsuga, Kashima City	36	4	140	37	3.8	B	b	I								
B35	Higashifukashiba, Kamisu City (Kashima Port)	35	55	140	40	5.7	B	a	R							0.8 1.6	T I
B36	Hasakishinko, Kamisu City (Shin Fishing Port)	35	45	140	51	3.3	A	a	I							0.4 0.5	P P
C1	Araoicho, Choshi City (Choshi Fishing Port)	35	44	140	50	2.7	A	a	I	0.6 0.95	0.7 1.05	5 5				0.5 0.5 0.6	P P T
C2	Inuwaka, Choshi City (Inuwaka Fishing Port)	35	42	140	51	4.8	A	a	I	2.09	2.19	5	2.14	msl (Togawa)	2.2	1.0 1.2	I I
C3	Shimonagai, Asahi City (Iioka Fishing Port)	35	42	140	44	3.2	A	a	I				3.5	tp	3.5	0.9 1.5	R R
C4	Hiramatsu, Asahi City	35	42	140	43	6.3	A	a	I	3.7	3.8	2					
C5	Ashikawa, Asahi City	35	42	140	40	5.1	A	a	I							0.3	P
C6		35	42	140	40	7.9	B	a	R								
C15	Koseki, Kujukuri Town (Katakai Fishing Port)	35	32	140	27	2.5	A	a	I	1.4	1.5	3				0.6 0.7 0.7 0.7 0.7	R R R R I
C18	Sendokyu, Ichinomiya Town	35	23	140	23	2.9	A	c	I	1.0	1.1	3	2	nt	2.0		
C22	Ohara, Isumi City (Ohara Fishing Port)	35	15	140	24	2.6	B	a	I	1.8 1.8	1.6 1.6	1 1	2-3	un		0.8	I
C26	Hama, Onjuku Town (Onjuku Fishing Port)	35	11	140	21	2.3	A	a	I	1.7	1.5	3				0.5 0.7 1.1	R R R
C27	Hamakatsuura, Katsuura City (Katsuura Fishing Port)	35	9	140	19	2.0	B	a	R	1.9 2.2	1.7 2.0	3 2	2.0-2.5	msl	1.8-2.3	0.6 0.8	P P

Table 2

continued

No.	Location Name	Latitude		Longitude		2011 height[a] (m)	rel.[b]	acc.[c]	type	1960 height						2010 height	
		deg	min	deg	min	(m)				CFI[d] (m)	CFI Corrected[e] (m)	CFI acc.[f]	JMA[g] (m)	JMA datum	JMA Corrected[e] (m)	Tsuji/Imai[h] (m)	type
C29	Kominato, Kamogawa City (Kominato Fishing Port)	35	7	140	12	1.7	B	a	P	1.5	1.3	3					
C30	Amatsu, Kamogawa City (Amatsu Fishing Port)	35	7	140	10	1.8	B	a	R	1.3	1.1	2					
C31		35	7	140	10	1.5	B	a	R								
C34	Wadacho-Wada, Minamiboso City (Wada Fishing Port)	35	2	140	1	1.9	A	a	I	1.0	0.8	2	0.7	ht	0.9		
C37	Chikuracho-Hedate, Minamiboso City (Chikura Fishing Port)	34	57	139	58	1.4	B	a	I	1.3 1.2	1.0 1.1	3 3	0.9	ht	1.1		
C38		34	57	139	58	0.9	B	a	P								
C39	Shirahamacho-Otohama, Minamiboso City (Otohama Fishing Port)	34	55	139	56	1.2	B	a	R	1.3	1.1	3					
C40	Shirahamacho-Shirahama, Minamiboso City (Nojimahigashi Fishing Port)	34	54	139	53	1.0	B	a	R	1.0	0.8	3[i]	2.0	da	1.0		
C43	Mera, Tateyama City (Tomisaki Fishing Port)	34	55	139	50	2.0	A	a	I	1.32	1.11	5	1.73	nt	1.7	0.7 0.8	P T

R runup height, *I* inundation height, *P* tsunami height in ports, *T* tide gauge, *da* double amplitude, *msl* tsunami heights above mean sea level, *dl* datum line, *tp* tsunami heights above TP, *nt* tsunami heights above sea level at time of maximum tsunami, *ht* tsunami heights above high tide level of the tsunami arrival date, *un* unknown, *x* eyewitness accounts

[a] 2011 heights above sea level at time of maximum tsunami

[b] Rel.: reliability, A: most reliable based on clear physical evidence or eyewitness account; B: mostly based on natural traces; C: least reliable based on equivocal evidence

[c] Acc.: accuracy; a: measurement error <0.2 m; b 0.2 ≤ error ≤ 0.5 m; c error >0.5 m

[d] Tsunami heights above Tokyo Peil (TP) taken from the Committee of Field Investigation of the Chilean Tsunami of 1960 (CFI 1961). The numbers in parentheses indicate questionable data

[e] Tsunami heights above sea level at time of maximum tsunami

[f] 5: Values observed by tide gauges; 4: values with highest accuracy; 3: values with moderate accuracy; 2: values with fair accuracy; 1: values obtained by other sources

[g] Tsunami heights from JMA (1961)

[h] Tsunami heights above sea level at the time of maximum tsunami taken from Tsuji *et al.* (2010) or Imai *et al.* (2010)

[i] Nojimanishi

heights were measured as 17 and 21 m (I133–I134). The 1960 tsunami heights were 3 and 4 m by CFI (1961), and 4 and 5 m by the JMA (1961), respectively.

5.2. *Northern and Southern Sanriku Coasts*

Along the northern Sanriku coast (north of latitude 40.2°N), the 2011 tsunami heights drastically decrease toward the north from 20 to 5 m. The two preceding Sanriku tsunami heights also become smaller toward the north (Fig. 14). The 1896 heights are similar to the 2011 heights (median ratio is 1.01; Table 3), while the 1933 heights are smaller (median

ratio 0.66). The 1960 and 2010 Chilean tsunami heights are more uniform throughout the Sanriku coast, and are smaller than the three Sanriku tsunamis (median ratio of 1960/2011 heights is 0.42, while one 2010 height is 0.15 of the 2011 height).

Along the southern Sanriku coast (south of latitudes 39.0°N), the 2011 tsunami heights range mostly from 10 to 20 m, and larger than all the previous tsunamis (median ratio is 0.29, 0.24, 0.28, 0.06 for 1896, 1933, 1960, and 2010 tsunami, respectively). The 1960 and 2010 tsunami heights are more uniformly distributed. The 1896 and 1933 tsunami heights drastically decrease toward the south

Table 3

Median ratios and correlation coefficients of the previous tsunami heights and the 2011 tsunami heights

Coast	1896/2011			1933/2011			1960/2011			2010/2011		
	Ratio (median)	Correlation	Number[a]	Ratio (median)	Correlation	Number[a]	Ratio (median)	Correlation	Number[a]	Ratio (median)	Correlation	Number[a]
N. Sanriku	1.01	0.63	11	0.66	0.65	13	0.42	–0.38	17	0.15	–	1
C. Sanriku	0.85	0.24	66	0.47	0.36	81	0.16	0.21	71	0.09	0.60	6
S. Sanriku	0.29	0.22	46	0.24	0.39	105	0.28	0.33	58	0.06	–0.12	12
All Sanriku	0.69	0.34	123	0.33	0.47	199	0.25	0.17	146	0.07	0.14	19
Ibaraki–Chiba	–	–	–	–	–	–	0.62	0.63	37	0.28	0.41	15

[a] Number of comparison; if multiple measurements are reported at one location, the median height is used for each report

and become smaller than the 1960 Chilean tsunami heights at the southernmost Sanriku coast.

At Utatsu-Tanoura in Minamisanriku Town, the 2011 tsunami was evidently much higher than the other historical tsunamis. Almost all the houses in low-lying areas were washed away, and large amounts of rubble, fishing boats, and equipment were transported by the tsunami. The 2011 runup heights were between 13 and 16 m (M21–M23), while the 1896 tsunami heights were 16 feet (5 m; Iki 1897) or 4–11 m (Matsuo 1933). The 1933 tsunami heights were 5–6 m (Matsuo 1933; Earthquake Research Institute 1934). The 1960 tsunami height was 3 m by CFI (1961) and the 2010 height was 1 m by Tsuji et al. (2010), respectively.

At Ogatsucho-Wakehama in Ishinomaki City, the 2011 tsunami height was the largest, followed by the 1960 Chilean tsunami height, which was larger than the 1896 and 1933 Sanriku tsunamis. The 2011 tsunami inundated areas up to a temple ∼250 m from the coast, and almost all houses were swept away. The 2011 runup height was 14 m (M59), while the reported 1896 and 1933 tsunami heights were ∼2 m and the 1960 tsunami height was reported as 3 m by CFI (1961).

5.3. Correlation of Tsunami Heights

The tsunami heights from the 1896 and 1933 Sanriku tsunamis and the 1960 and 2010 Chilean tsunamis were compared with the 2011 tsunami heights at the same locations (Fig. 19; Table 3). Along the Sanriku coasts, the 2011 tsunami heights are positively correlated with those from the 1896 and 1933 Sanriku tsunamis. The correlation coefficient of the 2011 and 1896 tsunamis is 0.34, and that for the 2011 and 1933 tsunamis is 0.47 for the entire Sanriku coasts. The correlation coefficients are much larger on the northern Sanriku coast (0.63 and 0.65 for 2011–1896 and 2011–1933, respectively), but smaller on the central coast (0.24 and 0.36) and southern coast (0.22 and 0.39). The positive correlation coefficients indicate that the tsunami height variation is similar for local tsunamis. On the other hand, the 2011 tsunami heights are weakly correlated with those from the Chilean tsunami heights; the correlation coefficients are 0.17 for the 2011 and

Figure 15
Tsunami stone transported by the 1896 Sanriku tsunami at Raga, Tanohata Village, Iwate Prefecture. The 2011 tsunami inundation limit (24.5 m; I37) is just below this stone

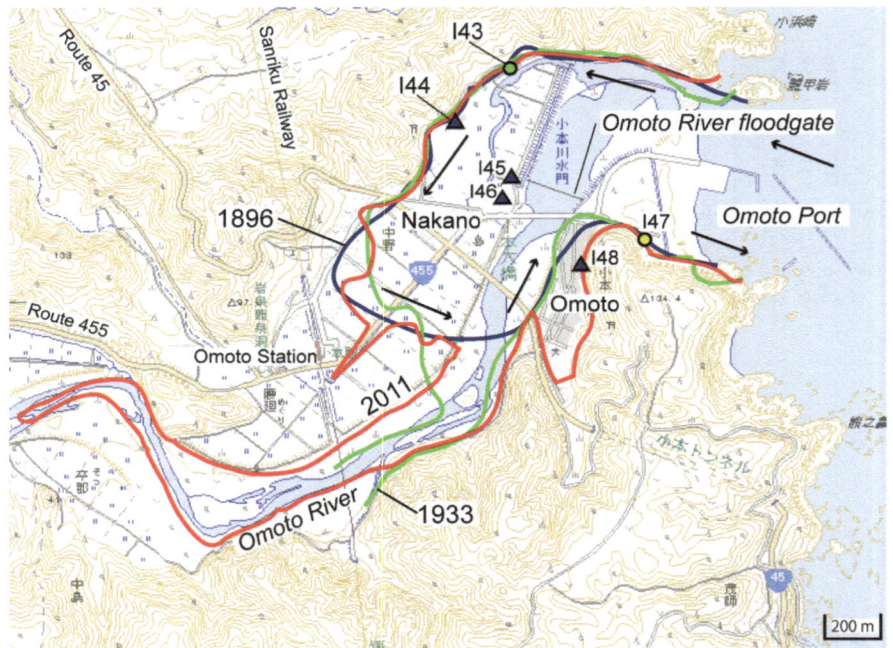

Figure 16
Inundation areas from the 1896, 1933, and 2011 tsunamis at Omoto, Iwaizumi Town, Iwate prefecture. The *blue curve* indicates the area severely damaged by the 1896 tsunami from Iki (1897). The *black arrows* show the direction of the 1896 tsunami reported by the Fudai Village chief in those days. The *green curves* show the inundation limits of the 1933 tsunami from the EARTHQUAKE RESEARCH INSTITUTE (1934). The *red curve* indicates the inundation limit of the 2011 tsunami by HARAGUCHI and IWAMATSU (2011). The *circles* and *triangles* indicate the measurement points of runup and inundation heights, respectively, with the same *color* code as Fig. 3

1960 tsunamis and 0.14 for the 2011 and 2010 tsunamis for the entire Sanriku coastline. They are negatively correlated on the northern Sanriku coast for the 1960 tsunami and on the southern coast for the 2010 tsunami. These indicate that tsunami height distribution of local tsunamis is different from that of trans-Pacific tsunamis.

Figure 17
Icehouse of the Japan Fisheries Cooperatives at Taro Fishing Port, which was severely damaged by the 2011 tsunami (14.8 m; I72). The two *white* markers on the back cliff show the heights of the 1896 (14.6 m; IKI 1897) and 1933 (10 m; EARTHQUAKE RESEARCH INSTITUTE 1934) Sanriku tsunamis. The tsunami height of 10 m for the 1933 tsunami was measured at a slightly different location; the tsunami heights at this location measured by EARTHQUAKE RESEARCH INSTITUTE (1934) and MATSUO (1933) were 7.0 and 6.4 m, respectively

Figure 18
Maximum tsunami height of 38.2 m from the 1896 Sanriku tsunami shown on a pole near the border of Okubo and Shirahama, Sanrikucho-Ryori, Ofunato City, Iwate Prefecture (labeled Ryori in Fig. 14). The 2011 tsunami with heights of 16.8 m (I133) and 21.2 m (I134) did not reach this location

5.4. Controlling Factors of Tsunami Heights and their Variations

The type, location, and magnitude of the 1896, 1933, and 2011 earthquakes are all different. The 1896 Sanriku earthquake (M_s 7.2) was an example of a "tsunami earthquake" (KANAMORI 1972) that produces weak ground shaking, but a very large tsunami. The 1896 tsunami source was estimated to be near the trench axis with a 210 km fault length and ~50 km fault width (Fig. 1; TANIOKA and SATAKE 1996; TANIOKA and SENO 2001). The seafloor deformation, landward subsidence, and seaward uplift were limited near the trench axis (Fig. 20). The 1933 Sanriku earthquake (M_s 8.5) was an outer-rise earthquake with a normal faulting mechanism (KANAMORI 1971). The seafloor deformation was dominantly subsidence (AIDA 1977). The 2011 Tohoku earthquake was the largest (M_w 9.0). The seafloor deformation extended much further than the above-mentioned Sanriku earthquakes. The largest slip occurred at around 38.3°N, 143.3°E, to the east of the epicenter, but the maximum tsunami heights were recorded on the central Sanriku coast ~100 km north of the largest slip. The 2011 Tohoku earthquake can be considered a combination of a great interplate earthquake and a "tsunami earthquake" (Fig. 1; FUJII *et al.* 2011; SATAKE *et al.* 2013), while OKAL (2013) argued that there is no evidence of 'slowness' in the earthquake source, and GRILLI *et al.* (2013) suggested additional tsunami generation mechanisms not represented in the coseismic sources (e.g., splay faults, sub-marine mass failure).

The tsunami heights from the three Sanriku tsunamis exhibit a large variation on the central Sanriku coast (between 39.0° and 40.2°N) regardless of the type, location, and magnitude of the earthquakes. The tsunami heights from the 1896 and 2011 earthquakes ranging 5–40 m are approximately similar on the central Sanriku coast, whereas the seismologically determined earthquake magnitudes of M_s 7.2 for the 1896 were much smaller than the 2011 earthquake (M_w 9.0). While the source region of the 2011 earthquake includes the rupture area of the 1896 earthquake, the 1896 tsunami heights are higher than the 2011 tsunami heights at some locations. These facts demonstrate that the coastal tsunami heights on the central Sanriku coast are not necessarily controlled by the location, type, or the magnitude of earthquake, and that the huge tsunami was not a surprise.

The Sanriku coast consists of numerous bays of various sizes and depths; it is called a ria coast, as it

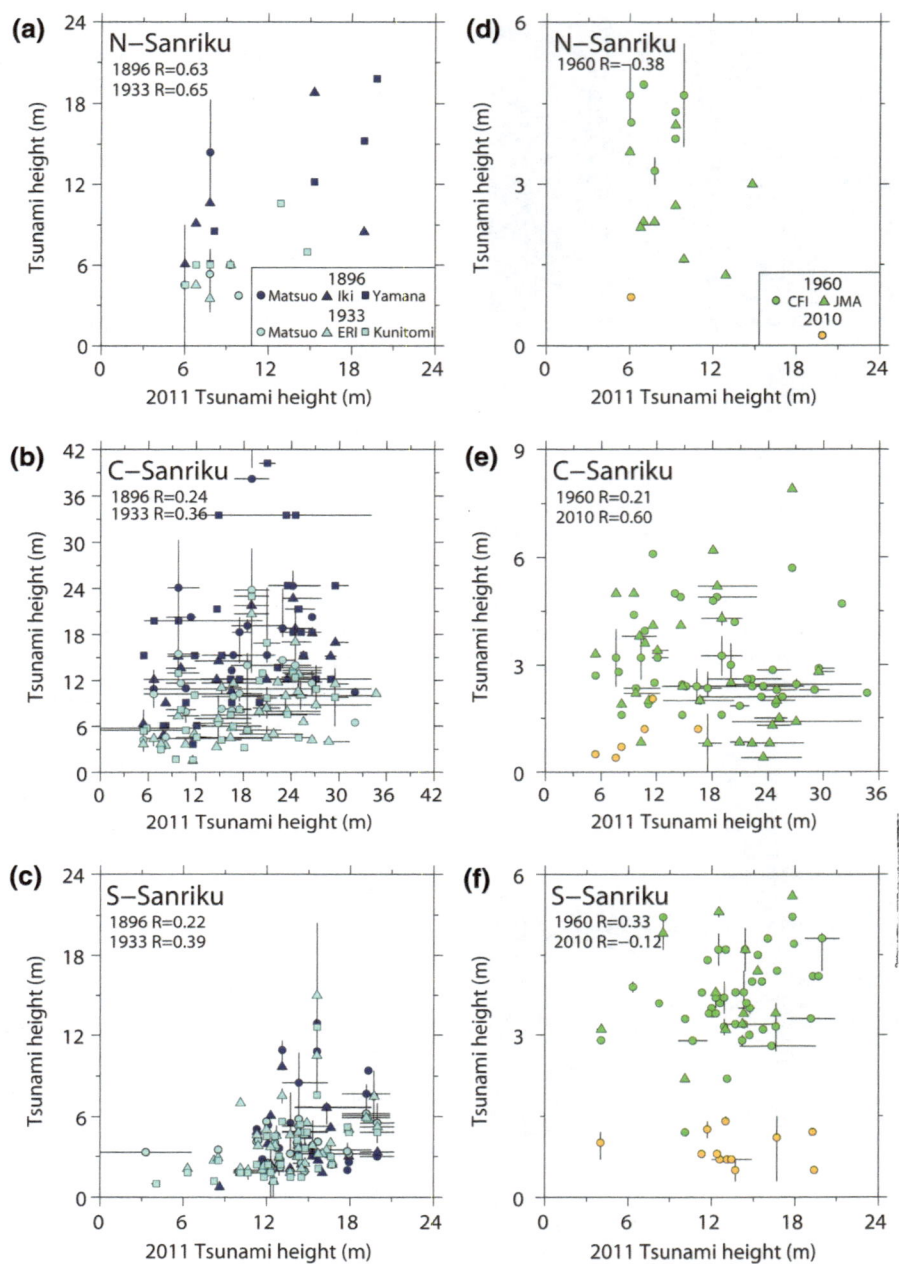

Figure 19

Comparison between the 2011 tsunami and the 1896 and 1933 Sanriku tsunamis for the northern (**a**), central (**b**), and southern Sanriku coasts (**c**). The similar comparison between the 2011 tsunami and the 1960 and 2010 Chilean tsunamis are shown in **d–f**. *Orange circles* indicate the 2010 tsunami heights from Tsuji *et al*. (2010). Other *symbols* are the same as in Fig. 14. Multiple height data at the same locations are represented as median values with ranges (shown as *bars*). The correlation coefficients of the previous tsunami heights and the 2011 tsunami heights are also shown

was created by the submergence and subsequent flooding of mountainous terrain (e.g., OTUKA 1934; KOIKE *et al.* 2005). The characteristic periods of sea level oscillations in bays are also variable; 55.2 min for Miyako Bay, the largest bay on the Sanriku coast, while 27.0 min for Otsuchi (HONDA *et al.* 1908).

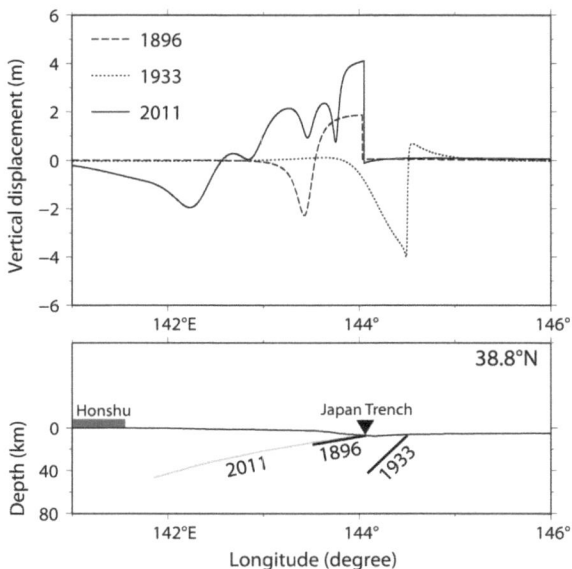

Figure 20
Seafloor deformation along the latitude of 38.8°N caused by the 1896 Sanriku (*dashed curve*), 1933 Sanriku (*dotted curve*), and 2011 Tohoku (*solid curve*) earthquakes. Cross sections of fault models for the 1896 Sanriku (TANIOKA and SENO 2001), 1933 Sanriku (AIDA 1977), and 2011 Tohoku (SATAKE et al. 2013) earthquakes are shown below. Locations of trench axis and land area are also shown

The period of an incoming tsunami wave is also a controlling factor. The dominant period for the 2011 event was estimated to be ∼45 min at DART 21418 and 33–66 min at DART 21413 (BORRERO and GREER 2013), while HEIDARZADEH and SATAKE (2013) estimated the two dominant periods of 37 and 67.4 min from multiple DART records. The dominant periods of the preceding two Sanriku tsunamis are expected to be shorter because their source dimensions were much smaller than that of the 2011 event (i.e., 210 km × 50 km for the 1896 tsunami; TANIOKA and SATAKE 1996, and 185 km × 50 km for the 1933 Sanriku tsunami; AIDA 1977). However, the dominant period of the 2010 Chilean tsunami was much longer (∼110 min) at DART 21413 (BORRERO and GREER 2013). The dominant period becomes longer after a tsunami propagates the Pacific Ocean because of the dispersion effect (WATADA et al. 2013).

For local tsunamis, the large variation of tsunami heights along the Sanriku coast are probably caused by matching the periods of incoming waves and the characteristic periods of some bays. On the other hand, tsunami heights are less sensitive to the coastal topography and show a more uniform distribution for trans-Pacific Chilean tsunamis, resulting from longer periods than the characteristic periods of bays along the Sanriku coast.

On the northern (north of 40.2°N) and southern (south of 39.0°N) Sanriku coasts, the local variations in tsunami height are much smaller. The tsunami heights from the three earthquakes were similar on the northern Sanriku coast, while the 2011 heights were much larger than those of the 1896 or 1933 Sanriku tsunamis on the southern Sanriku coast. The distance from the tsunami source and the earthquake magnitude control the tsunami heights on these coasts. The distances to the northern Sanriku coasts from the three sources were similar, while the 2011 tsunami source is much closer to the southern Sanriku coast (Fig. 1).

SUPPASRI et al. (2013) performed regression analyses between the earthquake magnitude and the maximum tsunami heights based on the historical tsunami trace database and the field survey of the 2011 Tohoku earthquake in each tsunami-affected location. They claimed that the earthquake magnitude is a major controlling factor in determining the maximum tsunami heights. However, the examples of the 1896 and 2011 tsunami heights are clear counterevidence for magnitude dependence. CHOI et al. (2012) approximated the distribution of tsunami heights along the coast by simple log-normal distributions, suggesting that the tsunami heights are controlled only by the distance from the source. However, the distribution of tsunami heights along the Sanriku coast clearly demonstrates a significant contribution by other factors such as irregular coastal topography.

6. Comparison of Tsunami Heights on Ibaraki and Chiba Coasts

For the Pacific coasts of Ibaraki and Chiba prefectures (between 34.9° and 36.9°N), TSUJI et al. (2011) measured 79 tsunami heights at 35 locations. Of these, tsunami heights from the 1960 and 2010 Chile earthquakes were also reported at 24 and 15 locations, respectively (Fig. 21; Table 2). The 1960

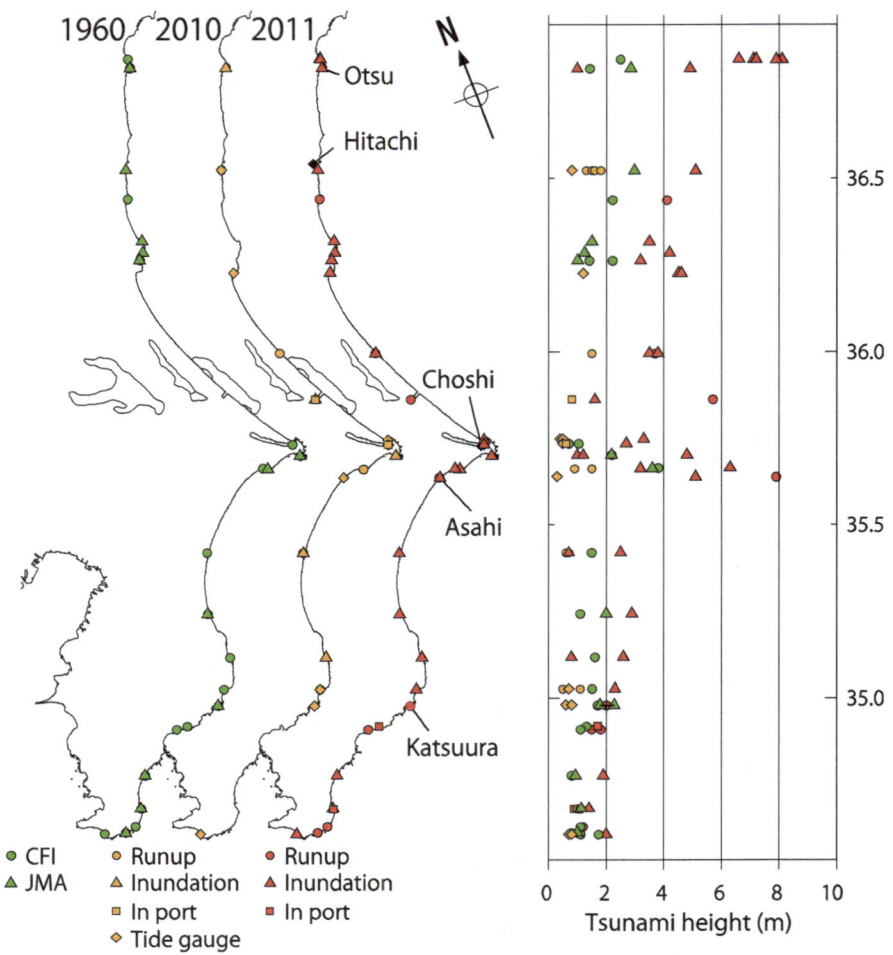

Figure 21
Comparison of 1960 and 2010 Chilean tsunami heights with 2011 Tohoku tsunami heights along the coasts of Ibaraki and Chiba prefectures. *Green circles* and *triangles* indicate 1960 heights from CFI (1961) and JMA (1961), respectively. *Orange circles, triangles, squares,* and *diamonds,* respectively, indicate runup heights, inundation heights, tsunami heights in ports, and tsunami heights from tide gauges from the 2010 Chile earthquake (Tsuji *et al.* 2010; Imai *et al.* 2010). *Red circles, triangles,* and *squares,* respectively, indicate 2011 runup heights, inundation heights, and tsunami heights in ports from Tsuji *et al.* (2011). The measurement locations are indicated on the maps on the *left*. Only data measured at the same locations for two or more tsunami are shown

tsunami heights are smaller than the 2011 tsunami heights (median ratio is 0.62; Table 3), and the 2010 heights are much smaller (median ratio 0.28).

6.1. Tsunami Heights

At Otsu in Kitaibaraki City, the measured inundation height from the 2011 Tohoku tsunami was ∼5 m (B6). The 1960 tsunami heights were 1 m (CFI 1961) or 3 m (JMA 1961), while the 2010 tsunami caused only a minor inundation with a height of 1 m.

Along the coast of Ibaraki and Chiba prefectures, the 2011 tsunami heights gradually decreased toward the south. The two Chilean tsunamis also showed similar tendencies, though the change is smaller. These similar variations are reflected in the positive correlation of tsunami heights (Table 3; Fig. 22). The correlation coefficient between the 2011 and 1960 tsunami heights is 0.63, and that for the 2011 and 2010 tsunamis is 0.41.

Local amplifications of tsunami heights around Asahi City were found for the 2011, 2010, and 1960

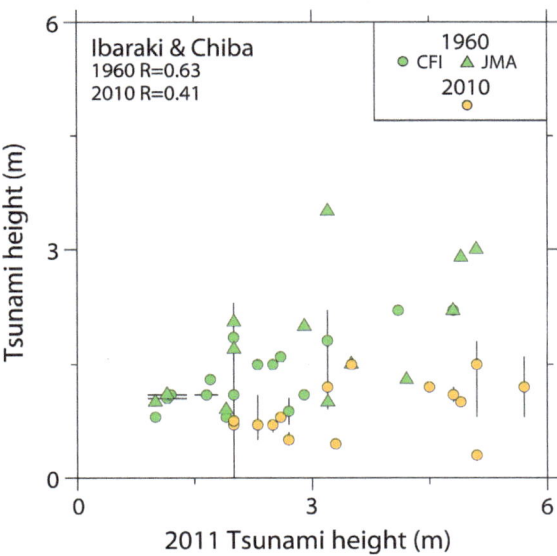

Figure 22

Comparison between the 2011 tsunami and the 1960 and 2010 Chilean tsunamis along the Ibaraki and Chiba coasts. *Green circles* and *triangles* indicate the 1960 Chilean tsunami heights from CFI (1961), and JMA (1961), respectively. *Orange circles* indicate the 2010 tsunami heights from Tsuji *et al.* (2010) or Imai *et al.* (2010). Multiple height data at the same locations are represented as median values with ranges (shown as *bars*)

tsunamis: tsunami heights were locally high (5–8 m) around Asahi City in 2011, while the 1960 and 2010 Chilean tsunami heights were 4 and 2 m, respectively. This local peak may be due to the local topography (the peninsula around Choshi and the local bathymetry off Asahi). The tsunami heights were also locally large around Katsuura, another gentle peninsula (Fig. 21). These common local variations also contributed to the large correlation coefficients.

6.2. *Factors Controlling Tsunami Heights and their Variability*

The 2011 tsunami heights generally decreased toward the south, away from the tsunami source. The 1960 and 2010 tsunamis, which were generated by the earthquakes in Chile (M_w 9.5 and 8.8, respectively) and propagated across the Pacific Ocean, show more uniform heights along the coasts, although they also decreased toward the south. Of the two Chilean tsunamis, the 2010 tsunami heights are consistently lower than the 1960 heights because of the smaller earthquake magnitude.

On a smaller scale, both near-field and transoceanic tsunamis show similar local variations. Local peaks around Choshi and Asahi, and near Katsuura were found for all the tsunamis, possibly as a result of the local topography, which consists of a number of small peninsulas. The peninsula is more distinct around Choshi and the tsunami heights show a more significant peak, while the peninsula near Katsuura is gentler and the peak in tsunami heights is less pronounced. This indicates that the local topography also affects local variation in tsunami height, although the main controlling factor is the source location, slip distribution, and the earthquake magnitude.

7. *Conclusions*

We summarized our 12 field surveys in which 296 tsunami heights accompanying the 2011 Tohoku earthquake were measured. The data and detailed locations of survey points and photographs (this paper and Tsuji *et al.* 2011) will be useful for modeling the 2011 tsunami source (e.g., Satake *et al.* 2013). We then compared tsunami heights for the 2011 Tohoku earthquake with those from past earthquakes: the 1896 and 1933 Sanriku earthquakes in Japan, and the 1960 and 2010 Chile earthquakes. Along the central Sanriku coast (between 39.0° and 40.2°N), the 2011 and 1896 tsunami heights ranged from 5 to 40 m, showing significant local variation. This may be due to the rugged and irregular coastline, indicating that local topography is a major factor in controlling tsunami height, together with the location, type, or magnitude of the earthquake. This is evident from the fact that the largest tsunami height was recorded at around 40.0°N for the 1896 and 2011 tsunamis, despite these tsunamis having different source locations. The local variations are much smaller on the northern and southern Sanriku coasts and the Ibaraki and Chiba coasts. The 2011 tsunami heights generally decrease toward the north and south, and also show local variations probably due to local topography. The 1960 and 2010 Chilean tsunami heights are more uniform. Both near-field and transoceanic tsunamis exhibit local peaks in their heights near peninsulas. Such local variations of tsunami heights may be helpful for educating coastal residents to reduce future tsunami disasters.

Acknowledgments

We thank Haeng Yoong Kim, Toshihiro Ueno, Satoko Murotani, Satoko Oki, Megumi Sugimoto, Jiro Tomari, Mohammad Heidarzadeh, Shingo Watada, Kentaro Imai, Byung Ho Choi, Sung Bum Yoon, Jae Seok Bae, Kyeong Ok Kim, Hyun Woo Kim, Makoto Yoshimizu, and Morio Koyama for helping to measure tsunami heights. We also thank Kentaro Imai for providing data on tsunami heights from the 2010 Chile earthquake. Valuable comments and suggestions from the guest editor, Dr. Jose C. Borrero, and two anonymous reviewers were very helpful in improving our manuscript. Most of the figures were generated using Generic Mapping Tools (WESSEL and SMITH 1998). We also used a digital topographic map with a scale of 1:25,000 from the Geospatial Information Authority of Japan. This study was partially supported by Grants-in-Aid for Scientific Research from the Ministry of Education, Culture, Sports, Science and Technology, Japan.

REFERENCES

ABE, K., 1994. *Instrumental magnitudes of historical earthquakes, 1892 to 1898*, Bulletin of the Seismological Society of America, *84*, 415–425.

AIDA, I, 1977. *Simulations of large tsunamis occurring in the past off the coast of the Sanriku district*, Bulletin of the Earthquake Research Institute, University of Tokyo, *52*, 71–101 (in Japanese with English abstract).

BORRERO, J.C. and GREER S.D., 2013. *Comparison of the 2010 Chile and 2011 Japan tsunamis in the far-field*, Pure and Applied Geophysics, *170*, 1249–1274, doi:10.1007/s00024-012-0559-4.

CFI (COMMITTEE for FIELD INVESTIGATION of the CHILEAN TSUNAMI of 1960), 1961. *Report on the Chilean tsunami of May 24, 1960, as observed along the coast of Japan*, Tokyo, Maruzen Co., 397 pp.

CHOI, B.H., MIN, B.I., PELINOVSKY, E., TSUJI, Y., and KIM, K.O., 2012. *Comparable analysis of the distribution functions of runup heights of the 1896, 1933 and 2011 Japanese Tsunamis in the Sanriku area*, Natural Hazards and Earth System Sciences, *12*, 1463–1467.

EARTHQUAKE RESEARCH INSTITUTE, 1934. *Field survey report of damage caused by the 1933 Sanriku earthquake*, Bulletin of the Earthquake Research Institute, University Tokyo (supplementary volume), -Papers and Reports on the Tsunami of 1933 on the Sanriku Coast, Japan, *1*, 9–139 (in Japanese).

FUJII, Y., SATAKE, K., SAKAI, S., SHINOHARA, M., and KANAZAWA, T., 2011. *Tsunami source of the 2011 off the Pacific coast of Tohoku Earthquake*, Earth, Planets and Space, *63*, 815–820.

GRILLI, S.T., HARRIS, J.C., BAKHSH, T.S.T., MASTERLARK, T.L., KYRIAKOPOULOS, C., KIRBY, J.T., and SHI, F., 2013. *Numerical simulation of the 2011 Tohoku tsunami based on a new transient FEM co-seismic source: comparison to far- and near-field observations*, Pure and Applied Geophysics, *170*, 1333–1359, doi:10.1007/s00024-012-0528-y.

HARAGUCHI, T. and IWAMATSU, A., 2011. *Detailed maps of the impacts of the 2011 Japan Tsunami, Vol. 1: Aomori, Iwate and Miyagi prefectures*, Kokon-Shoin, Publishers Ltd., Tokyo, 167 pp. (in Japanese).

HATORI, T., 1995. *Review of documents for the 1896 Meiji Sanriku tsunami along the coast of Iwate Prefecture*, Tsunami Engineering Technical Report, Disaster Control Research Center, Tohoku University, *12*, 59–65 (in Japanese).

HONDA, K., TERADA, T., YOSHIDA, Y., and ISITANI, D., 1908. *Secondary undulations of oceanic tides*, Journal of the College of Science, Imperial University, Tokyo, Japan, *24*, 1–113.

IKI, T., 1897. *A report of the field investigation of the tsunami of 1896 in the Sanriku District*, Reports of the Imperial Earthquake Investigation Committee, *11*, 5–34 (in Japanese).

IMAI, K., NAMEGAYA, Y., TSUJI, Y., FUJII, Y., ANDO, R., KOMATSUBARA, J., KOMATSUBARA, T., HORIKAWA, H., MIYACHI, Y., MATSUYAMA, M., YOSHII, T., ISHIBE, T., SATAKE, K., NISHIYAMA, A., HARADA, T., SHIGIHARA, Y., SHIGIHARA, Y., and FUJIMA, K., 2010. *Field survey for tsunami trace height along the coasts of the Kanto and Tokai districts from the 2010 Chile Earthquake*, Journal of Japan Society of Civil Engineers, Ser. B2 (Coastal Engineering), *66*, 1351–1355 (in Japanese with English abstract).

IMAMURA, A., 1934. *Past tsunamis of the Sanriku coast*, Japanese Journal of Astronomy and Geophysics, *11*, 79–93.

IMAMURA, F. and WATANABE, T., 1990. *Surveys of large Sanriku tsunamis at Taro, Iwate Prefecture*, Research Report of the Tsunami Disaster Prevention Laboratory, Civil Engineering, Tohoku University, *7*, 123–140 (in Japanese).

JMA (JAPAN METEOROLOGICAL AGENCY), 1961. *Report on the tsunami of the Chilean earthquake, 1960*, Technical Report of the Japan Meteorological Agency, *8*, 389 pp. (in Japanese).

JMA (JAPAN METEOROLOGICAL AGENCY), 2011. *Monthly report on earthquakes and volcanoes in Japan March 2011*, 321 pp. (in Japanese).

KANAMORI, H., 1971. *Seismological evidence for a lithospheric normal faulting—the Sanriku Earthquake of 1933*, Physics of the Earth and Planetary Interiors, *4*, 289–300.

KANAMORI, H., 1972. *Mechanism of tsunami earthquakes*, Physics of the Earth and Planetary Interiors, *6*, 346–359.

KOIKE, K., TAMURA, T., CHINZEI, K., and MIYAGI, T., editors, 2005. *Regional Geomorphology of the Japanese Islands Vol. 3 Geomorphology of Tohoku Region*, University of Tokyo Press, Tokyo, 355 pp. (in Japanese).

KUNITOMI, S., 1933. *Off Sanriku earthquake and tsunami on March 3, 1933*, Quarterly Journal of Seismology, *7*, 111–153.

MATSUO, H., 1933. *Report on the survey of the 1933 Sanriku tsunami*, Report of the Civil Engineering Laboratory, *24*, 83–136 (in Japanese).

MATSUO, H., 1934. *Report on the survey of the 1933 Sanriku tsunami (supplement)*, Report of the Civil Engineering Laboratory, *27*, 93–94 (in Japanese).

HEIDARZADEH, M., and SATAKE, K., 2013. *Waveform and spectral analyses of the 2011 Japan tsunami records on tide gauge and DART stations across the Pacific ocean*, Pure and Applied Geophysics, *170*, 1275–1293, doi:10.1007/s00024-012-0558-5.

MORI, N., TAKAHASHI, T., YASUDA, T., and YANAGISAWA, H., 2011. *Survey of 2011 Tohoku earthquake tsunami inundation and runup*, Geophysical Research Letters, *38*, L00G14, doi:10.1029/2011GL049210.

MORI, N., TAKAHASHI, T., and The 2011 TOHOKU EARTHQUAKE TSUNAMI JOINT SURVEY GROUP, 2012. *Nationwide post event survey and analysis of the 2011 Tohoku earthquake tsunami*, Coastal Engineering Journal, *54*(01), 1250001, doi:10.1142/S0578563412500015.

OKAL, E.A., 2013. *From 3-Hz P waves to $_0S_2$: no evidence of a slow component to the source of the 2011 Tohoku Earthquake*, Pure and Applied Geophysics, *170*, 963–973, doi:10.1007/s00024-012-0500-x.

OTUKA, Y., 1934. *Tsunami damages March 3rd, 1933, and the topography of Sanriku coast, Japan*, Bulletin of Earthquake Research Institute, University Tokyo (supplementary volume), - Papers and Reports on the Tsunami of 1933 on the Sanriku Coast, Japan, *1*, 127–151.

OZAWA, S., NISHIMURA, T., SUITO, H., KOBAYASHI, T., TOBITA, M., and IMAKIIRE, T., 2011. *Coseismic and postseismic slip of the 2011 magnitude-9 Tohoku-Oki earthquake*, Nature, *475*, 373–376.

SATAKE, K., FUJII, Y., HARADA, T., and NAMEGAYA, Y., 2013. *Time and space distribution of coseismic slip of the 2011 Tohoku earthquake as inferred from tsunami waveform data*, Bulletin of the Seismological Society of America, *103*, 1473–1492.

SHUTO, N. and GOTO, T., 1985a. *Trace surveys of the large Sanriku tsunamis—Raga, Hiraiga, and Shimanokoshi in Tanohata Village, Omoto and Shimokonari in Iwaizumi Town*, Research Report of the Tsunami Disaster Prevention Laboratory, Faculty of Civil Engineering, Tohoku University, *2*, 39–45 (in Japanese).

SHUTO, N. and GOTO, T., 1985b. *Trace surveys of the large Sanriku tsunami—Okirai in Sanriku Town*, Research Report of the Tsunami Disaster Prevention Laboratory, Faculty of Civil Engineering, Tohoku University, *2*, 46–53 (in Japanese).

SHUTO, N. and UNOHANA, M., 1984. *Traces on the 1983 Nihonkai-Chubu earthquake*, Research Report of the Tsunami Disaster Prevention Laboratory, Faculty of Civil Engineering, Tohoku University, *1*, 88–267 (in Japanese).

SUPPASRI, A., FUKUTANI Y., ABE, Y., and IMAMURA, F., 2013. *Relation between earthquake magnitude and tsunami height along the Tohoku coast based on historical tsunami trace database and the 2011 Great East Japan tsunami*, Report of Tsunami Engineering, *30*, 37–49.

TAKEDA A., 1987. *The tsunami stone at Raga—a result of leveling of the old tide trace due to the Meiji-Sanriku tsunami in 1896*, Report of the National Research Center for Disaster Prevention, *39*, 163–169 (in Japanese with English abstract).

TANIOKA, Y. and SATAKE, K., 1996. *Fault parameters of the 1896 Sanriku tsunami earthquake estimated from tsunami numerical modeling*, Geophysical Research Letters, *23*, 1549–1552.

TANIOKA, Y. and SENO, T., 2001. *Sediment effect on tsunami generation of the 1896 Sanriku tsunami earthquake*, Geophysical Research Letters, *28*, 3389–3392.

TSUJI, Y., OHTOSHI, K., NAKANO, S., NISHIMURA, Y., FUJIMA, K., IMAMURA, F., KAKINUMA, T., NAKAMURA, Y., IMAI, K., GOTO, K., NAMEGAYA, Y., SUZUKI, S., SHIROSHITA, H., and MATSUZAKI, Y., 2010. *Field investigation on the 2010 Chilean Earthquake Tsunami along the comprehensive coastal region in Japan*, Journal of Japan Society of Civil Engineers, Ser, B2 (Coastal Engineering), *66*, 1346–1350 (in Japanese with English abstract).

TSUJI, Y., SATAKE, K., ISHIBE, T., KUSUMOTO, S., HARADA, T., NISHIYAMA, A., KIM, H.Y., UENO, T., MUROTANI, S., OKI, S., SUGIMOTO, M., TOMARI, J., HEIDARZADEH, M., WATADA, S., IMAI, K., CHOI, B.H., YOON, S.B., BAE, J.S., KIM, K.O., and KIM, H.W., 2011. *Field surveys of tsunami heights from the 2011 off the Pacific Coast of Tohoku, Japan Earthquake*, Bulletin of Earthquake Research Institute, University of Tokyo, *86*, 29–279 (in Japanese with English abstract).

UNOHANA, M. and OTA, T., 1988. *Disaster records of Meiji Sanriku tsunami by Soshin Yamana*, Research Report of the Tsunami Disaster Prevention Laboratory, Faculty of Civil Engineering, Tohoku University, *5*, 57–379 (in Japanese).

WATADA, S., KUSUMOTO, S., and SATAKE, K., 2013. *Cause of travel-time difference between observed and synthetic tsunami waveforms at distant locations*, Abstract for IAHS-IAPSO-IASPEI Joint Assembly, SP1S1.03.

WESSEL, P. and SMITH, W.H.F., 1998. *New, improved version of the Generic Mapping Tools released*, EOS Transactions American Geophysical Union, *79*, 579.

YAMANA, S., 1896. *Disaster records of Meiji Sanriku tsunami, reproduced by Unohana and Ota (1988)*, Research Report of the Tsunami Disaster Prevention Laboratory, Faculty of Civil Engineering, Tohoku University, *5*, 57–379 (in Japanese).

YAMASHITA, F., 2003. *Brief history of "declaration of tsunami mitigation town" and large breakwaters in Taro Town, Sanriku coast*, Historical Earthquakes, *19*, 165–171 (in Japanese).

(Received September 15, 2013, revised January 8, 2014, accepted January 10, 2014, Published online March 19, 2014)

Reprinted from the journal

Pure Appl. Geophys. 171 (2014), 3217–3228
© 2014 Springer Basel
DOI 10.1007/s00024-014-0803-1

Short Wave Amplification and Extreme Runup by the 2011 Tohoku Tsunami

Takenori Shimozono,[1] Haiyang Cui,[2] Julie D. Pietrzak,[2] Hermann M. Fritz,[3] Akio Okayasu,[4] and
Andrew J. Hooper[5]

Abstract—Watermarks found during the post-event surveys of the 2011 Tohoku tsunami confirmed extreme runup heights at several locations along the central to northern part of the Sanriku coast, Japan. We measured the maximum height of nearly 40 m above mean sea level at a narrow coastal valley of the Aneyoshi district. Wave records by offshore GPS-buoys suggest that the remarkably high runup was associated with a leading, impulsive crest of the tsunami amplified by local bathymetry and topography. In order to elucidate the underlying amplification mechanism, we apply a numerical model to reproduce the measured distribution of tsunami heights along the target coastline. A series of numerical tests under different boundary conditions suggests that a spectral component with a dominant period of 4–5 min in the leading wave play a key role in generating the extreme runup. Further analyses focusing on the Aneyoshi district confirm that the short wavelength component undergoes critical amplification in a narrow inlet. Our findings highlight the importance of resolving offshore waveforms as well as local bathymetry and topography when simulating extreme runup events.

Key words: Tsunami, runup, amplification, topography, Sanriku coast.

1. Introduction

On March 11, 2011 at 2:46 PM local time (6:46 UTC), a Mw 9.0 earthquake occurred off the northern Pacific coast of Japan. The seismic rupture associated with the subduction of the Pacific plate extended over an area of about 500 km by 200 km along the Japan Trench. The maximum slip has been estimated to exceed 75 m with a significant fraction occurring all the way to the trench (Lay *et al.* 2011; Hooper *et al.* 2013). The ensuing massive tsunami struck the northern Pacific coast of Japan causing nearly 20,000 fatalities. The nationwide post-tsunami survey conducted by the Tohoku Tsunami Joint Survey Group (TTJSG, hereafter) revealed that maximum runup heights of >20 m were distributed along a 290 km stretch of the Pacific coast (Mori *et al.* 2011, 2012). The measured tsunami heights were remarkably large along the Sanriku coast located to the northwest of the epicenter (Fig. 1).

The GPS-buoy network system recorded various profiles of the tsunami as it approached the Pacific coast (Kawai *et al.* 2012). These surface buoys were deployed between the 100–200 m depth contour lines and transmitted signals of the tsunami-induced changes in water level to their land stations (see Fig. 1a for the buoy locations). In addition, the tsunami was also captured at deeper locations by two ocean-bottom pressure gauges, TM1 and TM2 (Maeda *et al.* 2011). Figure 2a shows sea level anomalies at the four GPS-buoys and the two pressure gauges off the Sanriku coast. Generated by complex rupture events, the wave records at different locations exhibit significant variation in their waveforms. Wave crest heights range from about 4–7 m with the maximum value at GPS-buoy 802.

[1] Department of Civil Engineering, The University of Tokyo, 7-3-1 Hongo, Bunkyo-ku, Tokyo 113-8656, Japan. E-mail: shimozono@coastal.t.u-tokyo.ac.jp
[2] Faculty of Civil Engineering and Geosciences, Delft University of Technology, Building 23, Stevinweg 1, 2628 CN Delft, The Netherlands. E-mail: h.cui@tudelft.nl; j.d.pietrzak@tudelft.nl
[3] School of Civil and Environmental Engineering, Georgia Institute of Technology, Atlanta, GA 30332, USA. E-mail: fritz@gatech.edu
[4] Department of Ocean Sciences, Tokyo University of Marine Science and Technology, 4-5-7 Konan, Minato-ku, Tokyo 108-8477, Japan. E-mail: okayasu@kaiyodai.ac.jp
[5] School of Earth and Environment, University of Leeds, Leeds LS2 9JT, UK. E-mail: a.hooper@leeds.ac.uk

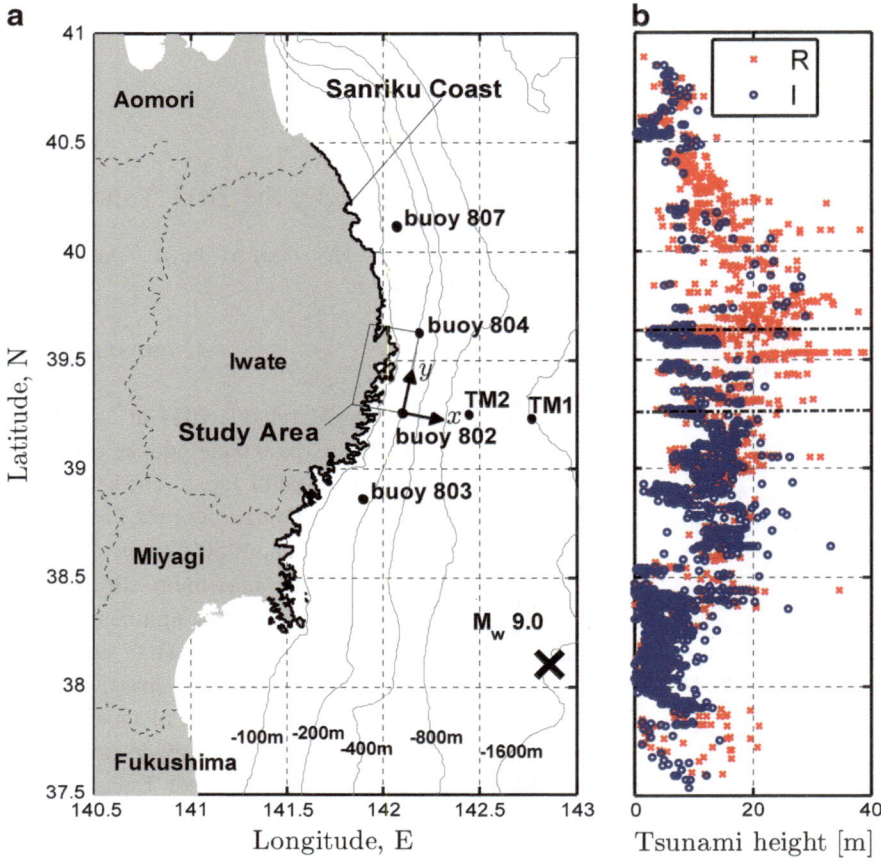

Figure 1
The Sanriku coast and measured tsunami heights. **a** Index map showing the study area, the Sanriku coast, the earthquake epicenter from the Japan Meteorological Agency and the locations of the GPS buoys. **b** Distribution of the measured runup and inundation heights along latitude from the TTJSG dataset

The regional tsunami behavior along the central Sanriku coast was analyzed based on numerical simulations and the model results were compared with field data (SHIMOZONO *et al* 2012; WEI *et al.* 2013). A comparison of the offshore wave records with the onshore tsunami heights suggested that the extreme runup heights (>25 m) were recorded along the northern Sanriku coast between 39.2° and 40.2°N rather than closer to the epicenter further south. The maximum amplification and runup on the coast may not be merely attributed to the coastal topography as the coastline complexity is similar along the northern and southern Sanriku coast (LIU *et al.* 2013). There should be some interactive mechanism for the drastic amplification between incoming wave properties and local topographic features.

Extreme runup heights have been recorded in past tsunami events such as the 1992 Flores tsunami (26.2 m at Riang-Kroko, TSUJI *et al.* 1995), the 1993 Okushiri tsunami (31.7 m at Monai, SHUTO and MATSUTOMI 1995; TITOV and SYNOLAKIS 1997), the 2004 Indian Ocean tsunami (31.0 m at Lhoknga, BORRERO *et al.* 2006) and the 2010 Chile tsunami (29.0 m at Constitución, FRITZ *et al.* 2011). They are characterized by unusual heights in comparison to their surroundings and commonly occur in coastal valleys and on coastal bluffs directly onshore from the tsunami source areas. The extreme runup heights from field surveys are often difficult to explain with existing runup theories or numerical simulation based on inferred fault models. The dynamic processes causing the extreme runup are poorly understood

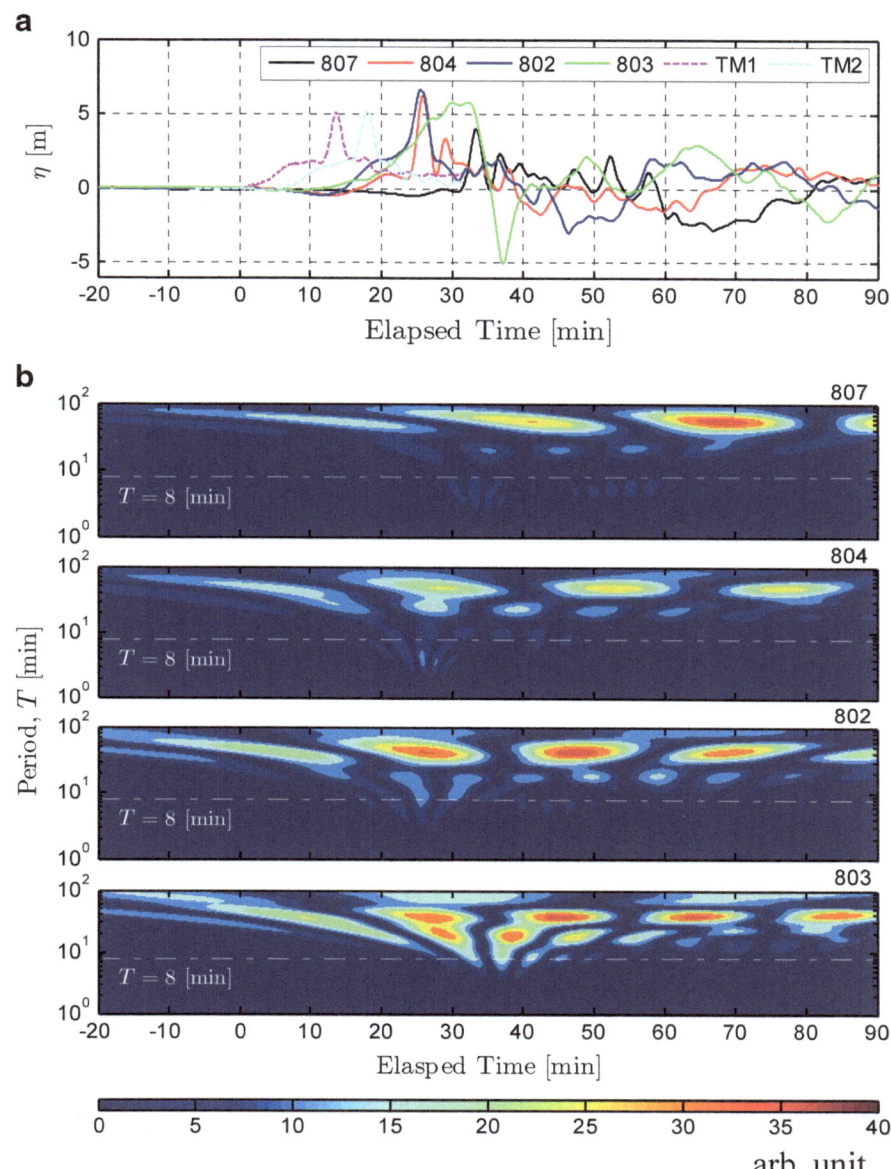

Figure 2

Offshore waveforms observed by the GPS buoys and the pressure sensors. **a** Water surface elevation at the six locations off the Sanriku coast. The origin of time axis is set to the start time of the earthquake. **b** Wavelet coefficients of the offshore waveforms by GPS-buoys based on the real Morlet wavelet

mainly because incoming wave properties such as wavelength, shape and amplitude remain uncertain in most cases.

The purpose of this paper is to investigate possible causes of the extreme runup generated by the 2011 Tohoku tsunami. The study focuses on a site in the Aneyoshi district, where runup heights of nearly 40 m were measured by some of the authors. This site marked one of the highest runup records in the TTJSG database. The study area is set to include the Aneyoshi district along the central Sanriku coast as indicated by a rectangle in Fig. 1a. Tsunami

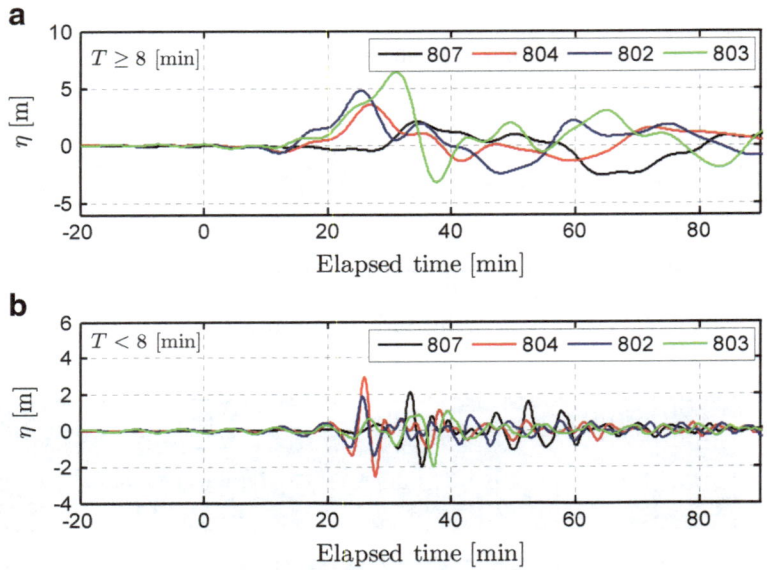

Figure 3
Long and short wavelength components at the GPS-buoys. **a** Wave components with periods longer than 8 min. **b** Wave components with periods shorter than 8 min

propagation and inundation in this range is simulated with a numerical model, using different offshore boundary conditions constructed from the GPS-buoy records. Through model applications, we finally elucidate the underlying mechanism to trigger the drastic amplification.

2. Extreme Runup

2.1. Study Area and Incident Waves

The characteristics of the incident tsunami waves approaching the study area can be inferred from the wave profiles at GPS-buoy 802 and 804 in Fig. 2a. The initial part of each profile represents an incident wave component of the tsunami that has propagated from the source area in the deep ocean, while a reflective component from the coast is only superimposed after a certain period of time. Tsunami travel time calculations between the GPS-buoys and coastline indicate that the reflective component becomes significant only after the arrival of the main initial crest. This argument is also supported by similarity of the initial waveforms at the three locations, GPS-buoy 802, TM1 and TM2 aligned in a cross-coast

line. The maximum runup heights on the coast are associated with the peaked crest through a specific amplification mechanism given the absence of similar scale successive waves observed on the coast (FRITZ et al. 2012).

The time–frequency characteristics of the four GPS-buoy recordings are represented by the wavelet spectrograms in Fig. 2b. A predominant period of 40–50 min is identified from the maximum values of the wavelet coefficients. The dominant periods of past tsunamis generated from a similar source area in 1896 and 1933 were estimated to be shorter (ABE 2005). The discrepancy may be attributed to a deep interplate slip that was not significant in the previous two events (FUJII et al. 2011). Clear differences in the time–frequency characteristics recorded at the four locations appear in a range of shorter periods. Large spectral peaks ranging from 8 to 25 min are commonly observed around the initial waves at all the locations. However, the magnitudes decrease from south to north, implying that larger seafloor displacements near the epicenter are responsible for these spectral peaks.

Besides the dominant spectral peaks, there are small, but distinct peaks in the period band of

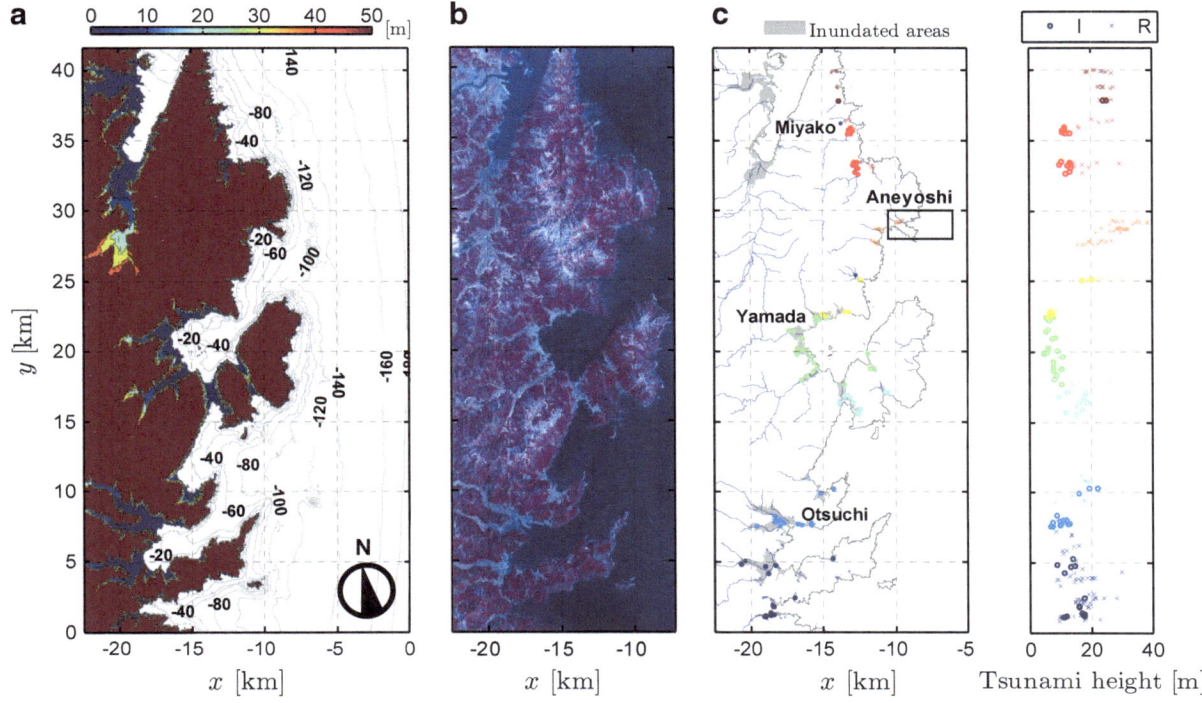

Figure 4

Close-up view of the study area. **a** Topographic–bathymetric map of the study area along the Sanriku coast (*upper left corner*: 39.6646°N, 141.9290°E, *bottom right corner*: 39.2586°N, 142.0969°E). **b** ASTER satellite image of the study area taken on March 19, 2011. **c** Surveyed locations and tsunami heights above the mean sea level by TTJGS. Inundated areas shown in *gray* were estimated from satellite and aerial images by the Geospatial Information Authority

<8 min in particular at the northern buoys. The spectral peaks localized in time are associated with the impulsive crests in the time domain signals. For clarification purpose, the four GPS-buoy data are divided into two components below and above a cut-off period of 8 min in Fig. 3. The amplitude of the remarkably short component is peaked at GPS-buoy 804. Thus, it was possibly caused by a local rupture in the northern part of the fault plane. It should be noted, however, that the large impulsive crests at GPS-buoy 804 and 807 have not been fully explained by estimated sources from existing tsunami-inversion studies (Fujii *et al.* 2011; Saito *et al.* 2011; Grilli *et al.* 2013). Some have conjectured that submarine mass failures generated an additional tsunami (Grilli *et al.* 2013), while others have conjectured that focusing of N-wave occurred in the corresponding area (Kanoglu *et al.* 2013). Hereafter, we refer to the wave component

with a dominant period of 4–5 min as the short wavelength component, as the period range is below typical tsunami periods.

2.2. Survey Results

Figure 4a shows a detailed map of the study area extending 42 km by 22 km. The *x*- and *y*-axes of an orthogonal coordinate system are defined with an origin at GPS-buoy 802 and the *y*-axis pointing to GPS-buoy 804 as indicated by two arrows in Fig. 1a. The topography and bathymetry are created from 10 m-grid topographic data of the Geospatial Information Authority of Japan and digital nautical charts of the Japan Hydrographic Association, respectively. An ASTER satellite image that covers the land area is shown in Fig. 4b. The mountainous coastal zone consists of several major bays of variable geometries with small plains at the head of the bays. A large portions of the coastal plains were inundated by the huge tsunami as indicated with gray areas in Fig. 4c.

The surveyed locations and corresponding tsunami heights above mean sea level classified into runup and inundation heights are also displayed in Fig. 4c. Detailed information on the field survey and data processing are available in MORI et al. (2011, 2012). We show only data flagged as "highly reliable" by respective surveyors from the TTJSG dataset, accounting for 90 % of the entire dataset. The tsunami heights are plotted by markers in different colors depending on the y-coordinate values. Because of the complex topographies, the runup and inundation heights of the tsunami exhibit a remarkable variation along the coastline. The most striking difference was observed between Yamada and Aneyoshi where the values vary from 8 to 40 m over a distance of about 10 km.

The maximum height of 38.8 m was measured in the Aneyoshi district and registered into the TTJSG dataset by the authors (OKAYASU et al. 2012). The Aneyoshi district is located at about 39.53°N and 142.05°E ($x = -10$ km and $y = 20$ km). An enlarged view of topography and bathymetry around the district is shown in Fig. 5a. A converging inlet with an entrance width of about 1 km and depth of about 40 m connects with a narrow, rocky canyon on land. Figure 5b represents a bed profile along the center line of the inlet. The water depth decreases almost linearly from 100 m depth with a mean slope of $\alpha = 0.042$. Despite the high runup of the tsunami, there was no tsunami victim at this site because the residential area had been relocated to higher ground after the disastrous Sanriku tsunami of 1933. The runup height of the 1933 tsunami was 21 m, which was high compared with surrounding areas, and only three people survived out of 92 residents (EARTHQUAKE RESEARCH INSTITUTE 1934).

A field survey was conducted on April 13, 2011, $\sim$ 1 month after the event, to measure tsunami trace heights above mean sea level. Later, the site was revisited by several independent survey teams confirming the extreme runup. The survey results are shown in Fig. 6a together with some photographs from the site. The limited low-lying area is flanked by steep slopes and constricted into a narrow canyon some 100–200 m inland from the shoreline (Fig. 6b,

c). Various tsunami traces such as rafted debris and knocked-down trees were found at corresponding elevations along the steep slopes (Fig. 6d, e), reliably indicating that the tsunami runup exceeded 38 m at several locations up the valley.

3. Numerical Study

3.1. Model Description

We perform 2D numerical simulations of tsunami inundation in the study area using a finite-volume model, H2Ocean (CUI et al. 2010, 2012). The model is based on the nonlinear shallow water equations with bottom friction terms based on the Manning formulation,

$$\frac{\partial \eta}{\partial t} + \nabla \cdot (h\mathbf{u}) = 0, \qquad (1)$$

$$\frac{\partial \mathbf{u}}{\partial t} + \mathbf{u} \cdot \nabla \mathbf{u} + g\nabla \eta + g\frac{n^2}{h^{4/3}}\mathbf{u}|\mathbf{u}| = 0, \qquad (2)$$

with time t, the water surface elevation η, the depth-averaged velocity $\mathbf{u}$, the total water depth h, the gravitational acceleration g and the Manning roughness n. The equations are discretized on an unstructured grid that enables an efficient refinement of the complex coastline. The model is capable of preserving mass and momentum in local cells as well as maintaining the positivity of the water depth in the case of wetting and drying. The costal structures and buildings on land are taken into account by the equivalent roughness. Because of the extensive area covered by the model, single values are given for the land and sea parts of the computational domain. Through preliminary tests, the value of n is determined at 0.02 s m$^{-1/3}$ on the sea floor and 0.05 s m$^{-1/3}$ on land. The set of values is typically employed to simulate tsunami flooding over a medium-density residential area (GOTO and SHUTO 1983). The computational grid is refined to cells of 20 by 20 m around the coastline without a significant difference in results by further refinements.

The computational domain is identical to the area shown in Fig. 4a, with an offshore boundary ($x = 0$) almost parallel to the 200 m depth contour line connecting GPS-buoy 802 and 804. The

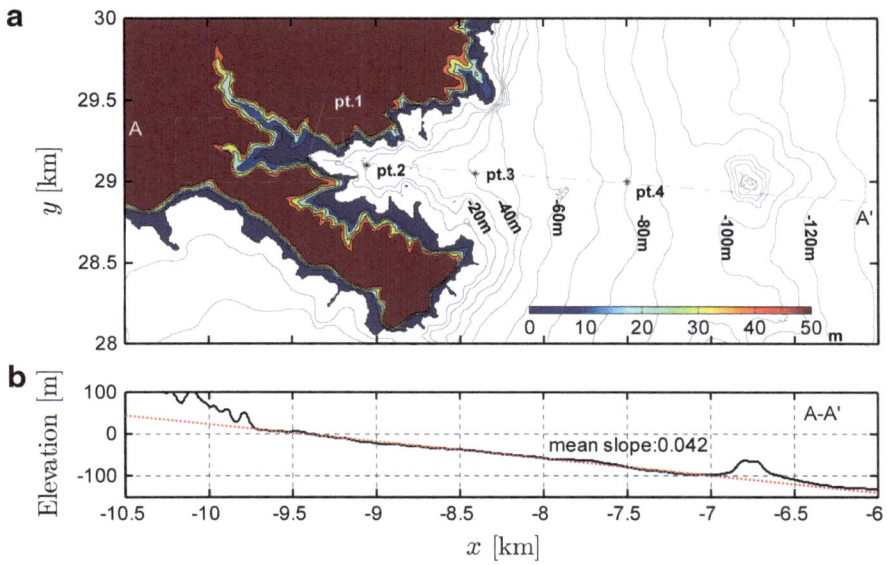

Figure 5
Detailed map of the extreme-runup site, Aneyoshi. **a** Topographic–bathymetric map around the Aneyoshi district. The area corresponds to a rectangle in Fig. 4c. **b** Bed profile along the inlet axis (A–A' cross-section)

tsunami originating from farther offshore, especially its long-wavelength component, travels in the direction perpendicular to the boundary due to refraction effect. The water surface elevation along the boundary can be, therefore, interpolated from the wave data at the two GPS-buoys, in the following manner,

$$\eta(0, y, t) = f(y)\{\eta_{802}(t) - \eta_{804}(t)\} + \eta_{804}(t) \quad (3)$$

with the interpolation function $f(y)$, the respective wave profiles η_{802} and η_{804}. This treatment of the boundary condition can be justified as far as the initial, dominant wave of the tsunami is concerned. For the interpolation function, we consider various types of monotonic functions on an assumption that there is no local maximum in η between the two locations. Here we bring three distinct cases into discussions, namely "linear interpolation" ($f = y/y_0$) where y_0 is a distance between the two buoys, "uniform η_{802}" ($f = 1$) and "uniform η_{804}" ($f = 0$). In order to prevent laterally reflected waves from immediately entering the study area, the northern and southern boundaries are offset outward by creating virtual computational regions on both sides. A series of computations were carried out with a time step

satisfying the CFL condition until the leading wave finishes flooding on the coast.

3.2. Model Validation

The computed and measured tsunami heights for the three different boundary conditions are compared in Fig. 7. These comparisons are made only at the surveyed locations with computed values at nearest neighbor grid points. Different marker colors indicate data locations, increasing latitude from cold to warm colors according to the data representation in Fig. 4c. The relative root mean square (rms) error, defined as the ratio of the residual rms to the rms of the measured values, is displayed in each panel as an indicator of the overall prediction error.

The model result agrees most closely with the field data when the offshore boundary condition is based only on η_{804}. The other two boundary conditions lead to systematic underestimation of tsunami heights exceeding 25 m, including the maximum value at Aneyoshi. Consequently, the overall prediction errors of the two cases become larger in spite of better agreements in the southern part of the domain.

Figure 6
Survey results and post-tsunami field observation scenarios in Aneyoshi. **a** Close-up view of the surveyed area in the Aneyoshi district. The extent of the area is indicated by a rectangle in Fig. 5a. **b** View of the coast from the sea side. **c** The narrow, rocky valley. **d** Debris brought by the extreme tsunami runup. **e** Trees on the slope felled by the tsunami. *Note* The power line poles were destroyed by the tsunami but restored on 13 April 2011

The results of the comparative study suggest that some wave properties specific to η_{804} are required to produce the extreme runup especially on the northern coast. Comparing the data of the two waves in Figs. 2 and 3, spectral components of η_{802} dominate those of η_{804} in a wide range of periods except in the range of $T < 8$ min. Thus, it is reasonable to conjecture that the short wavelength component plays a key role in the extreme runup.

It should be noted that the present model cannot fully reproduce the extreme runup of 38.8 m at Aneyoshi (the corresponding computed value is 32.5 m). Even with a finer grid, the runup height at the site does not increase significantly. Also, adding frequency dispersion to the model (using a non-hydrostatic version of the same model) results in only minor differences as the propagation distance of the tsunami is short in the computational domain. The most plausible reason for the under-estimation is that the monotonicity of η along the offshore boundary is not necessarily valid. This is especially true for the high-frequency component, as its wavelength is smaller than y_0, the distance between the two buoys. Nevertheless, the model generally accommodates the generation mechanism of the extreme runup, since the highest runup is marked at Aneyoshi in the computed results as well.

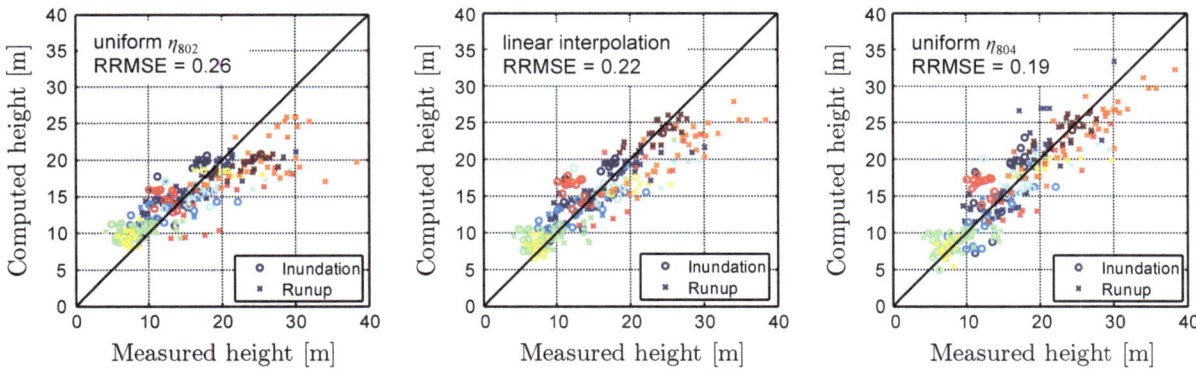

Figure 7

Comparisons of the computed and measured tsunami heights for the three different boundary conditions. The marker colors correspond to those in Fig. 4c. The RRMSE stands for the relative root mean square error

4. Discussion

We now focus on the extreme runup at Aneyoshi and apply the same model for further investigation of its possible cause. To examine the dynamic processes of the local amplification in details, temporal variations of the water surface at different depths, 0, 20, 40 and 80 m along the center line of the Aneyoshi inlet are plotted together in Fig. 8 (See Fig. 5a for these locations). The three panels, respectively, show the results by the different boundary conditions based on η_{802} and η_{804}. In these results, the incident waves are slightly amplified to point 3 at the inlet entrance where their original forms are mostly preserved. As the tsunami propagates further into the inlet, the short wavelength components with a period of 4–5 min are selectively enhanced in both flood and ebb directions. The ratio of wave heights at point 3–2 is almost identical to that from the Green's law on the assumption that both width and depth of the inlet decrease by half. The resulting waveforms in shallow water are asymmetrical about the horizontal axis due to strong nonlinearities. The wave height is larger in the lower panel simply because the corresponding spectral component is originally more significant in η_{804}.

In order to analyze the mechanism of the short wave amplification, we numerically force oscillations with different periods in the Aneyoshi inlet by transmitting waves from an offshore boundary placed in parallel to the y-axis at a depth of $h_0 = 100$ m. The

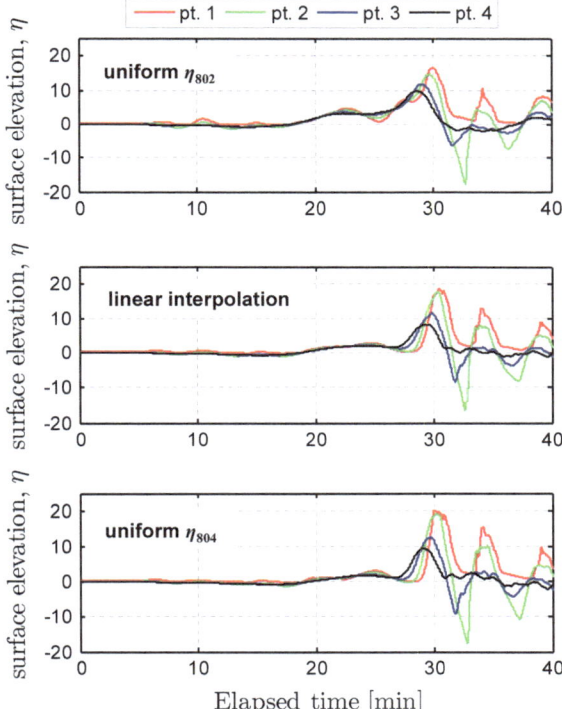

Figure 8

Temporal variations of water surface level at the four locations along the axis of the Aneyoshi inlet, computed with three different boundary conditions. The reference locations are indicated in Fig. 5a

response analysis is carried out using wavelets instead of monochromatic waves, since the short wavelength component of the current case is

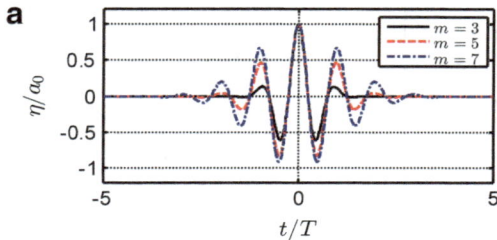

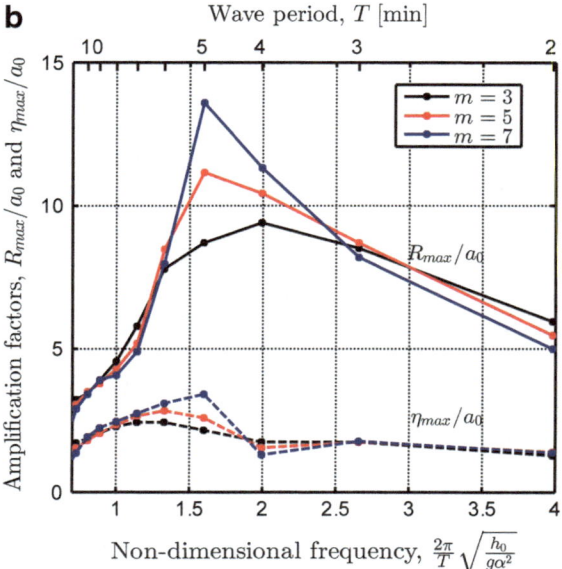

Figure 9

Results of the basin response analyses. **a** Wavelets with different wave numbers employed as incident waves. **b** Runup amplification factors and wave amplification factors at the inlet entrance (point 3) versus the non-dimensional frequency (wave period) for three different wave numbers

localized in time as shown in Fig. 3b. The wavelet is a plane wave modulated by a Gaussian envelope that is analogous to one used in the previous wavelet analysis,

$$\eta_0(t) = a_0 \exp\left(-\frac{\omega^2 t^2}{2m^2}\right)\cos(\omega t), \qquad (4)$$

with the amplitude a_0, the angular frequency ω and the wavenumber m, roughly corresponding to the number of oscillations in the wavelet (LIU 2000). If we let m go to infinity, the resulting wave will be a monochromatic wave with a period of T ($=2\pi/\omega$). Here we consider three cases with $m = 3$, 5 and 7 as depicted in Fig. 9a. The wave height $2a_0$ is fixed to be 1.0 m, while the wave period T is varied from 2 to 10 min in the short period range.

Figure 9b shows amplification factors in terms of the maximum runup height $R_{\max}/a_0$ and the maximum surface elevation at the inlet entrance $\eta_{\max}/a_0$ plotted versus the non-dimensional frequency $\omega(h_0/g\alpha^2)^{1/2}$ (and also the wave period T) for the three different values of m. The runup height $R_{\max}$ is defined as the maximum surface elevation in the hinterland. The runup amplification factors are sharply peaked around a period of 4–5 min, exhibiting high sensitivity of the runup height to the incoming wave period in this period range.

It is well known from the linear theory for wave runup over a plane beach that the runup amplification factor is an increasing function of the non-dimensional frequency (KELLER and KELLER 1964). Thus, the shorter waves result in higher runup in the linear regime. This relationship remains unchanged even in cases of wave runup in narrow bays, but the growth rate of the amplification factor against the non-dimensional frequency is increased remarkably due to the so-called funneling effect (ZAHIBO *et al.* 2006; DIDENKULOVA and PELINOVSKY 2011). Though in reality, the trend is inverted on the shorter-period side due to the occurrence of wave breaking, creating a peak in the amplification curve. Hence, an incoming wave with the peak period undergoes critical amplification in the inlet without being bounded by wave breaking.

The amplification curves by the different values of m illustrate that wave-to-wave interactions further strengthen the runup amplification around the peak period. The resonant period of the Helmholz mode for the Aneyoshi inlet is estimated at about 4 min using a semi-analytical formula for V-shaped bays (RABINOVICH 2009). The occurrence of wave resonance at the estimate period is not evident due to wave breaking, but a significant drop in $\eta_{\max}/a_0$ is indicative of the formation of a node at the inlet entrance. Indeed, wave amplification within the inlet, $R_{\max}/\eta_{\max}$ is sharply peaked at the estimated resonant period. In the actual event by the 2011 Tohoku tsunami, however, the strong wave resonance as in the case of $m = 7$ may not have been excited at the Aneyoshi inlet, judging from the actual waveforms in Fig. 3b. Nevertheless, the weak resonance through an interaction of a small preceding wave and the main crest may contribute to the extreme runup to some extent.

5. Conclusion

The strong and complex rupture of the 2011 To-hoku earthquake generated a huge tsunami that exhibited a variety of waveforms along the northern Pacific coast of Japan. The offshore waveforms of the tsunami approaching the present study area, the central Sanriku coast, with the extreme runup recordings were characterized by a small preceding wave followed by a sharp crest. The spectral component of a period of 4–5 min localized in the leading wave was found to play a key role in generation of the extreme runup on the intricate coastline. For the representative case of the Aneyoshi district, the short wavelength component went through critical ampli-fication due to the funneling effect, unsuppressed by wave breaking. The interaction of the forerunner wave with the main crest may have further contrib-uted to runup enhancement in the inlet. The Aneyoshi case represents a prominent example, but the same phenomenon could happen at other sites with similar topographic features.

The present study on the extreme runup by the 2011 Tohoku tsunami demonstrates that short wave-length components of a tsunami possibly undergo tremendous amplification in small bays or inlets. Certain topographic features are remarkably suscep-tible to this phenomenon. Consequently, a seemingly small tsunami may result in unexpectedly high runup. Some of the extreme runup in past, and other events may be generated through similar amplification mechanisms. The local enhancement of short wave-length components poses a difficulty in evaluating tsunami risks, since detailed waveforms of incoming tsunamis are not known a priori in most cases.

Acknowledgments

The post-tsunami survey was financially supported by the JST J-RAPID project (11103018). H.M.F. was supported by NSF RAPID award (CMMI-1135768).

REFERENCES

ABE, K. (2005), *Tsunami resonance curve from dominant periods observed in bays of northeastern japan*, Tsunamis: Case Studies and Recent Developments, edited by K. Satake, 97–113, Springer.

BORRERO, J. C., SYNOLAKIS, C. E. and FRITZ, H. M. (2006), *Northern Sumatra field survey after the December 2004 great Sumatra earthquake and Indian ocean tsunami*, Earthquake Spectra, 22(S3), S93–S104, doi:10.1193/1.2206793.

CUI, H., PIETRZAK, J. D. and STELLING, G. S. (2010), *A finite volume analogue of the P1NC-P1 finite element: With accurate flooding and drying*, Ocean Modelling, 35(1–2), 16–30.

CUI, H., PIETRZAK, J. D. and STELLING, G. S. (2012), *Improved efficiency of a non-hydrostatic, unstructured grid, finite volume model*, Ocean Modelling, 54–55, 55–67.

DIDENKULOVA, I. and PELINOVSKY, E. (2011), *Runup of tsunami waves in U-shaped bays*, Pure Appl. Geophys., 168(6–7), 1239–1249.

Earthquake Research Institute (1934), *Reports on the 3 March, Showa 8 Sanriku tsunami*, Bulletin of the Earthquake Research Institute, Tokyo Imperial University, Supplementary Vol.1, 248p., 1934 (in Japanese).

FRITZ, H.M., PETROFF, C.M., CATALAN, P., CIENFUEGOS, R., WINCKLER, P., KALLIGERIS, N., WEISS, R., BARRIENTOS, S. E., MENESES, G., VALDERAS-BERMEJO, C., EBELING, C., PAPADOPOULOS, A., CONTRERAS, M., ALMAR, R., DOMINGUEZ, J.C. and SYNOLAKIS C.E. (2011), *Field survey of the 27 February 2010 Chile tsunami*. Pure Appl. Geophys., 168(11), 1989–2010, doi:10.1007/s00024-011-0283-5.

FRITZ, H. M., PHILLIPS, D.A., OKAYASU, A., SHIMOZONO, T., LIU, H., MOHAMMED, F., SKANAVIS, V., SYNOLAKIS, C.E. and TAKAHASHI, T. (2012), *2011 Japan tsunami current velocity measurements from survivor videos at Kesennuma Bay using LiDAR*, Geophys. Res. Lett., 39, L00G23, doi:10.1029/2011GL050686.

FUJII, Y., SATAKE, K., SAKAI, S., SHINOHARA, S. and KANAZAWA, T. (2011), *Tsunami source of the 2011 off the Pacific coast of Tohoku, Japan earthquake*, Earth Planets Space, 63(7), 815–820.

GOTO, C. and SHUTO, N. (1983), *Effects of large obstacles on tsunami inundations*, Tsunamis - Their Science and Engineering, Terra Scientific Publishing Company, Tokyo, 511–525.

GRILLI, S.T., HARRIS, J.C., TAJALI BAKHSH, T.S., MASTERLARK, T.L., KYRIAKOPOULOS, C., KIRBY, J.T. and and SHI, F. (2013), *Numerical simulation of the 2011 Tohoku tsunami based on a new transient FEM co-seismic source: Comparison to far- and near-field observations*, Pure Appl. Geophys., 170(6–8), 1333–1359.

HOOPER, A., PIETRZAK, J., SIMONS, W., CUI, H., RIVA, R., NAEIJE, M., TERWISSCHA VAN SCHELTINGA, A., SCHRAMA, E., STELLING, G. and SOCQUET, A., (2013), *Importance of horizontal seafloor motion on tsunami height for the 2011 Mw = 9.0 Tohoku-Oki earthquake*, Earth planet. Sci. Lett., 361, 469–479.

KANOGLU U, TITOV VV, AYDIN B, MOORE C, STEFANAKIS TS, ZHOU H, SPILLANE M, SYNOLAKIS C.E., (2013), *Focusing of long waves with finite crest over constant depth*, Proc. R Soc., A 469: 20130015.

KAWAI, H., SATOH, M., KAWAGUCHI, K. and SEKI, K. (2012), *Recent tsunamis observed by GPS buoys off the Pacific coast of Japan*, Proc. 33rd International Conf. on Coastal Eng., currents.1.

KELLER, J. B. and KELLER, H. B. (1964), *Water wave run-up on a beach*, Office Naval Research Res. Rept., NONR-3828(00), Dept. Navy, Washington.

LAY, T., AMMON, C. J., KANAMORI, H., XUE, L. and KIM, M. J. (2011), *Possible large near-trench slip during the 2011 Mw 9.0 off the Pacific coast of Tohoku Earthquake*, Earth, Planets and Space, 63(7), 687–692.

Liu, H., Shimozono, T., Takagawa, T., Okayasu, A., Fritz, H. M., Sato, S. and Tajima Y. (2013), *The 11 March 2011 Tohoku tsunami survey in Rikuzentakata and comparison with historical events*. Pure Appl. Geophys., *170*(6–8), 1033–1046, doi:10.1007/s00024-012-0496-2.

Liu, P. C. (2000), *Wavelet transform and new perspective on coastal and ocean engineering data analysis*, Advances in coastal and ocean engineering, *6*, 57–102.

Maeda, T., Furumura, T., Sakai, S. and Shinohara, M. (2011), *Significant tsunami observed at ocean-bottom pressure gauges during the 2011 off the Pacific coast of Tohoku Earthquake*, Earth Plants Space, Vol.63, pp. 803–808.

Mori, N., Takahashi, T., Yasuda, T. and Yanagisawa, H. (2011), *Survey of 2011 Tohoku earthquake tsunami inundation and run-up*, Geophys. Res. Lett., *38*, L00G14, doi:10.1029/2011GL049210.

Mori, N., Takahashi, T. and the 2011 Tohoku Earthquake Tsunami Joint Survey Group (2012), *Nationwide Post Event Survey and Analysis of the 2011 Tohoku Earthquake Tsunami*, Coast. Eng. J. *54*(1), 1250001.

Okayasu, A., Shimozono, T., Sato, S., Tajima, T., Liu, H., Takagawa, T. and Fritz, H.M. (2012), *2011 Tohoku tsunami runup and devastating damages around Yamada Bay, Iwate: surveys and numerical simulation*, Proc. of 33rd International Conf. on Coastal Eng., currents.4.

Rabinovich, A. B. (2009), *Seiches and harbor oscillations*, Handbook of Coastal and Ocean Engineering, 193-236, edited by Y. C. Kim, World Scientific Publ., Singapore.

Saito, T., Ito, Y., Inazu, D. and Hino, R. (2011), *Tsunami source of the 2011 Tohoku-Oki earthquake, Japan: Inversion analysis based on dispersive tsunami simulations*, Geophys. Res. Lett., *38*, L00G19, doi:10.1029/2011GL049089.

Shimozono, T., Sato, S., Okayasu, Y., Tajima, Y., Fritz, H.M., Liu, H. and Takagawa, T. (2012), *Propagation and inundation characteristics of the 2011 Tohoku tsunami on the central Sanriku Coast*, Coastal Eng. J., *54*(1), 1250004.

Shuto, N., and Matsutomi, H. (1995), *Field survey of the 1993 Hokkaido Nansei-Oki earthquake tsunami*, Pure Appl. Geophys., *144*, 649–663.

Titov, V.V., and Synolakis, C.E. (1997), *Extreme inundation flows during the Hokkaido-Nansei-Oki tsunami*. Geophys. Res. Lett, *24*(11), 1315–1318.

Tsuji, Y., Matsutomi, H., Imamura, F., Takeo, M., Kawata, Y., Matsuyama, M., Takahashi, T., Sunarjo and Harjadi, P. (1995), *Damage to coastal villages due to the 1992 Flores Island earthquake tsunami*, Pure Appl. Geophys., *144*(3–4), 481–524.

Wei, Y., Chamberlin, C., Titov, V.V., Tang, L., and Bernard, E.N., (2013), *Modeling of the 2011 Japan tsunami: lessons for near-field forecast*, Pure Appl. Geophys. doi:10.1007/s00024-012-0519-z.

Zahibo, N., Pelinovsky, E., Golinko, V., and Osipenko, N. (2006), *Tsunami wave runup on coasts of narrow bays*, Int. J. Fluid Mechanics Research, *33*, 106–118.

(Received October 15, 2013, revised February 4, 2014, accepted February 16, 2014, Published online March 28, 2014)

Pure Appl. Geophys. 171 (2014), 3229–3240
© 2014 Springer Basel
DOI 10.1007/s00024-014-0809-8

▌Pure and Applied Geophysics

Delayed Survey of the 2011 Tohoku Tsunami in the Former Exclusion Zone in Minami-Soma, Fukushima Prefecture

SHINJI SATO,[1] AKIO OKAYASU,[2] HARRY YEH,[3] HERMANN M. FRITZ,[4] YOSHIMITSU TAJIMA,[1] and TAKENORI SHIMOZONO[1]

Abstract—Post-tsunami field surveys in the Minami-Soma exclusion zone in the Fukushima Prefecture were delayed for 15 months after the 2011 Tohoku tsunami. The area was subject to access restrictions until June 2012 due to high radiation levels caused by the meltdown at the Fukushima Dai-ichi Nuclear Power Plant. The distribution of the measured tsunami heights is presented in combination with observed infrastructure damage. The enhanced tsunami heights in the areas along the shoreline are attributed to wave reflection, funneling and splash-up at cliffs and seawalls, as well as the increased flow resistance as the tsunami plowed through coastal pine-tree forests. Consequently, large tsunami heights exceeding 10 m were limited to areas within 500 m from the shoreline. Onshore profiles of the maximum inundation levels were dependent on inland topography: tsunami heights increased inland in steep V-shaped valleys, while decaying with inundation distance along flat coastal plains. Tsunami flood levels in the coastal plains are affected by the extent of seawall damage: coastal flood levels are higher behind completely destroyed seawalls than behind partially damaged coastal defenses. Remnant seawalls provided valuable lessons to be implemented in future designs of tenacious structures based on the Japanese concept of 'nebari' representing resiliency to endure tsunami overflow as the original design height is exceeded.

Key words: Tohoku Tsunami, inland tsunami heights, tsunami disaster followed by a nuclear accident, seawall performance.

1. Introduction

At 14:46 on March 11, 2011 (JST), a magnitude 9.0 earthquake struck the northeast coast of Japan, resulting from thrust faulting along the subduction zone between the Pacific and the North American

Plates (SIMONS *et al.* 2011; FUJII *et al.* 2011). This earthquake generated catastrophic tsunamis, which severely impacted the Japanese Pacific coast. According to the Tohoku Earthquake Tsunami Joint Survey Group (TTSG 2013), the measured maximum tsunami run-up heights reached nearly 40 m. This tsunami represents the largest tsunami event possibly within the last 1,000 years in Japan. Significant casualties (15,884 fatalities with an additional missing 2,640 presumed dead as of January 10, 2014, according to the National Police Agency) and tremendous damage to coastal cities and infrastructure were caused by this tsunami event, while the earthquake accounts for <1 % of the fatalities. The tsunami inundation at the Fukushima Dai-ichi Nuclear Power Plant (NPP) resulted in the worst nuclear accident since the 1986 Chernobyl incident. The multiple meltdowns and release of radiation mandated an enforced exclusion zone around the Fukushima Dai-ichi NPP, which remained off-limits for the remainder of 2011 due to the area's elevated radiation levels.

Extensive post-tsunami surveys were conducted under the coordination of the Joint Survey Group (TTSG), which was established on the day after the event. Based on established protocols, tsunami surveys should be performed as quickly as possible without interfering with initial rescue operations, but prior to removing perishable evidence during recovery and reconstruction. Given the scale of this event with an affected area spanning more than 1,000 km, a comprehensive survey was facilitated through collaborative efforts by numerous teams involving various organizations, disciplines and countries. About 300 Japanese and international researchers from multiple disciplines participated in the Joint

---

[1] Department of Civil Engineering, The University of Tokyo, 7-3-1 Hongo, Bunkyo-ku, Tokyo, Japan. E-mail: sato@coastal.t.u-tokyo.ac.jp
[2] Department of Ocean Sciences, Tokyo University of Marine Science and Technology, 4-5-7 Konan, Minato-ku, Tokyo, Japan.
[3] School of Civil and Construction Engineering, Oregon State University, Corvallis, OR 97331-3212, USA.
[4] School of Civil and Environmental Engineering, Georgia Institute of Technology, Atlanta, GA 30322, USA.

Tsunami Survey Group (MORI *et al.* 2012). Rapid sharing of collected data enabled coordination among individual survey teams to complement and fill data gaps, while limiting redundant surveys. The evolving database updated on the Internet provided continuous public access for scientific research as well as rescue, recovery, and restoration efforts in the affected areas.

Here we report the findings from the first comprehensive tsunami survey in the southern portion of Minami-Soma, Fukushima Prefecture. Our field survey in this area had been delayed by the exclusion zone imposed after the multiple meltdowns and release of radiation at the Fukushima Dai-ichi NPP. We focus on the spatial distribution of tsunami inundation heights as well as damage to protective coastal structures such as seawalls. Since the nuclear accident was primarily caused by tsunami inundation, it is important to understand the detailed tsunami effects along this stretch of coastline. The Fukushima coast is distinguished by different topographic and bathymetric features from the Sanriku

and Sendai coasts further north. Almost all human beings, except a few workers at the nuclear power plants, were displaced from the area. Fortunately, many tsunami watermarks were preserved for over a year given the complete lack of human activities including postponed clean-up, recovery, and reconstruction efforts.

2. *Coastal Topography and Tsunami Inundation in Minami-Soma*

The coastal topography and tsunami watermark heights to the north of the Fukushima Dai-ichi NPP up to the Sendai Plain are shown in Fig. 1. The measured tsunami heights obtained by SATO *et al.* (2013) and those compiled by the Joint Survey Group to the north of the original 20 km exclusion zone are also included in Fig. 1 and compared with nearshore tsunami heights at a water depth of 10 m computed by SATO *et al.* (2013).

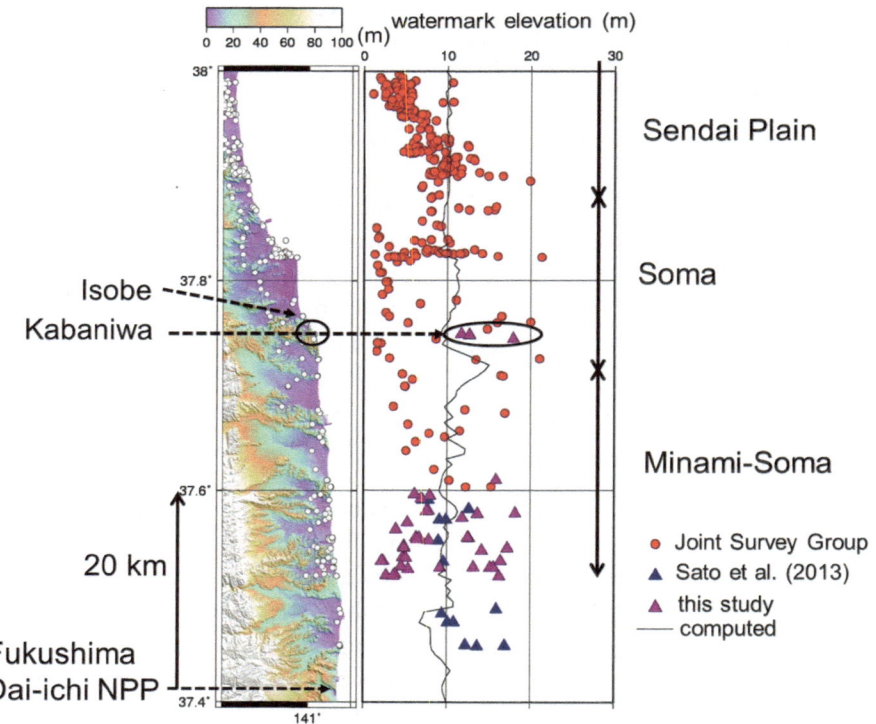

Figure 1
Distributions of measured tsunami heights around the study area. The *thin solid line* represents the nearshore tsunami height at a water depth of 10 m computed by SATO *et al.* (2013)

The shoreline of the Fukushima Prefecture is straight with a slightly convex curvature in contrast to the concave and low-lying bight along the Sendai Plain further north. The coastline in Minami-Soma is characterized by a repetitious pattern of coastal bluff formations (typically 20 m high) and small coastal plains (<2 km wide). Coastal cliffs reach heights up to 40 m above sea level along the central Fukushima Coast. Small rivers separate these hills and form confined low-lying coastal plains. The overall distribution of tsunami heights is influenced by the coastal topography (Fig. 1). The tsunami height is relatively low along the Sendai Plain, reflecting the divergence of wave rays by refraction associated with the concave bathymetry in combination with the wide continental shelf. The low-lying Sendai Plain allowed the tsunami inundation to penetrate several kilometers inland, while gradually attenuating its energy and height. In comparison, the tsunami heights along the Fukushima Coast increased due to the convex-shaped shoreline and the associated offshore bathymetry, which tend to focus tsunami energy (SATO et al. 2013).

Prior to the 2011 event, the coastal plains were protected by seawalls with a design height of 6 m T.P. Tokyo Peil (T.P.) is the standard topographic datum in Japan and represents the mean sea level in Tokyo Bay. Many seawalls collapsed due to the hydrodynamic loading from tsunamis with their onshore heights of 10–15 m, thereby far exceeding the seawall design height of 6 m. Co-seismic land-level subsidence of 0.5 m in the area further reduced the effective seawall height. Figure 2 shows large flooded areas 1 day after the tsunami attack. Inundated flat coastal plains with standing water affected by the land subsidence are separated by a series of coastal bluffs aligned to the shoreline. The following inundation areas are the focus of our investigation: from north to south, 'Obama' denotes the coastal plain around the Ohtagawa River, 'Tsukabara' denotes the coastal plain around the Odaka River, 'Murakami' denotes the isolated circular hill on the coast, 'Tsunobeuchi' denotes a relatively narrow coastal plain south of Murakami, and 'Idagawa' denotes a coastal plain around the Miyata River.[1]

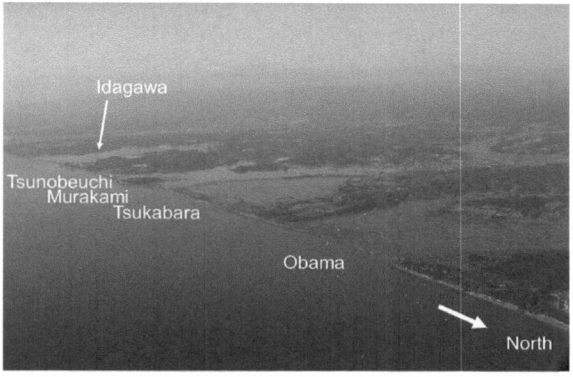

Figure 2
Aerial photo of Minami-Soma, March 12, 2011, courtesy Dr. K. Satake, UT

As noted earlier in Sect. 1, post-tsunami surveys in the southern portion of Minami-Soma were delayed owing to the nuclear accident, which isolated coastal areas within the exclusion zone. All human activities were restricted within a 20 km radius from the Fukushima Dai-ichi NPP after the infamous hydrogen explosions (the first one occurred on March 12, 1 day after the tsunami strike), and the residents were completely evacuated by the fifth day (March 16). Entry into the area was prohibited and access strictly controlled. In February 2012, <1 year after the tsunami event, the survey team, organized by the University of Tokyo and Fukushima Prefecture, was granted a special permit and conducted the first tsunami survey along the coast of the entire exclusion zone (SATO et al. 2013). In this paper, we report the findings of a follow-up survey performed on June 19, 20, 2012 as the access-restriction was lifted for the coastal areas in southern Minami-Soma. To our knowledge, this represents the only detailed tsunami survey conducted in the previously restricted areas of Minami-Soma.

As shown in Fig. 3, many clear tsunami watermarks, such as mudlines representing the inundation levels on buildings, were found intact even more than 1 year after the tsunami inundation. We could easily identify and record tsunami borne debris stranded in sheltered locations such as inside houses, tsunami-induced damage marks on communication towers, and breakage of tree limbs. (Note that all watermark photographs are archived and available on the Internet in the TSUNAMI JOINT SURVEY GROUP PHOTO ARCHIVE

---

[1] Obama (37°35.5′N, 141°01.5′E); Tsukabara (37°34.5′N, 141°01.5′E); Murakami (37°33.1′N, 141°01.6′E); Tsunobeuchi (37°32.6′N, 141°01.7′E); Idagawa (37°31.8′N, 141°01.8′E).

Figure 3
A clear mudline watermark around the exterior of a residential house located 1.2 km from the shoreline at Idagawa, Minami-Soma (tagged with RPDJ-0017 in TTSG database, June 19, 2012. Note that all watermark photographs are archived and available in the photo archives with the URL listed in the references)

2013). This highlights that some tsunami watermarks can remain preserved for an extended time under the absence of human activities such as clean-up, recovery and reconstruction.

Hydrodynamic tsunami modeling suggests focusing of tsunami wave energy towards Minami-Soma (SATO *et al.* 2013). The computation was based on linear long-wave theory with the use of the tsunami source estimated by FUJII *et al.* (2011). Figure 4 illustrates the distribution of maximum water level due to tsunami. The numerical simulation suggests that the refraction of the tsunami due to bathymetry played an important role in the propagation direction and the local tsunami amplification. Note that one of the local peaks appears to be headed towards Minami-Soma.

Based on videos and photos recorded by local residents, the tsunami reached the shore in the form

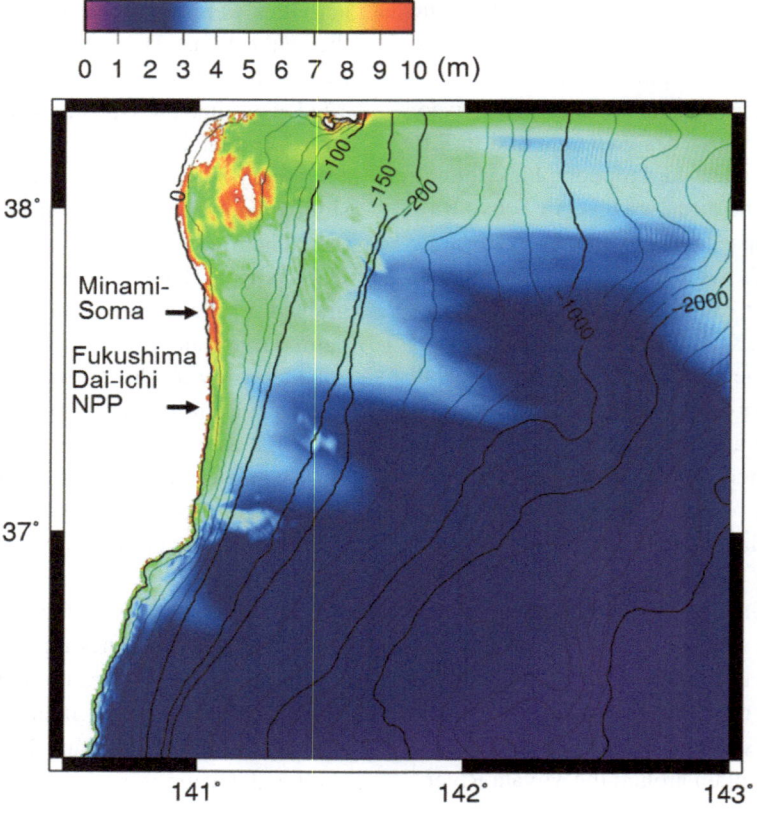

Figure 4
Maximum water level due to computed tsunami (SATO *et al.* 2013)

Figure 5
a–d Tsunami observed at Tsukabara, Minami-Soma (Photographs courtesy Mr. Sadatsugu Tomisawa)

of a breaking bore with a height of about 10 m (SATO *et al.* 2013; SANUKI *et al.* 2013). When the bore impacted coastal cliffs or seawalls, the deflected water splashed up to heights of 20–30 m. The formation of such a high splash-up near the shoreline is confirmed in a series of photographs recorded at Tsukabara (Fig. 5). Examining the successive series of photographs with a time interval of 0.5 s (only three snapshots are shown in Fig. 5a–c), the incident tsunami direction is inferred from the location of the two surviving pine trees (about 20 m high above sea level based on laser field measurements) marked by arrows in subsequent images in Fig. 5. The tsunami initially struck the area from the southeast (15:39:24, according to the camera time stamp) and seconds later also from the northeast (15:39:28 and 15:39:34).

SHIMOZONO *et al.* (2012) numerically simulated tsunami inundation processes for many localities affected by the 2011 Tohoku Tsunami and compared the results with field measurements, discussing the

effect of topographic slope on inland inundation heights. The topographic slope affects the energy dissipation in the tsunami runup process. The longer the tsunami penetration distance on a mild slope is, the greater the resulting energy loss. This trend is confirmed by the sharp contrast in inland tsunami behavior observed between the Sanriku and Sendai coasts. The steep slope topography along the Sanriku coast resulted in high tsunami runup near the maximum tsunami penetration, while the mild slope topography in the Sendai Plain exhibits tsunami heights decaying with distance from the shore. In the Sendai Plain, the runup heights at the inundation limit are significantly lower than corresponding tsunami heights near the shore. Funneling inland topography commonly found on the steep ria-type Sanriku coast further enhanced the amplification of tsunami runup (e.g., SHIMOZONO *et al.* 2014). Similar variation in onshore tsunami behavior was found along the Fukushima Coast.

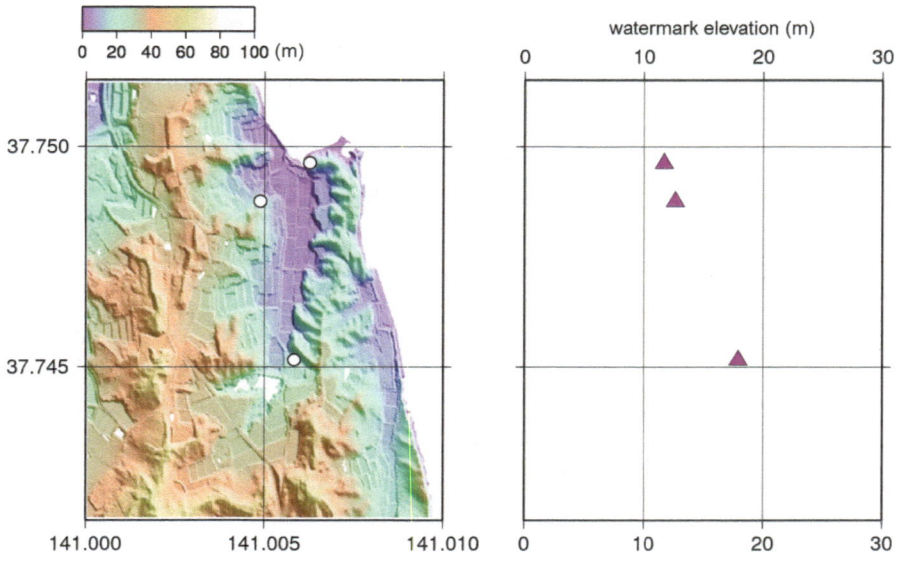

Figure 6
Distribution of measured tsunami heights in Kabaniwa, Soma

Figure 6 shows tsunami heights measured at Kabaniwa, Soma (see Fig. 1 for the location).[2] The inland topography of Kabaniwa is a V-shaped valley. The coastline was protected by a concrete rip-rap type mound seawall topped by a curved capping wall with the total height of 6 m T.P. The tsunami completely destroyed the seawalls to the ground level as shown in Fig. 7a. The maximum runup location is shown in Fig. 7b, where the tsunami height increased with the distance from the shore to the runup height at about 500 m inland (see Fig. 6).

The distribution of tsunami heights in Minami-Soma is shown in Fig. 8. Large variation in the measured heights is attributed to the coastal features in Minami-Soma characterized by alternating coastal cliffs and plains. The large variation in tsunami height can be described as follows: tsunami heights near geomorphological transitions between the coastal plain and the bluff are much higher than the heights measured within the plain. This is due to tsunami reflection on the bluff, resulting in wave amplification causing the converging flow from the amplified water level in front of the bluff to rush into the coastal plain. This is confirmed in Figs. 9

Figure 7
Photos of Kabaniwa, Soma, displaying **a** completely broken seawall and **b** location near the maximum runup (June 20, 2012)

and 10. Figure 9 shows the relationship between the tsunami height and the local ground elevation, whereas Fig. 10 shows the relationship between the tsunami height and the distance from the shoreline. Large

---

[2] Tsunami watermarks were measured by the RTK-GPS system with Virtual Reference Station mode with positional accuracy of a few centimeters.

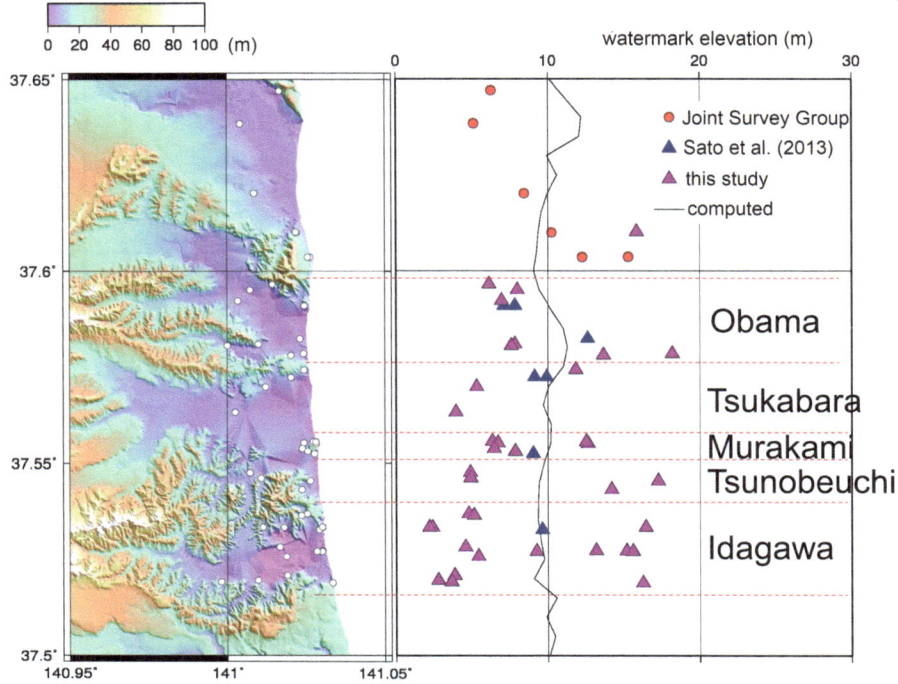

Figure 8
Distributions of measured tsunami heights in the study area

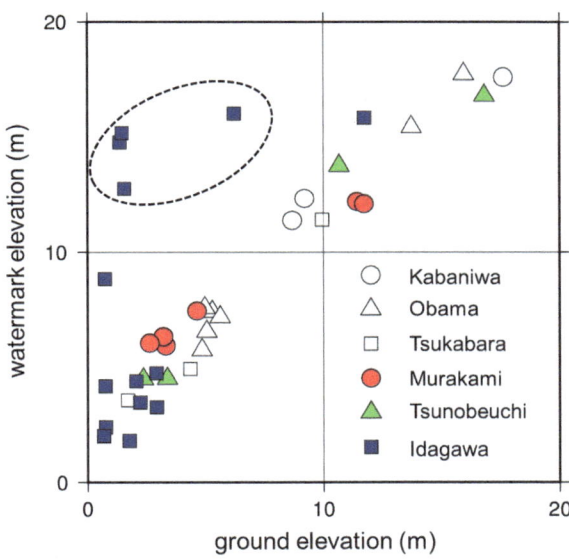

Figure 9
Relationship between tsunami heights and ground elevations

tsunami heights in excess of 10 m are generally observed at high ground elevations on the slopes of coastal bluffs with exception of a few points at Idagawa marked in Fig. 9. Large tsunami heights were observed

within 500 m from the shoreline (Fig. 10). However, inland tsunami inundation heights at different locations vary in spite of comparable tsunami heights near the shore. The tsunami heights more than 1.5 km inland at Obama are as high as 7 m, while those at Idagawa are smaller than 3 m. The reason for this significant difference remains unclear, but could be associated with the difference in damage levels to coastal structures (seawalls) as discussed later in Sect. 3.

Figure 11 shows aerial photos of Idagawa indicating the locations of the four measurements marked in Fig. 9. The photo taken before the tsunami (Fig. 11a) shows that the measurement points are located near or inside the coastal pine tree forest, which was approximately 100 m wide and used to exist behind the seawall. Most of the trees were completely washed away by the tsunami. Only six trees survived, which provided tsunami watermarks such as broken branches along with scratch marks on the bark of tree trunks. It is noted that devastating destruction of so-called tsunami control forests by the tsunami was also observed at Rikuzentakata, Iwate Prefecture (LIU *et al.* 2013).

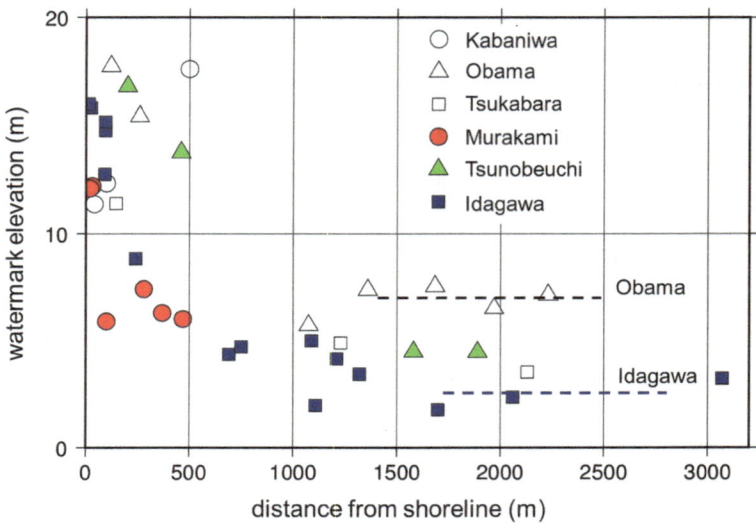

Figure 10
Relationship between tsunami heights and distance from the shoreline

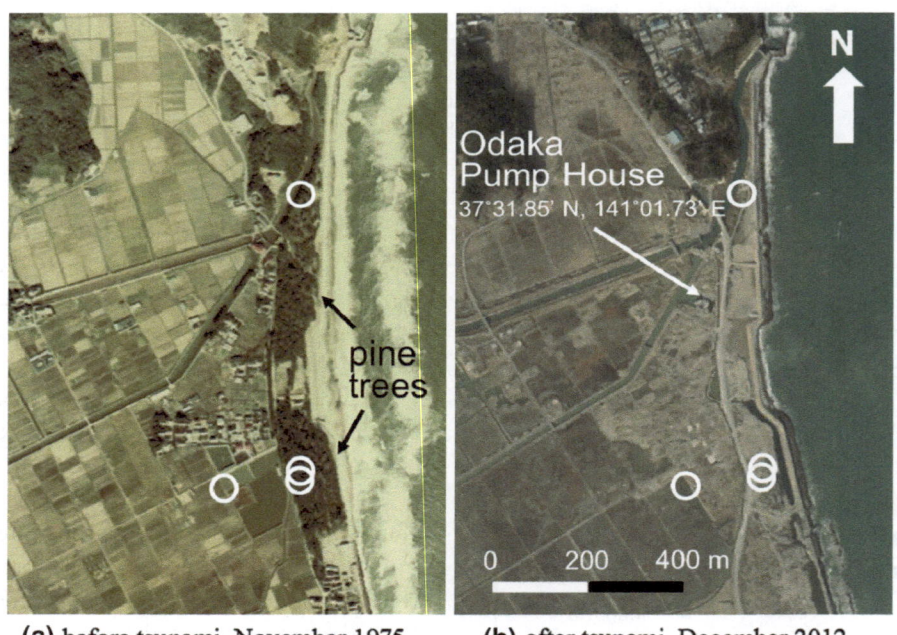

(a) before tsunami, November 1975   (b) after tsunami, December 2012

Figure 11
a, b Aerial photos before and after the tsunami showing a band of coastal pine trees (Photos are taken by Geospatial Information Authority Japan). Note that the forest remained unchanged in a satellite image taken in November 2009

Broken tree branches, as well as the damage marks on the two nearby communication towers, at Idagawa indicate that the tsunami reached 11–13 m above ground at a terrain elevation of about 1.5 m above sea level. Such large tsunami heights are attributed to the flow deflection by seawalls and the initially increased flow resistance induced by the dense coastal pine forest. Similarly at Futaba Beach, located 6 km south of Idagawa and 4 km north of the Fukushima Dai-ichi NPP, a 16.9 m high watermark

Table 1

Total and collapsed length of seawalls in Minami-Soma compiled by KATO et al. (2013)

Area	Total length (m)	Length (m) for 'complete collapse'	Ratio (%)
Obama	1,483	469	32
Tsukabara	2,041	99	5
Murakami	201	0	0
Tsunobeuchi	1,581	61	4
Idagawa	1,652	0	0

was measured in the forest while nearby watermarks adjacent to the forest were 12.2 m and 13.6 m (SATO et al. 2013).

3. Damage to Coastal Structures

Many seawalls were damaged along the Fukushima Coast, which hosts both the Fukushima Dai-ichi and Dai-ni nuclear power plants. Understanding the destruction mechanism of coastal structures in Fukushima is important since the heights of the tsunami exceeded those of the structures by several meters. Studying seawall overtopping flows is crucial to understand the vulnerability of structures as documented by SATO et al. (2012) in a field study at Nakoso located at the southern end of the Fukushima Prefecture.

KATO et al. (2013) compiled a comprehensive dataset of seawall destruction due to the 2011 Tohoku Tsunami. On the basis of visual inspection of aerial photographs and satellite images, they categorized the destruction of seawalls in 'complete collapse' when the seawall completely collapsed to the ground level, and 'partial collapse' when some part of the seawall remained. Table 1 summarizes the data for seawalls in Minami-Soma (locations are identified in Fig. 2). The ratio of 'complete collapse' is large at Obama, which may explain why the tsunami heights near the inundation limit at Obama were large compared to Idagawa (Fig. 9). This suggests that unless the protective coastal structures (seawalls) were completely destroyed, the remnant seawalls are capable of reducing the inland inundation level.

Based on the lessons learned from the Tohoku Tsunami, the Japanese government introduced two-

level tsunami hazards to deal with tsunami disaster mitigation. The first-level tsunami hazard (Level-1 tsunami) is defined as a frequent tsunami with a return period of 100 years. Shore protection structures are designed for a Level-1 tsunami, which dictates crown elevations along with other design criteria. On the other hand, the second-level tsunami hazard (Level-2 tsunami) represents the probable maximum tsunami and is used to establish evacuation strategies and plans. The return period of the Level-2 tsunami is set at 1,000 years. Partially damaged, protective coastal structures (e.g. seawalls) remain capable of reducing tsunami effects for tsunamis exceeding their design level (Level-1). Hence, the structures should remain partially effective for a Level-2 event, unless the structures were totally destroyed. Therefore, it is important to construct such protective structures with the consideration of tsunamis beyond the design level (Level-1). In other words, protective coastal structures like tsunami seawalls should be designed to maintain partial effectiveness even for 'beyond-the-design-basis' conditions. This design concept is called 'nebari' in Japanese, which implies tenacity, toughness and resilience. Careful field observations on damaged seawalls should lead to some insights for how to design seawalls using the 'nebari' concept. However, this concept should not be applied to critical infrastructure such as nuclear power plants where overtopping must be prevented.

Figure 12 shows a partially damaged seawall at Isobe in Soma City (see Fig. 1 for location). The gently sloped front face is intact but significant damage was found on the crown and along the landward slope. A large amount of water impounded on the lee side of the seawall filled a scour hole excavated by the extensive tsunami overtopping flow. The detached rubble mound breakwaters located about 100 m offshore of the seawall appeared to be intact.

The Tsukabara coast was also protected by seawalls with a crest height of 6 m T.P. Figure 13 shows upright concrete seawalls completely broken off at the ground level. The 3 m wide capping concrete plate and the landside concrete plate were displaced and transported inland by the overflowing tsunami. The sand-filled core inside the seawall was completely washed away. The progressive collapse and scour removed the backside

Figure 12
Seawall damage observed in Isobe, Soma (June 20, 2012)

Figure 14
A surviving seawall observed in Tsukabara, Minami-Soma (June 20, 2012)

Figure 13
Seawall damage observed in Tsukabara, Minami-Soma (September 3, 2012)

Figure 15
A surviving seawall connected with a completely broken section observed in Naraha (August 24, 2012) approximately 20 km south of the Fukushima Dai-ichi Power Plant

support leading to the fracture at ground level of the seaside concrete wall. Some of the concrete armor blocks, originally placed in front of the seawall, were found more than 300 m inland.

Figure 14 shows a surviving seawall on the south side of the Odaka River in Tsukabara. The gentle slope of the front face and the crown of the seawall are intact. The landward side is partly collapsed, but the gravel land fill which had been placed to cover the landward slope of the seawall appeared to minimize the scour there.

The foregoing observations on seawall failures suggest that design improvements for the crown portion and the lee side of the seawalls are important to enhance 'nebari' under overtopping flow. For example, a heavier crown cap combined with milder side slopes could help reduce damage to the seawall. Furthermore, a concrete apron extended landward

from the lee-side toe of the sea wall could reduce scour actions. Another example of a remnant seawall is shown in Fig. 15, indicating that the presence of partitioning walls was effective in supporting the seawall, even where a substantial amount of the fill material was scoured away. Black sand bags were placed in the section of total failure, where partitioning walls were not installed (Fig. 15). The foregoing and many other observations should be analyzed carefully to develop the designs of future protective coastal structures based on 'nebari.'

Figure 16 shows the destruction of the Odaka Pump House located close to the shoreline along the Idagawa coast (see Fig. 11 for the location). The pump house is a reinforced-concrete structure which failed

Figure 16
Devastating damage of the Odaka Pump House located close to the Idagawa coast (February 6, 2012)

completely as indicated by fractured walls and columns snapped off at their base. Observed scour holes around the foundation of the pump house did not appear to contribute to the structural failure. The seawalls initially blocked the incident tsunami flow allowing energy storage and accumulation until overtopping commenced. This stored energy—at least to the crown elevation (6.1 m) of the seawall—must have been released to the lee side. The overtopping flow released from the high elevation of the seawall crown created a high velocity (more than 11 m/s based on the Bernoulli theorem) onto the initially dry soil surface behind the seawall. The upper limit of the flow speed is estimated to be 17 m/s for a tsunami height of 15 m near the seawall at Idagawa (Fig. 9). Therefore, the flow speed behind the seawall in Idagawa is constrained within a range of 11–17 m/s, which is comparable to velocity measurements at Kesennuma (FRITZ *et al.* 2012). The corresponding stagnant pressure range is 61–145 kPa, which could explain the massive fluid forces applied to any trees and structures behind the seawall. The generation of such massive hydrodynamic forces by overtopping flows immediately behind the seawall can be considered one of the disadvantages of seawalls. Protective coastal structures limit the ocean view and the noise from an approaching tsunami, which in combination with psychological overconfidence in engineered defense structures, could mislead people to delay their evacuation. In Japan, signs on coastal structures remind residents to heed evacuations regardless of the engineered defenses.

4. Conclusions

This paper described the results of the detailed tsunami survey in Minami-Soma, where the survey was delayed due to the nuclear accident at the Fukushima Dai-ichi Nuclear Power Plant. The main findings are summarized as follows:

(1) Tsunami inundation heights in Fukushima Prefecture were relatively large owing to the macroscopic features in offshore bathymetry and the convex shape of the coastline. The variation in tsunami inundation height was significant in Minami-Soma reflecting alternating coastal bluffs and plains. Large tsunami heights exceeding 10 m were found within 500 m from the shoreline. Tsunami reflection by seawalls and cliffs, and increased flow resistance induced by coastal pine forest belts are considered to be the cause of local and temporal amplifications in areas near the shore.

(2) Distribution of the inland tsunami height profile was dependent on the inland topography. Steep V-shaped valleys produced landward increasing tsunami heights, while across flat coastal plains tsunami heights decayed with distance from the shoreline.

(3) Protective coastal structures (e.g. seawalls) as well as dense coastal forests temporarily slow down the incident tsunami, while absorbing and accumulating tsunami potential energy. The sudden release of the stored tsunami energy as the forests collapsed and walls were overtopped resulted in significant hydrodynamic forces along with debris impact and associated damming forces causing unanticipated structural damage and scour in the areas immediately behind the seawalls or coastal forests.

(4) Unless completely destroyed, seawalls were effective in reducing tsunami inundation levels. Some remnant seawalls provided valuable design lessons in enhancing the tenacity of seawalls against tsunami overflow. Gentle side slopes, landfill on the landward slope and placement of partitioning walls are considered among

enhancement factors to strengthen seawalls by using the concept of 'nebari', a Japanese word representing tenacity, toughness and resilience.

Acknowledgments

The authors of UT and TUMST were supported by Japan Science and Technology Agency through the J-RAPID program. H.Y. and H.M.F. were supported by the US National Science Foundation through the NSF RAPID award CMMI-1135768. The authors also acknowledge field survey assistance by several students of UT and TUMST.

REFERENCES

FRITZ, H.M., PHILLIPS, D.A., OKAYASU, A., SHIMOZONO, T., LIU, H., MOHAMMED, F., SKANAVIS, V., SYNOLAKIS, C.E., TAKAHASHI, T. (2012). *2011 Japan tsunami current velocity measurements from survivor videos at Kesennuma Bay using LiDAR*, Geophys. Res. Lett., *39*, L00G23, doi:10.1029/2011GL050686.

FUJII, Y., K. SATAKE, S. SAKAI, M. SHINOHARA and T. KANAZAWA: *Tsunami source of the 2011 off the Pacific coast of Tohoku Earthquake*, Earth Planets Space, Vol. *63* (No. 7), pp. 815–820, 2011.

KATO, F., Y. SUWA, K. WATANABE and S. HATOGAI: *Damages to shore protection facilities induced by the Great East Japan Earthquake Tsunami*, Journal of Disaster Research, Vol. *8* No. 4, 612–625, 2013.

LIU, H., SHIMOZONO, T., TAKAGAWA, T., OKAYASU, A., FRITZ, H.M., SATO, S., TAJIMA, Y.: *The 11 March 2011 Tohoku Tsunami survey in Rikuzentakata and comparison with historical events*. Pure Appl. Geophys., *170*(6–8):1033–1046, doi:10.1007/s00024-012-0496-2, 2013.

MORI, N., TAKAHASHI, T. and THE 2011 TOHOKU EARTHQUAKE TSUNAMI JOINT SURVEY GROUP: *Nationwide post event survey and analysis of the 2011 Tohoku Earthquake Tsunami*, Coastal Engineering Journal, JSCE, *54*(1), 1250001, 2012.

SANUKI, H., Y. TAJIMA, H. YEH and S. SATO: *Dynamics of tsunami flooding to river basin*, Proc. Coastal Dynamics, 2013.

SATO, S., S. TAKEWAKA, H. LIU and H. NOBUOKA: *Tsunami damages of Nakoso Coast due to the 2011 Tohoku Tsunami*, Proc. 33rd Intl. Conf. on Coastal Engineering, 2012.

SATO, S., H. YEH, M. ISOBE, K. MIZUHASHI, H. AIZAWA and H. ASHINO: *Coastal and nearshore behaviors of the 2011 Tohoku Tsunami along the central Fukushima Coast*, Proc. Coastal Dynamics, 2013.

SATO, S.: *2011 Tohoku Tsunami and future directions for tsunami disaster mitigation, Keynote lecture*, Proc. of 2013 IAHR Congress, Tsinghua University Press, 2013.

SHIMOZONO, T., S. SATO, A. OKAYASU, Y. TAJIMA, H. M. FRITZ, H. LIU and T. TAKAGAWA: *Propagation and inundation characteristics of the 2011 Tohoku Tsunami on the central Sanriku Coast*, Coastal Engineering Journal, JSCE, *54*(1), 1250004, 2012.

SHIMOZONO, T., H. CUI, J. D. PIETRZAK, H. M. FRITZ, A. OKAYASU and A. J. HOOPER: *Short wave amplification and extreme runup by the 2011 Tohoku Tsunami*, Pure Appl. Geophys., 2014 (submitted).

SIMONS, M., S.E. MINSON, A. SLADEN, F. ORTEGA, J. JIANG, S.E. OWEN, L. MENG, J.-P. AMPUERO, S. WEI, R. CHU, D.V. HELMBERGER, H. KANAMORI, E. HETLAND, A.W. MOORE, and F.H. WEBB: *The 2011 Magnitude 9.0 Tohoku-Oki Earthquake: Mosaicking the Megathrust from Seconds to Centuries*. Science, *332*:1421–1425. doi:10.1126/science.1206731, 2011.

TOHOKU TSUNAMI JOINT SURVEY GROUP (TTSG): Tsunami survey database, http://www.coastal.jp/tsunami2011/, referred on October 2013.

TSUNAMI JOINT SURVEY GROUP PHOTO ARCHIVE: http://grene-city.csis.u-tokyo.ac.jp/, referred on October 2013.

(Received November 4, 2013, revised February 19, 2014, accepted February 24, 2014, Published online March 29, 2014)

Pure Appl. Geophys. 171 (2014), 3241–3255
© 2014 Springer Basel
DOI 10.1007/s00024-013-0768-5

Seismic Regime in the Vicinity of the 2011 Tohoku Mega Earthquake (Japan, $M_w = 9$)

M. V. RODKIN[1,2] and I. N. TIKHONOV[2]

Abstract—The 2011 Tohoku mega earthquake ($M_w = 9$) is unique due to a combination of its large magnitude and the high level of detail of regional seismic data. The authors analyzed the seismic regime in the vicinity of this event using data from the Japan Meteorological Agency catalog and world databases. It was shown that a regional decrease in *b*-value and of the number of main shocks took place in the 6–7 years prior to the Tohoku mega earthquake. The space–time area of such changes coincided with the development of precursor effects in this area, as revealed by LYUBUSHIN (Geofiz Prots Biosfera 10:9–35, 2011) from the analysis of microseisms recorded by the broadband seismic network F-net in Japan. The combination of episodes of growth in the number of earthquakes, accompanied by a corresponding decrease in the b-value and average depth of the earthquakes, was observed for the foreshock and aftershock sequences of the 2011 Tohoku earthquake. Some of these anomalies were similar to those observed (also post factum) by KATSUMATA (Earth Planets Space 63:709–712, 2011), NANJO *et al.* (Geophys Res Lett 39, 2012), and HUANG and DING (Bull Seismol Soc Am 102:1878–1883, 2012), whereas others were not described before. The correlation of the periods of growth in seismic activity with the decrease of the average depth of earthquakes can be explained by the growth of fluid activity and the tendency of a penetration of low density fluids into the upper horizons of the lithosphere. The unexpectedly strong Tohoku mega earthquake with a rather small rupture area caused an unexpectedly high tsunami wave. From here it seems plausible that M9+ earthquakes with a large tsunami could occur in other subduction zones where such cases were suggested before to be impossible.

Key words: Japan region, earthquake catalog, seismic regime, mega earthquake, vicinity of mega earthquake, precursor effects, aftershock sequence.

1. Introduction

The Tohoku earthquake ($M_w = 9.0$) of 11 March 2011 was unexpected but the best instrumentally recorded great earthquake (RODKIN and TIKHONOV 2011, 2013; KAGAN and JACKSON 2013; LAY *et al.* 2013) to date. Seismic-hazard assessments based on instrumentally observed and historically documented events did not expect an earthquake of such magnitude in Japan. M9+ earthquakes were not recorded in the Japanese region before the 2011 Tohoku earthquake, either from instrumental data or from sketchy historical data available since the year 599 (USAMI 1979; UTSU 1979).

Moreover, it could be suggested that such great earthquakes are impossible in Japan because of the fragmentation of the lithosphere into different plates and an essential curvature of the deep trench in the Japanese region. There are not whole rectilinear segments of the deep trench which are able to accommodate an M9+ earthquake source on the order of 1,000 km long, as one might expect for a magnitude 9 earthquake. Hence, decisions about protective measures, such as heights of tsunami walls, have been based on maximum tsunami expectations from M8.5 earthquakes that proved to be inadequate when the 2011 Tohoku $M_w = 9.0$ earthquake occurred (LAY *et al.* 2013). Casualties were caused mainly by the tsunami impact on coastal towns, where tsunami heights significantly exceeded the protective harbor tsunami walls.

Despite the absence of records of M9+ earthquakes, a sizeable probability of the occurrence of an M9+ earthquake in Japan was indicated by PISARENKO *et al.* (2010), a result of the statistical analysis based upon the use of the limiting distributions of the theory of extreme values. The physical reasons that earthquakes of magnitude 9.0–9.7 can occur in the majority of subduction zones (Japan zone included) were presented in McCAFFREY (2008), and KAGAN and JACKSON (2013); moreover, it was argued that variations in seismic potential among different subduction zones are not significant.

---
[1] Institute of Earthquake Prediction Theory and Mathematical Geophysics, RAS, Moscow, Russia.
[2] Institute of Marine Geology and Geophysics, FEB RAS, Yuzhno-Sakhalinsk, Russia. E-mail: tikhonov@imgg.ru

Thus, the 2011 Tohoku earthquake is unique due to a combination of its surprisingly large magnitude and the high level of instrumental recording. It provides an opportunity for detailed examination of change in the seismic regime connected with such rare phenomenon as a mega earthquake occurrence.

Earthquakes of magnitude 9 affect large regions; therefore, a large number of smaller magnitude events occur in the region of influence of these great earthquakes. Given the growth in the available volume of data, one could expect that the situation for the prediction of great or mega earthquakes is better than that for earthquakes of smaller magnitude. However, this is not the case and there is not yet much success in the prediction of M9+ earthquakes.

According to the long-term major and great earthquake prediction given by the seismic gap method (FEDOTOV et al. 2011), although the focal zone of the Tohoku mega earthquake joins the zone of high seismic hazard from the south, the Tohoku earthquake occurred outside the zone of the greatest hazard. Moreover, during the Tohoku earthquake re-rupturing of several regions, which had experienced smaller events during the last century, took place (LAY et al. 2013); this rare style of rupture disagrees with the classical seismic gap scenarios.

No intermediate-term prediction of the most recent M9+ earthquakes (the 26 December 2004 $M_w = 9.3$ Sumatra–Andaman earthquake, and the 11 March 2011 $M_w = 9.0$ Tohoku earthquake) were made including the regularly used M8 and MSc algorithms of intermediate-term earthquake forecasting of KEILIS-BOROK and KOSSOBOKOV (1990), KOSSOBOKOV et al. (1990), and KOSSOBOKOV (2013). These algorithms have been used over the past two dozen years and have quite convincing statistics (KOSSOBOKOV 2013) on their success rates in comparison with the "seismic roulette forecasting method". The M8 algorithm was developed originally for the forecasting of the great (M8.0+) earthquakes (KEILIS-BOROK and KOSSOBOKOV 1986). With this algorithm, examined territory is scanned with overlapping circles, with the diameter depending upon the magnitude under forecasting. Within each circle the sequence of earthquakes without aftershocks is examined and a set of predictive functions is calculated (see KEILIS-BOROK and KOSSOBOKOV

1990 for details). If, for a given time interval, a definite number of these predictive functions exceeds the corresponding threshold levels, a large earthquake hazard is declared within this circle for the next 5-year period. The diameter of the circles representing the size of the zone of influence of the earthquake in preparation is rather large: 560, 854 and 1,333 km for magnitudes 7.0, 7.5 and 8.0, respectively.

The MSc algorithm allows the possibility of decreasing the size of the area of seismic hazard from that obtained by the use of the M8 algorithm (KOSSOBOKOV et al. 1990). The MSc algorithm was designed by retroactive analysis of seismicity prior to the Eureka earthquake (1980, $M = 7.2$) near Cape Mendocino in California. An application of the MSc algorithm requires a more detailed seismic catalog than the M8 algorithm.

Recent post factum studies have indicated, however, that a successful forecasting might be obtained if a proper renormalization procedure in the M8 and MSc algorithms were to be used (KOSSOBOKOV 2011; DAVIS et al. 2012).

In contrast with the failure of regularly used methods of earthquake forecasting, the location of the 2011 Tohoku earthquake was a priori specified from the analysis of low-frequency seismic noise recorded by the broadband seismic F-net network in Japan (LYUBUSHIN 2011). Post factum, a possibility of an intermediate-term prediction of the earthquake's time (LYUBUSHIN 2011) from the analysis of low-frequency noise was determined. The forecasting features, used in (LYUBUSHIN 2011), consist of the simplification of the structure of the seismic noise. Such simplification is similar to that occurring in a case of critical style behavior (MA 1976; HAKEN 1978; and others). The works of LYUBUSHIN (1998, 2011) provide a detailed description of the method. Note, however, that this method has no statistical validity. In fact, there is only one case of an incomplete successful prediction. There is no other evidence in support of this method.

SOBOLEV (2011) post factum showed that 1.5 months before the 2011 Tohoku earthquake the amplitude of seismic noise, with periods of minutes, had grown in the Tohoku focal zone. Similar results of an increase in amplitude of the low-frequency seismic noise in the Tohoku earthquake's epicentral zone were observed by MARSAN and ENESCU (2012).

In fact, an increase in amplitude of low-frequency seismic noise was previously observed before a few strong and major earthquakes (SOBOLEV and LYUBU-SHIN 2006, 2007).

One of the main problems arising in earthquake prediction is related to identification of earthquake precursors on the background of the strong stochastic changes of a seismic regime. However, a contribution of random components often decreases with an increase of available information. It is easy to see that the data volume used for the prediction of the 2011 Tohoku earthquake from the analysis of microseismic noise surpasses, by many orders of magnitude, the volume of data of the NEIC/USGS world catalog (http://earthquake.usgs.gov/regional/neic), which was used in earthquake prediction with M8 and MSc algorithms and in the seismic gap approach.

It is possible to assume that the contribution of random components in the result of analysis of seismic noise is small. Thus, the developing tendencies of a seismic regime are revealed more clearly, which have allowed the observation of precursory trends prior to the 2011 Tohoku earthquake.

The results of analysis of a seismic regime in the vicinity of earthquakes of magnitude $M_w \geq 6.5$ strong earthquake generalized vicinity (SEGV) testify in favor of the importance of the volume of data used (RODKIN 2008, 2012). The SEGV method is based upon an aggregation of information about a large number (several hundred, up to 1,000) of the strongest earthquakes. If the 500 strongest earthquakes from the Harvard CMT catalog are taken into account, the corresponding mb values change from mb = 6.3 until the maximum mb = 8.9 value, and the magnitude range increases correspondingly if the 1,000 strongest events are taken into consideration. Such aggregation of information has permitted us to reveal a complex of anomalies which indicate the preparation process of a strong earthquake. The clearly observable tendency for the growth of amplitude of these anomalies prior to the time of a generalized strong event allows one "to predict" the moment of a generalized strong earthquake.

In the problem of earthquake prediction, the task, in many respects, consists of the detection of the precursory effects against the strong background variability of a seismic regime. As a rule, the contribution of random components decreases with an increase in the volume of information available. Proceeding from there, it was suggested (RODKIN 2008; TIKHONOV and RODKIN 2011) that an increase in volume of seismic information, as used in the case of the SEGV approach, will facilitate a successful prediction of strong earthquakes. This work analyses the seismic regime of the Tohoku mega earthquake and discusses the possibility of obtaining a retrospective prediction of the Tohoku earthquake on the basis of data from the regional catalog of the Japan Meteorological Agency (JMA). The volume of data used significantly exceeds the volume of data of the NEIC/USGS world catalog for the same region (as used in major and great earthquake prediction with algorithms M8 and MSc), but this volume of data is much less than the volume of data on seismic noise registered by broad-band networks of Japan.

The questions under study are the following: does the increase of volume of the available data reveal the precursor features of the Tohoku mega earthquake? And do these features agree with those found recently (RODKIN 2008, 2012) in the generalized vicinity of strong earthquake?

When this plan of research was mainly fulfilled and a few such anomalies were found and even published in part (RODKIN and TIKHONOV 2013), it was found that similar precursor anomalies in the change of the rate of earthquakes and in b-values were found and published by NANJO et al. (2012), HUANG and DING (2012), KAWAMURA et al. (2013), and LAY et al. (2013). These results agree with our findings and will be discussed below.

2. Methods of Analysis and Data

The Introduction clearly defines the research problem: to try to use the most detailed regional catalog to enlarge the data volume used, and thus to reduce the noise contribution in seismic data, to detect clear signs of preparation of the Tohoku mega earthquake, and to compare the results with those obtained in examination of the strong earthquake generalized vicinity (SEVG) model.

The catalog used for this study is the JMA catalog. The completeness levels of the JMA regional catalog vary with time. The catalog is complete for M5+ earthquakes since 1930, complete for M3.5+ earthquakes since 1961, and complete for M3+ earthquakes since 1994. In this study we used earthquake data of M ≥ 3.5 from 1961 onwards and data of M ≥ 3.0 from 1994 onwards.

A considerable part of the random component of a seismic regime is connected with aftershock sequences. These sequences are secondary in relation to the size of released seismic energy and they alter the background catalog seismicity. They also provide a noticeable impact on other characteristics of a seismic regime. Because of this, we carried out the analysis of two datasets from the JMA catalog—(1) the dataset of all earthquakes and (2) the dataset of main shocks where the aftershock sequences were removed from the catalog.

The catalog of main shocks was formed with the help of a program (SMIRNOV 1997), which used the method by MOLCHAN and DMITRIEVA (1992) to remove aftershocks. The catalog subset of M ≥ 3 spans from January 1994 until 10 March 2011; there are 100,562 earthquakes with 52,071 main shocks among them. The aftershock sequence of the 2011 Tohoku mega earthquake was analyzed using the JMA catalogs until 11 September 2011.

We have used several methods of data analysis. In the framework of the first approach, sliding spatiotemporal cells were used and the number of events (N), released seismic energy (E), b-values, and fractal correlation dimensions (D) were calculated. The data used for these parameters was binned in time intervals of 2 years (± 1 year relative to midyear) and in overlapping spatial cells around integer values of latitude and longitude with a spatial cell size of 2 by 2 on both coordinates. A depth restriction of $H \leq 70$ km was used.

We have calculated the b-values according to the formula (UTSU 1965):

$$b = \log e/(M_{\mathrm{av}} - M_{\mathrm{z}}), \qquad (1)$$

where M_{z} is the threshold magnitude above which the data should be complete; in practice $M_{\mathrm{z}} = 3.0$ magnitude threshold was used here, and M_{av} is the average value of a magnitude in a sample. The b-values from (1) were estimated for the cases of more than 20 events in a cell. There were 1,949 such cells in the catalog of all events, and there were 1,652 such cells in the catalog of main shocks.

The value of the fractal correlation dimension, D (GRASSBERGER and PROCACCIA 1983), was determined from the plot of the logarithm of the number of couples of events located at a distance less than r from logarithm r. Calculation of D was carried out if a number of events in a cell exceed 100. There are 1,107 such cells in the catalog of all events, and 552 in the catalog of the main events.

The choice of the size of a space–time cell was chosen as a compromise between an increase in space–time resolution and of the accuracy of the estimates of the fractal correlation dimension and the b-value (D and b-value). We have suggested the estimates of D and b-value to be suitably valid if their time and spatial changeability significantly exceeded the random errors in estimation. A restriction on the number of events was chosen, taking into account the results of MANDAL and RODKIN (2011) and RODKIN and MANDAL (2012), where a minimum admissible number of events providing a suitable accuracy for b and D estimates were discussed; in this paper $n = 20$ and $n = 50$ were found correspondingly to be suitable in the cases of b and D estimates when the tendencies of change of these parameters rather than their exact values are examined. The uniform size of a space–time grid simplifies the comparison of different parameters of the seismic regime.

In the framework of the second approach the changes in the seismic parameters (N, E, b-value, D) in the vicinity of the 2011 Tohoku mega earthquake were analyzed. The radius of the epicentral area was chosen to be $R = 250$ km, half the size of the aftershock zone (HIROSE et al. 2011). We examined the change in character of seismic activity in the areas with a radius of $0.5R$, R, $2R$, $3R$, $4R$, $5R$ and $7R$ around the Tohoku mega earthquake epicenter.

In addition, the aftershock sequence of the 2011 Tohoku mega earthquake was analyzed. Aftershocks were chosen by the standard abovementioned method from the JMA catalog.

3. Calculation Results

In the framework of the first approach of sliding spatial–temporal cells, we have examined the characteristics of seismicity in space–time cells for the

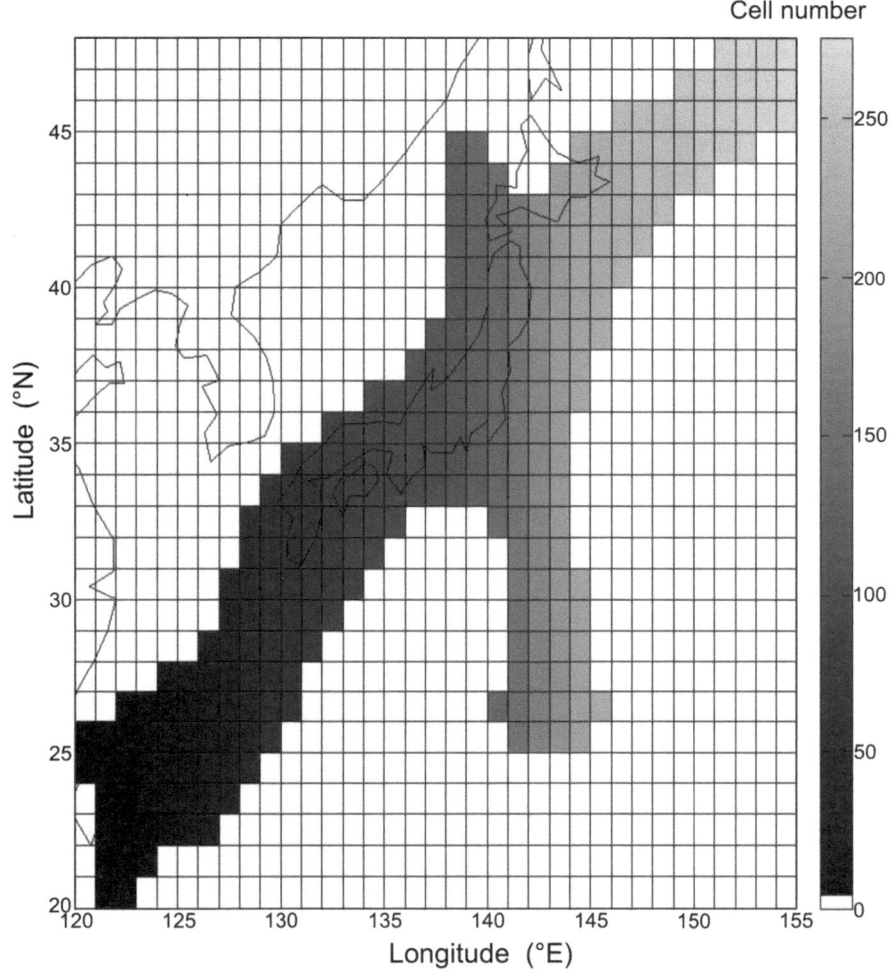

Figure 1
Location of averaging cells. The number of cells are given at the *right*; empty cells are shown as *white cells*

time period before the 2011 Tohoku earthquake using the JMA catalog. In order to highlight more clearly the time component in the change of the seismic parameters, we have used the normalized values of these parameters. Normalization of each of the parameters was obtained by subtracting the average value of that parameter and dividing by the normal deviation value for the given spatial cell. Normalization procedure permits us to minimize the influence of large spatial difference in level of seismic activity in different cells. In the case of seismic energy we have used the logarithm of the released energy. Figure 1 shows the spatial arrangement of cells. The numbers of cells increase in general from southwest

to northeast, the cells with a deficit of seismic information are shown as white cells.

In order to study the general tendencies in interrelation of space–time values of the seismic parameters (the number of all earthquakes and main shocks, released seismic energy, *b*-value and fractal dimension *D* value), the matrix of correlations between these parameters in different spatial–temporal cells (Table 1) was calculated. For the given sample size, if the correlation $|r| \geq 0.1$, this corresponds to tendencies that are statistically valid with a probability essentially more than 0.95, and if $|r| > 0.3$ then the relation is valid with probability more than 0.99.

Table 1

The correlation values between the four seismic parameters in different spatial–temporal cells

Parameters	Released seismic energy (E)	Number of main events (N)	b-value (b)	Fractal correlation dimensions (D)
E	1	0.38 (0.14)	−0.46 (−0.35)	0.15 (0.05)
N		1	0.16 (0.42)	0.53 (0.29)
b			1	−0.01 (0.25)
D				1

Values given in parentheses are those obtained for the normalized values (see text) of the parameters

Table 1 presents the correlation coefficients' values for the catalog of main shocks. The result of the calculation for all M3+ earthquakes gave similar, but lower correlation values. The results presented in Table 1 allow the following interpretation. A high negative correlation found between b-value and the released seismic energy E is quite expected because a decrease in b-value and an increase of the released seismic energy both correspond to an increase of proportion of moderate and strong earthquakes. Positive correlation of b-values with change in number of main shocks N will be discussed below.

Positive correlation of fractal dimension D values and a number of events is similar to that observed earlier in quite different cases of examination of seismicity and location of mineral deposits (CHEN *et al.* 2006; FORD and BLENKINSOP 2008; MANDAL and RODKIN 2011; RODKIN and MANDAL 2012). This tendency (MANDAL and RODKIN 2011; RODKIN and MANDAL 2012) is most likely an artifact of the method of estimation of D-value in the case of a shortage of available data. Positive correlation between b- and D-values relates to the well-known formula of proportionality of these values (AKI 1981); this correlation was repeatedly mentioned by different authors and can be considered typical of seismicity (CHEN *et al.* 2006; RODKIN and MANDAL 2012).

Thus, the revealed statistical interrelations turned out to be, in most cases, quite expected and natural. Further, we will be mainly interested in the time variability of the seismic parameters, which was seen

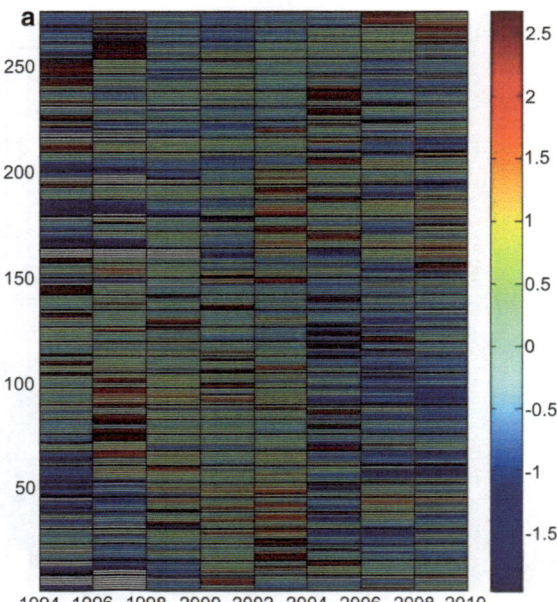

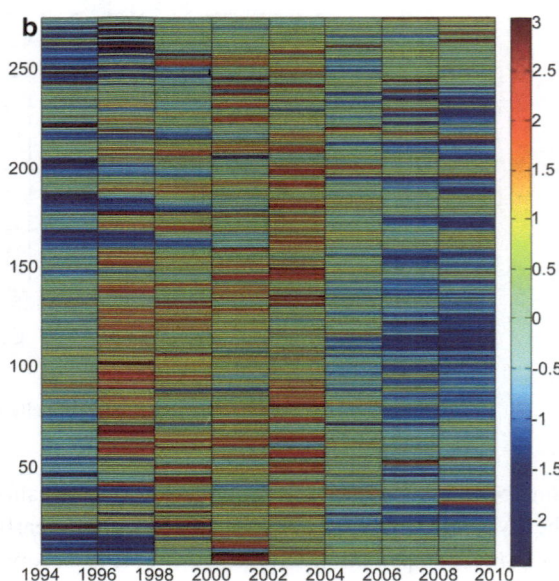

Figure 2

The normalized numbers of events (N) in different spatiotemporal cells for **a** all events and **b** main events. X-axis is time in years, Y-axis is the number of spatial cells (see Fig. 1 for the cells' location) (Color figure online)

in the result of analysis of the normalized characteristics of the seismic regime.

Figure 2 shows the variability of normalized numbers of M3+ shocks depending on a cell number (ordinate) and a time interval (abscissa) for all

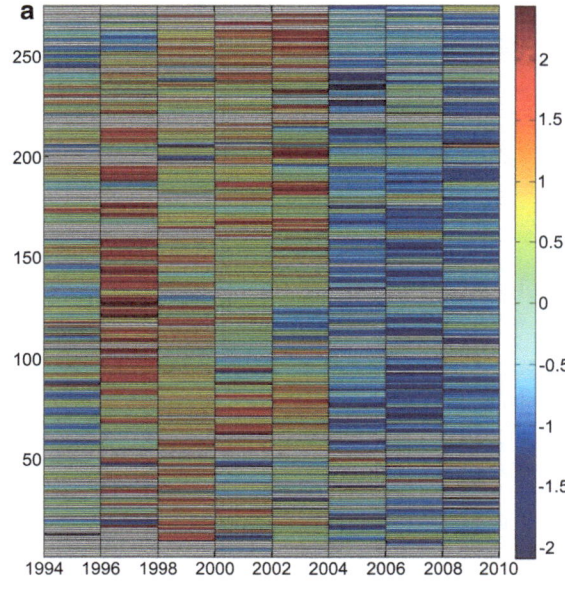

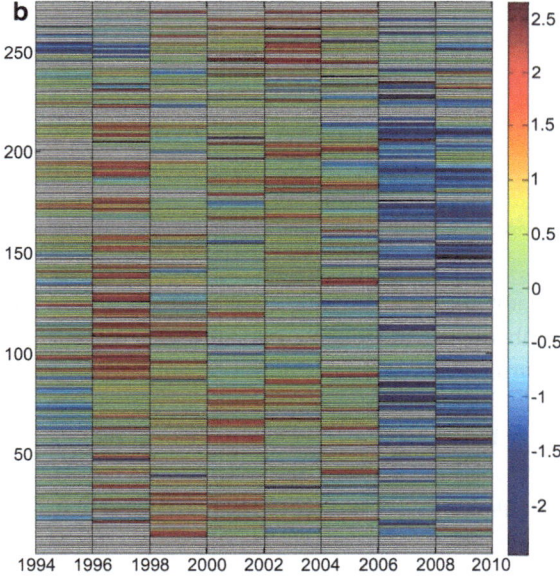

Figure 3

Change of normalized b-values for **a** all events and **b** main events. The X-axis is in years, the Y-axis is the number of spatial cells (see Fig. 1 for the cells' location). The numbers of cells are shown in Fig. 1; in general, the numbers increase from southwest to northeast (Color figure online)

earthquakes (Fig. 2a) and only for the main events (Fig. 2b). Recall that the normalized values were obtained for each cell by subtraction of average value and division by normal deviation of the given parameter value for the given cell. This way the temporal variability of the seismic parameters is made more clearly

for comparison. There is no obvious time variability in Fig. 2a. But Fig. 2b shows the clear tendency of a regional growth of the normalized numbers of the main events dataset from 1996 to 2004, and reduction in the number of main shocks from 2004 to the moment of occurrence of the 2011 Tohoku earthquake. Thus, in the change of the number of main shocks, one can see the effect of seismic quiescence arising 6–7 years before the Tohoku mega earthquake occurrence.

A similar effect of seismic quiescence prior to the 2011 M9.0 Tohoku earthquake was described by KATSUMATA (2011), and HUANG and DING (2012) where both possible approaches, with and without the declustering procedure of removing aftershocks, were used. Whereas the spatiotemporal characteristics of the zone of seismic quiescence differ in these papers (in accordance with the difference in the used methods), the very existence of the effect of intermediate-term seismic quiescence is quite clear.

An analogous procedure was applied to the examination of the b-values. The results are shown in Fig. 3, where empty (white) cells correspond to those with an insufficient number of events for b-value calculation (≤ 20 events). Figures. 2b and 3a, b clearly show time intervals of simultaneous change in the normalized number of main events and in the normalized b-values revealed almost for the whole area of the Japanese region. Such spatiotemporal variability specifies the character of correlation of N- and b-values, previously shown in Table 1.

Thus, we have that the time interval of 1994–1996 corresponds to a decrease in b-values and lower numbers of main shocks. A growth of these parameters then occurred in 1997–2004. From 2004 to 2010 lower values of N and b occurred again. These changes are seen clearly in the both figures for b-values (Fig. 3a, b). In the case of N, the number of events, this dependence is seen clearly only for the case of the main shocks (Fig. 2b).

The decrease of b-values corresponds to a growth in the probability of occurrence of a strong event. Actually, the period of regional decrease of N- and b-values in 2006–2010 had ended by the time of the 2011 Tohoku mega earthquake. The spatial location of this event coincided with the most significant and extended area of decrease in b-values (Fig. 4 shows the values of difference of normalized b-values for

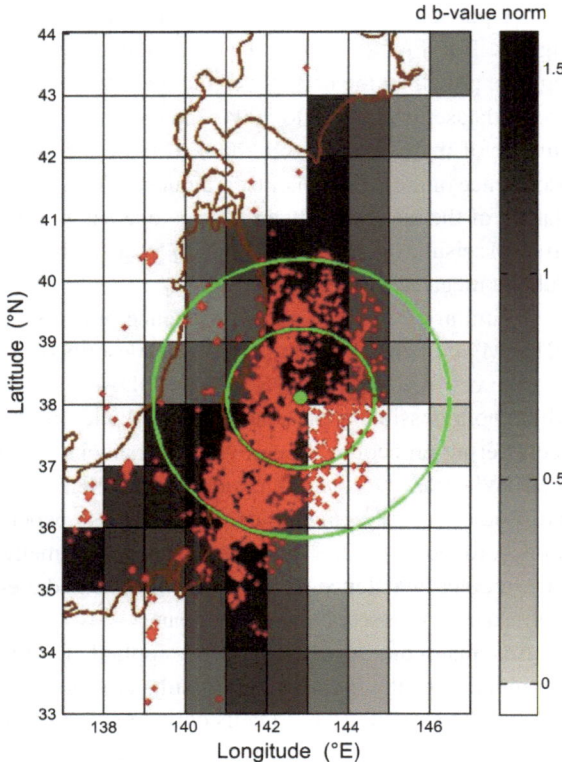

Figure 4
The difference of mean normalized *b*-values for 2006–2010 and 1996–2005 for the dataset of main events. *Dark colors* indicate bigger values of decrease of normalized *b*-values. The *green circle* is the epicenter of the 2011 Tohoku earthquake; the radii of the circles are equal to 0.5*R* and *R*, where *R* = 250 km. First week aftershocks are shown as *red points* (Color figure online)

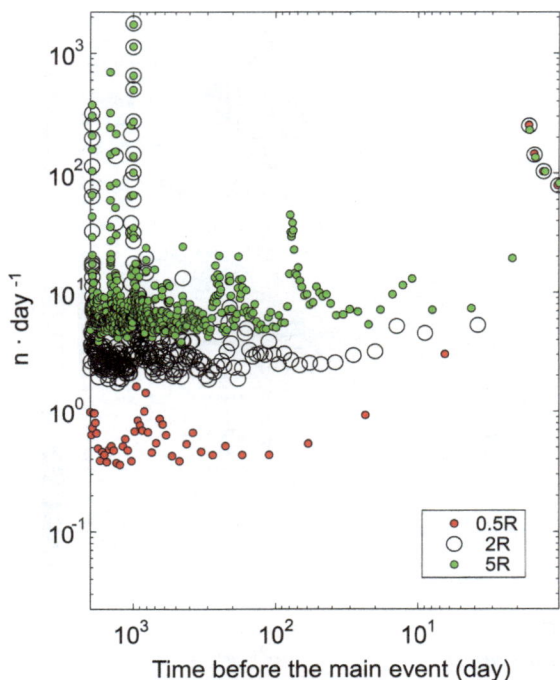

Figure 5
Change in the number of earthquakes in the vicinity of the 2011 Tohoku mega earthquake with radius 0.5*R* (*red solid circles*), 2*R* (*black open circles*) and 5*R* (*green solid circles*) for time sequential groups of 50 earthquakes with step 25 events. *R* = 250 km, half the length of the 2011 Tohoku rupture length (Color figure online)

the catalog of main events for 1996–2005 and 2006–2010). A similar decrease in *b*-values that occurred prior to the M9+ 2011 Tohoku and 2004 Sumatra quakes was presented by Nanjo *et al.* (2012).

We did not observe clear spatiotemporal changes in released seismic energy (*E*) and in fractal dimension values (*D*); correspondingly, these parameters are not mentioned below.

Previously (Rodkin 2008), the seismic analysis performed in the SEGV revealed two patterns of precursor activity: a weak long-term (up to a few years) growth of background seismic activity and a considerably stronger and short-term anomaly of an increase in foreshock numbers, which grew exponentially with time prior to the strong event. We have examined the vicinity of the Tohoku mega earthquake in an attempt to detect the same features but

did not find a systematic long-term increase in the number of earthquakes prior to the main shock occurrence. In contrast, in the Tohoku mega earthquake vicinity the effect of seismic quiescence was found for the declustered catalog both at the regional size scale and in the closer vicinity of the Tohoku earthquake. Katsumata (2011) and Huang and Ding (2012) had observed the effect of seismic quiescence prior to the 2011 Tohoku earthquake both for the declustered JMA catalog and the whole JMA catalog. It is unclear at this time, in which case seismic quiescence would take place and in which case the effect of background increase in earthquakes rates would take places.

We have observed a growth in the number of earthquakes in the close spatiotemporal vicinity of the Tohoku mega earthquake. This is the second type of precursor activity, typical of the SEGV, that was found. Figure 5 shows the change in the rate of events for sequential groups of earthquakes of 50

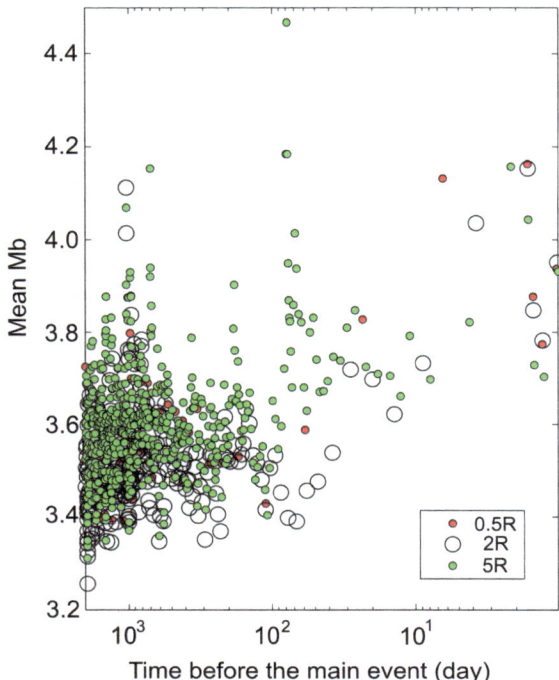

Figure 6
Change in the average values of magnitude *Mb* in the vicinity of the 2011 Tohoku mega earthquake of radius 0.5*R* (*red solid circles*), 2*R* (*black open circles*) and 5*R* (*green solid circles*) for time sequential groups of 50 earthquakes with steps of 25 events. *R* = 250 km (Color figure online)

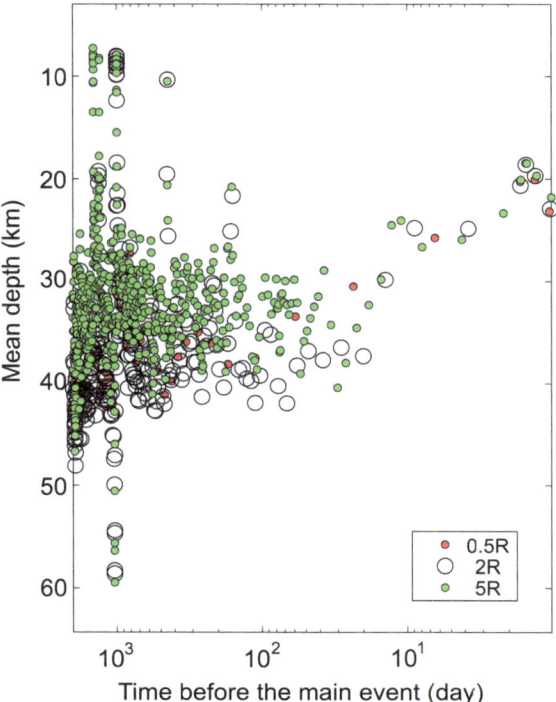

Figure 7
Change in the average depth of earthquakes in the vicinity of the 2011 Tohoku mega earthquake of radius 0.5*R* (*red solid circles*), 2*R* (*black open circles*) and 5*R* (*green solid circles*) for time sequential groups of 50 earthquakes with steps of 25 events. *R* = 250 km (Color figure online)

events with a step of 25 events. The results are presented in Fig. 5 for three distance ranges around the 2011 Tohoku epicenter with radii 0.5*R*, 2*R* and 5*R*, where *R* = 250 km is the radius of the Tohoku aftershock sequence (HIROSE *et al.* 2011). As can be seen from Fig. 5, the growth in number of events took place approximately 20 days prior to the Tohoku main shock in a small area with radius 0.5*R* from the epicenter. The overall growth in number of events, however, was not strong and the possibility of its identification in real time appears to be problematic.

Figure 6 similarly presents the data for the average values of magnitude *Mb* for the time sequential groups of 50 earthquakes over the distance ranges of 0.5*R*, 2*R* and 5*R*. There appears to be a tendency for the mean magnitude to grow in the 20 days immediately prior to the Tohoku main shock. For the JMA catalog at the M3 level, the growth in the mean magnitude, according to a formula (1), corresponds to the known precursory decrease in *b*-values.

Figure 7 shows changes in the average depth of earthquakes over time. It can be seen that immediately prior to the 2011 Tohoku earthquake the average depth of earthquakes decreases. Similar to the previous cases (Figs. 5, 6), the changes tend to concentrate in the 20 days prior the Tohoku earthquake.

Thus, several short-term precursor anomalies were found prior to the M9+ Tohoku earthquake as had been previously observed in the generalized vicinity of a strong earthquake (RODKIN 2008, 2012). In particular, an increase in the number of events, an increase in the mean magnitude, a decrease in *b*-value, and a decrease in mean depth value were observed.

Similar statistical interrelations between earthquake rates, *b*-values and average depth of earthquakes were observed in the 2011 Tohoku aftershock sequence. The time series of the aftershock rate, *b*-values, and mean depths for consecutive

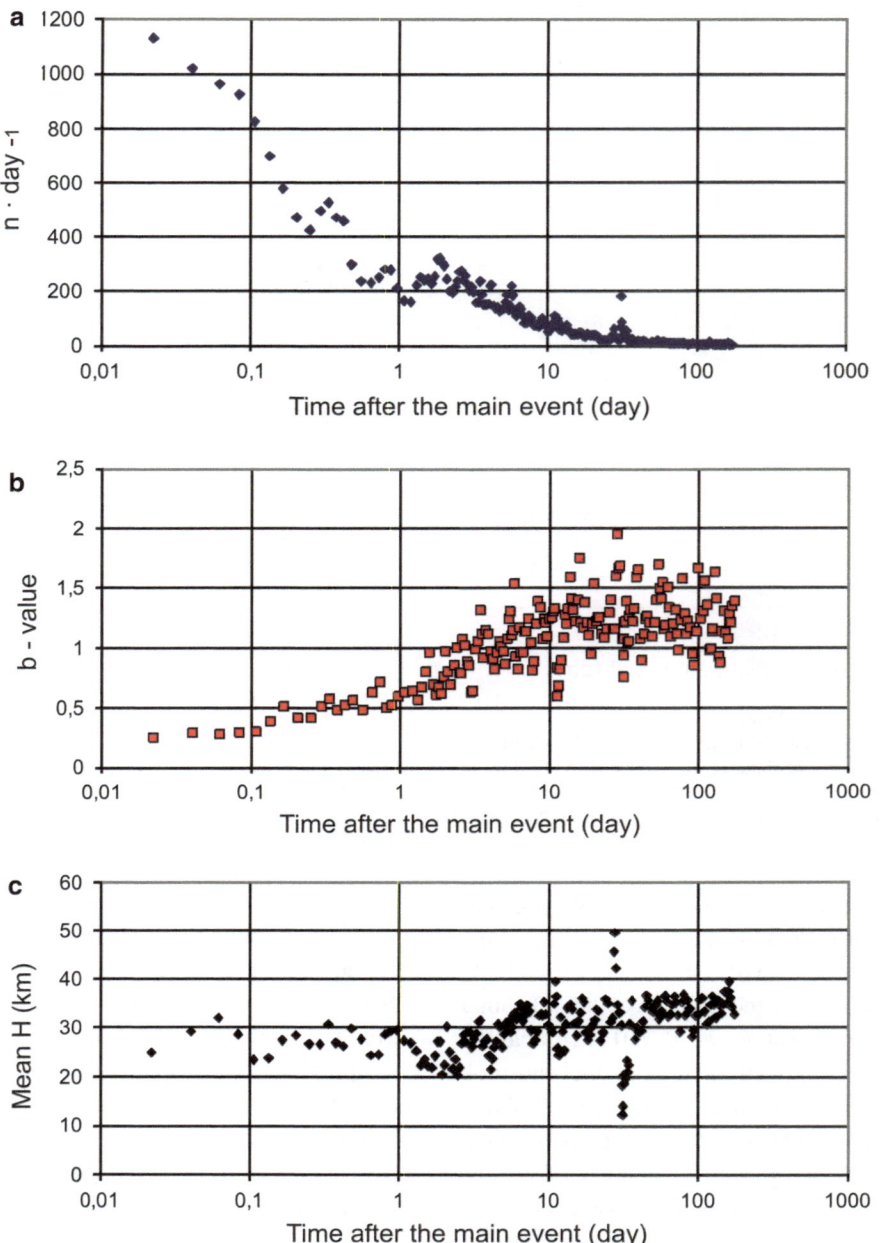

Figure 8

Changes in **a** the aftershocks' number for $M \geq 3.0$ (*blue points*), **b** b-value (*red points*), and **c** average depth of earthquake sources H (*black points*) from the JMA catalog data for Tohoku aftershock sequence (Color figure online)

groups of aftershocks of the Tohoku mega earthquake are presented in Fig. 8. It can be seen that an increase in earthquake rate corresponds in a number of cases, to episodes of decrease in b-values and in the mean depth of earthquakes. Similar anomalies were found in the aftershock sequence of the 2004, $M_w = 9.3$ Sumatra–Andaman earthquake by RODKIN and TIKHONOV (2011) and in the aftershock sequence of the 2001, M_w 7.7 Bhuj, India earthquake by RODKIN and MANDAL (2012).

4. Discussion

Regularly updated prognoses of strong seismicity carried out with the use of M8 and MSc algorithms were not successful in the case of the two recent M9+ earthquakes. The prediction of the location of the 2011 Tohoku earthquake in Japan was obtained timely only by LYUBUSHIN (1998, 2011) from analysis of seismic noise. A retrospective result in the 2011 Tohoku earthquake prediction was described by SOBOLEV (2011), by detecting the growth in amplitude of low-frequency seismic noise.

Previous studies (RODKIN 2008), based on the analysis of different parameters of the seismic regime in the SEGV, suggested that the efficient prediction of major and great earthquakes can be achieved with an increase (of 100-fold) in the available data volume. This growth in data volume was achieved by construction of the SEGV, where a number of predictive parameters were determined, and a "prediction" of a generalized strong earthquake was found to be possible. It was natural to ask the question whether a more detailed regional JMA catalog ($M \geq 3.0$) would provide an opportunity to observe additional precursor signs of a great earthquake. In order to reduce random fluctuations in seismicity, in addition to the catalog of all earthquakes we have analyzed the catalog of main events, declustered from aftershock sequences by the method of MOLCHAN and DMITRIEVA (1992) and implemented by the program of V. Smirnov.

From the examination of the JMA catalog (1994–2011, $M \geq 3.0$), periods of relatively low and high seismic activity were observed. The time interval beginning in 1996, of increased number of main shocks N and of the increase in mean b-values, was replaced in 2004 by a period of lower N- and b-values. A more noticeable reduction of the normalized N- and b-values occurred in 2006–2011. This time frame coincides with the beginning of a period of higher seismic hazard according to the results of a multi-fractal analysis of microseismic noise (LYUBUSHIN 1998, 2011).

Similar seismic hazard precursors were found post factum by HUANG and DING (2012) where seismic quiescence anomalies were found to arise around the 2011 Tohoku epicenter area between 2006 and 2008,

and by NANJO et al. (2012) where the decrease in b-value was detected in the same area in the 10 years prior to the main shock. The 2011 Tohoku earthquake took place in the area of the greatest reduction of normalized b-values (Fig. 4), the most simple structure of the microseismic field (LYUBUSHIN 1998, 2011), and an increase in the amplitude of seismic noise (SOBOLEV 2011).

The correlation of a normalized number of main shocks and b-values is not unexpected. It has been known that for time intervals in advance of large earthquakes a reduction of b-values, an increase in swarm activity of earthquakes, along with an increase in the numbers of aftershocks are typical. These effects are confirmed by experiments on rocks (LEI et al. 2003; SMIRNOV and PONOMAREV 2004; SOBOLEV et al. 1996). The mentioned changes in seismicity in the case of relatively stable levels of the seismic energy release should be accompanied by a decrease in the number of main shocks. This decrease was observed in this study.

The analysis of seismicity prior to the 2011 Tohoku earthquake allowed the observation of some anomalies in the 20 days before the main event. These anomalies showed a steady growth in amplitude prior to the Tohoku main shock. These anomalies were (1) an increase in the number of earthquakes, (2) an increase in the average magnitude of earthquakes, and (3) a decrease in the mean earthquake depth. The tendency for an increase in mean magnitudes value, in our case, is equivalent to a decrease in b-value. Concerning the decrease in mean earthquake depth, such an effect might be related to the growth of activity of deep fluid and an upward outburst of low density fluid.

A similar tendency of correlation in earthquakes rate, b-values, and mean earthquake depth was revealed in the aftershock sequence of the Tohoku mega earthquake (Fig. 8). A few episodes of an increase in aftershock activity coincide with a change in b-value and in the mean depth of aftershocks. Similar anomalies were found in the aftershock sequences of a few other major and great earthquakes by RODKIN and TIKHONOV (2011), and RODKIN and MANDAL (2012).

The character of change in seismicity in the close vicinity of the Tohoku mega earthquake agrees well with the variability revealed earlier in the strong

77

earthquake generalized vicinity (SEVG) constructed from an aggregation of data from hundreds of individual strongest earthquakes from the given catalog. The detection of a similar (but significantly more vague) complex of anomalies in the case of the 2011 Tohoku earthquake, unique by its magnitude and by the quality of a regional network of observation, confirms the earlier assumption that prediction of strong earthquakes can become possible with improved systems of seismic observations. To approach the number of earthquakes taken into account in the SEGV examination one would need rather detailed seismic information on approximately 10^4 events located in the spatial–temporal area of process of preparation of a future large earthquake. Analogously, it is suggested by VAN DER ELST *et al.* (2013) that an improvement in completeness magnitude from 3.7 to 2.0 would be needed to resolve the expected triggering signals in seismic regime on a regional scale.

The observed effect of a decrease in a mean earthquake depth for the Tohoku foreshocks (Fig. 7) and for the swarms in the Tohoku aftershock sequence (Fig. 8) can have a special interest. The same features were found in the aftershock sequences of a few recent large earthquakes (RODKIN and MANDAL 2012; RODKIN and TIKHONOV 2013). The similar case of a tendency for the direction of propagation of a set of earthquake foci was interpreted by MILLER *et al.* (2004), and INGEBRITSEN and MANNING (2010) as a sequence of the deep fluid front propagation. We agree that this interpretation appears to be mostly plausible. Moreover, we had found that this effect takes place in the generalized vicinity of a strong earthquake (submitted paper). We suggest that an increase in the deep fluid activity could be in direct connection with a large earthquake occurrence, and the change in mean earthquake depth could be used as a precursor of increase to a seismic activity.

It seems important that the rupture zone of the Tohoku $M_w = 9.0$ earthquake was found to be

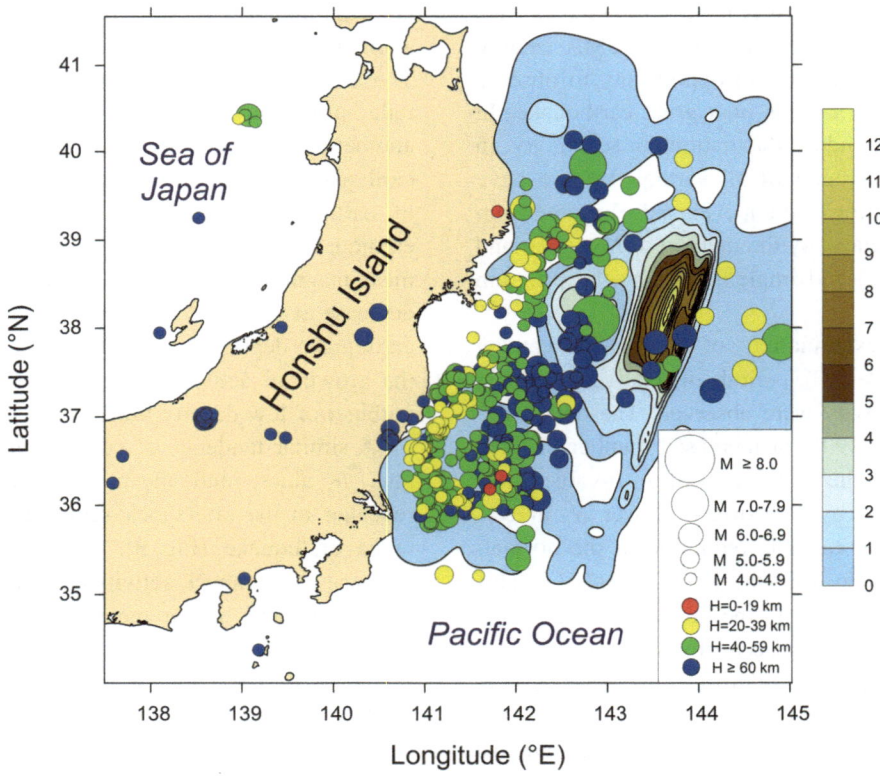

Figure 9
The model of vertical displacement of the sea surface after the 11 March 2011 earthquake in Japan and epicenters of the first week of aftershocks for $M \geq 4.0$ [sea surface displacement are taken from FINE *et al.* (2013) with authors' permission]

relatively small, ∼500 km, and the aftershock-zone length extended only 500–600 km along strike of the subduction zone (LAY *et al.* 2013), whereas the size of the M9+ mega earthquakes, which occurred prior to the Tohoku earthquake, were typically more than 1,000 km. This unexpectedly large earthquake with a rather small rupture area caused an unexpectedly high tsunami wave. Note that the size of area responsible for the highest mode of the 2011 Tohoku tsunami is also rather small, approximately 100 km (Fig. 9).

5. Conclusions

The data of the regional catalog of the JMA have shown that the 2011 Tohoku earthquake occurred after 6–7 years of regional decrease in mean *b*-values and in number of main shocks. Similar long-term precursors in the decrease in *b*-values and in the development of seismic quiescence were obtained by several authors (KATSUMATA 2011; NANJO *et al.* 2012; HUANG and DING 2012). The spatiotemporal area of maximum development of these anomalies agrees well with the area of development of predictive anomalies obtained in results of the analysis of the character of seismic noise (LYUBUSHIN 1998, 2011; SOBOLEV 2011).

The short-term anomalies of an increase in the number of foreshocks and in the average earthquake magnitude value along with the decrease in mean earthquake depth were found in a zone with a radius of 120 km from the epicenter of the Tohoku main event. These anomalies grew in amplitude to the occurrence of the Tohoku mega earthquake and are detectable post factum from about 20 days before the event.

The revealed complex of short-term precursor anomalies of the Tohoku mega earthquake is similar (but essentially less statistically valid) to the character of anomalies revealed in the result of the examination of the generalized vicinity of a strong earthquake obtained as an aggregation of data on hundreds of individual strong and major earthquakes from the Harvard CMT and other catalogs. The received result supports the assumption (RODKIN 2008) that a short-term (10–100 days) prediction of strong earthquakes may be possible in the case of the

improvement of seismic networks resulting in a radical increase of volume of seismic data available in the procedure of earthquake forecasting. To approach the number of earthquakes taken into account in the SEGV examination, one would need rather detailed seismic information on approximately 10^4 events located within the spatial–temporal area of process of preparation of the future strong earthquake.

The rupture zone of the Tohoku $M_w = 9.0$ earthquake was found to be relatively small ∼500 km, and the aftershock zone length extended only 500–600 km along strike of the subduction zone. An unexpectedly great earthquake with a rather small focal size causes an unexpectedly high tsunami wave. The size of area being responsible for the highest mode of the 2011 Tohoku tsunami is also rather small, only about 100 km in length. This result supports the idea that mega (M9+) earthquakes with a huge tsunami wave could take place in a majority of subduction zones where such cases were suggested before to be impossible.

Acknowledgments

The authors are grateful to the Japan Meteorological Agency for providing the seismic catalog data. We would like to thank Dr. Alexei Ivashchenko (Institute Oceanology, RAS, Moscow) and Dr. Taimi Mulder (Pacific Geoscience Centre, Sidney, BC, Canada) for their helpful comments and suggestions that significantly improved this paper. We are also extremely grateful to Dr. Mulder for her tremendous editorial work in approving our English. We thank Dr. A. Loskutov for his help in preparing Fig. 9. M.V. Rodkin was partly supported by the Russian Foundation for Basic Research, project 11-05-00663.

REFERENCES

AKI, K. (1981), *A probabilistic synthesis of precursory phenomena.* In *Earthquake Prediction: an International Review* (eds. SIMPSON, D.W., and RICHARDS, P.G.), Maurice Ewing Ser. 4, Am. Geophys. Union, Washington, D.C. 566–574.
CHEN, C.C., WANG, W.C., CHANG, Y.F., WU, Y.M., and LEE, Y.H. (2006), *A correlation between the b-value and the fractal dimension from the aftershock sequence of the 1999 Chi–Chi, Taiwan, earthquake,* Geophys. J. Int. *167*, 1215–1219.

DAVIS, C., KEILIS-BOROK, V., KOSSOBOKOV, V., and SOLOVIEV, A. (2012), *Advance prediction of the March 11, 2011 Great East Japan Earthquake: A missed opportunity for disaster preparedness*, Int. J. Disaster Risk Reduction *1*(1), doi.org/10.1016/j.ijdrr.2012.03.001.

FEDOTOV, S.A., SOLOMATIN, A.V., and CHERNYSHEV, S.D. (2011), *A Long-Term Earthquake Forecast for the Kuril-Kamchatka Arc for the Period from September 2011 to August 2016. The likely location, time, and evolution of the next great earthquake with $M \geq 7.7$ in Kamchatka*, Volcanology and Seismology. *6*(2), 65–88.

FINE, I.V., KULIKOV, E.A., and CHERNIAWSKY, J.Y. (2013), *Japan's 2011 tsunami: characteristics of wave propagation from observations and numerical modelling*, Pure Appl. Geophys. *170*, 1297–1307.

FORD, A., and BLENKINSOP, T.G. (2008), *Combining fractal analysis of mineral deposit clustering with weights of evidence to evaluate patterns of mineralization: application to copper deposits of the Mount Isa Inlier, NW Queensland, Australia*, Ore Geology Reviews *33*, 435–450.

GRASSBERGER, P., and PROCACCIA, I. (1983), *Measuring the strangeness of strange attractors*, Physica D 9, 189–208.

HAKEN, H., *Synergetics* (Springer-Verlag, Berlin Heidelberg 1978).

HIROSE, F., MIYAOKA, K., HAYASHIMOTO, N., YAMAZAKI, T., and NAKAMURA, M. (2011), *Outline of the 2011 off the Pacific coast of Tohoku Earthquake (Mw 9.0). Seismicity: foreshocks, mainshock, aftershocks, and induced activity*, Earth Planets Space *63*, 513–518.

HUANG, Q., and DING, X. (2012), *Spatiotemporal Variations of Seismic Quiescence prior to the 2011 M 9.0 Tohoku Earthquake Revealed by an Improved Region–Time–Length Algorithm*, Bull. Seismol. Soc. Am. *102*(4), 1878–1883.

INGEBRITSEN, S.E., and MANNING, C.E. (2010), *Permeability of the continental crust: dynamic variations inferred from seismicity and metamorphism*, Geofluids *10*, 193–205.

KATSUMATA, K. (2011), *A long-term seismic quiescence started 23 years before the 2011 off the Pacific coast of Tohoku Earthquake (M = 9.0)*, Earth Planets Space. *63*, 709–712.

KAGAN, Y., and JACKSON, D.D. (2013), *Tohoku Earthquake: a surprise?*, Bull. Seismol. Soc. Am. *103*(2B), 1181–1194.

KAWAMURA, M., WU, Y., KUDO, T., and CHEN, C. (2013), *Precursory migration of anomalous seismic activity revealed by the pattern informatics method: a case study of the 2011 Tohoku Earthquake, Japan*, Bull. Seismol. Soc. Am. *103*(2B), 1171–1180.

KEILIS-BOROK, V.I., and KOSSOBOKOV, V.G. (1986), *Time of Increased Probability for the Largest Earthquakes of the World*. In *Mathematical Methods in Seismology and Geodynamics*, Comp. Seismol., Nauka, Moscow. *19*, 48–57 (in Russian).

KEILIS-BOROK, V.I., and KOSSOBOKOV, V.G. (1990), *Premonitory activation of earthquake flow: algorithm M8*, Phys. Earth Planetary Inter. *61*(1–2), 73–83.

KOSSOBOKOV, V. (2011), *Are mega earthquakes predictable?*, Izv. Atmosph. Oceanic Physics *46*(8), 951–961.

KOSSOBOKOV, V. (2013), *Earthquake prediction: 20 years of global experiment*, Nat. Hazards 69, 1155–1177, doi:10.1007/s11069-012-0198-1.

KOSSOBOKOV, V.G., KEILIS-BOROK, V.I., and SMITH, S.W. (1990), *Localization of intermediate-term earthquake prediction*, J. Geophys. Res 95(B12), 19763–19772.

LAY, T., FUJII, Y., GEIST, E., KOKETSU, K., RUBINSTEIN, J., SAGIYA, T., and SIMONS, M. (2013), *Introduction to the Special Issue on the 2011 Tohoku Earthquake and Tsunami*, Bull. Seismol. Soc. Am. *103*(2B), 1165–1170.

LEI, X., KUSUNOSE, K., SATOH, T., and NISHIZAWA, O. (2003), *The hierarchical rupture process of a fault: an experimental study*, Phys. Earth Planetary Int. *137*, 213–228.

LYUBUSHIN, A.A. (1998), *Analysis of low-frequency multidimensional time series for geophysical monitoring and earthquake prediction*, J. Earthquake Prediction Res. 7(4), 496–509.

LYUBUSHIN, A.A. (2011), *Seismic Catastrophe in Japan on March 11, 2011: Long Term Prediction on the Basis of Low-Frequency Microseisms*, Geofiz. Prots. Biosfera *10*(1), 9–35.

MA, S.-K. (1976), *Modern Theory of Critical Phenomena*, 561 pp., Benjamin, Reading, Mass.

MANDAL, P., and RODKIN, M.V. (2011), *Seismic imaging of the 2001 Bhuj Mw7.7 earthquake source zone: b-value, fractal dimension and seismic velocity tomography studies*, Tectonophysics *512*, 1–11.

MARSAN, D., and ENESCU, B. (2012), *Modeling the foreshock sequence prior to the 2011, MW9.0 Tohoku, Japan, earthquake*, J. Geophys. Res. *117*(B06316), doi:10.1029/2011JB009039.

McCAFFREY, R. (2008), *Global frequency of magnitude 9 earthquakes*, Geology 36(3), 263–266.

MILLER, S.A., COLLETTINI, C., CHIARALUCE, L., COCCO, M., BARCHI, M., KAUS, B.J.P. (2004), *Aftershocks driven by a high pressure CO_2 source at depth*, Nature *427*, 724–727.

MOLCHAN, G.M., and DMITRIEVA, O.E. (1992), *Aftershock identification: methods and new approaches*, Geophys. J. Int. *109*(3), 501–516.

NANJO, K. Z., HIRATA, N., OBARA, K., and K. KASAHARA. (2012), *Decade-scale decrease in b value prior to the M9-class 2011 Tohoku and 2004 Sumatra quakes*, Geophys. Res. Lett. *39*(L20304), doi:10.1029/2012GL052997.

PISARENKO, V.F., SORNETTE, D., and RODKIN, M.V. (2010), *Distribution of maximum earthquake magnitudes in future time intervals: application to the seismicity of Japan(1923–2007)*, Earth Planets Space 62, 567–578.

RODKIN, M. V. (2008), *Seismicity in the Generalized Vicinity of Large Earthquakes*, Volcanology and Seismology 2(6), 435–445.

RODKIN, M.V. (2012), *Patterns of seismicity found in the generalized vicinity of a strong earthquake: Agreement with common scenarios of instability development*, In Extreme Events and Natural Hazards: The Complexity Perspective, Geophys. Monogr. Ser. *196*, edited by A. S. SHARMA et al. 27–39, AGU, Washington, D. C., doi:10.1029/2011GM001060.

RODKIN, M.V., and MANDAL, P. (2012), *A possible physical mechanism for the unusually long sequence of seismic activity following the 2001 Bhuj Mw7.7 earthquake, Gujarat, India*, Tectonophysics *536–537*, 101–109.

RODKIN, M.V., and TIKHONOV, I.N. (2011), *Megaearthquake of March 11, 2011, in Japan: the event magnitude and the character of the aftershock sequence*, Izv. Atmospheric Oceanic Physics *46*(8), 941–950.

RODKIN, M.V., and TIKHONOV, I.N. (2013), *On the seismic regime of the Japan region before the Tohoku mega earthquake (Mw = 9)*, Volcanology and Seismology 7(4), 243–251.

SMIRNOV, V.B. (1997), *Experience of Estimating the Completeness of Earthquake Catalogues*, Volcanology and Seismology 4, 93–105 (in Russian).

SMIRNOV, V.B., and PONOMAREV, A.V. (2004), *Patterns in the Relaxation of Seismicity from Field and Laboratory Observations*, Izv. RAN, Earth Physics *10*, 26–36 (in Russian).

SOBOLEV, G.A., PONOMAREV, A.V., KOLTSOV, A.V., and SMIRNOV, V.B. (1996), *Simulation of trigger earthquakes in the laboratory*, Pure Appl. Geophys. *147*(2), 345–355.

SOBOLEV, G. A., and LYUBUSHIN, A.A. (2006), *Microseismic impulses as earthquake precursors*, Izv. Phys. Solid Earth *42*(9), 721–733.

SOBOLEV, G. A., and LYUBUSHIN, A.A. (2007), *Microseismic anomalies before the Sumatra earthquake of 26 December 2004*, Izv. Phys. Solid Earth *43*(5), 3–41, doi:10.1134/S1069351307050011.

SOBOLEV, G.A. (2011), *Low frequency seismic noise before a magnitude 9.0 Tohoku earthquake on March 11*, Izv. Phys. Solid Earth *47*(12), 1034–1044.

TIKHONOV, I.N. and RODKIN, M.V. (2011), *The current state of art in earthquake prediction, the typical precursors, and the experience in the earthquake forecasting at the Sakhalin Island and the surrounding areas*, Earthquake Research and Analysis. Statistical Studies, Observations and Planning. Rijeka, Croatia. 43–78.

USAMI, T. (1979), *Study of historical earthquakes in Japan*, Bull. Earthquake Res. Inst. Univ. Tokyo *54*(3–4), 399–439.

UTSU, T. A. (1965), *Method for determining the value of b in a formula log n = a−bM showing the magnitude–frequency relation for earthquakes*, Geophysical Bulletin Hokkaido University *13*, 99–103 (in Japanese).

UTSU, T. (1979), *Seismicity of Japan from 1885 through 1925—a new catalog of earthquakes of M 6 felt in Japan and smaller earthquakes which caused damage in Japan*, Bull. Earthquake Res. Inst. Univ. Tokyo *54*(2), 253–308.

VAN DER ELST, N.J., BRODSKY, E.E., and LAY, T. (2013), Remote triggering not evident near epicenters of impending great earthquakes, Bull. Seismol. Soc. Am. *103*(2B), 1522–1540.

(Received May 18, 2013, revised December 24, 2013, accepted December 26, 2013, Published online January 21, 2014)

Pure Appl. Geophys. 171 (2014), 3257–3278
© 2013 Springer Basel
DOI 10.1007/s00024-013-0677-7

▌Pure and Applied Geophysics

The Great Tohoku-Oki Earthquake and Tsunami of March 11, 2011 in Japan: A Critical Review and Evaluation of the Tsunami Source Mechanism

GEORGE PARARAS-CARAYANNIS[1]

Abstract—The great Tohoku-Oki earthquake of March 11, 2011 generated a very destructive and anomalously high tsunami. To understand its source mechanism, an examination was undertaken of the seismotectonics of the region and of the earthquake's focal mechanism, energy release, rupture patterns and spatial and temporal sequencing and clustering of major aftershocks. It was determined that the great tsunami resulted from a combination of crustal deformations of the ocean floor due to up-thrust tectonic motions, augmented by additional uplift due to the quake's slow and long rupturing process, as well as to large coseismic lateral movements which compressed and deformed the compacted sediments along the accretionary prism of the overriding plane. The deformation occurred randomly and non-uniformly along parallel normal faults and along oblique, en-echelon faults to the earthquake's overall rupture direction—the latter failing in a sequential bookshelf manner with variable slip angles. As the 1992 Nicaragua and the 2004 Sumatra earthquakes demonstrated, such bookshelf failures of sedimentary layers could contribute to anomalously high tsunamis. As with the 1896 tsunami, additional ocean floor deformation and uplift of the sediments was responsible for the higher waves generated by the 2011 earthquake. The efficiency of tsunami generation was greater along the shallow eastern segment of the fault off the Miyagi Prefecture where most of the energy release of the earthquake and the deformations occurred, while the segment off the Ibaraki Prefecture—where the rupture process was rapid—released less seismic energy, resulted in less compaction and deformation of sedimentary layers and thus to a tsunami of lesser offshore height. The greater tsunamigenic efficiency of the 2011 earthquake and high degree of the tsunami's destructiveness along Honshu's coastlines resulted from vertical crustal displacements of more than 10 m due to up-thrust faulting and from lateral compression and folding of sedimentary layers in an east-southeast direction which contributed additional uplift estimated at about 7 m—mainly along the leading segment of the accretionary prism of the overriding tectonic plate.

Key words: Japan, Honshu, Sanriku, great 2011 Tohoku-Oki earthquake, Japan seismotectonics, tsunami, source-mechanism, tsunamigenic efficiency, Japan Trench.

1. Introduction

The most powerful earthquake in Japan in recent years occurred on March 11, 2011 off the coast of Sanriku (which includes the Aomori, Iwate, and Miyagi prefectures) (Fig. 1). It was one of five great earthquakes in the world since 1900. It generated a Pacific-wide, tsunami, which was particularly devastating and anomalously high along the northeast coast of Honshu. Warnings were issued for more than 20 countries and Pacific islands.

In Japan, both the earthquake and the tsunami caused extensive and severe damage to roads and railways, ignited fires and triggered a dam collapse. Many electrical generators were taken down. Most of the destruction and deaths in Japan were caused by the tsunami. As of 13 February 2013, the death toll in Japan was 15,880, another 2,694 people remained missing and another 136,481 were displaced (Japan's National Police Agency). There was extensive destruction at the Fukushima-Daiichi nuclear plant. The disaster left about 4.4 million homes in northeastern Japan without electricity and 1.4 million without water. There were power outages for about 4 million homes in Tokyo and the surrounding areas. Early estimates indicated that the monetary losses would exceed $100 billion.

The objectives of the present study are: (a) to provide a critical review and evaluation of the surveys undertaken immediately after the great 2011 Tohoku-Oki earthquake in Japan and of subsequent research work on the anomalous and extremely destructive tsunami it generated; (b) to review the seismotectonics of the region and of the earthquake's focal mechanism, rupture patterns and spatial and temporal sequencing and clustering of major aftershocks—the latter defining limits of crustal

[1] Tsunami Society International, Honolulu, HI, USA. E-mail: drgeorgepc@yahoo.com

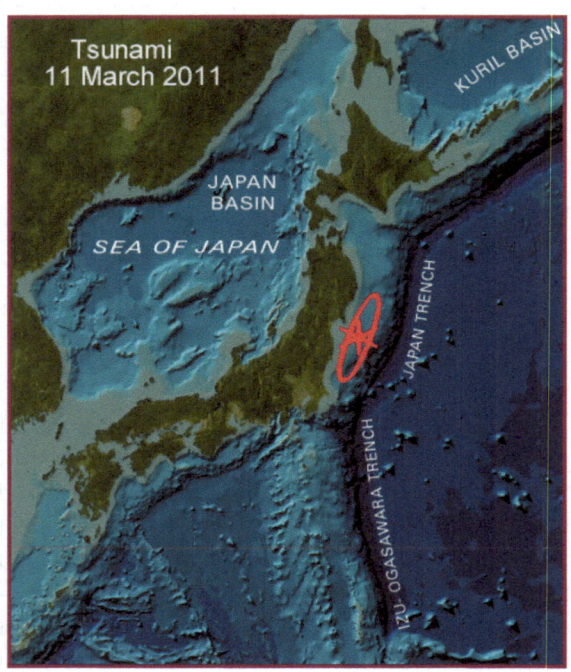

Figure 1
Epicenter of the March 11, 2011 earthquake; tsunami generating
area; major basins and trenches

displacements and the amount of energy release; (c) to understand the tsunami's generation mechanism and the cause of the extreme wave heights along northeast Honshu; (d) to evaluate the tsunamigenic efficiency of this event, based on a review of the combined earthquake rupturing impact on both the subducting oceanic lithosphere and on the overriding plate, as well as on examination of other large co-seismic, vertical and lateral displacements that contributed to the tsunami's anomalous height; and (e) to further evaluate the temporal elastic deformations caused by faulting and the collateral impact of co-seismic lateral compression on the sediments on the accretionary wedge near the trench axis.

To accomplish these objectives—and in addition to a critical overview of published reports and data of the earthquake event—the present study examines: (a) the foreshocks, aftershocks, rupture process, rupture duration and speed, pulses of seismic energy release and the co-seismic crustal movements of the great Tohoku-Oki earthquake of 2011 on land and offshore; (b) proposes a hypothesis for the 2011 earthquake's sequential rupture process that can

reasonably explain the source mechanism and efficiency of this destructive tsunami and reconciles the observed high run-up along the Sanriku coast (attributed to a mixed co-seismic and seabed failure) with the seismic source analysis based on seismic waveform inversion; and (c) makes a comparison of the 2011 tsunami source mechanism with those of the 1896 and 1933 events for the purpose of evaluating similarities of factors that contributed to the enhancement of the tsunami's height and destructiveness.

2. The Earthquake and The Tsunami

2.1. The Earthquake

The great Tohoku-Oki earthquake of March 11, 2011 occurred at 05:46 UTC, 14:46 JST (local time). The quake epicenter was at 37.68°N; 143.03°E (USGS–NEIS; HAYES, 2011) was about 373 km (231 miles) away from Tokyo, about 130 km (81 mi) off the east coast of Oshika Peninsula and about 150 km west of the tectonic boundary of the Eurasian and Pacific plates, characterized by the Japan Trench (Fig. 1). Strong ground motions were felt as far away as Tokyo.

There were differences in the estimates of the quake's moment magnitude—which was initially reported at $M_w = 8.9$ (USGS) but later revised upward to $M_w = 9.0$. Based on a seismic moment estimate of 4.2×10^{22} Nm, yielded an $M_w = 9.0$ (YOKOTA et al., 2011). However, based on long period surface waves (ranging from 166 to 333 s), the total seismic moment was recalculated to be about 5.6×10^{22} Nm, corresponding to a moment magnitude of 9.1—almost as much as that of the 2004 Sumatra earthquake (M_w 9.15). The focal depth of the quake was 15.2 miles (24.4 km) (USGS). Focal mechanism analysis indicated a low angle nodal plane with a strike of 199°, a dip angle of 10° and a slip angle of 92° (SHAO et al., 2011).

Similarly, using even longer period data (longer than 300 s) from the Global Seismographic Network retrieved in near-real time and the inversion algorithm employed in the Global CMT Project, NETTLES et al. (2011) calculated that the centroid moment tensor

(CMT) of this event had a moment of 5.3×10^{29} dyne-cm, and a geometry that indicated thrust motion of the Pacific Plate beneath the island of Honshu on a plane dipping 10°. This large scalar moment, translated to moment magnitude $M_w = 9.1$, established the Tohoku-Oki earthquake of March 11, 2011 as the fourth largest earthquake in the last 100 years.

2.2. The Tsunami

2.2.1 Near and Far Field Tsunami Impact

The great Tohoku-Oki earthquake of 2011 caused a devastating tsunami from the Tohoku to Kanto districts on the east coast of Honshu Island. The tsunami was widely recorded by GPS buoys, wave gauges and ocean bottom pressure sensors around the source and by tide gauges throughout the Pacific. Immediately following the disaster, researchers from throughout Japan and around the world, joined the 2011 Tohoku Earthquake Tsunami Joint Survey Group (2011) and other survey groups and conducted extensive in situ surveys of the near field tsunami inundation and run-up along 2,000 km of the Japanese coastline (MORI et al., 2011; Tohoku Earthquake Tsunami Information, 2011). The teams surveyed more than 5,300 locations and generated the most extensive tsunami survey dataset in the world. The extent of tsunami inundation and the run-up heights were graphically displayed at the website of the Joint Survey Group (http://www.coastal.jp/tsunami2011/). The results of the joint surveys and analyses were subsequently published (MORI et al., 2011, 2012). Also, the near and far field effects of the tsunami and the quantitative run-up heights were reported extensively in the scientific literature and in Internet summaries (PARARAS-CARAYANNIS, 2009, 2011).

2.2.1.1 Near-Field Effects Many of the tide gauge stations operating along the Tohoku coast were damaged or destroyed by the tsunami. However, two cabled ocean-bottom sensors (of OBPG) installed off the Kamaichi coast recorded the tsunami waveform just above the rupture area. According to MAEDA et al. (2011) records exhibited a two-stage tsunami development sequence: a smoothly increasing tsunami amplitude from 0 to 2 m during the first 800 s from the earthquake origin time, and a short-period impulsive tsunami with a peak of more than 5 m in the following 200 s.

Waves began striking the shores of Sanriku a few minutes after the quake. The impact was particularly devastating along coastal areas of northeastern coastal areas of Honshu, about 50–200 km north of Sendai, where the irregular coastline and the numerous narrow bays amplified and focused the tsunami wave heights, thus resulting in the largest inundation and run-up heights (MORI et al., 2011). The survey data clearly indicates this region's high tsunami vulnerability.

Hardest hit was the Miyagi Prefecture (Fig. 2). In some areas the waves inundated as far as 10 km (6 miles) inland. On the Sendai Plain, the maximum inundation height was 19.5 m, and the tsunami propagated like a bore for more than 5 km inland (MORI et al., 2011). Further south, at the Fukushima Daiichi Nuclear Power Plant, a 19-foot high protective levee was overtopped by the tsunami, which

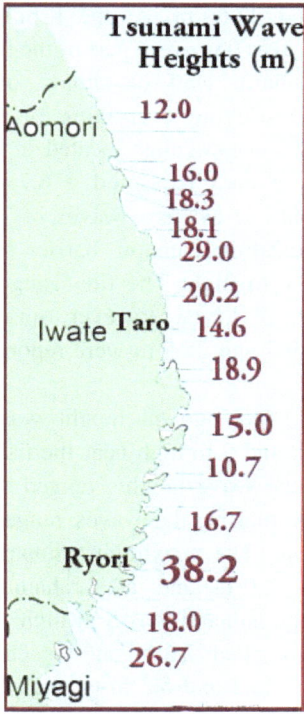

Figure 2
Reported maximum wave heights in meters of the 2011 tsunami along the Sanriku coast

submerged the lower height structures including the diesel generators and knocked out regular and backup cooling systems. The maximum tsunami wave height was ~46 feet. A 150-foot high splash was photographed as the tsunami impacted the turbine building of the plant, passing over its roof and striking the adjacent reactor building. Four of the plant's six reactors suffered damage to their radioactive cores. A state of emergency was declared which required massive evacuation of more than 200,000 residents living within a 20 km (12 mi) radius.

Reported run-up heights ranged up to 25.5 m, but at Koborinai the maximum runup reached 37.9 m (124 feet). A fire was ignited at an oil refinery in Chiba Prefecture near Tokyo. At Rikuzentakata, the maximum tsunami wave was 13 m high and overtopped the existing protective tsunami seawall, which was 6.5 m high. The tsunami reshaped the entire coastline and flooded the agricultural fields further north. Along the Taro District the tsunami overtopped the protective seawall and caused extreme destruction. Reported run-up heights ranged from 19.5 to 25.5 m. At Ryori Bay-Shirahama, the tsunami run-up height reached 23.60 m. At the fishing village of Ryoishi the waves destroyed part of the 9-m (30-foot) protective tsunami wall or simply overtopped it completely, destroying everything in the way. At Iwate, the GPS ocean gauge located in 204 m depth in the offshore area measured a 6.7 m (22 feet) tsunami height. At Miyako, waves of 11.5 m overtopped the existing tsunami barrier and seawall, which was 7.6 m high. The tide gauge recorded a tsunami height of 8.5 m. However, run-up heights of as much as 19.5 and 25.5 m were reported from this area.

At Natori, the tsunami height was 12 m near Sendai airport and 9 m high near the fishing port. At Kesennuma, the wave heights ranged from 9.10 to 14.7 m. At Kamaishi, the waves ranged from 7 to 9 m in height. The maximum tsunami height at Ofunato was 8.0 m and at Arahama 9.3 m. At Ishinomaki, the tsunami was 5 m high in the harbor but run-up reached 16 m at Ogachi-machi. At Kashima Port, 4.22 and 5.2 m tsunami heights were observed. At Fudai, 3,000 residents survived because of a 51-foot (15.5 m) floodgate. However, the tsunami run-up at the towers of the floodgate reached

66 feet (20 m). At Ishinomaki City the tsunami was over 10 feet high and washed homes away.

2.2.1.2 Far-Field Effects The tsunami's energy flux radiated across the Pacific and caused extensive destruction at distant shores in the Hawaiian Islands, California, Oregon, Chile and elsewhere (Fig. 3). The following are some of the reported measurements, observations and impacts of the tsunami in the Pacific.

2.2.1.3 Vanuatu The tide gauge measured 1.88 m (6.2 feet).

2.2.1.4 Hawaiian Islands Maximum runup heights ranged from 2 to 3 m (7–11 feet) on the islands of Maui and Hawaii (the Big Island). Four waves struck Midway Atoll at the Northwest end of the Hawaiian Islands. The highest wave reached a height of nearly 5 feet and completely washed out the reef and Spit Island, the smallest in the atoll. The tide gauge measured 1.27 m (4.2 feet). According to reports the tsunami killed hundreds of birds and swept away nests protecting seabird chicks, which were unable to fly. Reportedly, 110,000 Layson and black-footed albatross chicks were killed, along with 2,000 adult birds.

Damage to boats on the island of Oahu was extensive. Kahului on the Island of Maui suffered the worst damage. The tide gauge there measured 1.74 m (5.7 feet). On the Island of Hawaii, there was flooding and minimal damage of a hotel lobby near

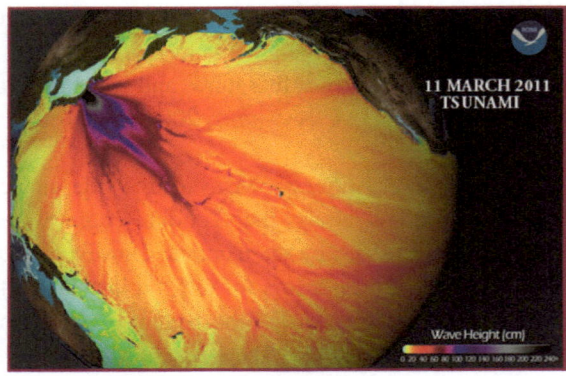

Figure 3
Wave heights and energy propagation of the 2011 tsunami across the pacific (NOAA graphic)

Kealakekua Bay. The Hilo tide gauge measured a 0.69 m (2.3 feet) high wave.

2.2.1.5 California   There was damage to docks and boats. At Crescent City, tsunami waves did extensive damage to port docks and severely damaged 35 boats. The reported maximum wave height was estimated at 2.5 m. The tide gauge measured a wave of 2.02 m (6.6 feet) high. The tsunami caused also extensive damage at Santa Cruz Harbor, estimated at more than 2 million dollars.

2.2.1.6 Oregon   Wave heights were relatively small, ranging from 0.90 to 1.20 (3–4 feet). Wave periods ranged from 10 to 15 min.

2.2.1.7 Chile   There were reports of major damage. Maximum-recorded tsunami runup at Coquimbo was 2.55 m, at Caldera 2.43 m and at Talcahuano 2.15 m.

3. Seismotectonics of the Region

Active tectonic convergence of the Pacific plate with the Eurasian and North American plates has created an extensive and complex tectonic plate margin along the Japanese island group. The following is a brief overview of the seismotectonics of the region.

3.1. Tectonic Evolution

Japan was originally the coastal part of Eastern Eurasia. However, many hundreds of millions of years ago (from mid-Silurian to the Pleistocene) oceanic crust movements caused by subduction processes began pulling Japan away from the continental block. About 15 million years ago these processes begun to open the Sea of Japan—a complex basin between Japan and the Korea/Okhotsk Sea Basin—which represents another sub plate with apparent counter clock rotational movement as it interacts against the Okhotsk plate along the inland sea boundary of the Hidaka Collision Zone (HCZ) (Fig. 4). Thus, a separate Amurian microplate has been postulated (WEI and SENO, 1998), which appears to be rotating in a counterclockwise direction. Two

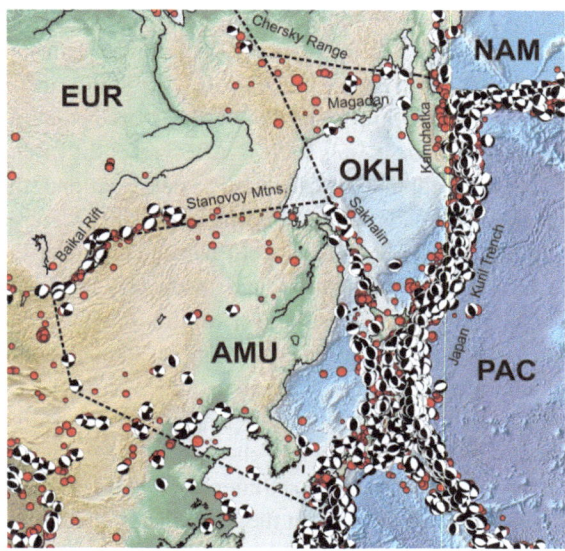

Figure 4
The postulated Amurian (AMU) microplate in relation to the pacific (PAC) and the Okhotsk (OKH) plates (graphic of seismo.berkeley.edu pertaining to the Okhotsk plate modeling)

earthquakes in 1983 and 1993 in the eastern boundary of this microplate generated destructive tsunamis in the Sea of Japan basin. Furthermore, Sakhalin Island, north of Hokkaido, which separates the Sea of Japan from the Sea of Okhotsk, is probably the result of transpressional tectonics along the North America-Eurasia boundary (PARARAS-CARAYANNIS, 1983, 1994). However, whether the Okhotsk plate and Northern Honshu are part of the North American plate or not, has not been ascertained (SENO et al., 1981). Similarly, it has not been determined whether the island of Honshu is part of the North America plate, of the Eurasian plate or an independent microplate.

In brief, Japan is a mature island arc. Subduction of the Philippine Sea Plate beneath the continental Amurian Plate, the Okinawa Plate to the south and of the Pacific Plate under the Okhotsk Plate to the north, continues to the present day and is the cause of frequent earthquakes, tsunamis and of occasional volcanic eruptions. The convergence rate across the boundary between the Pacific and Eurasian tectonic plates along the east side of Honshu Island varies from about 8 to 9 cm/year (3.1–3.5 in). The 2011 Tohoku-Oki earthquake occurred on the megathrust where the Pacific Plate subducts below Japan at an average rate of about 8–8.5 cm/year (SIMONS *et al.*, 2011). As the Pacific plate subducts under Honshu, the high convergence results in the build-up of stresses. Arc stresses contribute to back-arc spreading (SENO and YAMANAKA, 1998). After reaching aseismically a threshold of elastic deformation, the stresses are suddenly released by earthquakes. Large tsunamigenic earthquakes occur periodically near the eastern tectonic boundary characterized by the Japan Trench, as the western edge of the Pacific Plate subducts under Honshu. Destructive tsunamis are generated from large earthquakes either on the outer ridge of the subducting plate or on the overriding plate, west of the Japan Trench—where the extensive forearc and accretionary sedimentary wedge seem to have significant effects on the type of boundary slips that can be expected and on the frequency and intensity of tsunami-earthquakes.

3.2. *Seismicity of Japan*

Japan accounts for about 20 % of the world's earthquakes. Its high seismicity (Fig. 5) results from compression along the Pacific-North America subduction zone, from outer rise and intra-plate events and from magmatic effects of plumes or super plumes which may have hydrated the subducting oceanic lithosphere (PARARAS-CARAYANNIS, 1994; SENO and YAMANAKA, 1996). Usually, shallow normal faulting occurs in the trench-outer rise region, as well as on the overriding Eurasian plate and the outer slope of the Japan Trench (SENO and GONZALEZ, 1987).

3.2.1 *Seismicity of Honshu*

Along the northern segment of the Honshu arc, there is a triple seismic zone with variable degrees of seismicity due to regional stress fields (KAWAKATSU and SENO, 1983; SENO and TAKANO, 1996; SENO, 1999b) (Fig. 6). As stated previously, it has been proposed that northern Honshu may be a separate microplate (SENO, 1999a). Also, it has been postulated that the subduction zone off Miyage Prefecture—where the 2011 tsunami was anomalously high—has a double zone of seismicity (SENO and PONGSAWAT, 1981; SENO and KROEGER, 1983).

Small earthquakes occur frequently along the east side of Honshu. Stronger earthquakes of M 6.0 and large earthquake M 7.0 occur less frequently and intermittently. Such earthquakes usually result from normal-types of faulting in areas where the shear stress is predominant—thus, less likely to generate

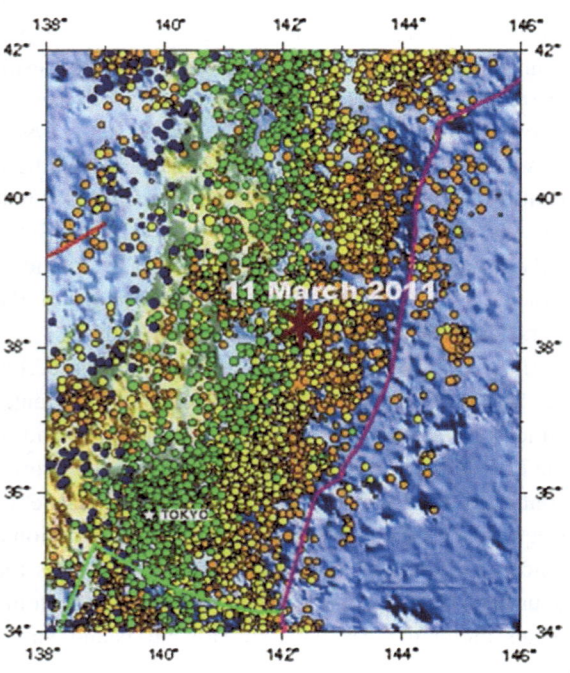

Figure 5
Seismicity of Japan. Epicenter of the 11 March 2011 earthquake (modified USGS map). Epicenter of the 11 March 2011 earthquake

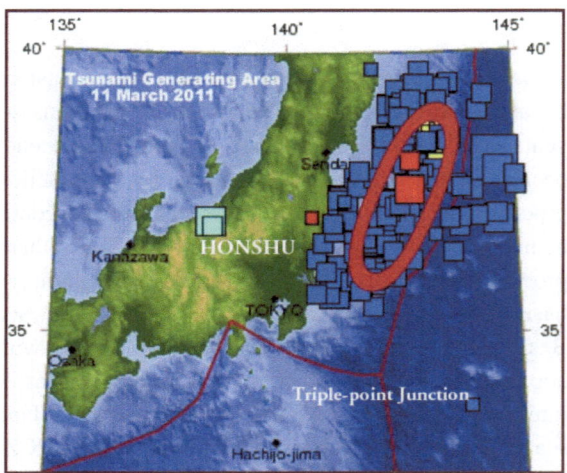

Figure 6
Earthquake epicenter (large *red block*) major aftershocks and tsunami generating area (modified USGS map)

great tsunamis. Earthquakes of M_w 8.0 or greater occur infrequently—on the average every 50–100 years. Such mega-thrust earthquakes involve mainly a mechanism where the compressional stress is predominant and their large vertical and horizontal displacements are the cause of destructive tsunamis. The great Meiji Sanriku earthquake of 1896 off the Tōhoku region was such an event. It generated a destructive tsunami with waves reaching as high as 38 m along the Sanriku coast. Another earthquake known as the 1933 (Shōwa) Sanriku Earthquake, generated another destructive tsunami. The March 11, 2011 earthquake (M_w 9.0) had many similarities with the 1896 event (PARARAS-CARAYANNIS, 2011). All such tsunamigenic events in the past were shallow (about 20 km focal depth) and involved a thrust mechanism of compressional stress, which resulted in the uplift of the overriding tectonic plate, as well as in great horizontal movements that disturbed upwards the sedimentary layers on the accretionary prism.

4. Past and Recent Destructive Tsunamis in Japan

The Sanriku coast in particular and other areas in the Tōhoku region have been impacted by numerous large tsunamigenic earthquakes. The historic record shows that a total of 65 destructive tsunamis struck Japan between A.D. 684 and 1960 (IIDA et al., 1967; PARARAS-CARAYANNIS, 2000). As early as 18 July 869 (also reported as July 13, 869), an earthquake with an estimated magnitude of 8.3 generated a tsunami along the Sanriku coast, which resulted in the loss of 1,000 lives and the destruction of hundreds of villages. On 3 August 1361, another tsunami destroyed 1,700 houses in this same area and killed a large number of people. On 20 September 1498, 1,000 houses were washed away and 500 deaths resulted from a tsunami, which struck the Kii peninsula. Kyushu was struck by a destructive tsunami in September 1596. Great loss of life occurred on 31 January 1596 from a tsunami on the island of Shikoku, affecting also a number of regions in Honshu.

On 2 December 1611 another destructive tsunami struck Keichō killing almost 3,000 people along the Sanriku coast. The same tsunami killed more than 3,000 people in the Nanbu-Tsugaru area. Another earthquake on 17 February 1793 generated a tsunami that struck the Sanriku coast killing a number of people. On August 23, 1856, a strong offshore earthquake off the Sanriku coast generated another destructive tsunami.

The great Meiji Sanriku earthquake of 15 June 1896 generated a tsunami, which resulted in 27,122 deaths, thousands of injuries and the loss of over 10,000 structures and of more than 7,000 boats and ships. On 3 March 1933 the Shōwa Sanriku Earthquake generated another destructive tsunami that struck the Sanriku region. The maximum wave height of this tsunami at Ryōri Bay was 28.7 m. The waves killed more than 3,000 people, injured hundreds more and destroyed approximately 9,000 homes and 8,000 boats. In December 1944, another offshore earthquake near central Honshu generated a tsunami that caused almost 1,000 deaths and the destruction of over 3,000 houses. The Nankaido tsunami, on 21 December 1946, resulted in 1,500 deaths and the destruction of 1,151 houses (IIDA et al., 1967). The 11 March 2011 Tohoku-Oki earthquake generated the most destructive tsunami in recent times in the same general area. Given the history of catastrophic tsunamis along the Sanriku region and the rates of tectonic plate collision, this latest event was expected to occur—although its timing could not be forecasted.

5. Tsunami Source Mechanism

The March 11, 2011 quake had characteristics of severity of tsunami generation usually associated with slow rupture velocity within compacted, sedimentary layers. Because of their tsunamigenic efficiency, such events are known as tsunami-earthquakes. The 2011 earthquake had a complex focal mechanism, which involved mainly reverse thrusting and compression, but also multiple parallel ruptures, as well as extensive lateral and vertical sediment displacements—which contributed to the tsunami's severity.

As stated, the present evaluation includes a review of the combined rupturing impact on both the subducting Pacific oceanic lithosphere and on the overriding Eurasian tectonic plate, as well as on the large vertical and horizontal tectonic crustal

89

displacements. Additionally examined are the spatial and time sequence of foreshock and aftershock distribution, the clustering of aftershocks, the three-dimensional dynamics of shallow and deeper sub-duction processes, the effects of temporal elastic deformation caused by faulting and the collateral impact on the sediments of the accretionary prism. Finally, a comparison is undertaken of source char-acteristics of the destructive tsunamis of 1896 and 1933 in the same general area off Sanriku's coast-lines. Although the 2011 earthquake occurred slightly to the south of the 1896 event, it had many similar source characteristics. The similarities and differ-ences are discussed in a subsequent section.

5.1. Examination of Foreshocks and Aftershocks

5.1.1 Foreshocks

Many large earthquakes are known to be preceded by one or more foreshocks, although it is unclear how such foreshocks relate to the nucleation process of the main shock (KATO et al., 2012). The mainshock of the great 2011 Tohoku-Oki earthquake was preceded by a foreshock sequence that lasted two days which was examined by IDE et al. (2011). On 9 March 2011, a large M_w 7.2, shallow (less than 30 km), foreshock occurred at 38.42°N, 42.83°E, a little north of the subsequent great earthquake of March 11. The large foreshock was followed by three more with magni-tudes greater than 6. IDE et al. (2011) determined that the larger foreshock of M_w 7.2 had a similar low-angle thrust mechanism as that of the subsequent main event and that their epicenters were separated by only ~45 km. The relative proximity and the low-angle thrust mechanism of the foreshock may have resulted in subsequent nucleation that triggered the main shock.

Similarly, on the basis of a waveform correlation technique, KATO et al. (2012) identified two distinct sequences of foreshocks which migrated at rates ranging from 2 to 10 km/day along the axis of the Japan Trench towards the epicenter of the main 2011 Tohoku-Oki earthquake. Based on the time history of quasi-static slip along the plate interface—as indi-cated by the small repeating earthquakes that were part of the migrating seismicity—the same study

suggested that there were two sequences involving slow-slip transients, which propagated toward the initial rupture point of the main event. The second sequence, which involved large slip rates, may have caused substantial stress loading, prompting the unstable dynamic rupture of the main shock (KATO et al., 2012). Subsequent aftershocks prior to the main earthquake, spread to the north, but several of the larger events appeared to have migrated towards the eventual nucleation region of the 11 March 2011 main earthquake, at 38.30°N, 142.34°E (USGS). None of the foreshocks generated a tsunami.

The hypothesis of the initial rupturing process and of the subsequent rupture on the other side of the main 2011 Tohoku-Oki earthquake's hypocenter—as proposed and discussed in a subsequent section of this report—perhaps is related to such a nucleation effect due to the above described sequences of foreshocks, which may be characteristic for future large earthquakes on the overriding tectonic plate in this segment of the Japan Trench.

5.1.2 Aftershocks

Examination of aftershock distribution indicates that the seismic energy of the March 11, 2011 earthquake was mainly released about 100 km off the coast of Miyagi and Fukushima Prefectures. About 50 min following the main quake on March 11, there were a large number of aftershocks, the largest having a magnitude of 7.1 (Table 1). Shortly afterwards, 35 more aftershocks larger than magnitude 5.0 and 14 larger than magnitude 6 were recorded (USGS). By mid-March 2011 more than 250 aftershocks with magnitudes of over 5.0 had occurred and 25 of these

Table 1

Main earthquake and major aftershocks in the first 71 min (M_w 8.9 assigned to main earthquake—revised later to M_w 9 and M_w 9.1)

UTC date-time 2011/03/11	Mag	Lat (deg)	Long (deg)	Depth (km)
05:46:24	9	38.322	142.369	24.4
06:06:11	6.4	39.025	142.316	25.1
06:07:22	6.4	36.401	141.862	35.4
06:15:46	6.8	36.126	140.234	30.2
06:25:51	7.1	38.106	144.553	19.7
06:48:47	6.3	37.993	142.764	22.3
06:57:15	6.3	35.758	140.992	30.2

had magnitudes over 6.0. Strong, shallow aftershocks with magnitudes greater than 6 were recorded on March 27 and 28.

5.1.3 Clustering of Aftershocks

A cluster algorithm analysis of aftershocks in chronological sequence undertaken as part of the present study, determined a big cluster of 260 events; a second cluster of 120 events and a third cluster of 60 events, as well as 20 very small clusters—typically one or two events each (Fig. 7).

Many of the aftershocks may have occurred on unmapped, minor faults both in the intra-plate region as well as on the outer rise of the subducting plate. Plotting the aftershock focal depths along eastern Honshu indicated that there was a spectacular peak at a focal depth of about 24–25 km (Fig. 8). The significance of this to tsunami generation was evaluated in terms of regional, spatial subduction geometry, slip, crustal movements and sediment displacements. Buckling of the crust due to subduction friction probably activated many minor, normal faults, which gave rise to subsequent aftershocks—even outside the tsunami generating region on the outer ridge of the subducting plate. Indeed the

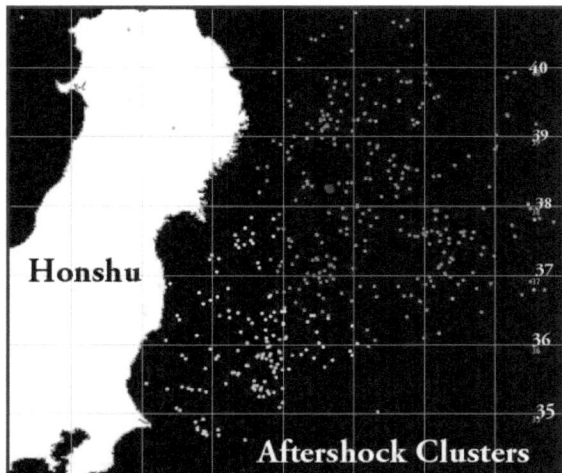

Figure 7
Three major clusters of aftershocks. Big cluster of 260 events (in *red dots*), followed by a second cluster of 120 events (in *blue dots*) and by third cluster of 60 events (in *yellow dots*)—as well as 20 very small clusters of tensional aftershocks along the outer ridge of the subducting Pacific plate

aftershock distribution was extensive, covering an area that was approximately 500 km long and 300 km wide.

5.1.4 Examination of the Earthquake's Rupture Propagation and Duration

The significance of effects on sediments from either aseismic or seismic subduction as well as variations in earthquake rupture velocities and tsunami generation for other subduction regions—such as Makran in the North Arabian Sea, the Sunda Trench segment in the Andaman Sea and the Mid-America Trench— have been examined in the past (PARARAS-CARAYANNIS, 1992, 2005, 2006). In all these regions of subduction, block motions of consolidated sediments were associated with bookshelf faulting, which contributed to slow-rupturing, silent and deadly tsunami-earthquakes. The reason is that within consolidated, dewatered, or lithified sedimentary layers, there is a lot more shear, thus the earthquake ruptures tend to be much slower and the tsunamigenic efficiency much greater. In all of these cases, the degree of sediment consolidation along the plate boundary appeared to have been a key factor in locking slippage on the megathrust region of the tectonic boundary, then releasing greater energy when the stress thresholds were exceeded (PARARAS-CARAYANNIS, 2006).

Apparently, the region where the 11 March 2011 Tohoku earthquake occurred had reached a very high level of stress. However, since subduction near Honshu does not follow a straight fault line along the tectonic boundary as defined by the Japan Trench, most large tsunamigenic earthquakes along the San-riku region—even the most destructive—involve relatively short ruptures, but proportionately large slips. Although their seismic energy release may be quite high, the affected crustal blocks tend to be smaller because of existing asperities and crustal heterogeneities.

The 2011 earthquake—like most large earthquakes—had a very complex rupture, which exhibited a variation in velocity. The rupture propagation, duration and displacements were investigated by numerous researchers using different models and techniques (WANG and MORI, 2011;

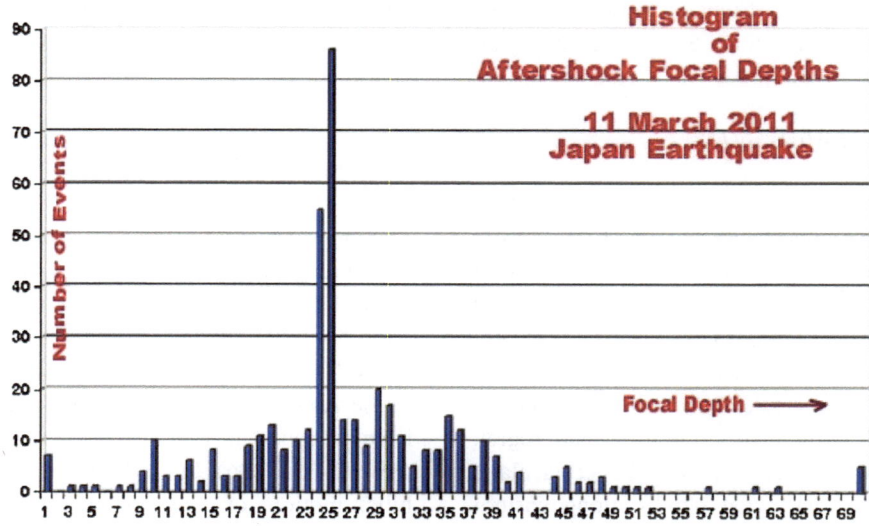

Figure 8
Note spectacular peak at about 24–25 km depth

SHAO *et al.*, 2011; AMMON *et al.*, 2011; IDE *et al.*, 2011; HAYES, 2011; HAYES *et al.*, 2011). Accordingly, the earthquake ruptured the intra-plate boundary offshore east Honshu, with fault displacements of up to 40 m, variable rates of propagation and a rupture duration which was estimated to range from 150 to 170 s. The estimates were relatively consistent. For example, data recorded by the dense USArray network in Japan, indicated that the quake exhibited a variable rupture propagation, which ranged from about 1.0 to 3.0 km/s for the high-frequency radiation and lasted approximately 170 s. Based on finite-source imaging, IDE *et al.* (2011) concluded that the rupture consisted of a small initial phase, deep rupture for up to 40 s, extensive shallow rupture at 60–70 s, and subsequently continuing deep rupture, which lasted more than 100 s. Significant heterogeneous changes of physical properties along the fault plane(s) may be the reason for the variability in the velocity of rupture propagation. The overall rupture length was estimated to be about 450 km long (WANG and MORI, 2011).

5.1.5 Examination of Seismic Energy Release

There were different estimates of the seismic energy release. Inversion of teleseismic P waves and broad-band Rayleigh wave observations with high-rate GPS recordings indicated a moment of 3.9×10^{22} Nm (M_w 9.0) and a centroid time of 71 s. (AMMON *et al.*, 2011). Teleseismic body and surface wave analysis of both broadband body waves and long period seismic waves (SHAO *et al.*, 2011) determined the total seismic moment to be 5.8×10^{22} Nm. Furthermore, the resulting rupture models showed a steady increase of moment rate for the first 80 s and an initial rupture speed of 1.5 km/s mainly in a northwesterly direction. Subsequent rupturing in a southwestward direction continued at a speed estimated at about 2.5–3.0 km/s. As stated, changes in physical properties of crustal and sedimentary materials may be responsible for the variation in speed. Usually, areas of low rupture velocity are associated with large energy release along the fault plane, while high rupture speed may be associated with lower energy release.

5.1.6 Examination of Crustal Movements

The March 11, 2011 Tohoku-Oki earthquake was a megathrust event, with the Pacific plate moving underneath the Eurasian plate. Apparently, there was great accumulation of strain in the region over a period of several decades. The earthquake caused subsidence and extensive crustal movements on land and in the offshore region. Such movements and large slip variations near a trench are not uncommon for great subduction earthquakes of $M_w >8$. Specifically during this 2011 quake, the landmass of Honshu Island moved in an east-southeasterly direction and there were significant crustal movements in the offshore region that generated the destructive tsunami. The large tsunami is believed to have been caused by a fault rupture extending to a shallow part of the subduction zone at the Japan Trench. The present study paid special attention to the investigations of co-seismic crustal and sediment movements, because of their significance in evaluating the earthquake's tsunamigenic efficiency and the anomalously high waves.

Several studies were conducted in an effort to understand the 2011 Tohoku-Oki earthquake and tsunami, particularly the extent of the co-seismic crustal movements (Ozawa *et al.*, 2011; Ito *et al.*, 2011; Kido *et al.*, 2011; Sato *et al.*, 2011; Fujiwara *et al.*, 2011; Ammon *et al.*, 2011; Yamazaki *et al.*, 2011; Gusman *et al.* 2012). Studies of co-seismic and post-seismic movements and slip included land and seafloor geodetic measurements using the Global Positioning System (GPS) and acoustic techniques (Kido *et al.*, 2011; Sato *et al.*, 2011; Ozawa *et al.*, 2011; Yamazaki *et al.*, 2011), multi-beam bathymetric surveys before and after the earthquake (Fujiwara *et al.*, 2011) and ocean-bottom pressure gauge data (Ito *et al.*, 2011). Although there were quantitative differences in the estimates, all such studies determined that very large sea-floor movements were associated with this event near and directly above the focal region.

There was good agreement on the measurements of uplift and subsidence on the coastal area. Based on the GPS, the Geospatial Information Authority in Tsukuba, Japan, stated that there were land mass movements in many areas of Honshu, from the northeastern region of Tohoku to the Kanto region, including Tokyo. The largest displacement on land occurred at the Oshika Peninsula near the epicentral area where there was movement of a little over 5 m (17 feet) eastward and subsidence by a little over 1 m (4 feet). As expected—there were some variations in the estimates of crustal movements in the offshore source region, depending on the specific techniques that were used.

Studies of co-seismic and post-seismic slip distributions were determined from ground displacements, which were detected by space geodetic techniques using a network of the GPS (Ozawa *et al.*, 2011). For the offshore source region this study concluded that the co-seismic slip area extended for approximately 400 km along the Japan Trench, matching the area of the pre-seismic, locked, zone, and that the after-slip begun to overlap the co-seismic slip area and extended into the surrounding region.

Based on seafloor geodetic measurements and the GPS/acoustic technique just above the rupture area of the 2011 Tohoku-Oki earthquake, a study by Kido *et al.* (2011) revealed strong trench-normal variation in the horizontal crustal displacement associated with the earthquake. The GPS/acoustic technique had high resolution and accuracy. One set of observations, at a site only 50 km away from the trench, yielded the most trench-ward data ever reported with an accuracy better than 1 m. It showed a co-seismic displacement of up to 31 m, whereas another set of observations 100 km away from the trench showed a 15-m displacement. The horizontal data, as well as the vertical data from pressure gauges and tsunami observations, strongly indicated that indeed the seismic slip reached the trench.

Another study (Ito *et al.*, 2011) based on measurements of ocean-bottom pressure gauges—which had been installed before the 2011 Tohoku-Oki earthquake—also determined the extent of frontal wedge deformation near the quake's source region. Specifically this particular study reported on an uplift of 5 m and on a co-seismic, horizontal displacement of more than 60 m along the frontal wedge near the Japan trench—the latter determined by measurements which used local benchmark displacements obtained by acoustic ranging before and after the earthquake. The same study concluded that there were significant horizontal and vertical deformations in the source region, that the co-seismic slip beneath the frontal wedge of the plate boundary had average displacements of about 58 m east and 74 m east-southeast, and that the estimated magnitude of the slip along the main fault was as much as 80 m near the trench. Thus, by using the ocean-bottom gauge data, there was good estimation of co-seismic vertical movements, whereas the GPS/acoustic technique provided good estimates of co-seismic horizontal movements along the frontal wedge of the overriding tectonic plate in the source region along the Japan Trench. However, it should be pointed out that the measurements obtained by the cable pressure sensors TM1 and TM2 offshore of Iwate for the 2011 tsunami are indicative of the deep-water height of the tsunami at these locations—which may have been overestimated. At this location, according to GSI measurements (Geospatial Information Authority of Japan 2011), the deformation should have been −1.19 and −1.4 m, respectively, (ANNUNZIATO, 2012, proposing −0.82 and −0.96 m to respect the Okada model). According to ANNUNZIATO (2012), in order to have a coherent set of measurements that can allow for the correct estimation of the 2011 tsunami height at the source region, the measurements by these gauges—and also the other off-shore or tide gauges used for the inversion—should be corrected for the amount of local subsidence—something that was not done. The reason given is that the inversion techniques tend to minimize the difference between the measured signals and the calculated values. Thus, without considering a correction for subsidence, the initial period and tsunami wave peak, would be estimated to be much higher at the source—perhaps the tsunami height would be overestimated by as much as 1.4 m.

Slip and fault displacement were further estimated by another study (AMMON *et al.*, 2011), which concluded that these displacements near the quake's source region were up to 40 m. However, direct measurement of seafloor deformation near the trench axis by JAMSTEC (Japan Agency for Marine Earth Science and Technology) indicated that the average seafloor displacement was 50 m within the 40 km west of the trench axis.

A study by SATO *et al.* (2011)—also combining GPS and acoustic data—determined that very large sea-floor movements occurred directly above the focal region. Specifically, this investigation stated that an area with more than 20 m of horizontal displacement, stretched for several tens of kilometers along the trench and reached the largest amount of about 24 m toward the east-southeast just above the quake's hypocenter. It was concluded that the movements were four times larger than those on land and that there was about 3 m of vertical uplift along the

frontal edge—contrary to the lesser subsidence observed on land.

Another investigation by FUJIWARA *et al.* (2011), estimated the up-dip limit and quantified the co-seismic sea-floor displacement caused by the 2011 Tohoku-Oki earthquake, by using multi-beam bathymetric survey data. Differences between bathymetric data acquired before (in 1999 and 2004) and after the earthquake, were compared to conclude that indeed the displacements extended out to the axis of the Japan Trench—also suggesting that the fault rupture reached it. According to this study, the sea floor on the outermost landward area moved by about 50 m horizontally in an east-southeast direction and about 10 m upward. This study concluded that the large horizontal displacement lifted the sea floor by a total of up to 16 m.

In conclusion, all of the above reviewed studies determined large crustal movements. The existence of thrust earthquakes in this segment off Honshu indicates that either the sediments along the plate boundary become sufficiently well consolidated and dewatered at about 70 km from the deformation front, or that older, lithified rocks are present within the forearc so that stick–slip sliding behavior becomes possible when the stress exceeds a critical level. This has been indicated for the Eastern Makran subduction zone in the Northern Arabian Sea (PARARAS-CARAYANNIS, 2006). Similar changes in the physical properties of the subducted sediments along Sumatra and the Andaman Sea were responsible for the two different rupture velocities and slip displacements of the great 2004 earthquake.

Figure 9 illustrates that the maximum slip of the 2011 Tohoku-Oki earthquake occurred along a relatively shallow region ranging 140–180 km in length, in both directions above the quake's hypocenter. Presently, there is no sufficient data or surveys to fully evaluate the effect of sediments on the subduction dynamics off the northeast coast of Honshu. None of the studies of the co-seismic displacements—using the various seismic and geodetic inversion procedures mentioned in this report—determined an accurate up-dip limit or the extent of tsunami height boost by up-thrusting of the sediments due to the lateral compressional forces. Further studies are likely to contribute quantitatively to better

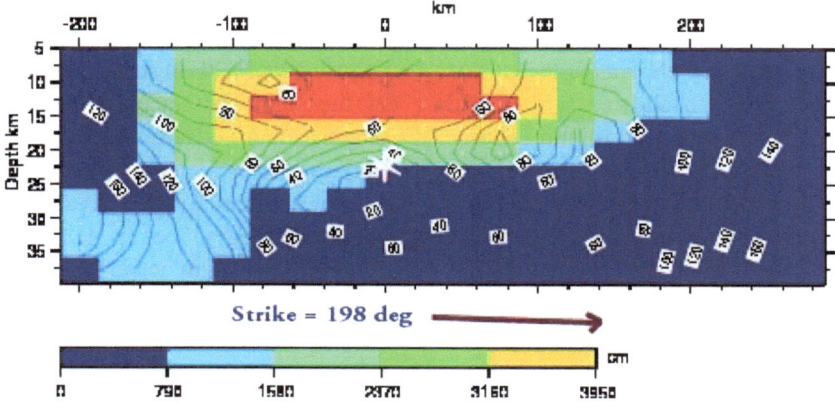

Figure 9

Cross-section of the quake's slip distribution. The strike direction of the fault plane is indicated by the red arrow and the hypocenter location and depth of the March 11, 2011 earthquake are denoted by the white star. Contours show the rupture initiation time in seconds. (modified after SHAO *et al.*, 2011)

estimates of the co-seismic behavior of the frontal wedge of the overriding plate in the source region along the Japan Trench. However, based on the kinematic rupture history and review of data reported in the literature, the present study estimates that the vertical uplift by sediments was up to 10 m on the east side of the fault and nearer to the Japan Trench axis—thus contributing significantly to the anomalously destructive tsunami.

5.1.7 Tsunami Source Area

As stated, the 2011 Tohoku-Oki earthquake packed a great deal of energy. Several studies were conducted in trying to determine the tsunamigenic earthquake's source process and area by using seismic wave inversion, focal mechanism analysis, aftershock distribution, ocean floor static changes, parameters of rupture and co-seismic slip (IDE *et al.*, 2011; YOKOTA *et al.*, 2011; YOSHIDA *et al.*, 2011; SIMONS *et al.* 2011; FUJII *et al.*, 2011).

Apparently, strong variation of the rupture characteristics controlled both the intensity of ground motions as well as the size of the source area where the tsunami was generated. However, the source area of the 2011 tsunami was relatively small compared to those of the great Sumatra (2004) and Chile (2010) earthquakes—both of which had long ruptures and much larger tsunami source areas (PARARAS-CARAYANNIS, 2005, 2010). The Sumatra earthquake

had a rupture that propagated also at varying speeds along two segments of the Sunda tectonic boundary and its overall tsunami source area was almost 1,300 km long, about 300 km wide and involved a large slip. By comparison, the 2011 Tohoku-Oki earthquake had an extensive aftershock region that was almost 450-km long and 200-km wide, but the source region that generated the large tsunami was smaller and rather compact, but involved large crustal displacements and uplift of sediments.

The 2011 quake's main rupture propagated not only in the strike direction but also in the dip direction, included both the deep area known as the Miyagi-oki region and the compact shallow area near the Japan Trench (YOKOTA *et al.*, 2011). Based on finite-source imaging, the study by IDE *et al.* (2011) concluded that the rupture consisted of a small initial phase, deep rupture for up to 40 s, and extensive shallow rupture at 60–70 s and continuing deep rupture lasting more than 100 s. Also, the same investigation concluded that a combination of a shallow dipping fault and a compliant hanging wall may have enabled large shallow slip near the trench and that normal faulting aftershocks in the area of high slip, suggested a dynamic overshoot on the fault. Such dynamic overshoot could have increased the area of the tsunami source towards the Japan Trench—as most of the other studies also indicated.

Based on the study of seismic waveform data of teleseismic P waves and on regional strong motion

95

data, YOSHIDA *et al.* (2011), concluded that the initial rupture gradually expanded near the hypocenter in the first 40 s, that the main rupture was to the east of the initial break point (the shallower side of the quake's hypocenter), and that the rupture subsequently propagated both southward and northward for more than 150 s. The same study also estimated that the maximum slip amounts to be more than 25 m and further concluded that the impacted area was about 450 km in length and 200 km in width.

The maximum co-seismic slip was estimated at 35 m according to a study by YOKOTA *et al.* (2011), which performed a quadruple joint inversion of all the data resulting from separate inversions to determine a source model most suitable for explaining all the separate datasets. However, other studies estimated the co-seismic slip to be as much as 50 m. An investigation by SIMONS *et al.* (2011) indicated that the Tohoku-Oki earthquake was relatively deficient in high-frequency seismic radiation—a difference attributed to its relatively shallow focal depth. However, based on sources of high-frequency seismic waves, the edges of the deepest portions of the co-seismic slip were delineated, although they did not correlate with the locations of maximum slip. The study concluded that models of the 2011 earthquake indicated that the distribution of co-seismic fault slip exceeded 50 m in places. Such large fault slip and expansion of source area is supported by other studies. Based on observations from the two cabled ocean-bottom sensors off Kamaichi, MAEDA *et al.* (2011) suggested the lack of any sea floor upheaval in these particular locations during the earthquake but concluded that an extremely large slip had occurred in the shallow portion of the subducting Pacific Plate near the trench axis. The source model that the investigation derived from the offshore gauge tsunami records indicated that a very large slip of 57 m had occurred off the Miyagi near the trench axis, south of the 1896 Meiji Sanriku tsunami earthquake. The investigation concluded that this was the major source of the highly destructive tsunami.

In conclusion, focal mechanism analysis indicated dipole crustal movements involving both subsidence and uplift in the tsunami source region. Apparently, most of the positive vertical static displacements occurred closer to the Japan Trench on the

accretionary prism (Fig. 10). Also, teleseismic P waves and broadband Rayleigh wave observations (AMMON *et al.*, 2011) support the conclusion that most of the significant displacements that generated the larger tsunami occurred in the first 80 s after rupture initiation.

To summarize, the overall tsunami source area was estimated to be about 300–350-km long and about 150–175-km wide. However, the spatial and temporal sequencing and clustering of major aftershocks and the energy release indicate that the main source area that generated the higher tsunami was about 120–140 km long and about 80 km wide (Fig. 11). What contributed to the higher tsunami heights were the earthquake's up-dip rupture expansion, mainly above the quake's hypocenter (at

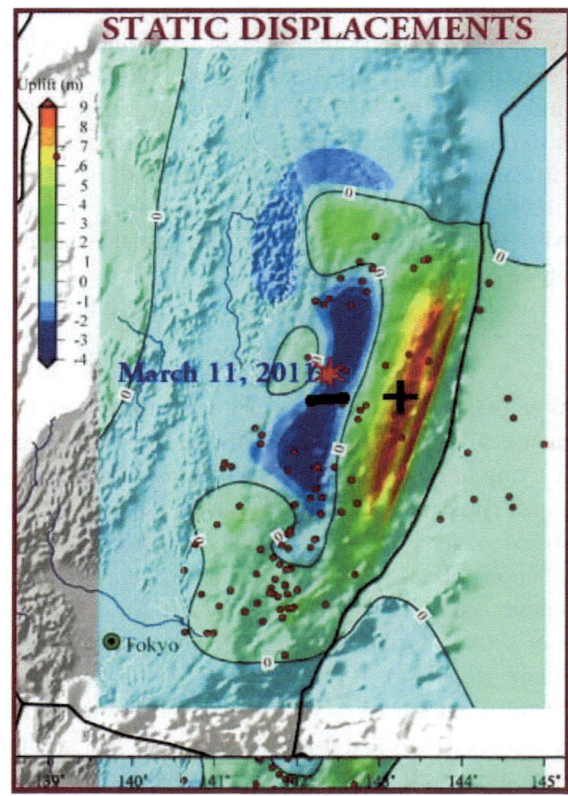

Figure 10
Predicted vertical static displacements (positive and negative) based on the inverted slip model (cross-section of slip distribution shown, indicating maximum tectonic uplift of almost 10 m (graphic modified after SHAO *et al.*, 2011—preliminary result for rupture process. *Red star* indicates the epicenter location and *red dots* are aftershocks ($M > 5$), from USGS)

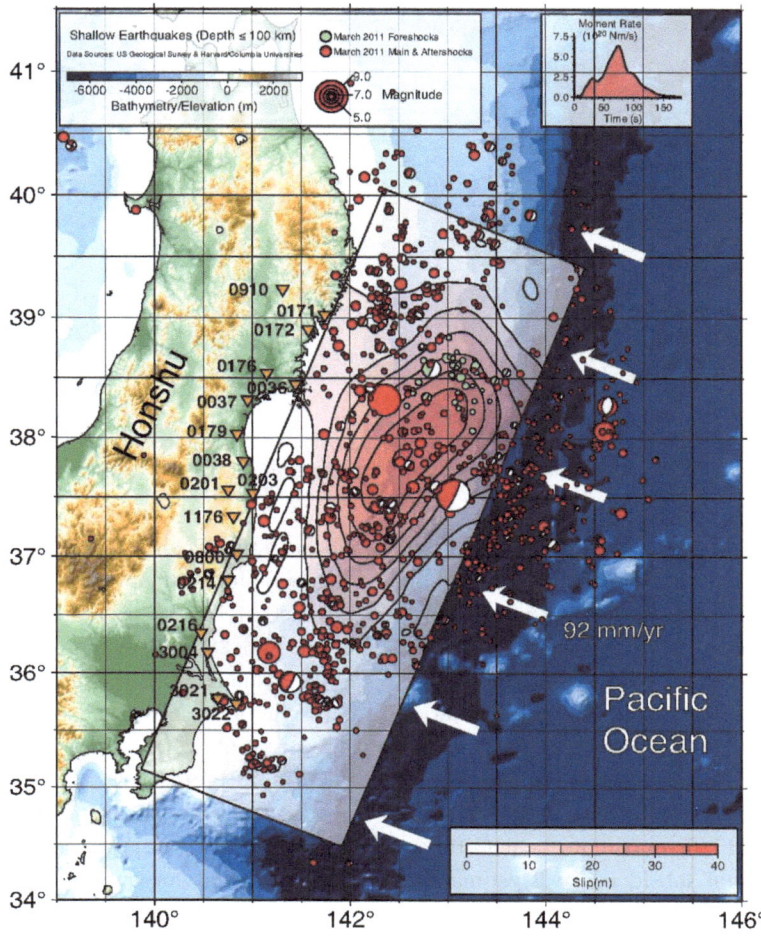

Figure 11
Tsunami generating area showing the epicenter of the main earthquake on March 11, 2011, a major aftershock with magnitude 7.1, 50 min later and 172 aftershocks which were recorded by the USGS by March 12, 02:04:53 UTC 201 and slip estimates (after Ammon *et al.*, 2011)

20–24 km depth), which resulted in extensive additional vertical and horizontal movements and uplift of sediments on the overriding plate—extending almost to the edge of the Japan Trench. The rupture propagated not only in the strike direction but also in the dip direction and included both the deep area known as the Miyagi-oki region and the compact shallow area near the Japan Trench.

5.1.8 Tsunami Source Mechanism

Subduction of the Pacific tectonic plate beneath the Eurasian plate resulted in the great Tohoku-Oki earthquake of March 11, 2011. The earthquake involved primarily reverse thrust compression, as well as strike slip shear and east-southeast trending lateral movement of the overriding Japan volcanic arc of the Eurasian plate. The following source mechanism scenario is proposed to explain the larger height tsunami that struck northeastern coastal areas of Honshu such as Miyagi and Fukushima Prefectures (Fig. 12). However, according to the field survey, the maximum tsunami heights were measured along the coast of Iwate Prefecture (around 39°–40°N).

Initial rupturing begun at the earthquake hypocenter focal depth of 24.4 km and expanded upward on the accretionary prism towards the ocean floor, first through well-compacted non-hydrated sedimentary layers and subsequently through fully hydrated layers. Rupturing continued in this manner for the

97

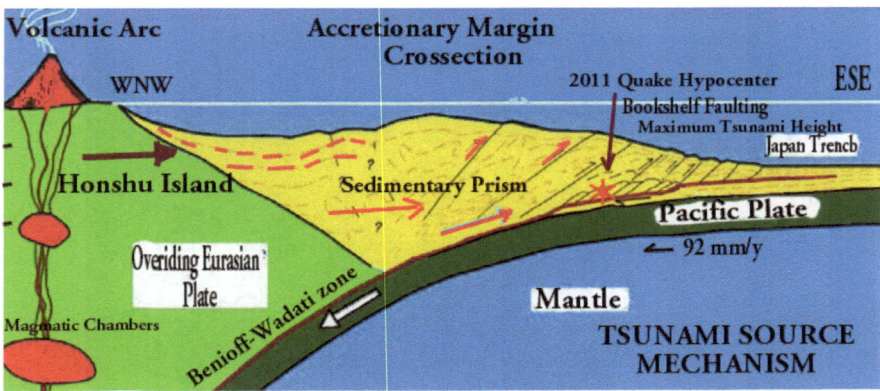

Figure 12
Postulated crosssection of the accretionary margin east of Honshu Island. Compression of the sedimentary prism and subsequent normal and bookshelf faulting contributed to the tsunami's source mechanism and to greater tsunamigenic efficiency by uplift of sediments along the frontal edge and up to the Japan Trench

first 80 s—propagating at the rate of about 1.5 km/s for an approximate distance of about 120 km in a northwesterly direction from the epicenter location. The initial combination of reverse thrust compression, strike slip shear and east-southeast lateral movement and compression of the sedimentary prism, resulted in slip and fault displacements estimated to be up to 50 m. The maximum eastward movement of land mass was estimated to be as much as 5 m at the surface (17 ft. maximum near Oshika Peninsula) and the coastal subsidence was estimated to be about 1.2 m.

At first, the sediments on the accretionary prism along both east and west of the hypocenter compressed elastically. However, the elastic deformation was short-lived, as in the next few seconds the rupturing process nucleated existing normal faults on the continental shelf on both sides of the rupture, which also begun to fail in sequence. Additionally, the reverse thrust motions and lateral compression ruptured the sedimentary layers of the accretionary prism, which begun failing sequentially in a bookshelf fashion creating several parallel and en-echelon thrust faults along a much wider zone of deformation that extended westward toward the Japan Trench. Because of the lateral compression of the sedimentary layers and the subsequent failure due to bookshelf faulting, large volumes of sediments begun thrusting upward with fault dips becoming progressively steeper on the east side of the initial rupture and along the eastern zone of the accretionary prism

closer to the Japan Trench. In this eastern region, the greater volume of up-thrusted sediments contributed significantly to the generation of the higher tsunami in the first 80 s. Thus, and as stated previously, the larger tsunami was generated along a zone of deformation that was about 120–140 km long and about 80 km wide off the coast of the Miyagi and Fukushima Prefectures—the regions that also experienced the greater tsunami devastation.

Subsequent rupturing in a southwestward direction continued for approximately 100 s at a speed estimated at about 3.0 km/s. for a total distance of about 300 km. This region did not experience as much bookshelf faulting and there was lesser upward displacement of sediments and a smaller offshore tsunami. Thus, the tsunami that struck coastal sites along the Ibaraki Prefecture was not as high. Studies of the rupture process in space and time (HONDA et al., 2011) also confirm the present study's hypothesis of tsunami source mechanism and conclusions.

6. Evaluation of Tsunamigenic Efficiency

As stated earlier, W-phase inversion indicated a moment of 3.9×10^{22} Nm (M_w 9.0) and a centroid time of 71 s. (AMMON et al., 2011). However, based on long period surface waves (ranging from 166 to 333 s), the total seismic moment was recalculated to be about 5.6×10^{22} Nm, corresponding to a moment magnitude of 9.1. The back-projection method—which used data recorded on the USArray network (WANG and MORI, 2011)—estimated a rupture propagation with variable speed ranging from about 1.0 to 3.0 km/s for about 450 km in length in approximately 170 s.

The variable rupture propagation and the change in directionality indicate that the tsunami generation area can be divided into two distinct segments along a wide zone of deformation. Apparently, the most significant disturbance of sediments occurred in the first segment when the rupture speed was slower and thus the tsunami height was greater. Since the two initial pulses occurred at 30 and 80 s in a rupture segment located 50–70 km northwest of the epicenter, we can conclude that the higher tsunami was generated along this segment during that time interval following the main earthquake. The third impulse which occurred about 250 km southwest of the epicenter about 180 s after the main shock, is indicative of higher rupture velocity in a southwestward direction from the quake's hypocenter and of a second region of tsunami generation of much lesser height with lesser sediment contribution. The observed three-pulse energy release is also supported by the cluster algorithm analysis of the aftershocks, which indicates the segmentation of the tsunami generating area. As previously stated, there was a big cluster of 260 events, followed by a second cluster of 120 events and by third cluster of 60 events—as well as 20 very small clusters, most outside the tsunami generation area—some on the outer rise of the subducting tectonic plate (Fig. 7). These variations indeed reflect differences in the strength of physical properties on both the subducting and overriding tectonic plates—such as rigidity, compaction and degree of hydration/serpentinization—along the 450 km fault(s) that ruptured sequentially when the 2011 Tohoku earthquake struck.

Based on the above-described evaluation, we conclude that the great height of the 2011 tsunami along the Honshu's coastlines was caused by the crustal displacements due to up-thrust faulting and by the displacement and excessive uplift of sediments along the accretionary prism of the overriding tectonic plate, as it thrusted east-southeast towards the Japan Trench by as much as 5 m (about 17 feet) in

certain areas on land and probably substantially more at the décollement depth.

To what extent and what volume of sediments were uplifted cannot be estimated with any certainty since the displacements were non-uniform over the entire length of the two main segments of the rupture zone. In view of this uncertainty, a comparison and evaluation of the 2011 earthquake and tsunami with past events is given in the following section. However, we can conclude that most of the sediment displacements occurred mainly along the segment of the fault off the Miyagi and Fukushima Prefectures where the seismic energy release was greater (Fig. 11). As stated, there was lesser seismic energy release along the segment off the Ibaraki Prefecture.

7. Evaluation and Comparison of the 1896, 1933 and 2011 Sanriku Earthquakes and Tsunamis

The 2011 Tohoku-Oki earthquake was one of the largest and generated one of the most destructive tsunamis. Two other great earthquakes of 2004 in Indonesia and of 2010 in Chile generated destructive tsunamis. All of these recent great earthquakes had rupture zones that extended for several hundred of kilometers and slips, which were 10 m of more. In the following section, an evaluation and brief comparison is made of the two great earthquakes and tsunamis that struck the same Tohoku region of Japan in 1896 (Fig. 13) and 1933 with that of 2011, in terms of destruction, seismic intensities, source mechanisms, sediment uplift mechanisms (Fig. 14) and tsunami run-up heights.

7.1. Destruction and Fatalities

Both earthquakes of 1896 and of 2011 generated extremely destructive tsunamis along the Sanriku coast. The Shōwa Earthquake of 1933 also generated a destructive tsunami on the Sanriku coasts (IIDA et al., 1967; HATORI et al., 1982; PARARAS-CARAYANNIS, 1969, 2005).

There were many similarities in the height of the tsunami waves and, thus, to the degree of destructiveness and the number of fatalities for all three events. For example, 30–60 min following the 1896

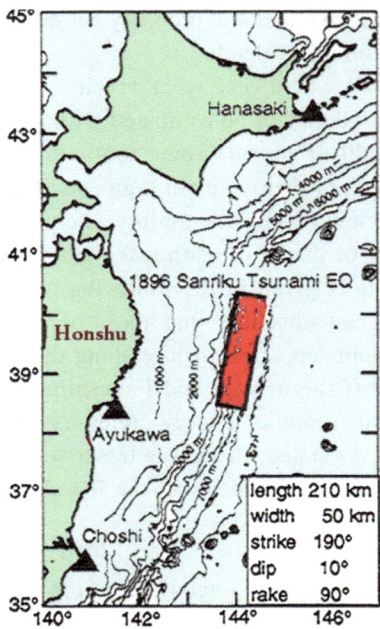

Figure 13
Estimated earthquake fault parameters and generation area of the 1896 tsunami as determined by reverse wave refraction from three tide gauges in the region (modified original graphic by TANIOKA and SATAKE, 1996; also shown in TANIOKA and SENO, 2001 with a change of fault dip from 20° to 10°). The sum of elastic deformation and the additional uplift of sediments were used with three different models of displacement to estimate total ocean floor deformation

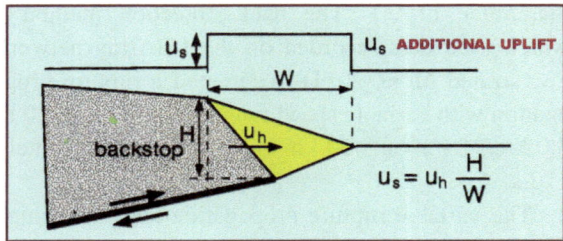

Figure 14
Uplift of sediments by compressional forces. One of three models of displacements along the leading edge of the accretionary prism which was used to estimate additional contribution to tsunami height from a sediment uplift process for the 1896 event (modified after TANIOKA and SENO, 2001)

earthquake, tsunami waves begun to strike the Sanriku coastal region as well as the southern coasts of Hokkaido. The death toll of the tsunami was 26,360. At the village of Tarō, only 36 people of its total population of 2,000 survived while all infrastructure and most of the houses were destroyed (NAKAO, 2009). The 1933 tsunami was equally devastating. A total of 3,064 people were killed and 1,092 were injured. At Yoshihama, close to Ryori Bay, the 1933 tsunami was responsible for 982 deaths. The death toll of the 2011 tsunami was given as 13,843 at the end of April 2011, with another 14,030 people missing. The impact of the Fukushima nuclear disaster may have a long-term collateral impact and add to the death toll.

7.2. Seismic Intensities

The great earthquakes of 1611 (Keichō), of 1896 (Meiji Sanriku) and of the 2011 (Tohoku-Oki) were

not associated with strong ground motions over large areas but generated extremely devastating tsunamis. All three involved reverse faulting with slow rupturing within thick sedimentary layers. By contrast, the 1933 "Showa" quake occurred on a normal fault.

The ground motions of the 1896 earthquake were not substantial and seismic intensities ranging from 2 to 4 were assigned to this event (Japan Meteorological Observatory scale) for a relatively small area of Sanriku (Fig. 15). The quake's rupture velocity was relatively slow, indicating the presence of compacted sedimentary layers in the source region. However, the generated tsunami was extremely high, as this was a distinct tsunami-earthquake with a slower fault slip than that which usually occurs during normal earthquakes. The ground motions of the 1933 earthquake were much stronger and were assigned an intensity of five. However, the tsunami it generated was not as high as that of 1896. Although the 2011 event was also characterized as a tsunami-earthquake, strong ground motions were felt as far away as Tokyo. The stronger ground motions were probably generated along the second segment of the earthquake's rupture, which was associated with higher propagation velocity of up to 3 km/s.

7.3. Source Mechanisms of the Tohoku-Oki 2011 and "Meiji" 1896 Earthquakes

The estimated source region of the 1896 tsunami was somewhat north of that of 1833 and of 2011. Both Tohoku-Oki 2011 and "Meiji" 1896 earthquakes occurred on reverse faults and had characteristics of

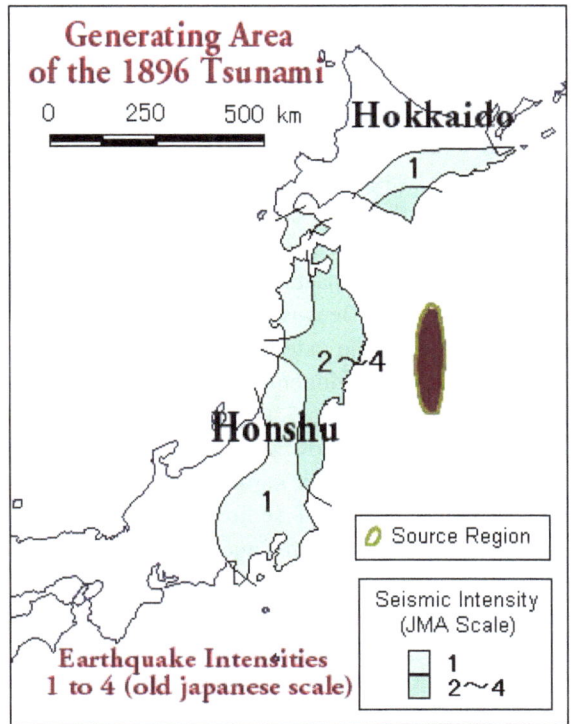

Figure 15
Approximate generating area of the tsunami (modified after Japan Meteorological Observatory, 1896)

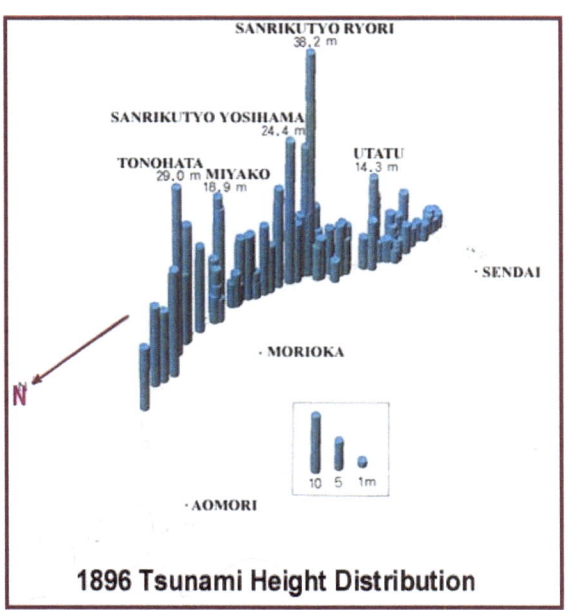

Figure 16
Tsunami height distribution north of Sendai (modified after HATORI et al., 1982)

severity of tsunami generation usually associated with slower initial fault slips and slow rupture velocities within compacted sedimentary layers. Both had slow rupture velocities initially—mainly within compacted sedimentary layers—thus, resulting in greater horizontal and vertical displacements of sediments along the accretionary prism near the Japan Trench. Both events can be characterized as "tsunami-earthquakes". By contrast and as stated, the 1933 "Showa" quake occurred on a normal fault. Although this was also a great earthquake and generated a very destructive tsunami, its impact was not as severe as that of the 1896 and 2011 events.

7.4. Comparison of Run-up Heights of the 1896 and 2011 Tsunamis

The March 11, 2011 tsunami impact on Honshu's coasts was similar to those of 1896 and 1933. All three tsunamis reached the shores of Honshu within 30–40 min after the main shocks were felt. The maximum heights for all three tsunamis occurred at Ryōri Bay in Sanriku, Iwate Prefecture. The height of the 1896 tsunami was 38.2 m, the highest ever in Japan since the tsunami of 1868 (Fig. 16). The maximum height of the 2011 tsunami was roughly the same as that of 1896, approximately 38 m at the village of Ryōri. The 1933 tsunami in the same area reached a height of 23.0 m.

The far field impact of the 1896 tsunami was greater than that of 2011. In 1896, waves up to 9 m (30 feet) struck the Hawaiian Islands, causing extensive destruction to wharves, boats and houses (PARARAS-CARAYANNIS, 1969). The waves of the 2011 tsunami were destructive but ranged from 2 to 3 m (7–11 feet). Similarly in California, a wave of about 3 m (9.5 feet) was observed in San Francisco in 1896, but the 2011 was less. In Crescent City the maximum run-up height of the 2011 tsunami was about 2.5 m.

7.5. Evaluation and Comparison of Sediment Uplift

The amount of additional sediment uplift of the 1896 earthquake (Fig. 13) was examined in the past (TANIOKA and SATAKE, 1996; TANIOKA and SENO,

2001). This event was characterized as a tsunami-earthquake because of its slow rupturing velocity and weak ground motions along the Sanriku coast. In trying to estimate the slip for this 1896 event by modeling, it was determined that a postulated 20° dipping fault along the top of the subducting plate, offered a match for both the observed seismic response and the tsunami, but required an estimated slip displacement of 10.4 m. However, when a shallower fault dip of 10° was used and the additional uplift of sediments was taken into account, a slip ranging from 5.9 to 6.7 m was obtained, which was in better agreement with the tsunami's waveforms recorded at three tide stations.

By revising the fault modeling, TANIOKA and SENO (2001) estimated the magnitude of the 1896 earthquake to be $M_w = 8.0$–8.1. However, this is believed to be an underestimate as the magnitude of the 1896 earthquake was probably similar or even greater to that of 2011. This also supported by the much greater far-field tsunami run-up heights that were observed for the 1896 event. Also, of the three displacement models TANIOKA and SENO (2001) considered for the contribution of sediment uplift to tsunami generation, the one shown in Fig. 14—involving only the leading edge of the accretionary prism—is more realistic, but would still underestimate the actual sediment uplift of the 1896 tsunami. Apparently, the 2011 tsunami involved sediment uplift over a much wider area on the accretionary prism and thus it would be difficult to estimate quantitatively the effect on ocean surface displacement and wave heights at the source region.

8. Summary and Conclusions

The anomalously high 2011 tsunami which occurred along Honshu's east coast resulted from a combination of crustal deformations of the ocean floor due to the upthrust tectonic motions of the earthquake, augmented by additional uplift due to large coseismic lateral movement which compressed and deformed sediments along the accretionary prism on the overriding tectonic plane near the Japan Trench. The event was a "tsunami-earthquake" in the sense that it was mostly associated with a slow rupturing process and lateral movement within shallow and highly compacted sedimentary layers.

The deformation occurred randomly along parallel and en-echelon faults, which failed in a sequential bookshelf manner. Most of the energy release and deformations that generated the huge tsunami occurred along the shallow eastern segment of the fault off the Miyagi Prefecture, while the segment off the Ibaraki Prefecture—where the rupture process was rapid—released minor seismic energy and resulted in lesser compaction and deformations of the sedimentary layers. Because of the complexity of the rupturing process, the extent of additional uplift due to buckling of the sediments in the tsunami generation area off the Miyagi segment of the fault is difficult to estimate. However, both the 1992 Nicaragua and 2004 Sumatra earthquakes demonstrated that bookshelf failure of sedimentary layers could generate anomalously high tsunamis. Apparently, the same mechanism was responsible for the high tsunami generated in the offshore area off Honshu when the March 11, 2011 Tohoku-Oki earthquake struck.

The great 1896 Meiji earthquake was also a tsunami-earthquake with a similar mechanism of tsunami generation, enhanced by sediment deformation and uplift. Finally, the March 2011 event may have released most of the stress that had accumulated—thus ending a seismic cycle for this particular region off the Sanriku coast. However, due to energy transference, a new seismic cycle of stress has begun for the adjacent regions, which will culminate in another large tsunamigenic earthquake in the near future—perhaps in the next 2–4 years. A large tsunamigenic earthquake with moment magnitude up to M_w 8 can be expected to occur either to the north closer to Hokkaido and the Kurile Islands, or to the south closer to the Izu-Ogasawara Trench area.

Acknowledgment

The author thanks Peter Zhol for his assistance in providing earthquake aftershock clustering data.

REFERENCES

AMMON C. J., THORNE L., KANAMORI H., and CLEVELAND M., 2011, "A rupture model of the great 2011 Tohoku earthquake". Earth Planets Space, 63(7), 693–696.

ANNUNZIATO, A. (2012), *Sea Level Signals Correction for the 2011 Tohoku Tsunami*. Science of Tsunami Hazards, Vol *31* No. 2, pp 99–111.

FUJII, Y., SATAKE, K., SAKAI, S., SHINOBARA, M., and KANAZAWA, T. (2011), *Tsunami source of the 2011 Off the Pacific Coast of Tohoku earthquake*, Earth Planets Space *63*, no. 7, 815–820.b.

FUJIWARA, T., KODAIRA, S., NO, T., KAIHO, Y., TAKAHASHI, N., KANEDA, Y. (2011), *The 2011 Tohoku-Oki Earthquake: Displacement Reaching the Trench Axis*, Science *334*, no. 6060, 1240–1240; doi:10.1126/science.1211554.

Geospatial Information Authority of Japan (2011), *The 2011 off the Pacific coast of Tohoku Earthquake: Crustal deformation and fault model (preliminary)*, http://www.gsi.go.jp/cais/topic110313-indexe.html.

GUSMAN, ADITYA, R., TANIOKA, Y., SAKAI, S., TSUSHIMA, H. (2012), *Source model of the great 2011 Tohoku earthquake estimated from tsunami waveforms and crustal deformation data*, Earth and Planetary Science Letters, *341–344*: 234–242.

HATORI, T., AIDA, I, KOYAMA M., and T. HIBIYA (1982), *Field Survey of the Tsunamis In Inundating Ofunato City—The 1960 Chile and 1933 Sanriku Tsunamis*. Bull. of the Earthquake Research Institute, Vol. *57*, pp. 133–150.

HAYES, G. (2011), *Rapid source characterization of the 03-11-2011 Mw 9.0 off the Pacific coast of Tohoku earthquake*, Earth Planets Space, Special Issue: First Results of the 2011 Off the Pacific Coast of Tohoku Earthquake, *63*(7), 525–528.

HAYES, G.P., EARLE, P.S., BENZ, H.M., WALD, D.J., BRIGGS, R., and the USGS/NEIC Earthquake Response Team (2011), *88 hours: the U.S. Geological Survey National Earthquake Information Center response to the March 11, 2011 Mw 9.0 Tohoku earthquake*, Seismol. Res. Lett., *82*(4), 481–493, doi:10.1785/gssrl.82.4.481.

HONDA, R., YUKUTAKE, Y., ITO, H., HARADA, M., AKETAGAWA, T., YOSHIDA, A., SAKAI, S., NAKAGAWA, S., HIRATA. N., OBARA. K., and KIMURA H. (2011), *A complex rupture image of the 2011 Tohoku earthquake revealed by the MeSO-net*. TERRAPUB Report (Received April 10, 2011; Revised May 18, 2011; Accepted May 29, 2011 and published Online).

IDE, S., BALTAY, A., BEROZA, G.C. (2011), *Shallow dynamic overshoot and energetic deep rupture in the 2011 Mw 9.0 Tohoku-Oki earthquake*, Science *332*, no. 6036, 1426–1429. doi: 10.1126/science.1207020.

IIDA, K., COX D.C., and PARARAS–CARAYANNIS, G. (1967), *Preliminary Catalog of Tsunamis Occurring in the Pacific Ocean*. Data Report No. 5. Honolulu: Hawaii Inst. Geophys. Aug. 1967.

ITO, Y., TSUJI, T., OSADA, Y., KIDO, M., INAZU, D., HAYASHI, Y., TSUSHIMA, H., HINO, R., and H. FUJIMOTO (2011), *Frontal wedge deformation near the source region of the 2011 Tohoku-Oki earthquake*, Geophys. Res. Lett. *38*, L00G05, doi:10.1029/2011GL048355.

KATO, A., OBARA, K., IGARASHI, T., TSURUOKA, H., NAKAGAWA, S., HIRATA, N. (2012), *Propagation of Slow Slip Leading Up to the 2011 Mw 9.0 Tohoku-Oki Earthquake*, Science, 10 February 2012, Vol. *335* no. 6069, pp. 705–708. doi:10.1126/science.1215141.

KAWAKATSU, H., and T. SENO (1983), *Triple seismic zone and the regional variation of seismicity along the northern Honshu arc*. J. Geophys. Res. *88*, 4215–4230, 1983.

KIDO, M., OSADA, Y., FUJIMOTO, H., HINO, R., ITO, Y. (2011), *Trench-normal variation in observed seafloor displacements associated with the 2011 Tohoku-Oki earthquake*, Geophys. Res. Lett. *38*, L24303, doi:10.1029/2011GL050057.

MAEDA, T., FURUMURA, T., SAKAI, S., and SHINOHARA, M. (2011), *Significant tsunami observed at ocean-bottom pressure gauges during the 2011 off the Pacific coast of Tohoku earthquake*, Earth Planets Space, *63*(7), 803–808.

MORI, N., TAKAHASHI T., YASUDA T., and H. YANAGISAWA (2011), *Survey of 2011 Tohoku earthquake tsunami inundation and run-up*, Geophysical Research Letters, *38*, L00G14, doi:10.1029/2011GL049210.

MORI, N., TAKAHASHI T., and The 2011 Tohoku Earthquake Tsunami Joint Survey Group (2012), *Nationwide Post Event Survey and Analysis of the 2011 Tohoku Earthquake Tsunami*, Coastal Engineering Journal, Vol. *54*, Issue 4, 1250001, 27 p.

NAKAO, M. (2009), *The Great Meiji Sanriku Tsunami June 15, 1896 at the Sanriku coast of the Tohoku region*. Retrieved 2009-10-18.

NETTLES, M., EKSTROM, G., and KOSS, H.C. (2011), *Centroid-momnet-tensor analysis of the 2011 off the Pacific coast of Tohoku Earthquake and its larger foreshocks and aftershocks*, Earth Planets Space, *63*(7), 519–523.

OZAWA, S., NISHIMURA, T., SUITO, H., KOBAYASHI, T., TOBITA, M., and IMAKIIRE, T. (2011), *Coseismic and postseismic slip of the 2011 magnitude-9 Tohoku-Oki earthquake*, Nature *475*, no. 7356, 373–376 (21 July 2011). doi:10.1038/nature10227.

PARARAS-CARAYANNIS, G. (1969), *Catalog of Tsunamis in the Hawaiian Islands*. World Data Center A- Tsunami, U.S. Dept. of Commerce Environmental Science Service Administration— Coast and Geodetic Survey, May 1969.

PARARAS-CARAYANNIS, G. (1983), *The Earthquake and Tsunami of 26 May 1983 in the Sea of Japan*. http://www.drgeorgepc.com/Tsunami1983Japan.html.

PARARAS-CARAYANNIS, G. (1992), *The Earthquake and Tsunami of 2 September 1992 in Nicaragua*. http://drgeorgepc.com/Tsunami1992Nicaragua.html.

PARARAS-CARAYANNIS, G., (1994), *The Earthquake and Tsunami of July 12, 1993 in the Sea of Japan/East Sea—The Hokkaido "Nansei-Oki" Earthquake and Tsunami*. http://www.drgeorgepc.com/Tsunami1993JAPANOkushiri.html.

PARARAS-CARAYANNIS, G., (2000), *Major Earthquakes in Japan in the 20th Century*. http://drgeorgepc.com/EarthquakesJapan.html.

PARARAS-CARAYANNIS, G. (2005), *Earthquake and Tsunami of December 26, 2004, in Indonesia*. http://drgeorgepc.com/Tsunami2004Indonesia.html.

PARARAS-CARAYANNIS G., (2006), *Potential of Tsunami Generation along the Makran Subduction Zone in the Northern Arabian Sea, Case Study: The Earthquake and Tsunami of November 28, 1945*, Science of Tsunami Hazards, Vol. *24*, No. 5, pp 358–384. http://drgeorgepc.com/Tsunami1945Pakistan.html.

PARARAS-CARAYANNIS, G., (2009), *Earthquake and Tsunami of 3 March 1933 in Sanriku, Japan*. http://drgeorgepc.com/Tsunami1933JapanSanriku.html.

PARARAS-CARAYANNIS, G. (2010), *Earthquake and Tsunami of 27 February 2010 in Chile—Evaluation of Source Mechanism and of Near and Far-field Tsunami Effects*. Science of Tsunami Hazards, Vol. 29, No. 2. 2010. Summary at http://www.drgeorgepc.com/Tsunami2010Chile.html.

PARARAS-CARAYANNIS, G. (2011), *The Great Tsunami of March 11, 2011 in Japan—Analysis of Source Mechanism and Tsunami-genic Efficiency*. OCEANS 11, MTS/IEEE Proceedings, 2011.

SATO, M., ISHIKAWA, T., UJIHARA, N., YOSHIDA, S., FUJITA, M., MOCHIZUKI, M., and ASADA, A. (2011), *Displacement above the* hypocenter of the 2011 Tohoku-Oki earthquake, Science, 332, 1395.

SENO, T., and GONZALEZ, D. G. (1987), Faulting caused by earthquakes beneath the outer slope of the Japan Trench. J. Phys. Earth 35, 381–407, 1987.

SENO, T. (1999a), Is northern Honshu a microplate? Tectonophysics 115, 177–196, 1985.

SENO, T. (1999b), Syntheses of the regional stress fields of the Japanese islands. The Island Arc 8, 66–79, 1999.

SENO, T., and Y. YAMANAKA (1996), Double seismic zones, compressional deep trench - outer rise events and superplumes in Subduction Top to Bottom, edited by G. E. Bebout, D. W. Scholl, S. H. Kirby, and J. P. Platt Geophys. Monogr. 96, 347–355, 1996.

SENO, T., and Y. YAMANAKA (1998), Arc stresses determined by slabs: Implications for mechanisms of back-arc spreading. Geophys. Res. Lett. 25, 3227–3230, 1998.

SENO, T., SAKURAI T, and S. STEIN (1981), Can the Okhotsk plate be discriminated from the North American plate? J. Geophys. Res. 101, 11305–11315, 1996.

SENO, T., and B. PONGSAWAT (1981), A triple-planed structure of seismicity and earthquake mechanisms at the subduction zone off Miyagi Prefecture, northern Honshu, Japan Earth Planet. Sci. Lett. 55, 25–36, 1981.

SENO, T., and G. C. KROEGER (1983), A reexamination of earthquakes previously thought to have occurred within the slab between the trench axis and double seismic zone, northern Honshu. J. Phys. Earth. 31, 195–216, 1983.

SENO, T., and T. TAKANO (1996), Seismotectonics at the trench–trench-trench triple junction off central Honshu. Pure Appl. Geophys. 129, 27–40, 1989.

SHAO, G., LI, X., JI, C., and T. MAEDA (2011), Focal mechanism and slip history of 2011 Mw 9.1 off the Pacific coast of Tohoku earthquake, constrained with teleseismic body and surface waves, Earth Planets Space, 63, 559–564, 2011.

SIMONS, M., MINSON, S.E., SLADEN, A., ORTEGA, F., JIANG, J., OWEN, S.E., MENG, L., AMPUERO, J-P., WEI, S., CHU, R., HELMBERGER, D.V., KANAMORI, H., HETLAND, E., MOORE, A.W., and

WEBB, F.H. (2011), The 2011 magnitude 9.0 Tohoku-Oki earthquake: Mosaicking the megathrust from seconds to centuries, Science, 332(6036), 1421–1425.

TANIOKA, Y., and K. SATAKE (1996), Fault parameters of the 1896 Sanriku tsunami earthquake estimated from tsunami numerical modeling, Geophys. Res. Letters, 23–13, 1549–1552.

TANIOKA, Y., and SENO, T. (2001), Sediment effect on tsunami generation of the 1896 Sanriku tsunami earthquake. Geophysical Research Letters 28(17): 3389–3392.

The 2011 Tohoku Earthquake Tsunami Joint Survey Group (2011), Nationwide Field Survey of the 2011 Off the Pacific Coast of Tohoku Earthquake Tsunami, Journal of Japan Society of Civil Engineers, Series B, Vol. 67 (2011), No. 1, pp. 63–66.

Tohoku Earthquake Tsunami Information, The 2011 Tohoku Earthquake Tsunami Joint Survey Group. http://www.coastal.jp/tsunami2011/.

WANG, D., and MORI, J. (2011), Rupture Process of the 2011 off the Pacific Coast of Tohoku Earthquake (Mw 9.0) as Imaged with Back-Projection of Teleseismic P-waves. Earth Planets Space, 1–5, 2011.

WEI, D. and SENO T., 1998, Determination of the Amurian Plate Motion, In Mantle Dynamics and Plate Interactions in East Asia, edited by Martin Flower, GeoDynamics Series., AGU, 1998.

YAMAZAKI Y., LAY T., CHEUNG, K.F., YUE, H., and KANAMORI, H. (2011), Modeling near-field tsunami observations to improve finite-fault slip models for the 11 March 2011 Tohoku earthquake, Geophys. Res. Lett., 38, L12605, doi:10.1029/2011GL047508.

YOSHIDA, Y., UENO, H., MUTO, D., and S. AOKI (2011), Source process of the 2011 Off the Pacific Coast of Tohoku earthquake with the combination of teleseismic and strong motion data, Earth Planets Space 63, no. 7, 565–569.

YOKOTA, Y., KOKETSU, K., FUJII, Y., SATAKE, K., SAKAI, S., SHINOHARA, M., and T. KANAZAWA (2011), Joint inversion of strong motion, teleseismic, geodetic, and tsunami datasets for the rupture process of the 2011 Tohoku earthquake, Geophys. Res. Lett. doi:10.1029/2011GL050098.

(Received December 17, 2012, revised April 28, 2013, accepted April 29, 2013, Published online May 17, 2013)

Pure Appl. Geophys. 171 (2014), 3279–3280
© 2014 Springer Basel
DOI 10.1007/s00024-014-0992-7

Erratum

Erratum to: The Great Tohoku-Oki Earthquake and Tsunami of March 11, 2011 in Japan: A Critical Review and Evaluation of the Tsunami Source Mechanism

GEORGE PARARAS-CARAYANNIS[1]

Erratum to: Pure Appl. Geophys.
DOI 10.1007/s00024-013-0677-7

Figure 2 in the body of the original paper regarding inundation and wave heights of the 2011 Tohoku tsunami along the coastline of Eastern Honshu was taken from an earlier Japanese publication which was erroneously translated, thus does not represent the correct values. This Fig. 2 has been substituted by a new Fig. 2 which represents the actual corrected values of inundation and wave heights of the 2011 Tohoku tsunami along the coastline of Eastern Honshu as determined by the comprehensive survey summarized by MORI et al. (2011) and by MORI et al. (2012) or other investigations. Therefore, the caption of the new Fig. 2 has been corrected to reflect these changes and the text in the section of the original paper with heading of "Near and Far Field Effects" needs to include the reference MORI et al. (2012).

Nothing more in the subsequent text of the original paper refers to Fig. 2 and none of the subsequent descriptions of tsunami effects, impact, mechanism or conclusions contradict the findings of surveys as summarized by MORI et al. (2011) and by MORI et al. (2012) or other investigations.

The online version of the original article can be found under doi:10.1007/s00024-013-0677-7.

[1] Tsunami Society International, Honolulu, HI, USA. E-mail: drgeorgepc@yahoo.com

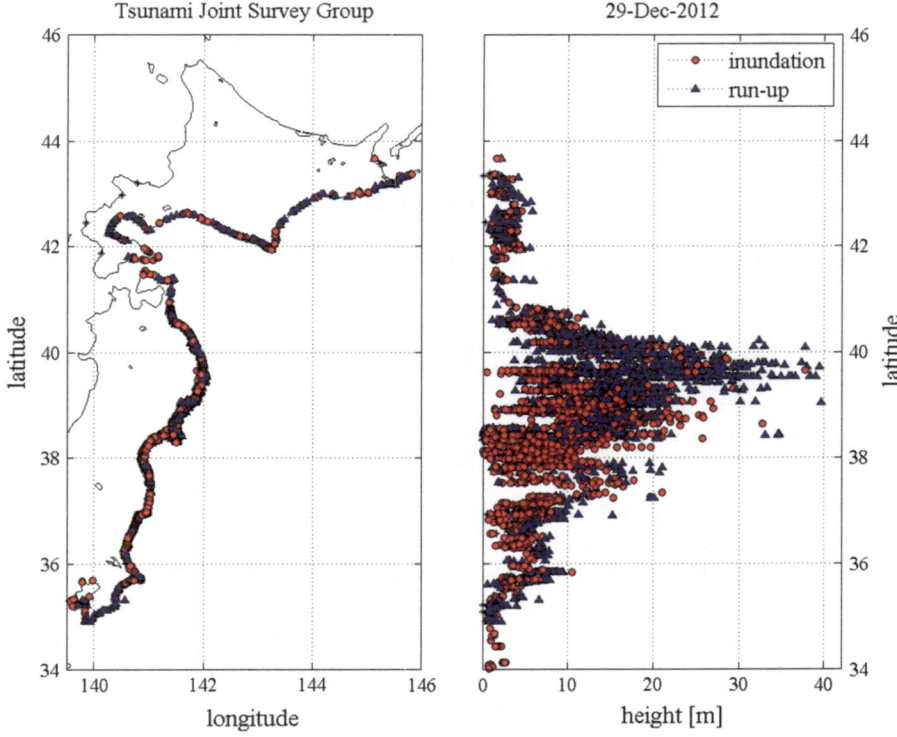

Figure 2
Inundation and wave heights in meters of the 2011 Tsunami along the Tohoku coast as determined by a comprehensive post event survey
(Mori *et al.*, 2012) (The date 29 Dec 2012 on the *top right side* of the figure above, refers to the completion of curation of the 2011 tsunami
data)

REFERENCE

Mori, N., Takahashi T. and The 2011 Tohoku Earthquake Tsu-
nami Joint Survey Group, 2012. *The 2011 Tohoku Earthquake
Tsunami Joint Survey Group (2012), Nationwide Post Event
Survey and Analysis of the 2011 Tohoku Earthquake Tsunami*,
Coastal Engineering Journal, Vol. *54*, Issue 4, 1250001, 27 p,
doi:10.1142/S0578563412500015

Pure Appl. Geophys. 171 (2014), 3281–3305
© 2014 Springer (outside the USA)
DOI 10.1007/s00024-014-0777-z

Tsunami Forecast by Joint Inversion of Real-Time Tsunami Waveforms and Seismic or GPS Data: Application to the Tohoku 2011 Tsunami

Yong Wei,[1,2] Andrew V. Newman,[3] Gavin P. Hayes,[4] Vasily V. Titov,[1] and Liujuan Tang[1,2]

Abstract—Correctly characterizing tsunami source generation is the most critical component of modern tsunami forecasting. Although difficult to quantify directly, a tsunami source can be modeled via different methods using a variety of measurements from deep-ocean tsunameters, seismometers, GPS, and other advanced instruments, some of which in or near real time. Here we assess the performance of different source models for the destructive 11 March 2011 Japan tsunami using model–data comparison for the generation, propagation, and inundation in the near field of Japan. This comparative study of tsunami source models addresses the advantages and limitations of different real-time measurements with potential use in early tsunami warning in the near and far field. The study highlights the critical role of deep-ocean tsunami measurements and rapid validation of the approximate tsunami source for high-quality forecasting. We show that these tsunami measurements are compatible with other real-time geodetic data, and may provide more insightful understanding of tsunami generation from earthquakes, as well as from nonseismic processes such as submarine landslide failures.

Key words: Tsunameter, GPS, finite-fault solution, tsunami, inversion, tsunami forecast, runup, inundation, modeling, near field.

1. Introduction

Rapid seafloor deformation caused by earthquake slip is the dominant source of large long-wavelength transoceanic waves driven by gravity, termed tsunami. Determination of tsunami source characteristics in the early stages of a progressing tsunami is crucial for accurate and useful inundation and runup forecasting. At present, deep-ocean tsunameters, seismographic networks, and high-rate GPS stations are probably the most capable instruments for providing real-time observations of earthquake rupture and tsunami propagation. These observations can be used for rapid estimation of a slip-distribution-based tsunami source, which can in turn be used as direct input for model forecasts of tsunami impact in real time (Tang *et al.*, 2012; Wei *et al.*, 2013) or near real time (Hayes *et al.*, 2011).

A tsunami source, for which the water column is instantaneously disturbed by the deformed ocean bottom, can be inferred from a variety of tsunami measurements provided by deep-ocean tsunameters (Wei *et al.*, 2003, 2008; Titov *et al.*, 2005; Titov, 2009; Percival *et al.*, 2010), cabled ocean-bottom pressure sensors (Tsushima *et al.*, 2009), satellite altimetry (Arcas and Titov, 2006; Hirata *et al.*, 2006), and/or tide gage data (Satake, 1987; Satake and Kanamori, 1991). An indirect measure of a tsunami source can come from the estimation of coseismic displacements resulting from fault slip, assuming that the resulting vertical disturbance of the ocean floor is instantaneously transferred to the ocean surface (Kajiura, 1970). Traditionally, the approximate location, depth, fault orientation, and seismic moment of an event can be approximated from a centroid moment tensor (CMT) solution within 8–15 min after large earthquakes (Whitmore, 2009). However, since they assume a point source, CMT inversions do not adequately describe the spatial extent of rupture, causing tsunami forecasts to be inaccurate for coastal communities at risk (Government Accountability Office, 2006). In the last decade, finite-fault inversion algorithms using globally distributed broadband seismic waveforms have

---

[1] Joint Institute for the Study of the Atmosphere and Ocean, University of Washington, Seattle, WA, USA. E-mail: Yong.Wei@noaa.gov
[2] Pacific Marine Environmental Laboratory, National Oceanic and Atmospheric Administration, Seattle, WA, USA.
[3] School of Earth and Atmospheric Sciences, Georgia Institute of Technology, Atlanta, GA, USA.
[4] United States Geological Survey, National Earthquake Information Center, Golden, Colorado, USA.

Reprinted from the journal

become a widely used technique to constrain the detailed spatial dimensions and slip distribution of the seismic source (hereafter "earthquake source dimensions"; e.g., JI et al., 2002). This methodology has been used as source input for models of tsunami from some recent large earthquakes (FRITZ et al., 2011; NEWMAN et al., 2011; YAMAZAKI et al., 2011; WEI et al., 2013; MACINNES et al., 2013). Global Positioning System (GPS) data offer precise measurements of ground displacements from an earthquake rupture, which have been shown to facilitate computation of earthquake slip parameters shortly after large events (BLEWITT et al., 2006; PIETRZEK et al., 2007; SIMONS et al., 2011; VINGY et al., 2011; SONG et al., 2012).

Each method has its own merits and limitations. Tsunameters provide direct measurements of the water pressure changes caused by tsunami waves, which are in general the earliest direct tsunami observations available during an event. Different from the water level registered at tide gages, deep-ocean tsunami data are free of interference from harbor and local shelf effects, yet are not sensitive to short-wavelength wind-driven waves, leaving clean signatures (resolution to the sub-cm level) of the tsunami. Because tsunami propagation in the deep ocean follows linear wave dynamics, pressure information from tsunameters allows for rapid tsunami source inversions (SATAKE, 1987; SATAKE and KANAMORI, 1991; WEI et al., 2003; PERCIVAL et al., 2010). Real-time tsunami measurements in the deep ocean are the core component of the NOAA's tsunami forecast system, and have led to many successful real-time forecasts in the last decade, especially after the 2004 Indian Ocean tsunami (GONZÁLEZ et al., 2005; TITOV, 2009; TANG et al., 2008, 2012; WEI et al., 2008, 2013). Tsunameters are predominantly sited near subduction zones—the environment responsible for most tsunamigenic earthquakes. Instruments are placed at distances equivalent to 30–60 min of tsunami wave travel time from expected tsunamigenic earthquake sources, and seaward of the trench to avoid effects from coastal and shallow-water interference that complicate signals by causing nonlinearity and dispersion of waves (SPILLANE et al., 2008). Unfortunately, such tsunameter data are particularly critical in the first 30 min for near-field

tsunami warning and forecasting. Several, more proximal tsunameters were recently deployed offshore of Japan and provided useful data within 10–20 min after the 7 December 2012 moment magnitude (M_w) 7.2 Japanese earthquake, which generated a small tsunami (BERNARD et al., 2013). This improvement shows that tsunami detection times can be significantly shortened when tsunameters are located closer to potential sources (WEI et al., 2013).

Inversion of seismic waves via finite-fault modeling is a method used to rapidly characterize an earthquake source in terms of its spatiotemporal slip distribution (JI et al., 2002; HAYES, 2011, and references therein). Following the 2011 Japan earthquake, the US Geological Survey (USGS) National Earthquake Information Center (NEIC) provided a quick finite slip model within several hours of the earthquake origin time (HAYES, 2011). This method is robust for capturing the broad characteristics of an earthquake's slip distribution, but can be limited by the initial assumption of the fault geometry (HAYES, 2011). For tsunami modeling, a major concern when using such fault slip models is that the energy conversion process—from fault rupture to ocean water—remains one of the most difficult geophysical problems to solve. For great earthquakes, such as the 2004 Sumatra, Indonesia and 2011 Tohoku-Oki, Japan events, one of the biggest challenges is to quickly determine an accurate magnitude for the earthquake, since the time available for analysis of the energy content in long-period seismic waves is limited to before waves arrive at the closest shorelines (WHITMORE, 2009). In the past, moment variations have been as large as a factor of ten between initial and final estimations (TANG et al., 2012). In the future, analyses of the seismic W-phase (DUPUTEL et al., 2012), particularly at regional distances (RIVERA et al., 2011), will help to improve both the accuracy and speed of rapid magnitude estimates.

Ground deformation monitors such as GPS provide accurate measurements of fault movement that can be used to reconstruct the detailed structure of the earthquake source. However, except for a very few underwater measurements during the Tohoku earthquake (SATO et al., 2011), which were not available in real time, GPS instruments are mostly tied to land-based observations because the radar signals used cannot penetrate water, and thus the results are inadequate for characterizing earthquake slip that occurs predominantly offshore and near the trench, where it is most effective at tsunami generation (NEWMAN, 2011). Therefore, the exclusive use of land-based GPS measurements may lead to poorly constrained estimates of tsunami sources. One solution is to combine coseismic measurements made on land with deep-ocean tsunameter measurements. GUSMAN et al. (2010) performed a retrospective joint inversion of tsunami waveforms and InSAR data to understand the magnitude and spatial extent of the 2007 Bengkulu earthquake. YOKOTA et al. (2011) reassessed the 2011 Tohoku earthquake source via a joint inversion of strong motion, teleseismic, geodetic, and tsunami datasets. YAMAZAKI et al. (2011) applied a perturbation method to tune the tsunami model results using tsunami measurements to improve the finite-fault inversions.

The deadly 11 March 2011 Japan tsunami is likely the fourth largest documented in history from the perspective of tsunami energy (TANG et al., 2012). It is also the largest event to have been widely recorded by tsunameter, seismic, and GPS networks. Despite the early tsunami warning in Japan 3 min after the earthquake based on a preliminary magnitude of M_w 7.9 event (OZAKI, 2012), the earliest estimate of the tsunami source itself came from NOAA's inversion model using waveforms recorded at a tsunameter (also called DART® in the USA as an acronym for Deep-ocean Assessment and Reporting for Tsunamis) 500 km east of the epicenter about an hour after the earthquake (TANG et al., 2012; WEI et al., 2013). The earliest estimates of the earthquake source useful for tsunami modeling were the USGS CMT and rapid finite-fault model solutions (HAYES, 2011; HAYES et al., 2011), both inferred from seismological data within minutes to hours of the mainshock. GPS stations throughout Japan recorded measurements of coseismic offsets within minutes of the earthquake, and these measurements, if made available in real time, could have aided a very rapid estimate of earthquake source dimensions (OHTA et al., 2012).

In the following sections, we assess source models inferred from different real-time measurements for prediction of the 11 March 2011 Japan tsunami (Fig. 1). We use these source models as input for a

tsunami model, and compare the modeling results with real-time measurements and survey results for the 2011 Tohoku tsunami. This comparative study provides an overview of the advantages and limitations of different real-time measurements potentially useful for early tsunami warning in the near and far fields.

2. Methodology and Tsunami Models

In this study, we use the tsunami inundation models developed in WEI et al. (2013) to simulate the 2011 Tohoku tsunami for all source models. These inundation models are based on the Method of Splitting Tsunami (MOST), a suite of numerical simulation codes capable of estimating tsunami generation, transoceanic propagation, and inundation (TITOV et al., 1997; TITOV and SYNOLAKIS, 1998). MOST uses nonlinear shallow-water (NSW) equations with bottom friction and a moving boundary to compute the tsunami inundation on land. We note here that, in NSW, wave breaking manifests itself as a discontinuity of the solution (TITOV and SYNOLAKIS, 1998). Because mass and momentum are conserved in MOST, the computed solution reflects the overall evolution of tsunami inundation without reproducing the details of the breaking front. With proper setup between the spatial and temporal time steps, the MOST model can numerically mimic the physical wave dispersion (BURWELL et al., 2007). ZHOU et al. (2012) demonstrated that predictions generated using MOST, a shallow-water model with numerical dispersion, and a fully dispersive model developed by ZHOU et al. (2011) all perform comparably in predicting the leading waves of a tsunami. We note here that a two-dimensional (2D) shallow-water model with numerical dispersion such as MOST cannot fully reproduce the physical dispersion of a tsunami as a Boussinesq (dispersive NSW) model does (LOVHOLT et al., 2010). The most significant discrepancies exist in the short trailing waves, and may be more noticeable at distant destinations, during a dispersive tsunami (ZHOU et al. 2011). However, from the perspectives of runup/rundown, wave breaking, and computational efficiency, MOST is more robust, efficient, and flexible as a real-time computational

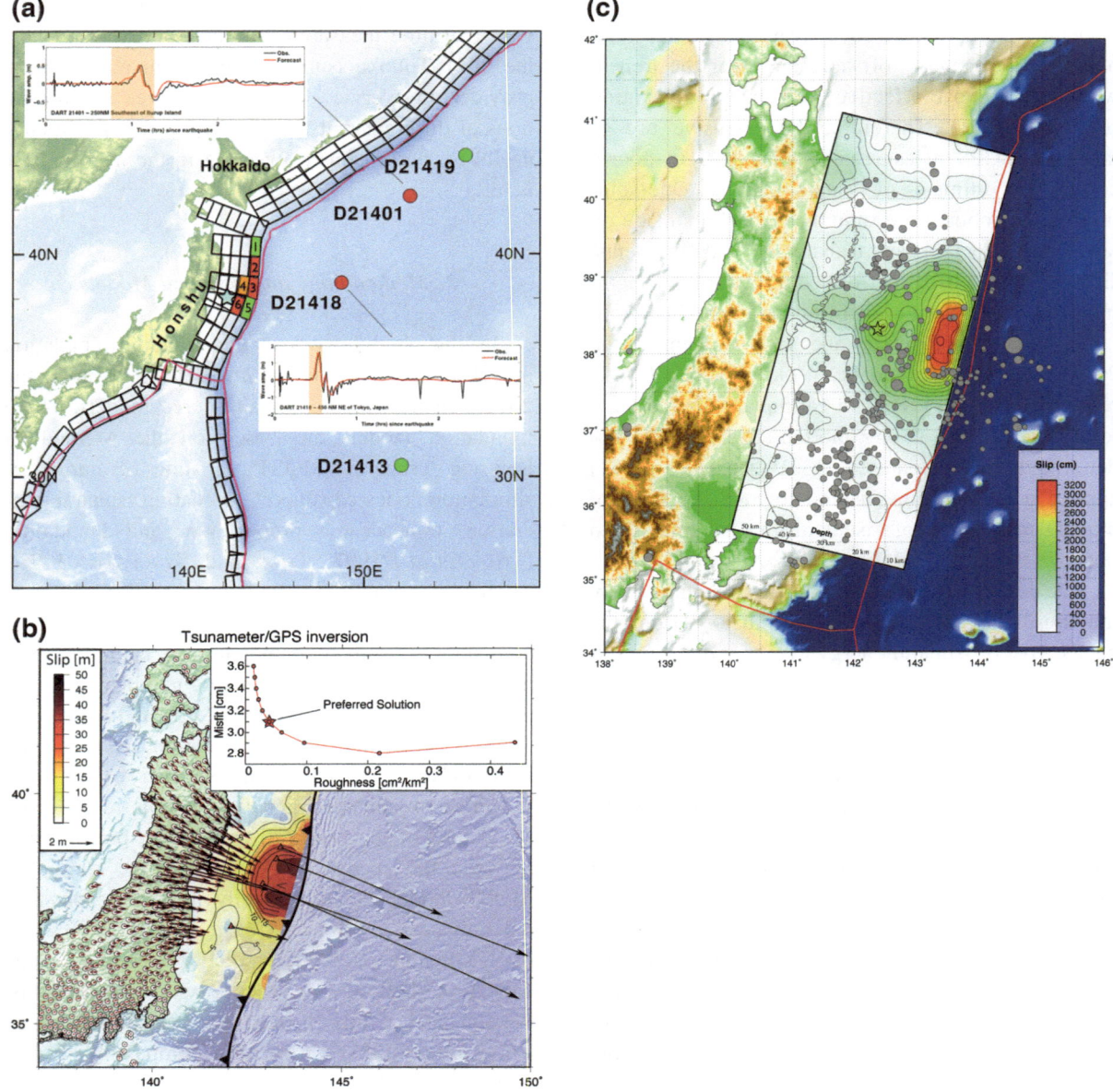

Figure 1

a The 11 March 2011 Japan tsunami source inferred from tsunameter measurements. This inversion was based on real-time measurements from tsunameters D21418 and D21401, where the *orange area* indicates the segment of the time series used in the inversion. The *cyan star* is the USGS epicenter for the 2011 Tohoku earthquake. The *purple line* indicates the plate boundaries. The *black boxes* are the 100 km × 50 km tsunami unit sources precomputed in NCTR's database. The *filled boxes* reflect the inversion results with *color* indicating the slip amount (details in Table 1). **b** GPS/tsunameter inversion for slip along a single 12° dipping plane (*color contours*). Onland GPS data (*black arrows*) are mimicked by the inversion prediction of ground displacement (*red arrows* behind). Seafloor horizontal displacements from SATO *et al.* (2011), which were not used in the inversion, are shown for comparison. An *inset* shows the relative trade-off between decreased roughness (increased smoothness) and misfit. The preferred slip solution is chosen near an inflection point where further reductions in model roughness significantly increase misfit. **c** Surface projection of the slip distribution superimposed on GEBCO bathymetry. *Red lines* indicate major plate boundaries (BIRD, 2003). *Gray circles*, if present, are aftershock locations, sized by magnitude (http://earthquake.usgs.gov/earthquakes/eqinthenews/2011/usc0001xgp/finite_fault.php)

tool compared with a fully dispersive model (Lov-HOLT et al., 2010). MOST employs the elastic model of OKADA (1985) to compute the initial seafloor deformation resulting from a source model, predictions that are then used directly as the initial deformation of the ocean surface in the tsunami inundation models (KAJIURA, 1970). The MOST model has been tested against laboratory experiments and benchmarks (SYNOLAKIS et al., 2008). Since the 2004 Indian Ocean tsunami, MOST has been validated against many modern tsunamis, and used to forecast the tsunami waves at many harbors along US coastlines.

MACINNES et al. (2013) used tsunami inundation models to evaluate source dimensions for nine earthquakes solely or jointly from tsunami measurements, seismic data, and GPS data. Using relatively low-resolution computational grids (30 arc s, ~1 km), they found significant resolution loss compared with high-resolution runs given the data available. High-resolution bathymetric and topographic grids are needed to differentiate a variety of source models. The tsunami inundation models used in this study compute the tsunami flooding with telescoping grids of increasing resolution; the finest grid uses a resolution of 2 arc s (~60 m) to compute the tsunami inundation. This approach also addresses the low-resolution problem raised by MACINNES et al. (2013). It is worth noting that the seafloor deformation offshore of the Sanriku Coast of Japan identified by the real-time tsunameter inversion of the NOAA Center for Tsunami Research (NCTR) (TANG et al., 2012; WEI et al., 2013) was further confirmed by MACINNES et al. (2013), who concluded that an additional source of tsunamigenic energy is needed to explain high tsunami runup along Japan's east coast between latitude 39°N and 40°N. Similarly, a recent study by GRILLI et al. (2013a, b) found evidence of a seismically triggered seafloor failure that may be the mechanism responsible for high tsunami runup in the north.

In this study, a total of 11 tsunami inundation models were used to cover the coastline of Japan between latitude 34.5°N and 44.0°N (Fig. 2a). Each of these models contains three telescoped grids with increasing spatial resolution of 2 arc min (~3.6 km), 15 arc s (~450 m), and 2 arc s (~60 m), to compute the tsunami wave dynamics from offshore to onshore.

TANG et al. (2009) studied 14 historical events and showed through model validation that a 7–8× variation in grid size provides optimized accuracy and speed for a tsunami forecast model. The bathymetric and topographic grids used in these models were obtained from a combination of the 1-arc-min global relief model of earth surface (ETOPO1), Japan Oceanographic Data Center JODC-expert Grid data for Geography—500 m (J-EGG500), and the GeoSpatial Information Authority of Japan (GSI) 50-m digital elevation model. The GSI digital elevation models also contain 5-m Light Detection and Ranging (LIDAR) data for several small areas. These datasets have different limitations in data quality and data density that may degrade the model accuracy. WEI et al. (2013) showed that the large error in the USGS Shuttle Radar Topography Mission 3-s topography resulted in significant underestimation of the inundation limit in the Sendai Plain.

Tsunami impact in the near field, compared with the far field, is more dependent on source geometry (OKAL and SYNOLAKIS, 2004). This comparative study allows us to examine how source models inferred from different methods fit the measured tsunami impact along Japan's coastline. It focuses on how to utilize these methods to achieve rapid and accurate tsunami hazard assessments in the near field, provided these measurements are available in real time for future earthquakes.

Lastly, this study provides discussion on the utilization of rapidly available measurements that may lead to improved real-time tsunami model forecasts, addressing: (1) What are the characteristics of the source? (2) What are the strengths and limitations of these methods? and (3) How could these methods be utilized to expedite tsunami warning, especially in the near field?

3. Source Models of the 11 March 2011 Japan Tsunami

3.1. Tsunami Source Derived from Tsunameter Measurements

Located 500 km east of the epicenter, tsunameter D21418 measured a 1.8-m-high pulse within 30 min of the earthquake (Fig. 1a). This is the largest tsunami

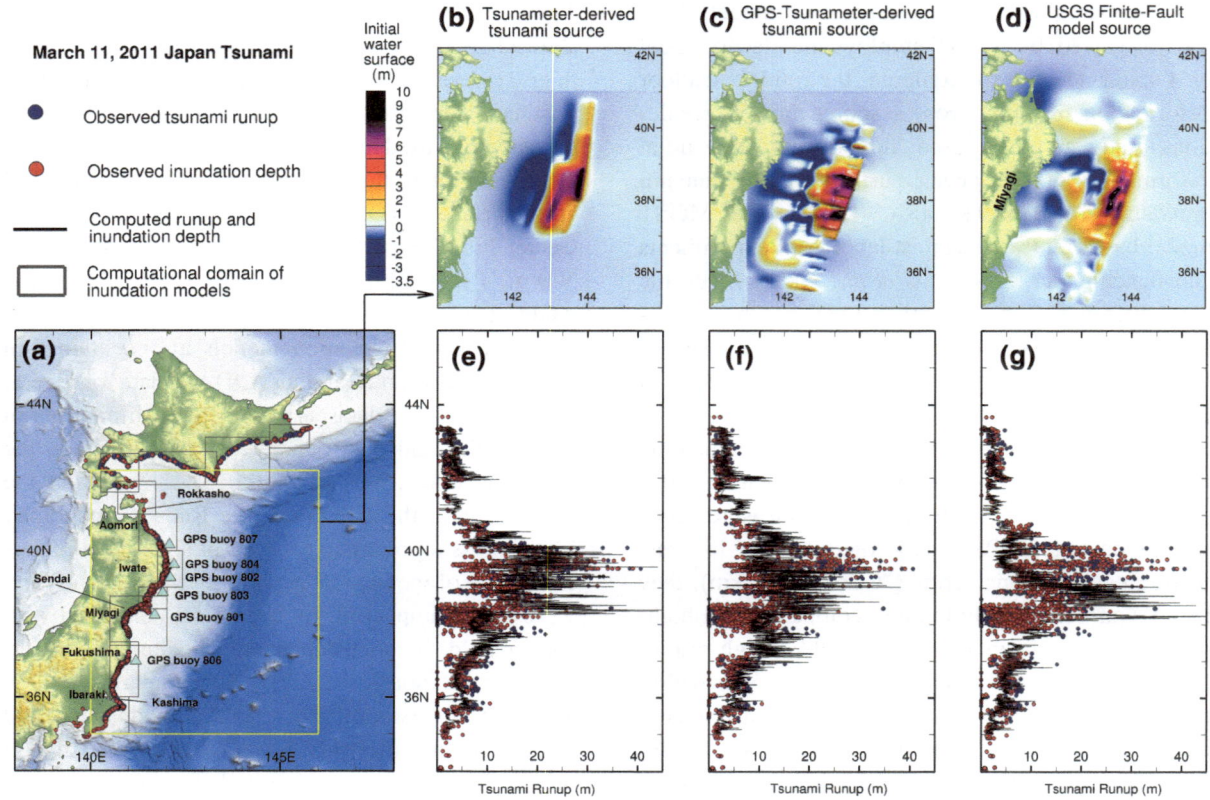

Figure 2
Model–observation comparison of 11 March 2011 Japan tsunami runup height. **a** Model setup along Japan's Pacific coastline, where the *yellow box* corresponds to the domain boundary of the tsunami source computation, and the *cyan triangles* indicate the locations of the Japan Nationwide Ocean Wave information network for Ports and Harbors GPS buoys. *Upper panel* **b** tsunami source derived from tsunameter measurements; **c** tsunami source derived from GPS and tsunami measurements; **d** tsunami source obtained based on USGS finite-fault solution. *Lower panel* **e** comparison between computed and observed tsunami runup for the tsunameter-derived source; **f** comparison between computed and observed tsunami runup for the tsunameter/GPS-derived source; **g** comparison between computed and observed tsunami runup for the USGS finite-fault solution

wave ever recorded by a deep-ocean tsunameter (TANG *et al.*, 2012). The tsunami forecast system of NCTR used records at tsunameters D21418 and D21419 to estimate the tsunami source in a rapid inversion process developed by PERCIVAL *et al.* (2010). Rapid forecasts based on this method rely on developing an a priori database of precomputed tsunamis from a network of subfault patches parameterized to represent global subduction zones (GICA *et al.*, 2008). With rupture dimensions of 100 km × 50 km and 1-m slip, each of these unit sources represents the tsunami source generated by an M_w 7.5 earthquake. MOST computes the tsunami propagation with an ocean basin grid at 4-arc-min resolution for all unit sources. All the

computational results are stored in the tsunami propagation database of NCTR. The precomputed tsunamis are used with real-time tsunami observations from bottom-pressure gages in the deep sea to estimate the approximate tsunami source (WEI *et al.*, 2008; TANG *et al.*, 2009; TITOV, 2009). This provides a quick estimate of the seafloor deformation over the tsunami generation area. During the Tohoku event, NCTR first computed a preliminary inversion of the tsunami source 56 min after the earthquake initiation with just station D21418. Upon the passage of the first tsunami peak at station D21401, 990 km northeast of the epicenter, a second inversion was carried out 90 min after the earthquake using the half-wave period

Table 1

Tsunami forecast source constrained from deep-ocean tsunami measurements at tsunameters D21418 and D21401

Unit source	Location	Strike	Dip	Rake	Depth (km)	Source coefficient (m)
1	143.5273E, 40.3125N	185	19	90	5.0	4.66
2	143.4246E, 39.4176N	185	19	90	5.0	12.23
3	143.2930E, 38.5254N	188	19	90	5.0	21.27
4	142.7622E, 38.5837N	188	21	90	21.3	26.31
5	143.0357E, 37.6534N	198	19	90	5.0	4.98
6	142.5320E, 37.7830N	198	21	90	21.3	22.75

Each unit source has dimensions of 100 km length and 50 km width. GICA *et al.* (2008) provide details of the acronyms for all unit sources. The numbering of the unit sources corresponds to that shown in Fig. 1a. The location of each unit source represents the center of the deformation rectangle side parallel to the foot wall (on the subsidence side of the rectangle)

measured from both stations D21418 and D21401. This provided a refined estimate of the tsunami source (Table 1) and indicated that the initially disturbed water surface was from a source with as much as 10 m of sea surface uplift (Fig. 2). Additionally, this model showed that the seafloor deformation was best described by a source rupture 400 km along strike and 100 km landward from the Japan Trench. Based on this tsunameter-derived source, real-time model inundation forecasts for 30 coastal communities in US territories were accomplished within 30 min of obtaining the tsunami source (i.e., 2 h after the earthquake), nearly 5 h before the tsunami struck the Hawaiian Islands, and 7 h before hitting the West Coast of the continental USA (TANG *et al.*, 2012). The same source was used for real-time model prediction of the tsunami heights in New Zealand (BORRERO *et al.*, 2013). The local civil defense increased the hazard level there after model-predicted wave heights were shown to be larger for the tsunameter inverted source than for a uniform slip case. We emphasize that the data (from two tsunameters) used to determine this tsunami source are only a small subset of tsunami records that we use in this study. TANG *et al.* (2012) provided validation of this tsunami source using measurements at 28 additional tsunameters throughout the Pacific. BORRERO *et al.* (2013) reported that their real-time model predictions along the coast of New Zealand were accurate despite the coarse nearshore bathymetry used in the assessment. The tsunami wave measurements and surveyed runup and inundation data in Japan (MORI *et al.*, 2011) provide data to further validate this source in the near field, as discussed in Sect. 4.

3.2. Combined Tsunameter/GPS Determination of Earthquake Rupture

Hundreds of GPS stations operated by the GPS Earth Observation Network (GEONET) of Japan recorded the onland deformation caused by the 11 March 2011 Tohoku earthquake. Computation of the raw GPS signals from the Geospatial Information Authority (GSI) of Japan, using preliminary satellite orbit estimates by the Advanced Rapid Imaging and Analysis (ARIA) project at the NASA Jet Propulsion Laboratory and Caltech (OWEN *et al.*, 2011), indicated large-scale ESE seaward displacements as large as 5.2 m horizontally and 1.1 m vertically downward (Fig. 1). These data indicated a very large coseismic rupture offshore and were subsequently corroborated by later available seafloor geodetic observations using the GPS/acoustic combination technique at five sites, which measured between 5 and 24 m of east-southeast horizontal motion and between −0.8 and 3 m of uplift (SATO *et al.*, 2011).

In order to evaluate the earthquake rupture extent and tsunami potential geodetically, we took the unique approach of combining the earliest available ARIA GPS displacement solutions with predictions of the seafloor vertical displacement from the early MOST tsunameter inversions using predefined Green's functions for an existing fault database. The tsunami-predicted vertical seafloor motion as estimated using OKADA (1985) was used instead of the predicted thrust on individual segments because the a priori subduction interface along the Japan Trench did not well represent the current state of knowledge about the slab geometry, and dipped about twice as steeply as the currently accepted value of 10–12° dip as described in Slab 1.0 (HAYES et al., 2012).

To evenly distribute the data selection between the seafloor uplift estimates (Fig. 2b) and the onland GPS (Fig. 1b), we selected a grid of 77 seafloor estimates along a 0.5° × 0.5° grid along the rupture area and the surrounding region including sections seaward of the trench (142–145°E, 36–41°N). Each of these 77 seafloor vertical estimates was given a factor of 3 weighting over the onland GPS stations to account for the comparable loss in components (1 component offshore, 3 components onland). We used 366 GPS stations nearest the rupture area (138–142°E, 35–42°N) and assigned error estimates to be 10 cm in each horizontal component and 20 cm in the vertical. The offshore vertical estimates were all assigned 20 cm error. The selection for errors, weighting, and discretization of seafloor vertical estimates is somewhat arbitrary, but corresponds to our normal relative confidence between horizontal and vertical GPS data, as well as our sensitivity to seafloor changes in tsunami excitation. Overall, the effects of errors on model predictions are directly dependent on the choice of weighting that we used. While more precise station-dependent and location-weighted errors could be constructed through substantial postprocessing, such details are not readily available for rapid assessments and hence are not considered here.

The onland GPS data and seafloor vertical displacement predictions were inverted using the geometry shown in Fig. 1b (520-km-long fault striking S15W from 144.4°E, 40°N, and dipping 10° west, including 16 along-strike and 8 down-dip segments from 0 to 50 km depth) and a smoothness trade-off with misfit as described in CHEN et al. (2009). The preferred slip solution was found at an inflection point at which decreased roughness (increased smoothing) began to substantially increase misfit. The final solution (Fig. 1b) agrees well with onland GPS and seafloor vertical estimates (RMS = 2.8 cm in horizontal and 3.9 cm in vertical). Likewise, the horizontal projection of the thrust component agrees well with horizontal seafloor displacements later determined by SATO et al. (2011). Because the near-field tsunameter inversion results (Fig. 2b) were inputs to this inversion, and the final results were

perturbed by onland GPS data and model smoothing, the newly predicted tsunami runup and open-ocean tsunami waves were somewhat degraded (peaks not as well fit). However, the results still match well, and are in almost complete agreement with the onland GPS data.

One interesting feature of this GPS/tsunameter solution is that it shows the utility of the method regardless of whether the predefined faults for the original tsunami inversion were accurate (i.e., whether their assumed geometry reflects the true geometry of the source fault). However, one could argue that, for such solutions in the future, predefined MOST tsunami solutions need not go back to a priori fault characterizations, but instead should just describe discrete pistons of vertical motion, which can be used directly with available onland GPS data to obtain the full cross-shore geodetic solution.

3.3. Finite-Fault Source Derived from Seismological Data

The rapid finite-fault source model for the Japan earthquake was constrained from Global Seismographic Network broadband waveforms based on an inversion algorithm developed by JI et al. (2002), and is discussed in greater detail in HAYES (2011) and HAYES et al. (2011). This approach inverts teleseismic body- and surface-wave data for the slip amplitude, direction, rise time, and rupture initiation time of a collection of subfaults that make up the fault surface. The fault planes used in the inversion were defined using the US Geological Survey (USGS) National Earthquake Information Center (NEIC) W-phase moment tensor solution, adjusted to match local slab geometry (HAYES et al., 2012). The initial fault size used in the inversion procedure is established using empirical relations between magnitude and fault length and width (e.g., BLASER et al., 2010). Subfault source time functions are modeled with an asymmetric cosine function (JI et al. 2002, 2003), and the velocity model used for Green's function computation is based on a combination of Preliminary Reference Earth Model (DZIEWONSKI and ANDERSON, 1981) and Crust 2.0 (BASSIN et al., 2000).

The preliminary solution, made publicly available several hours after the earthquake, predicted a maximum 17.5 m of slip on a nodal plane striking 195° and dipping 14°, and a corresponding seismic moment of 4.0×10^{29} dyne cm (M_w 9.0) (http://earthquake.usgs.gov/). An updated solution resolved nearly twice the amount of maximum slip—33 m—on a nodal plane striking in the same direction but slightly more shallowly at 10° (Fig. 1c). The seismic moment of the updated solution is 4.9×10^{29} dyne cm (M_w 9.1), an approximately 20 % increase over the preliminary solution. For tsunami modeling, the source parameters of the finite-fault solution are used as input in MOST to compute the ocean surface deformation for each subfault, which first describes the seafloor vertical motion using OKADA (1985). They are then linearly combined to obtain the initial disturbance to the ocean surface (i.e., the tsunami source of the finite-fault solution).

3.4. Characteristics of the Initial Source Deformation

The three source deformations illustrated in Fig. 2, although derived from different methods, share some important features: all source models estimate up to 10 m of vertical displacement near the trench; all models infer the largest displacement to have occurred in the segment between latitude 37.5°N and 39°N; and all models indicate a similar east–west rupture width of about 100 km.

One of the main differences among the source models in Fig. 2a–c lies in the along-strike slip distribution. NOAA's tsunameter-derived source (Fig. 2a) suggests that the water surface along the trench between latitude 39°N and 41°N, at the northern end of the rupture, was uplifted by up to 6 m. This northern extent of the water surface disturbance was not evident in source models derived from seismic and/or GPS data alone. WEI et al. (2013) indicate that this northern disturbance was responsible for the high tsunami along Sanriku's coastline. After examination of nine source models, MACINNES et al. (2013) confirmed that such an additional source of tsunamigenic energy is needed to explain the high tsunami runup along Japan's east coast between latitude 39°N and 40°N. GUSMAN et al. (2012) inferred that this additional uplift originated from motion of a sedimentary wedge that also caused

the large tsunami associated with the 1896 Sanriku earthquake. A recent study by GRILLI et al. (2013a) investigated a possible seismically triggered seafloor failure in the same region, which could also explain this water surface disturbance. One can also see that the tsunameter-derived source estimates an initial subsidence of up to 3.5 m, which decreases westward to less than 0.5 m when reaching the coastline of Honshu (Fig. 2a). The GPS/tsunameter source shows similar results for the largest subsidence, with scattered uplift over the shelf and coastline. The finite-fault source, however, indicates a main subsidence between 38°N and 39.5°N, and a secondary uplifting crustal movement over the shelf offshore of Miyagi (i.e., fault slip and thus seafloor uplift extending farther downdip).

Another difference among the source models is that both the GPS/tsunameter source model and the finite-fault model show 10–15 m of slip along the updip portion of the southern rupture area, where the tsunameter-derived source model indicates no disturbance. This is probably due to the size of the unit sources in the tsunameter model (100 km × 50 km), which usually filters out negligible water surface disturbances when they are linearly combined for an inversion process. The small subfaults employed in the GPS/tsunameter and finite-fault methods are useful for resolving detailed slip distributions of earthquake rupture. We note though that the increase in degrees of freedom resulting from small subfaults adds more uncertainty to the inversion solution, and thus may delay the dissemination of a valid tsunami warning and forecast. HAYES (2011) discussed that additional constraints, such as a priori fault geometry, and bounds to rupture velocity and peak slip, can speed up finite-fault inversions.

It is worth noting that, in the three source models that we compare in this study, the number and azimuthal distribution of observing stations are very different. The tsunameters are sparsely, but optimally, placed in the Pacific Ocean. During an event, the tsunameter inversion process starts after the earliest observation (most likely at the closest station to the source), such as 21418 in Fig. 1. However, measurements from a second tsunameter located off the main focus of tsunami energy, such as 21401, will improve the inversion results. Ideally, three tsunameters encompassing the main focus of the tsunami energy (such as 21418, 21401, and 21413) provide a robust estimate of the tsunami source. Tang et al. (2012) compared the inversion results of the tsunami source (used in this study) based on one (21418), two (21418 and 201401), and three tsunameters (21418, 21401, and 21413). They showed that the inversion using two tsunameters improved the tsunami energy estimate by 20 % from the inversion using only one tsunameter, while little more information was gained (6 %) from the inversion using three tsunameters. For the joint GPS/tsunameter solutions, resolution in the portion that most affects tsunamis is poorly constrained by GPS alone, given that the nearest stations are approximately 200 km from the updip edge of the slip models, the region most responsible for tsunami generation. In inversions tested using only GPS data we found similar downdip slip values; however, the more distal updip region had a maximum of only 33 m of thrust motion. Determining the true offshore patch-size resolution of this model is difficult because seafloor uplift estimates use predictions reconstructed from tsunami-derived interface slip models. Thus, for these results the farthest offshore points will have resolution patches comparable in size to the surface expression of the tsunami slip patches (100 km along-strike and about 50 km downdip). For the seismic solution, data are selected such that they are evenly distributed azimuthally around the source, and thus provide an optimal configuration for minimizing azimuthal bias on the model solution, though observations are inherently farther from the source because of the teleseismic nature of the inversion. Hayes (2011) discussed resolution tests of the seismic models; we refer readers there for more detail.

4. Model Results

The posttsunami survey by Japanese scientists provided data of tsunami runup (or inundation depth) and inundation limits at nearly 3,000 locations along the Pacific coastline of Honshu and Hokkaido (Mori et al., 2011; Fig. 2). The largest tsunami runup they identified was 39.7 m in the Omoe Aneyoshi District

of Miyako, Iwate Prefecture (Mori et al., 2011). The latitudinal distribution of tsunami runup and inundation depth shows that high runup ($\geq$20 m) was mostly focused along the coastline between 39°N and 40°N, where many narrow valleys are found that may have funneled the tsunami waves to much higher elevations. Runup height along the coastline changes abruptly from greater than 30 m to less than 10 m, at 40.2°N in the north and 36.8°N in the south, which approximately corresponds to the northern and southern extent of the high-uplift zone.

4.1. Tsunami Runup Height and Inundation Depth along Japan's East Coast

Wei et al. (2013) and MacInnes et al. (2013) both pointed out the importance of using high-resolution model grids to reproduce the high runup between 39°N and 40°N. MacInnes et al. (2013) showed that a grid resolution of 30 arc s ($\sim$900 to 1,000 m) was not enough to resolve the detailed coastal features, and led to an underestimation (by 10 to 20 m) of the runup height and inundation depth north of 39°N. Therefore, we have applied the same tsunami inundation models as in Wei et al. (2013) with grid resolution of 2 arc s ($\sim$60 m) for all three source models. All models reproduce the smaller runup well, but show differences in matching the high runup values, especially between 39°N and 40°N (Fig. 2e–g). The runup computed from the tsunameter-derived source model fits well with the measurements for both high and low runup, and also reproduces the sharp changes in runup height at the ends of the rupture well (Fig. 2a). The tsunameter-derived source (Fig. 2e) can also successfully reproduce the highest runup of 39.7 m at Omoe Aneyoshi (39.5337°N, 142.046°E), though the model slightly underestimates the runup heights and inundation depths along the coastline of Sendai (between 38°N and 38.3°N). Compared with the tsunameter-derived source model, the GPS/tsunameter-derived model reproduces observations better in the low-runup segments, but underestimates runup heights by almost 10 m in the high-runup area between 39.5°N and 40°N (Fig. 2f). This model also matches the tsunami measurements in the Sendai Plain well. The updated USGS finite-fault source model also reproduces low runup well, but significantly underestimates the high-runup zone between 39°N and 40°N (Fig. 2f). This indicates an inadequate tsunami source representation in the northern portion of the model. Despite issues in matching the highest runup in the north, the overall distributions of the modeled runup heights along the coastline from all three models are encouraging, especially for the two source models that used the tsunameter measurements (Fig. 2a, b).

4.2. Deep-Ocean Propagation

The time series recorded by the tsunameters are used not only for quick constraint of the tsunami source, but also for providing validation of the model forecast for distant coastlines in real time. The model–data comparisons provided in Fig. 3 show that all source models were able to achieve excellent matches to tsunami arrival times at stations D21418 and D21413. The leading peak wave computed from the GPS/tsunameter source and the finite-fault solution are slightly out of phase with the observations at stations D21401 and D21419. The GPS/tsunameter source suggests an arrival time approximately 3 min earlier than observed, and the finite-fault solution implies an arrival 5 min earlier than observed. These phase mismatches indicate a range of 10 to 20 km error (less than 5 % error with respect to a 400-km rupture length) in prediction of the high-slip-patch location from different source models. The excellent agreement with the time series at all four tsunameters shows the credibility of the tsunameter-derived source (Fig. 3a), beyond the two stations used to produce the model. Table 2 indicates that the tsunameter-derived source produces a model accuracy of 91.6 % in predicting the maximum tsunami wave amplitude at the four deep-ocean tsunameters. The GPS/tsunameter source model also shows very good agreement between the model and measured time series, with the exception of a 41 % underestimate of the first peak recorded at station D21418 (Fig. 3b). This source has a model accuracy of 76.8 % in predicting the maximum tsunami wave amplitude at the tsunameters (Table 2). When comparing the computed maximum tsunami wave amplitudes in the northwestern Pacific Ocean, it is clear that station D21418 is at a location that receives higher tsunami energy from the tsunameter-derived

source model (Fig. 3a) than from the GPS/tsunameter-derived source model (Fig. 3b). Again, we attribute this difference to the higher uplift in the northern portion of the rupture in the tsunameter-derived source model than was present in the other two. The seismic source model (which did not include any tsunami data in its inversion) also provides excellent estimations of the first peak at all four tsunameters, as indicated by the model accuracy of 91.6 % for the maximum tsunami wave amplitude (Table 2). This source model implies a second positive pulse that does not match the observed depression after the first peak, which is likely caused by the secondary uplift to the east of the main rupture area (Fig. 3c). The corresponding maximum wave amplitude map indicates that this source model produces the least tsunami energy among the three, whereas the tsunameter-derived source produces the most.

To better quantify the model results, we compared the wave spectrum of the modeled time series against the observations at all tsunameters. Figure 4 shows that all source models give reasonable prediction of the observations in the frequency domain. At tsunameters D21401, D21419, and D21413, all source models produce results concordant with the observations for wave periods up to 40 min. Wave periods greater than an hour registered at these deep-ocean tsunameters may belong to the background noise rather than the tsunami signatures themselves (ZHOU et al., 2014). It is worth noting that all modeling results provide good agreement at high frequencies (wave period between 1 and 10 min), suggesting that the modeling results also predict the trailing waves well. Modeling results from the tsunameter-derived source show the best agreement with the spectral energy at wave periods between 10 and 40 min, whereas the other two sources slightly underestimate for wave periods of 20 min and greater. At station D21418, the tsunameter-derived source presents better agreement for wave periods less than 10 min relative to the other two source models.

One would expect good agreement between the computed and observed time series at tsunameters D21418 and D21401, as they were used to directly obtain the tsunami source. Merely retrieving the various time series used in the joint inversion is certainly not proof of better forecasting ability of the tsunameter method. We therefore compared the computed results at other tsunameters (e.g., 21413

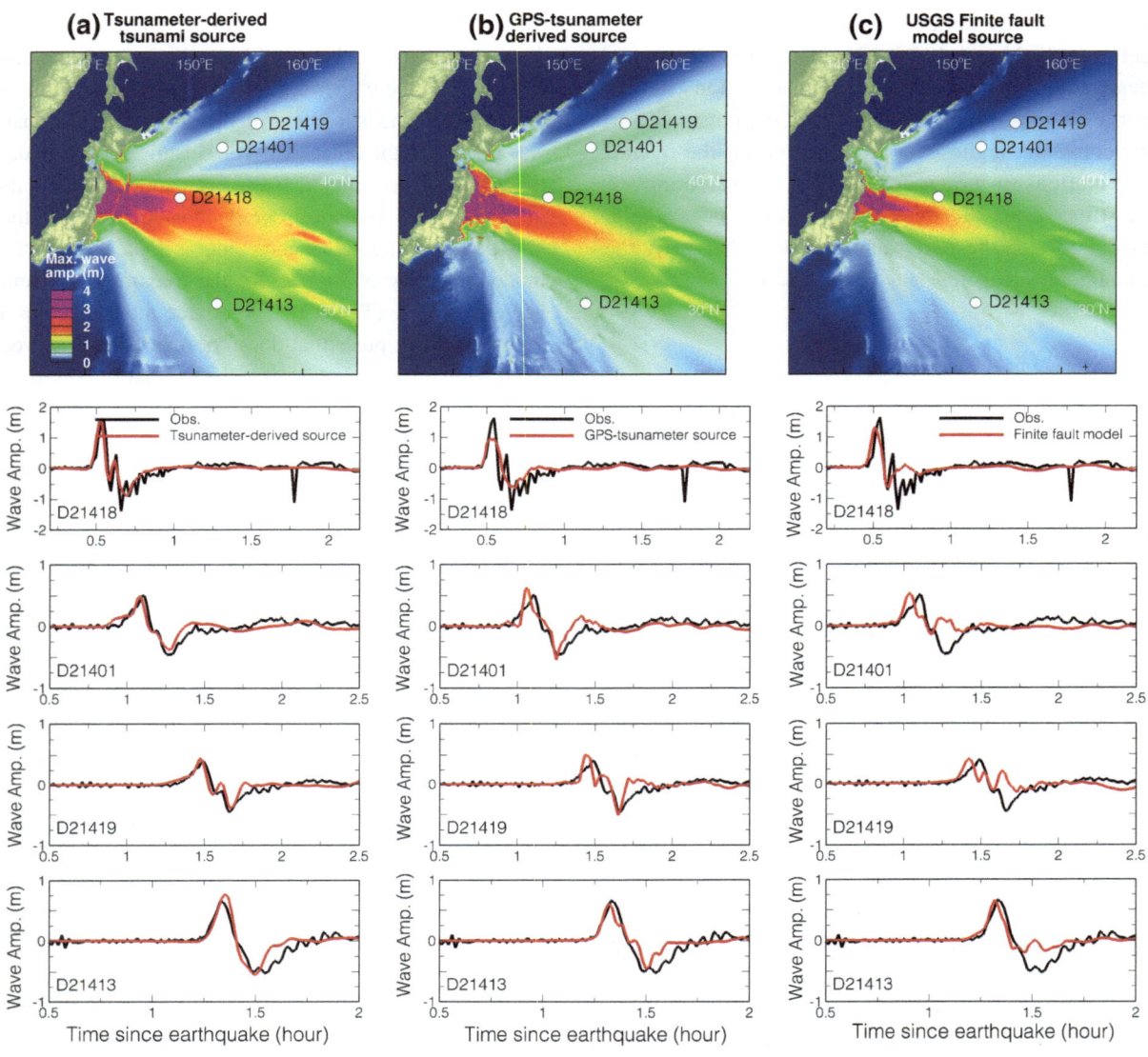

Figure 3

Model–observation comparison of 11 March 2011 Japan tsunami waveforms at deep-ocean tsunameters. **a** *Upper panel* maximum tsunami amplitude obtained using the tsunameter-derived source; *lower panel* comparison between computed and observed tsunami waveforms, where model results are obtained using the tsunameter-derived source. **b** *Upper panel* maximum tsunami amplitude obtained using the GPS/ tsunameter-derived source; *lower panel* comparison between computed and observed tsunami waveforms, where model results are obtained using the GPS/tsunameter-derived source. **c** *Upper panel* maximum tsunami amplitude obtained using the finite-fault model source; *lower panel* comparison between computed and observed tsunami waveforms, where model results are obtained using the finite-fault model source

and 21419 in Fig. 3) to verify the source derivation based on tsunameters D21418 and D21401. Tang *et al.* (2012) presented comparisons at 30 tsunameters throughout the Pacific Ocean. In this study, we provide further evidence of the successful performance of this method using model-independent results such as nearshore measurements (Sect. 4.3), runup (Sect. 4.1), and inundation (Sect. 4.4).

4.3. Nearshore Transformation of the Tsunami Waves

The GPS buoys deployed by the Japan Nation-wide Ocean Wave information network for Ports and Harbors (NOWPHAS) (Kato *et al.*, 2000, 2008) provided real-time measurements of tsunami propagation over the continental shelf between the Japan

Table 2

Comparison of maximum tsunami wave amplitude between models and observations at deep-ocean tsunameters and nearshore Japan Nationwide Ocean Wave information network for Ports and Harbors GPS buoys

	Station	Observed max. tsunami amp., A_{obs} (m)	Modeled max. tsunami amp. (m) and E (%)		
			Tsunameter-derived source	GPS/tsunameter source	Seismic source
Deep ocean	D21418	1.63	1.57 (−3.7 %)	0.96 (−41.1 %)	1.28 (−21.0 %)
	D21401	0.50	0.50 (0 %)	0.61 (+22.0 %)	0.53 (+6.0 %)
	D21419	0.40	0.44 (+10.0 %)	0.50 (+25.0 %)	0.42 (+5.0 %)
	D21413	0.65	0.78 (+20.0 %)	0.62 (−4.6 %)	0.66 (+1.5 %)
Model accuracy			91.6 %	76.8 %	91.6 %
Nearshore	GPS807	4.02	3.50 (−12.9 %)	2.92 (−27.4 %)	1.18 (−70.6 %)
	GPS804	6.30	5.42 (−14.0 %)	2.52 (−60.0 %)	1.68 (−73.3 %)
	GPS802	6.17	5.64 (−8.6 %)	3.07 (−50.2 %)	2.80 (−54.6 %)
	GPS803	5.18	5.28 (+1.9 %)	3.75 (−27.6 %)	7.80 (+50.6 %)
	GPS801	4.78	7.55 (+57.9 %)	6.17 (29.1 %)	7.80 (+63.2 %)
	GPS806	2.12	1.33 (−37.3 %)	2.71 (27.8 %)	1.82 (−14.2 %)
Model accuracy			77.9 %	63.0 %	45.6 %

The error E is calculated as $E = (\eta_m - \eta_{obs})/\eta_{obs} \times 100$ %, where η_{obs} is the observed maximum tsunami amplitude at the station and η_m is the computed maximum tsunami amplitude. The model accuracy is calculated as $1 - (\sum |E|)/n$, where n is the number of stations used for error calculation

Trench and the Honshu coastline. If one considers the tsunameter comparison as a validation of the tsunami propagation on the uplifted side of the source, then the NOWPHAS buoy (Fig. 2) measurements allow us to look closely at the tsunami wave dynamics on the subsided side of the tsunami source. Since all NOWPHAS buoys are deployed at water depths between 100 and 250 m, their measurements are not affected by the rugged coastline (which becomes particularly complex in Iwate). These buoy measurements contain direct information about the origin of the tsunami. The spatial coverage of these buoys from Iwate in the north to Miyagi in the south also offers a good calibration of the slip distribution along the length of the rupture. In Fig. 5, one can see that the pattern of GPS buoy observations from north to south is consistent with that shown by the runup height and inundation depth along the coast: large offshore waves (up to 6.3 m) corresponding to high tsunami runup were recorded between latitude 38°N and 40°N, and decayed further north and south. All three source models produce good estimates of the offshore waveforms, including the trailing waves, at the three NOWPHAS GPS buoys in the south (GPS buoys 803, 801, and 806) in terms of wave amplitude and period. The largest difference among the source model

predictions is seen for GPS buoys 804 and 802 located offshore of Iwate. The tsunameter-derived source model produces excellent agreement of the time series at these two buoys, especially the 6-m-high amplitude registered for the first wave. This source produces modeling results with small errors, about 14 % at GPS804 and 8.6 % at GPS buoy 802, in comparison with the observations. The GPS/tsunameter-derived source model gives a 50 % lower estimate for the first wave at GPS buoy 802, and a 60 % lower estimate for the first wave at GPS buoy 804. The seismic source model underestimates by about 73.3 and 54.6 % the first wave amplitude at GPS buoys 804 and 802, respectively, indicating that its tsunami source estimate in the north was not sufficient where a nonseismic seafloor failure might be the dominant tsunami generation mechanism (GRILLI *et al.*, 2013b). Table 2 shows that the average model accuracies (in terms of maximum tsunami wave amplitude) at all six GPS buoys are 77.9, 63.0, and 45.6 %, respectively, for the tsunameter-derived, GPS/tsunameter-derived, and seismic sources, respectively. Again, the initial water surface deformations in Fig. 2b–d clearly show the differences between the three source model predictions between 39°N and 40°N: (1) The tsunameter-derived

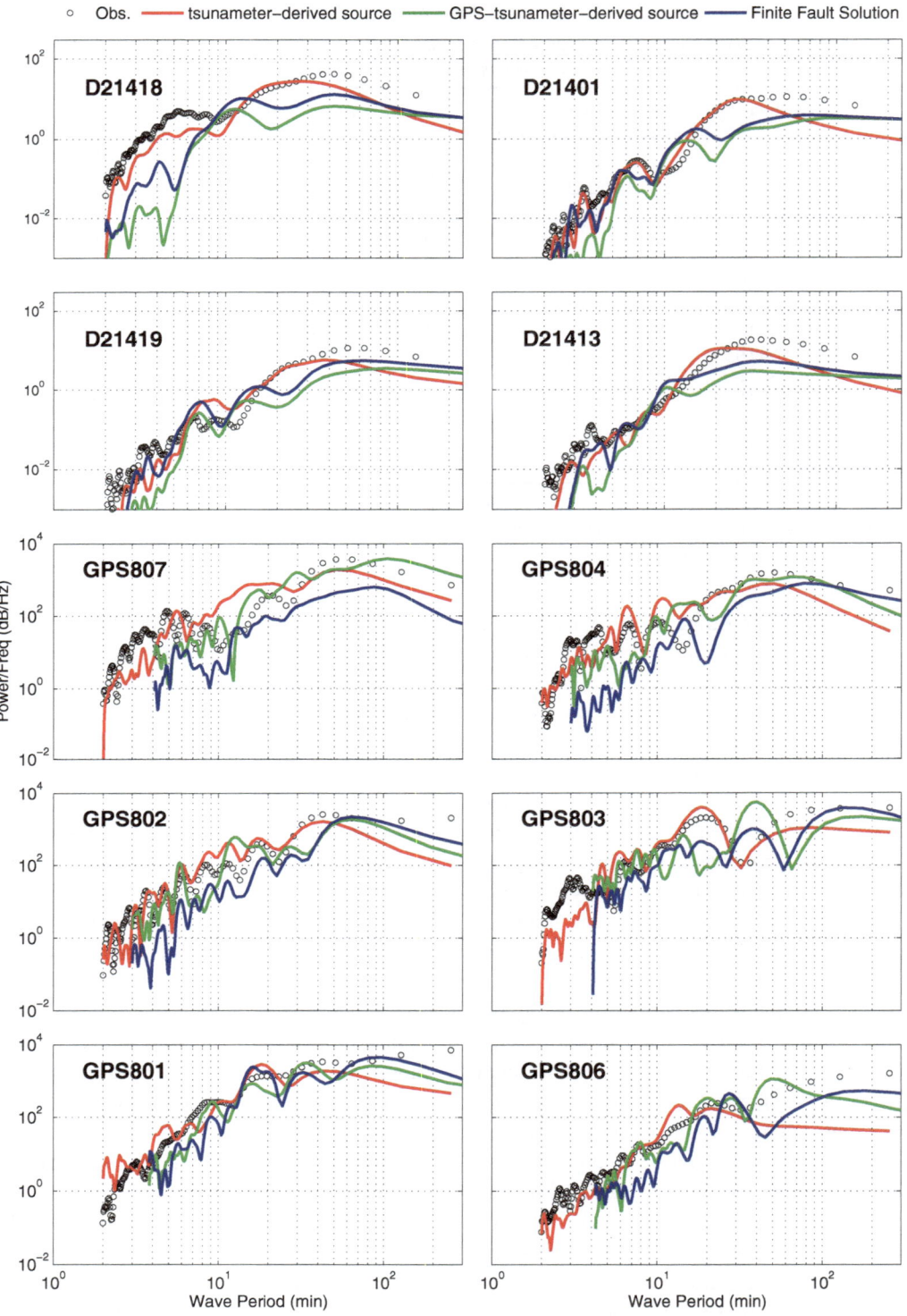

Figure 4
Spectral analysis of modeling results against observations at open-ocean tsunameters and nearshore GPS buoys

source indicates up to 5 m of initial surface deformation, whereas the other two source models predicted less than 1 m. (2) Similar to the tsunameter-derived source, the GPS/tsunameter-derived source predicts the greatest surface deformation—about 10 m (black color in Fig. 2b–d)—between 38.5°N and 39°N, whereas the seismic model's largest surface deformation is concentrated further south near 38°N. The GPS/tsunameter-derived source produces better predictions of the tsunami height at GPS buoys 807 and 802 compared with the seismic model. The underestimation by both of these models at GPS buoy 804 can probably be attributed to the underprediction of the surface deformation between 39°N and 40°N (Fig. 2c). Despite their different estimates of the amplitude of the first wave, all source models provide very good predictions of the trailing waves at all GPS buoys. The spectral analysis in Fig. 4 also shows that the high-frequency components (wave periods between 1 and 10 min) in the trailing waves are captured by the modeling results at all GPS buoys. This phenomenon probably indicates that these trailing waves were more dominated by wave interaction—such as wave shoaling, reflection, and refraction—with the local bathymetry and coastal features, rather than by initial source deformation. WEI et al. (2013) specifically pointed out that these complex wave interactions may not be accurately resolved by a simple semiempirical formulation or coarse grid resolution, and thus a high-resolution nonlinear tsunami inundation model is needed for accurate modeling forecasts. All source models produce very encouraging results for spectral energy, matching the observations (Fig. 4). The tsunameter-derived source and GPS/tsunameter-derived source give more similar results in terms of spectral energy than the seismic source. Although the seismic source slightly underestimates the spectral energy at GPS buoys 807 and 804, it provides good prediction at the other four GPS buoys. The tsunameter-derived source provides good estimation of the waves with period up to 40 min at all GPS buoys. However, it slightly underestimates the energy at wave periods greater than 1 h, especially at GPS buoy 806.

We note here that these forecast models can finish up to an 8-h simulation of tsunami inundation within 20 min, and this can also be significantly shortened to less than a couple of minutes with new computing technology, such as graphics processing unit (GPU) computing or parallel computing on a high-performance computer.

4.4. Tsunami Inundation Along Japan's East Coast

The tsunami inundation surveyed by Japanese scientists (MORI et al., 2011) provides additional validation for the reference source models. The most severe tsunami flooding occurred along the coastline between Sendai and Soma, and farther south to the Fukushima Daiich Nuclear Power Plant (NPP). Here we use the inundation model covering this section of Japan's coastline to illustrate differences among the three source models in predicting the tsunami flooding. As shown in Fig. 6, the inundation computations from all three source models provide very good estimation of the inundation extent along the coast between Soma and the Fukushima Daiich NPP. The GPS/tsunameter combined source model presents the largest inundation extent overall (Fig. 6b), especially in the southern part of Sendai (between latitudes 37.9°N and 38.1°N), where the tsunameter-derived source model slightly underestimates the inundated areas. The inundation distance computed using the seismic source model is the smallest among the three source models. The predictions of the maximum tsunami amplitude offshore from each model clearly correlate with the different estimates of tsunami inundation. In Fig. 2c, the GPS/tsunameter-derived source model produces the smallest uplift along the trench offshore of the Sendai Plain, but higher uplift offshore of the Soma area. This subtle difference leads to distinctly different offshore tsunami amplitude predictions along the Sendai Plain and Soma areas: smaller wave amplitude offshore of Sendai Plain, but larger wave amplitude offshore of Soma. It is worth noting that the GPS/tsunameter-derived source model produces better agreement with the observations from NOWPHAS buoy GPS801 (Fig. 5), implying that this source model may provide better estimates of the initial source deformation for the rupture segment offshore of the Sendai Plain.

To illustrate how the source models vary in predictions of the tsunami impact farther north along Japan's coast, Fig. 7 shows the computed inundations

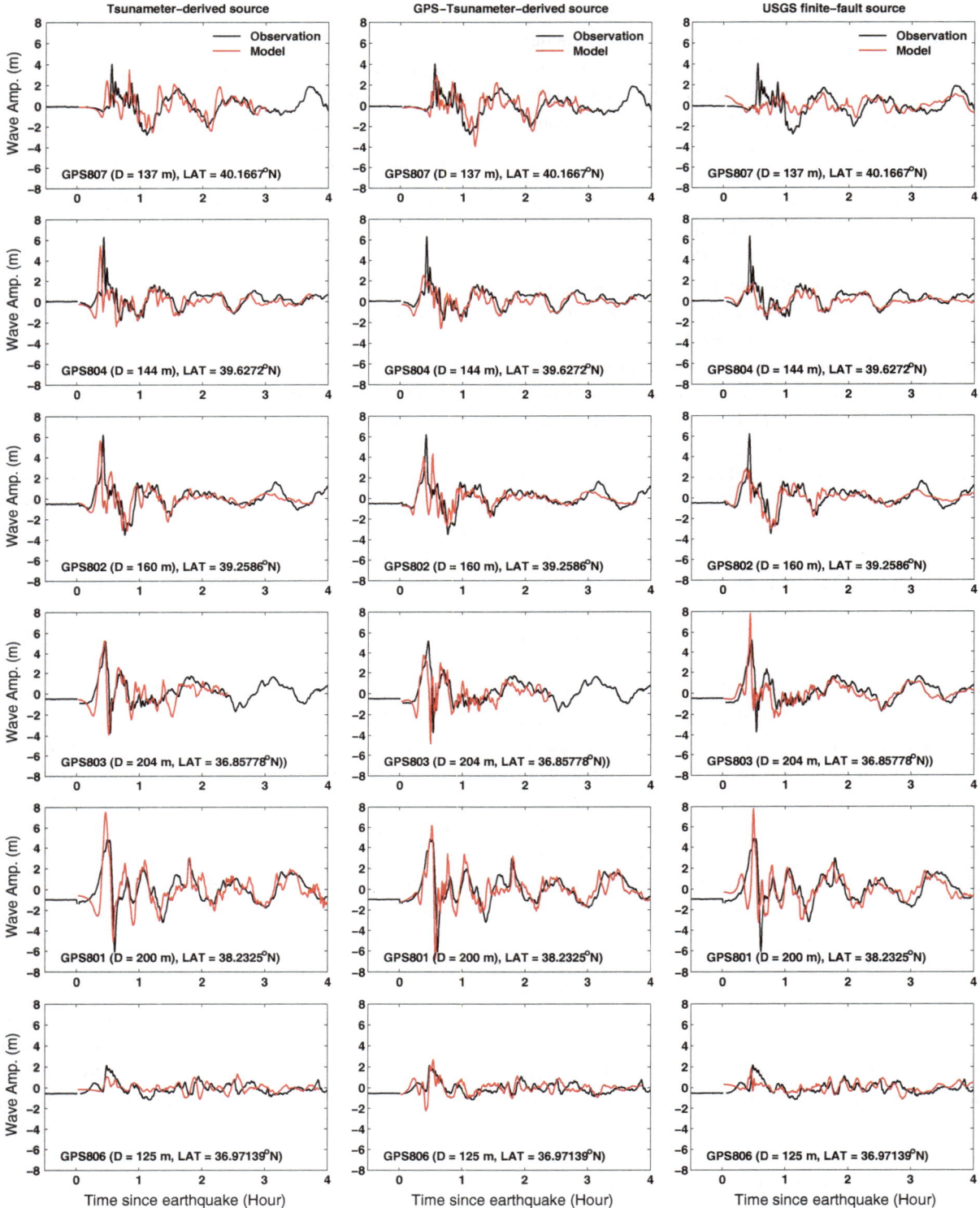

◄

Figure 5

Model–observation comparison of the 11 March 2011 Japan tsunami at NOWPHAS GPS buoys in order from north to south (Fig. 2). *Left panel* comparison between computed and observed tsunami inundation, where model results are obtained using the tsunameter-derived source. *Central panel* comparison between computed and observed tsunami inundation, where model results are obtained using the GPS/tsunameter-derived source. *Right panel* comparison between computed and observed tsunami inundation, where model results are obtained using the finite-fault model source

at three locations from north in Kuji and Noda (Fig. 7a–c) to south in Miyako (Fig. 7d–f) and in Ofunato and Rikuzentakata (Fig. 7g–i). At Rikuzentakata and Ofunato (Fig. 7g–i), all source models predict similar inundation extent. The maximum tsunami water level predicted by the tsunameter-derived source model is about 22 m at Rikuzentakata, whereas the GPS/tsunameter-derived source model and the seismic source model both predict maximum water levels of approximately 15 m

in the same area. These differences lead to contrasting predictions of the inundation extent in the Kesen River Valley. The tsunameter-derived source model provides the best match with the surveyed inundation limit, particularly in the valley west of Kesen River. Farther north at Miyako (Fig. 7d–f), it is clear that the seismic source model significantly underestimates the inundation extent, with only a 2.6 m maximum wave amplitude in the open ocean and up to 4 m in the bay area of Miyako. The tsunameter-derived source model predicts the highest maximum water level along Miyako's coastline of 30 m near cliffs to the east of Miyako, and 26 m along the coastline north of Miyako. Predictions from both the tsunami-derived and GPS/tsunameter source models show good agreement with the observed inundation line. Similar results are seen in the Kuji and Noda areas (Fig. 7a–c), where the seismic source model predicts the lowest estimates of both offshore and onshore water levels, and underestimates the inundation extent at

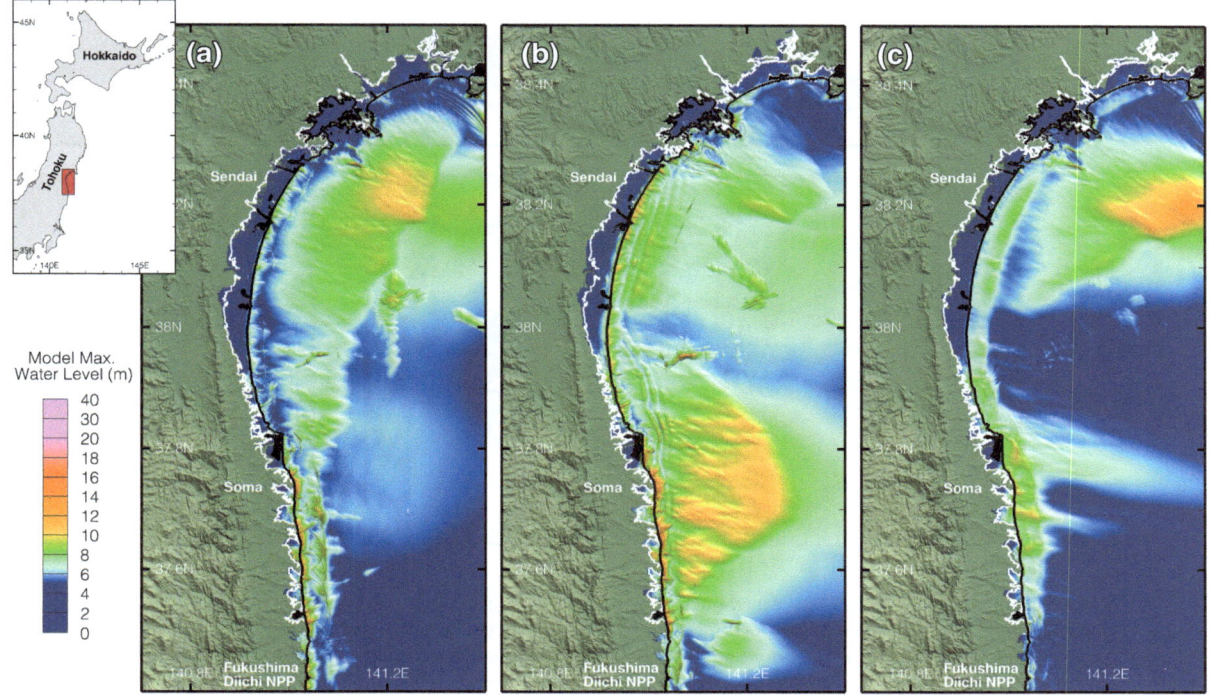

Figure 6

Model–observation comparison of 11 March 2011 Japan tsunami inundation in the areas of Sendai, Soma, and Fukushima, where the *thick white line* indicates the observed inundation limit, and the *color* represents the computed maximum water level. **a** Comparison between computed and observed tsunami inundation, where model results are obtained using the tsunameter-derived source. **b** Comparison between computed and observed tsunami inundation, where model results are obtained using the GPS/tsunameter-derived source. **c** Comparison between computed and observed tsunami inundation, where model results are obtained using the finite-fault model source

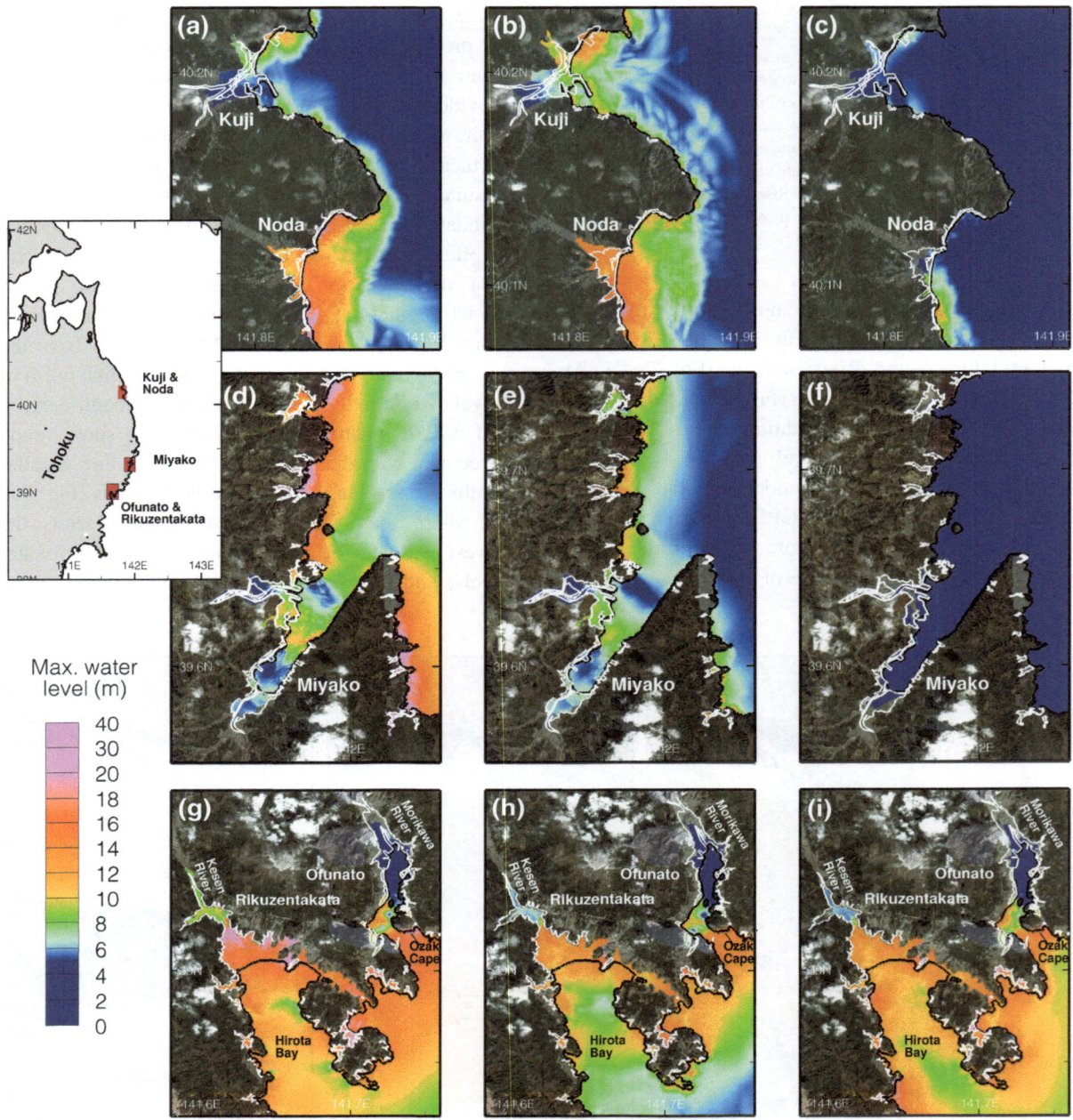

Figure 7

Model–observation comparison of 11 March 2011 Japan tsunami inundation along the coastline in the north of Tohoku Island, where the *thick white line* indicates the observed inundation limit, and the *color* represents the computed maximum water level. **a** Comparison between computed and observed tsunami inundation at Kuji and Noda for the tsunameter-derived source; **b** comparison between computed and observed tsunami inundation at Kuji and Noda for the GPS/tsunameter-derived source; **c** comparison between computed and observed tsunami inundation at Kuji and Noda for the USGS finite-fault solution; **d** comparison between computed and observed tsunami inundation at Miyako for the tsunameter-derived source; **e** comparison between computed and observed tsunami inundation at Miyako for the GPS/tsunameter-derived source; **f** comparison between computed and observed tsunami inundation at Miyako for the USGS finite-fault solution; **g** comparison between computed and observed tsunami inundation at Ofunato and Rikuzentakata for the tsunameter-derived source; **h** comparison between computed and observed tsunami inundation at Ofunato and Rikuzentakata for the GPS/tsunameter-derived source; **i** comparison between computed and observed tsunami inundation at Ofunato and Rikuzentakata for the USGS finite-fault solution

Table 3

Comparison of tsunami inundation accuracy measured by inundated area during the 11 March 2011 Japan tsunami

Prefecture	IA_m (km^2)	Tsunameter-derived source IA_t (km^2)	GPS/tsunameter source IA_g (km^2)	Finite-fault source IA_f (km^2)
Aomori (south of Rokkasho)	19	52	67	11
Iwate	58	58	79	71
Miyagi	327	345	372	174
Fukushima	112	131	155	234
Ibaraki (north of Kashima)	17	24	34	45
RMS (km^2)		19.1	37.2	88.7
L2 accuracy (%)		98.5	94.4	68.2

L2 accuracy = $\sum(I_m^2)/\sum(I_o^2) \times 100$ %, where I_m represents the model-computed inundated area for each prefecture, and I_o represents the observed inundated area in each prefecture

IA_m measured inundated area provided by the Geospatial Information Authority of Japan, IA_t computed inundated area based on tsunameter-derived source, IA_g computed inundated area based on GPS/tsunameter-derived source, IA_f computed inundated area based on USGS finite-fault solution, *RMS* root mean square

both locations. The GPS/tsunameter source model shows slightly higher maximum water levels than the tsunameter-derived source model, especially for off-shore wave amplitudes near Kuji Harbor. Wei *et al.* (2013) specifically discussed the model limitation at a 60-m grid resolution due to a lack of high-resolution topography for these shallow regions, and showed that the hilly bathymetry surrounding Noda was probably more dominant than the seawall in forming focused wave energy in the narrow valleys.

We note that the elastic earthquake deformation formula implemented in the MOST model is based on a rectangular fault model with a uniform slip distribution, and may not completely and precisely predict seafloor displacements from more complicated slip models. In addition, the transformation of the coseismic displacement into ambient water body deformation is one of the most understudied problems in tsunami science to date. The seismic source model also requires further transformation, from fault slip to seafloor deformation. These limitations may result in biased estimates of the tsunami source (or of the initial disturbed water surface). However, the theory of Kajiura (1970), of an instantaneous transfer of the vertical disturbance at the seafloor to the ocean surface, has been widely accepted and coupled with standard shallow-water models of tsunami wave dynamics, and has performed well in numerous other tsunami model simulations (e.g., Titov and Synolaski, 1998; Wang and Liu, 2007; Yamazaki *et al.*,

2011; MacInnes *et al.*, 2013). On the other hand, Grilli *et al.* (2013a) used a three-dimensional (3D) nonhydrostatic model to generate the initial tsunami source as a function of the space- and time-dependent seafloor deformation. They compared the subsequent model to one generated using the more typical approach of Kajiura (1970), specifying the maximum seafloor deformation as the initial free-surface deformation, in the tsunami model and found significant differences. Therefore, to avoid the uncertainties involved in converting a seismic source to a tsunami source, especially for the purpose of tsunami forecasts, deriving a tsunami source from direct tsunami wave measurements can significantly improve the efficiency and accuracy of tsunami forecasts by reducing the number of unknowns. It is worth noting that this tsunami source may not fully represent the physical processes (e.g., the earthquake rupture process) that are responsible for its generation.

Another interesting aspect of the inundation modeling arises from comparisons of the predicted total inundated area with survey results, which present an overall picture of how each source model performs. It is necessary to perform such an analysis region by region, to account for variations in the source model characteristics and coastal topographic features along different parts of Japan's coast. Table 3 presents the total computed inundated areas for each prefecture and comparison with observations provided by the Ministry of Land, Infrastructure,

Table 4

Time of earliest measurements made available for source estimation, first source estimation, and refined source estimation for all three models

Source model	Time of earliest measurements made available for source estimation	Time of first source estimation	Time of refined source estimation
Tsunameter-derived source	30 min	0.9 h	1.5 h
USGS finite-fault source	5–15 min	1.75 h	7 h
GPS/tsunameter source	Several hours[a]	Hindcast	

[a] While rapid "precise point position" solutions were used in this model, immediate real-time kinematic solutions would provide similar quality data and could become available before the earliest tsunami data

Transport, and Tourism (MLIT) of Japan (MLIT, 2011a, b). In a slightly different approach to WEI *et al.*, (2013), here we apply an L2-norm to better describe the model accuracy in each prefecture. Table 3 lists L2-norm accuracies of 98.5, 94.4, and 68.2 %, corresponding to root-mean-square errors of 19.1, 37.2, and 88.7 km^2, for the models derived from tsunameter data, GPS and tsunameter data, and seismic data, respectively. One can see that the tsunameter-derived source model and the GPS/tsunameter source model provide close estimates of the inundated area in all prefectures affected. The modeling results from these two source models also match the spatial change of the inundated area well, i.e., a maximum inundated area in Miyagi Prefecture, decreasing to the north and south. However, the inundated areas computed from the seismic source model overestimate the observations in Fukushima Prefecture, and underestimate those in Miyagi and Aomori (south of Rokkasho). These discrepancies, coupled with the above-mentioned results, indicate that the rapid USGS seismic source model could be further improved, and that including deep-ocean tsunami measurements in the inversion may lead to better prediction of the observed tsunami source. A similar approach is described in YAMAZAKI *et al.* (2011).

5. Discussion

Among the three source models described in this study, both the tsunameter-derived source and the finite-fault seismic model were rapidly created while the 2011 Japan tsunami was ongoing. The GPS/tsunameter source model was developed afterward

but using information that in the future could become available as soon as it is collected; tsunameter data are already relayed in real time, while a number of methods are available for obtaining real-time kinematic GPS solutions with cm-level accuracy relative to nearby, undeformed stations. The seismically defined finite-fault source model employed data from globally available seismic stations. As such, the solution is sensitive to heterogeneity in local velocity structure, and may be biased by uncertainties in the earthquake hypocenter; hence, the models may not be optimal for rapid tsunami forecasts when used alone—particularly in cases where nonseismic seafloor deformations (e.g., landslides or slumps) contribute to the tsunami.

Table 4 presents the temporal availability of the preliminary and refined source models relative to the onset of the 2011 Japan earthquake. While real-time kinematic GPS solutions could yield rapid coseismic rupture propagation and surface displacement fields, no such automated system was available at that time, nor was the implementation of codes for rapid GPS-based source inversions. Moving forward, GPS-based solutions should be shown to be robust, and preferably demonstrated in a large earthquake, before being considered as highly useful real-time tsunami warning tools.

Since the 2011 Japan tsunami, observation and modeling technologies have been improving in terms of the rapidity of solutions. Deep-ocean tsunameters can be deployed closer to the source region to detect tsunami signals quickly, in 10–20 min. Some rapid data products were made available with 2-h latency after the September 2012 Nicoya, Costa Rica earthquake (YUE *et al.*, 2013). Networks are being established to provide real-time kinematic GPS data for Cascadia (RABAK et al., 2010) and Southern California (LANGBEIN and BOCK, 2004; CROWELL et al., 2012). The problem with GPS networks remains that they are currently limited to onshore, and usually lack the necessary geometry to reproduce the true source with fidelity. Over the coming decade, teleseismic finite-fault models will likely remain the best first information about the rupture details of an earthquake. However, the inclusion of real-time GPS data, when geographically available, is invaluable to help constrain the lateral position and dip of a rupturing fault. Rapid deep-ocean tsunami measurements can help resolve the location, and particularly the near-trench extent, of seismically derived finite-fault models, which may be poorly resolved due to weak radiation of body waves. A comprehensive system that properly integrates seismic, tsunami, and geodetic measurements for the development of rapid earthquake rupture and tsunami excitation models would be a major improvement to providing timely and accurate tsunami forecasts for coastlines at risk from tsunami inundation. The development of robust tsunameter, geodetic, and seismic networks, communications, automated modeling, and decision-making procedures is essential.

6. Conclusions

Data collected from the 11 March 2011 earthquake and tsunami, including regional and global seismometers, GPS land stations, and deep-ocean tsunameters, are the earliest available observations useful for real-time or near-real-time tsunami source inversions. Measurements provided by these devices were used to estimate the earthquake and tsunami sources. In this study, we explore the near-field impact of the 11 March 2011 Japan tsunami via comparisons of three source models inferred from tsunameter measurements alone, from GPS and tsunami measurements combined, or from seismic waveforms. These source models are used as inputs to the MOST tsunami modeling software to simulate the generation, propagation, and inundation of the Japanese tsunami. Results permit comparative study to further understand the strengths and limitations of each modeling approach from the perspective of local tsunami predictions.

This study compares modeling results with deep-ocean and nearshore tsunami waveform observations, surveyed tsunami runup heights and inundation heights, and inundation limits and areas. Among the three source models, the tsunameter-derived real-time source, though an unexceptional characterization of the earthquake rupture geometry, presents the most complete agreement with tsunami observations, and in particular provides the best comparison with the high tsunami runup and inundation depths along the Sanriku Coast. The GPS/tsunameter and seismic source models perform well in the southern region (south of Sendai) of the earthquake rupture. The GPS/tsunameter source model underestimates by about 50 % the maximum runup of 30 to 40 m along the coastline between latitude 39°N and 40°N. The seismic source model predicts wave heights of just 10 m in the same location. A land-failure-induced disturbance in the northern rupture area, not captured by either the seismometer networks or the GPS stations, may explain the underestimation of tsunami runup in those models.

These model studies highlight the critical role of deep-ocean tsunami measurements—the earliest available tsunami data during the Japan event—for tsunami modeling and prediction. The source models generated using these measurements, such as the tsunameter-derived source model and the GPS/tsunameter-derived source model, have led to high-quality tsunami modeling results validated by real-time and posttsunami observations. Direct use of tsunami measurements in such source inversions reduces the ambiguity of the interaction between fault rupture and tsunami generation, and thus is the key for efficient and accurate tsunami forecasts. These direct tsunami measurements help to capture, in real time, the tsunami characteristics that are not directly related to the earthquake rupture and cannot be explained by available seismic data. Reducing the lag time between collection and availability and use of these measurements for near-field warning and inundation forecasts is an issue that should be addressed in the near future. GPS measurements, when made available more rapidly after an earthquake, and seismic models, when generated sooner without significantly sacrificing accuracy, would help reduce the lag time to create more effective assessment of the impending tsunami hazard in the near field. However, we suggest that these data be combined with tsunami measurements to produce the best estimates of tsunami impact along near-field coastlines. Recent deployment of deep-ocean tsunameters offshore of Japan, between the trench and the coastline, has shortened the earliest direct tsunami observations and detection times to less than 20 min, depending on source location.

With the exception of a very few underwater observation points (e.g., SATO *et al.*, 2011), GPS devices are predominantly restricted to land areas. International scientific communities are now stressing the need for improved underwater deformation monitoring methods and increased data collection and access. This study shows that deep-ocean tsunami measurements are highly complementary to onland GPS for estimating the earthquake rupture area, leading to more complete understanding of both the earthquake source properties and the tsunami generation process. In the coming years, combining such GPS and tsunami geodetic data with seismic data (particularly where GPS and tsunameter data are scarce) may lead to the best early estimates of earthquake rupture and tsunami-generating processes useful for both real-time forecasting as well as early post-event disaster relief.

We also stress that the short-term forecasting of tsunamis described in this study is different from the long-term assessment of tsunami hazards, which prepares a coastal community for maximal probable tsunami scenarios. A long-term tsunami hazard assessment may include predicting future hazards based on available knowledge or some assumptions of the physical origin of a tsunami. It can also include the identification of new physical processes responsible for generating large tsunamis, and forecasting future events based on such new findings. It is worth noting that the methods and tools developed for short-term forecasts, such as those discussed in this paper, are very useful for helping to improve long-term tsunami hazard assessments. Both approaches need to be carried out, continuously and interactively, to better mitigate future tsunamis.

Acknowledgments

The authors would like to thank William Barnhart of the U.S. Geological Survey and two anonymous reviewers for their thorough reviews and constructive remarks, which helped improve this manuscript. This publication is partially funded by the Joint Institute for the Study of the Atmosphere and Ocean (JISAO) under NOAA cooperative agreement NA100AR4320148; JISAO contribution 2137; PMEL contribution 4014. Any use of trade, product, or firm names is for descriptive purposes only and does not imply endorsement by the U.S. Government.

REFERENCES

ARCAS, D., and TITOV, V. (2006). *Sumatra tsunami: Lessons from modeling*, Surv. Geophys., *27*(6), doi:10.1007/s10712-006-9, 679–705.

BIRD, P. (2003). *An updated digital model of plate boundaries*, Geochem. Geophys. Geosys., *4*(3), 1027, doi:10.1029/2001GC000252.

BASSIN, C., LASKE, G. and MASTERS, G., (2000). *The current limits of resolution for surface wave tomography in North America*, EOS Trans. AGU, *81*, Fall Meet. Suppl., Abstract S12A-03.

BERNARD, E.N., TANG, L., WEI, Y., and TITOV, V.V. (2013). *Impact of near-field, deep-ocean tsunami observations on forecasting the December 7, 2012 Japanese tsunami*, submitted to this issue, in press (this issue), doi:10.1007/s00024-013-0720-8.

BLASER, L., KRÜUGER, F., OHRNBERGER, M. and SCHERBAUM, F., (2010). *Scaling relations of earthquake source parameter estimates with special focus on subduction environment*, Bull. Seismol. Soc. Am., *100*, 2914–2926.

BLEWITT, G., KREEMER, C., HAMMOND, W.C., PLAG, H.P., STEIN, S. and OKAL, E. (2006). *Rapid determination of earthquake magnitude using GPS for tsunami warning system*, Geophys. Res. Lett., *33*, L11309, doi:10.1029/2006GL026145.

BORRERO, J., BELL, R., CSATO, C., DELANGE, W., GREER, D., GORING, D., PICKETT, V. and POWER, W. (2013). *Observations, effects and real time assessment of the March 11, 2011 Tohoku-oki tsunami in New Zealand*, Pure Appl. Geophys., *170*, 1229–1248, doi:10.1007/s00024-012-0492-6.

BURWELL, D., TOLKOVA, E. and CHAWLA, A. (2007). *Diffusion and dispersion characterization of a numerical tsunami model*. Ocean Model., *19*(1–2), 10–30.

CHEN, T., NEWMAN, A.V., FENG, L., FRITZ, H.M. (2009). *Slip distribution from the 1 April 2007 Solomon Islands earthquake: a unique image of near-trench rupture*, Geophys. Res. Lett., *36*, L16307, doi:10.1029/2009GL039496.

CROWELL, B. BOCK, W.Y., MELGAR, D. (2012). *Real-time inversion of GPS data for finite fault modeling and rapid hazard assessment*, Geophys. Res. Lett., *39*(9), doi:10.1029/2012GL051318.

DUPUTEL, Z., RIVERA, L., KANAMORI, H., and HAYES, G. (2012). *W phase source inversion for moderate to large earthquakes (1990–2010)*, Geophys. J. Int. *189*, 1125–1147.

DZIEWONSKI, A.M. and ANDERSON, D.L., 1981. *Preliminary reference Earth model*, Phys. Earth Planet. Inter., *25*, 297–356.

FRITZ, H.M., BORRERO, J.C., SYNOLAKIS C.E., OKAL, E.A., WEISS, R., TITOV, V.V., JAFEE, B.E., FOTEINIS, S., LYNETT, P.K., CHAN, I-C., and LIU, P.L-F. (2011). *Insights on the 2009 South Pacific* tsunami in Samoa and Tonga from field surveys and numerical simulations, Earth-Sci. Rev., *107*, 66–75.

GICA, E., SPILLANE, M., TITOV, V.V., CHAMBERLIN, C. and NEWMAN, J.C. (2008). *Development of the forecast propagation database for NOAA's Short-term Inundation Forecast for Tsunamis (SIFT)*, NOAA Tech. Memo. OAR PMEL-139, pp 89.

GONZÁLEZ, F.I., BERNARD, E.N., MEINIG, C., EBLE, M., MOFJELD, H.O., and STALIN, S. (2005). *The NTHMP tsunameter network*, Nat. Hazards, *35*(1), Special Issue, U.S. National Tsunami Hazard Mitigation Program, 2005, 25–39.

GOVERNMENT ACCOUNTABILITY OFFICE (2006), "U.S. Tsunami Preparedness — Federal and state partners collaborate to help communities reduce potential impacts, but significant challenges remain," Report to Congressional Committees and Senator Dianne Feinstein, United States Government Accountability Office, pp 60.

GRILLI, S.T., HARRIS, J.C., KIRBY, J.T., SHI, F., and MA, G., MASTERLARK, T., TAPPIN, D.R. and TAJALI-BAKHSH, T.S. (2013a). Modeling of the Tohoku-Oki 2011 tsunami generation, far-field and coastal impact: A mixed co-seismic and SMF source. In Proc. 7th Intl. Conf. on Coastal Dynamics (Arcachon, France, June 2013) (ed. P. Bonneton), 749–758, paper 068.

GRILLI, S.T., HARRIS, J.C., TAJALIBAKHSH, T., MASTERLARK, T.L., KYRIAKOPOULOS, C., KIRBY, J.T. and SHI, F. (2013b). *Numerical simulation of the 2011 Tohoku tsunami based on a new transient FEM co-seismic source: comparison to far- and near-field observations*. Pure Appl. Geophys., *170*, 1333–1359.

GUSMAN, A.R., TANIOKA, Y., KOBAYASHI, T., LATIEF, H. and PANDOW, W. (2010). *Slip distribution of the 2007 Bengkulu earthquake inferred from tsunami waveforms and InSAR data*, J. Geophys. Res., *115*, B12316, doi:10.1029/2010JB007565.

GUSMAN, A.R., TANIOKA, Y., SAKAI, S., TSUSHIMA, H. (2012). *Source model of the great 2011 Tohoku earthquake estimated from tsunami waveforms and crustal deformation data*. Earth Planet. Sci. Lett., *341–344*, 234–242.

HAYES, G.P. (2011). *Rapid source characterization of the 2011 Mw 9.0 off the Pacific Coast of Tohoku earthquake*, Earth Planets Space, *63*, 529–534.

HAYES, G.P., EARLE, P.S., BENZ, H.M., WALD, D.J., BRIGGS, R.W., and the USGS/NEIC Earthquake Response Team (2011). *88 Hours: The U.S. Geological Survey National Earthquake Information Center response to the 11 March 2011 Mw 9.0 Tohoku earthquake*, Seis. Res. Lett. *82*(4), 481–493.

HAYES, G.P., WALD, D.J., JOHNSON, R.L. (2012). *Slab 1.0: A three-dimensional model of global subduction zone geometries*, J. Geophys. Res.: Solid Earth, *117*, B01302, doi:10.1029/2011JB008524.

HIRATA, K., SATAKI, K., TANIOKA, Y., KURAGANO, T., HASEGAWA, Y., HAYASHI, Y., and HAMADA, N. (2006). *The 2004 Indian Ocean tsunami: Tsunami source model from satellite altimetry*, Earth Planets Space, *58*, 195–201.

JI, C., WALD, D.J., and HELMBERGER, D.V. (2002). *Source description of the 1999 Hector Mine, California earthquake; part I: wavelet domain inversion theory and resolution analysis*, Bull. Seismol. Soc. Am., *92*(4), 1192–1207.

JI, C., HELMBERGER, D.V., WALD, D.J. and MA, K.-F., 2003. *Slip history and dynamic implications of the 1999 Chi–Chi, Taiwan, earthquake*, J. Geophys. Res., *108*, doi:10.1029/2002JB001764.

KAJIURA, K., (1970). *Tsunami source, energy and the directivity of wave radiation*, Bull. Earthq. Res. Inst., Univ. Tokyo, *48*, 835–869.

KATO, T., TERADA, Y., KINOSHITA, M., KAKIMOTO, H., ISSHIKI, H., MATSUISHI, M., YOKOYAMA, A., and TANNO, T. (2000). *Real-time observation of tsunami by RTK-GPS*, Earth Planets Space, *52*, 841–845.

KATO, T., TERADA, Y., NAGAI, T., SHIMIZU, K., TOMITA, T. and KOSHIMURA, S. (2008). Development of a new tsunami monitoring system using a GPS buoy, American Geophysical Union, Fall Meeting 2008, abstract #G43B-03.

LANGBEIN, J. and BOCK, Y (2004). *High-rate real-time GPS network at Parkfield: utility for detecting fault slip and seismic displacements*, Geophys. Res. Lett., *31*(15), doi:10.1029/2003GL019408.

LOVHOLT, F., PEDERSON, G. and GLIMSDAL, S. (2010). *Coupling of dispersive tsunami propagation and shallow water coastal response*. Open Oceanogr. J., *4*, 71–82.

MACINNES, B.T., GUSMAN, A.R., LEVEQUE, R.J., and TANIOKA, Y. (2013). *Comparison of earthquake source models for the 2011 Tohoku-oki event using tsunami simulations and near field observations*. Bull. Seismol. Soc. Am., *103*, no. 2B, 1256–1274, doi:10.1785/0120120121.

Ministry of Land, Infrastructure, Transport, and Tourism of Japan (2011a). Status survey reports of the East Japan Earthquake, Ministry of Land, Infrastructure, Transport, and Tourism Press Release (first report), pp 16.

Ministry of Land, Infrastructure, Transport, and Tourism of Japan (2011b). The Great East Japan Earthquake (107th report): Outline, Ministry of Land, Infrastructure, Transport, and Tourism, pp 1.

MORI, N., TAKAHASHI, T., YASUDA, T.and YANAGISAWA, H. (2011). *Survey of 2011 Tohoku earthquake tsunami inundation and run-up*, Geophys. Res. Lett., *38*, L00G14, doi:10.1029/2011GL049210.

NEWMAN, A.V. (2011). *Hidden depth*, Nature, *474*, 441–443.

NEWMAN, A.V., HAYES, G., WEI, Y. and CONVERS, J. (2011). *The 25 October 2010 Mentawai tsunami earthquake, from real-time discriminants, finite-fault rupture, and tsunami excitation*. Geophys. Res. Lett., *38*(5), L05302, doi:10.1029/2010GL046498.

OHTA, Y., KOBAYASHI, T., TSUSHIMA, H., MIURA, S., HINO, R., TAKASU, T., FUJIMOTO, H., IINUMA, T., TACHIBANA, K., DEMACHI, T., SATO, T., OHZONO, M., and UMINO, N. (2012), *Quasi real-time fault model estimation for near-field tsunami forecasting based on RTK-GPS analysis: application to the 2011 Tohoku-Oki earthquake (Mw 9.0)*, J. Geophys. Res. *117*, B02311, doi:10.1029/2011JB008750.

OKAL, E.A. and SYNOLAKIS, C.E. (2004). *Source discriminants for near-field tsunamis*, Geophys. J. Int., *158*, 899–912, doi:10.1111/j/1365-246X.2004.02347.x.

OKADA, M. (1985). *Surface deformation due to shear and tensile faults in a half-space*. Bull. Seismol. Soc. Am., *75*(4), 1135–1154.

OWEN, S.E., WEBB, F., SIMONS, M., ROSE, P.A., CRUZ, J., YUN, S., FIELDING, E.J., MOORE, A.W., HUA, H., and AGRAM, P.S. (2011). The ARIA-EQ project: Advanced Rapid Imaging and Analysis for Earthquakes, *American Geophysical Union*, Fall Meeting 2011, abstract #1N11B-1298.

OZAKI, T. (2012). *JMA's tsunami warning for the 2011 great Tohoku earthquake and tsunami warning improvement plan*. J. Disaster Res., *7*(7), 439–445.

PERCIVAL, D.B., DENBO, D.W., EBLE, M.C., GICA, E., MOFJELD, H.O., SPILLANE, M.C., TANG, L., and TITOV, V.V. (2010). *Extraction of tsunami source coefficients via inversion of DART® buoy data*, Nat. Hazards, doi:10.1007/s11069-010-9688-1.

PIETRZEK, J., SOCQUET, A., HAM, D., SIMONS, W., VIGNY, C., LABEUR, R.J., SCHRAMA, E., STELLING, G., and VATVANI, D. (2007). *Defining the source region of the Indian Ocean tsunami from GPS, altimeters, tide gauges and tsunami models*, Earth Planet. Sci. Lett., *261*(1–2), 49–64.

RABAK, I., MELBOURNE, T. I., SANTILLAN, M., SCRIVNER, C.W., KINKAID, K., STAHL, R., Rapid assessment and mitigation of Cascadia earthquakes using the combined PANGA and PBO real-time GPS networks, *Abstract G11C-04, presented at the 2010 Fall Meeting, AGU, San Francisco, Calif.*, 2010.

RIVERA, L.A., KANAMORI, H., and DUPUTEL., Z. (2011). W phase source inversion using high-rate regional GPS data of the 2011 Tohoku-oki earthquake, *Abstract G33C-04 presented at the 2011 Fall Meeting, AGU, San Francisco, Calif.*, 5–9 Dec.

SATAKE, K. (1987). *Inversion of tsunami waveforms for the estimation of a fault heterogeneity: Methods and numerical experiments*, J. Phys. Earth, *35*, 241–254.

SATAKE, K. and KANAMORI, H. (1991). *Use of tsunami waveforms for earthquake source study*, Nat. Hazards, *4*, 193–208.

SATO, M., ISHIKAWA, T., UJIHARA, N., YOSHIDA, S., FUJITA, M., MOCHIZUKI, M., and ASADA, A. (2011). *Displacement above the hypocenter of the 2011 Tohoku-Oki earthquake*, Science, *332*(6036), 1395, doi:10.1126/science.1207401.

SIMONS, M., MINSON, S.E., SLADEN, A., ORTEGA, F., JIANG, J., OWEN, S.E., MENG, L., J-P. AMPUERO, S. WEI, R. CHU, D.V. HELMBERGER, H. KANAMORI, E. HETLAND, A.W. MOORE, and F.H. WEBB (2011), *The 2011 magnitude 9.0 Tohoku-Oki earthquake: Mosaicking the megathrust from seconds to centuries*, Science, *332*(6036), 2011, pp. 1421–1425.

SONG, Y.T., FUKUMORI, I., SHUM, C.K., and YI, Y. (2012). *Merging tsunamis of the 2011 Tohoku-Oki earthquake detected over the open ocean*. Geophys. Res. Lett., *39*, L05606, doi:10.1029/2011GL050767.

SPILLANE, M.C., GICA, E., TITOV, V.V., and MOFJELD, H.O. (2008) *Tsunameter network design for the U.S. DART® arrays in the Pacific and Atlantic Oceans*, NOAA Tech. Memo. OAR PMEL-143, pp 165.

SYNOLAKIS, C.E., BERNARD, E.N., TITOV, V.V., KÂNOĞLU, U., and GONZÁLEZ, F.I. (2008). *Validation and verification of tsunami numerical models*, Pure Appl. Geophys., *165*(11–12), 2197–2228.

TANG, L., TITOV, V.V., BERNARD, E.N., WEI, Y., CHAMBERLIN, C., NEWMAN, J.C., MOFJELD, H., ARCAS, D., EBLE, M., MOORE, C., USLU, B., PELLS, C., SPILLANE, M.C., WRIGHT, L.M., and GICA, E. (2012). *Direct energy estimation of the 2011 Japan tsunami using deep-ocean pressure measurements*. J. Geophys. Res., *117*, C08008, doi:10.1029/2011JC007635.

TANG, L., TITOV, V.V., and CHAMBERLIN, C.D. (2009). *Development, testing, and applications of site-specific tsunami inundation models for real-time forecasting*, J. Geophys. Res., *114*, C12025, doi:10.1029/2009JC005476.

TANG, L., TITOV, V.V., WEI, Y., MOFJELD, H.O., SPILLANE, M., ARCAS, D., BERNARD, E.N., CHAMBERLIN, C., GICA, E., and NEWMAN, J. (2008). *Tsunami forecast analysis for the May 2006 Tonga tsunami*, J. Geophys. Res., *113*, C12015, doi:10.1029/2008JC004922.

TITOV, V.V. (2009). Tsunami forecasting, Chapter 12 in *The Sea, Volume 15: Tsunamis*, Harvard University Press, Cambridge, MA and London, England, 371–400.

TITOV, V., and GONZÁLEZ, F.I. (1997). Implementation and testing of the Method of Splitting Tsunami (MOST) model, *NOAA Tech. Memo. ERL PMEL-112 (PB98-122773)*, NOAA/Pacific Marine Environmental Laboratory, Seattle, WA, 11 pp.

TITOV, V.V., GONZÁLEZ, F.I., BERNARD, E.N., EBLE, M.C., MOFJELD, H.O., NEWMAN, J.C., and VENTURATO, A.J. (2005). *Real-time tsunami forecasting: challenges and solutions*, Nat. Hazards, *35*(1), Special Issue, U.S. National Tsunami Hazard Mitigation Program, 41–58.

TITOV, V.V., and SYNOLAKIS, C.E. (1998). *Numerical modeling of tidal wave runup*, J. Waterw. Port Coastal Ocean Eng., *124*(4), 157–171.

TSUSHIMA, H., HINO, R., FUJIMOTO, H., TANIOKA, Y., and IMAMURA, F. (2009). *Near-field tsunami forecasting from cabled ocean bottom pressure data*, J. Geophys. Res., *114*, B06309, doi:10.1029/2008JB005988.

VIGNY, C., SOCQUET, A., PEYRAT, S., RUEGG, J.C., MÉTOIS, M., MADARIAGA, R., MORVAN, S., LANCIERI, M., LACASSIN, R., CAMPOS, J., CARRIZO, D., BEJAR-PIZARRO, M., BARRIENTOS, S., ARMIJO, R., ARANDA, C., VALDERAS-BERMEJO, M.C., ORTEGA, I., BONDOUX, F., BAIZE, S., LYON-CAEN, H., PAVEZ, A., VILOTTE, J.P., BEVIS, M., BROOKS, B., SMALLEY, R., PARRA, H., BAEZ, J.C., BLANCO, M., CIMBARO, S., KENDRICK, E. (2011). *The 2010 Mw 8.8 Maule megathrust earthquake of central Chile, monitored by GPS*, Science, 332, 1417–1421.

WANG, X. and LIU, P.L.-F. (2007), *Numerical simulations of the 2004 Indian Ocean tsunamis—coastal effects*, J. Earthq. Tsunami, *1*(3), 273, doi:10.1142/S179343110700016X.

WEI, Y., BERNARD, E., TANG, L., WEISS, R., TITOV, V.V., MOORE, C., SPILLANE, M., HOPKINS, M.,.and KÂNOĞLU, U. (2008). *Real-time experimental forecast of the Peruvian tsunami of August 2007 for U.S. coastlines*, Geophys. Res. Lett., *35*, L04609, doi:10.1029/2007GL032250.

WEI, Y., CHAMBERLIN, C., TITOV, V.V., TANG, L. and BERNARD, E.N. (2013). *Modeling of the 2011 Japan tsunami: Lessons for near-field forecast*, Pure Appl. Geophys., *170*, 1309–1331.

WEI, Y., CHEUNG, K.F., CURTIS, G.D., and MCCREERY, C.S. (2003). *Inversion algorithm for tsunami forecast*, J. Waterw. Port Coastal Ocean Eng., *129*, 2003, pp. 60–69.

WHITMORE, P.M. (2009). Tsunami warning systems, Chapter 13 in *The Sea, Volume 15: Tsunamis*, Harvard University Press, Cambridge, MA and London, England, 401–442.

YAMAZAKI, Y., LAY, T., CHEUNG, K.F., YUE, H., and KANAMORI, H. (2011). *Modeling near-field tsunami observations to improve finite-fault slip models for the 11 March 2011 Tohoku earthquake*. Geophys. Res. Lett., *38*(7), doi:10.1029/2011GL049130.

YOKOTA, Y., KOKETSU, K., FUJII, Y., SATAKE, K., SAKAO, S., SHINOHARA, M., and KANAZAWA, T. (2011). *Joint inversion of strong motion, teleseismic, geodetic, and tsunami datasets for the rupture process of the 2011 Tohoku earthquake*, Geophys. Res. Lett., *38*, L00G21, doi:10.1029/2011GL050098.

YUE, H., LAY, T., SCHWARTZ, S. Y., RIVERA, J., PROTTI, M., DIXON, T., OWEN, S., NEWMAN, A. (2013). *The 5 September 2012 Nicoya, Costa Rica Mw 7.6 earthquake rupture process from joint inversion of high-rate GPS, strong-motion, and teleseismic P wave data and its relationship to adjacent plate boundary interface properties*, J. Geophys. Res. Solid Earth, *118*(10), 5453–5466.

ZHOU, H., MOORE, C., WEI, Y.and TITOV, V.V. (2011). *A nested-grid Boussinesq-type approach to modeling dispersive propagation* and runup of landslide-generated tsunami. Nat. Hazards Earth Syst. Sci., *11*(10), 2677–2697, doi:10.5194/nhess-11-2677-2011.

ZHOU, H., WEI, Y., WRIGHT, L., and TITOV, V.V (2014). *Waves and currents in Hawaii waters induced by the dispersive 2011 Tohoku tsunami.* Pure Appl. Geophys., doi:10.1007/s00024-014-0781-3

ZHOU, H., WEI, Y., and TITOV, V.V. (2012). *Dispersive modeling of the 2009 Samoa tsunami.* Geophys. Res. Lett., *39*(16), L16603, doi:10.1029/2012GL053068.

(Received September 28, 2013, revised January 7, 2014, accepted January 9, 2014, Published online April 19, 2014)

Pure Appl. Geophys. 171 (2014), 3307–3328
© 2014 Springer Basel
DOI 10.1007/s00024-014-0794-y

| Pure and Applied Geophysics

The Tohoku Tsunami of 11 March 2011: The Key Event to Understanding Tsunami Sedimentation on the Coasts of Closed Bays of the Lesser Kuril Islands

N. G. Razjigaeva,[1] L. A. Ganzey,[1] T. A. Grebennikova,[1] E. D. Ivanova,[1] A. A. Kharlamov,[2] V. M. Kaistrenko,[3] Kh. A. Arslanov,[4] and S. B. Chernov[4]

Abstract—The Tohoku tsunami of 11 March 2011 manifested in the region of the South Kuril Islands, although, as a rule, the run-up heights in this region did not exceed 3 m. In closed bays that were covered with ice before the tsunami, the eroding capacity of tsunami waves was aggravated by the ice fragments they carried. Here, mud sheets formed, reaching up to 106 m inland. The 2012 studies have shown well-preserved tsunami deposits, evident 1.5 years after the event. A comparative analysis of tsunami deposits from the periphery and from the near-field area close to the tsunami source was performed; this was important for understanding the deposition mechanism during the event, as it had different strengths on different shores. The difference in run-up heights determined the considerable differences in erosion, sedimentation, distribution of tsunami deposits, the formation of sedimentary structures, grain-size composition, and diatom and foraminifera assemblages. The sources of the material also varied significantly from each other: the material came from offshore in closed bays located in the tsunami source periphery, while in the near-field region close to the epicenter, the most active erosion occurred in the inundation area. In the latter area, the main sources of sand were beaches and dunes, while soil erosion was the source of mud. Studies of the Tohoku tsunami on the coasts of the Lesser Kuril Islands demonstrated that mud layers in the sections of coastal lowlands in closed bays could contain preserved detailed geological records of paleotsunamis, even those with a small-height run-up. In the sections of coastal peatlands of closed bays on Shikotan Island, up to 7–9 layers of mud and silty sands were found, these can easily be traced for more than 500 m inland. The grain-size composition of the mud is similar to the deposits of the 2011 Tohoku tsunami. The marine origin of these deposits is confirmed by the diatom analysis data.

Key words: 2011 Tohoku tsunami deposits, mud, grain size, diatoms, foraminifera, lesser Kuril Islands, Pacific Ocean.

[1] Pacific Geographical Institute, FEB RAS, Vladivostok, Russia. E-mail: nadyar@tig.dvo.ru

[2] P.P. Shirshov Institute of Oceanology, RAS, Moscow, Russia. E-mail: harl51@mail.ru

[3] Institute of Marine Geology and Geophysics, FEB RAS, Yuzhno-Sakhalinsk, Russia. E-mail: victor@imgg.ru

[4] Department of Geography and Geoecology, St. Petersburg State University, St. Petersburg, Russia. E-mail: arslanovkh@mail.ru

1. Introduction

The Tohoku tsunami of 11 March 2011 was one of the strongest events to have ever occurred in the Pacific Ocean; it was recorded along the entire Kuril Islands from the Lesser Kuril Islands in the south up to Paramushir Island in the north (Kaistrenko *et al.* 2011, 2013; Shevchenko *et al.* 2014). The Kuril Islands were in the periphery zone of the main tsunami propagation flux. The Tohoku tsunami run-up heights generally did not exceed 3 m on the coasts of the South Kuril Islands, located 750–800 km away from the source area. A particular feature of this event on the coasts of the Kuril Islands was heavy sea ice moving with the water (Kaistrenko *et al.* 2013). The sedimentation during such an exclusive and intensive event requires special investigation. The water coming with the tsunami waves was filled with ice fragments, which increased the wave erosive capacity; this effect was especially strong in closed bays, where a mud cover was found to have formed (Razjigaeva *et al.* 2013). In contrast, on the shores of open bays, the waves did not cause active erosion and left almost no deposits. The absence of deposits cannot be explained by the erosion during back flow, because it was not intensive. In all bays on the Pacific side of the island, the grass blades in the inundation zone were oriented landward, in accordance with the inflow. Locally, where the outflow was dominant, the grass blades were turned seaward and traces of soil erosion were found.

Some specific features of sedimentation during a tsunami with ice were reconstructed by Minoura *et al.* (1996) for the 1923 Kamchatka event based on deposits collected 70 years after the tsunami. The study of the 2011 Tohoku tsunami deposits in the region of the South Kuril Islands gives us a unique opportunity to examine the sedimentation of a tsunami with ice shortly after the event.

The study of the 2011 Tohoku tsunami in the South Kuril Islands also has another interesting aspect. In order to perform the reconstruction of transoceanic tsunamis of the past and to estimate their recurrence period, it is important to understand the difference between deposits formed in the periphery of the tsunami propagation and in the near-field zone of the highest run-up. The coastal area of eastern Japan was a region of very intensive research on the 2011 Tohoku tsunami deposits (ABE et al. 2012; GOTO et al. 2011, 2012; INUI et al. 2012; NAKAMURA et al. 2012; NARUSE et al. 2012; RICHMOND et al. 2012; SZCZUCIŃSKI et al. 2012; TAKASHIMIZU et al. 2012). The comparison of the characteristics of the 2011 tsunami deposits in the Lesser Kuril Islands with those collected on Honshu Island can explain the differences in the deposition mechanism in near-field and far-field areas.

The present paper is a continuation of our previous study, R2013, of the 2011 Tohoku tsunami deposits on the coasts of the Lesser Kuril Islands (RAZJIGAEVA et al. 2013). In R2013, we described the first deposit data collected from ice immediately after the event (i.e., in March 2011), and during the field survey in August–September 2011. The main purpose of R2013 was to examine the factors controlling tsunami sedimentation patterns in a variety of locations along the coast. We found that the characteristics of tsunami deposition varied significantly, essentially depending on local coastal topography; large differences were observed between the coasts of open and closed bays. Continuous mud deposits were found to have formed only on the coasts of closed bays. Our findings indicate that one of the crucial factors controlling the formation of the mud covers, even for coasts with insignificant run-up heights, was the presence of the ice. The main source of the tsunami mud deposits were the bottom inshore and offshore mud flows in closed bays with relatively low wave energy. However, we assume that if a tsunami of this scale occurred at an ice-free time of year, no mud covers could be formed.

One of the objectives of R2013 was to find differences of the 2011 Tohoku tsunami deposits from other coastal facies. The study of the composition of benthic foraminifera and diatoms indicated that the tsunami wave had entrained material from the upper shoreline. Considerable differences in the grain-size composition were found between the 2011 Tohoku tsunami deposits and the deposits associated with the 1994 Shikotan tsunami (IVANOV 1997; RAZZHIGAEVA et al. 2007). Additionally, it was discussed that tsunami sedimentation is significantly different from that of extreme storms, even with similar run-up values. The manifestation of the 2011 Tohoku tsunami in the Southern Kuril Islands was also examined by KAISTRENKO et al. (2013), who collected eyewitness reports and photos of tsunami effects on the coasts of these islands, in particular the results of interaction of tsunami waves with ice. They estimated maximum run-up heights and inshore inundation distances in various bays, however, they did not discuss tsunami sedimentation. SHEVCHENKO et al. (2014) examined coastal and offshore instrumental records of the 2011 tsunami. At Shikotan Island (Lesser Kuril Islands), the maximum recorded trough-to-crest wave height was almost 3 m, in close agreement with the field survey results (KAISTRENKO et al. 2013).

The purpose of the present paper is to analyze the specifics of the 2011 Tohoku tsunami sedimentation in the periphery zone, summarizing the 2011 observations and the new data obtained during the 2012 field survey. One of the objectives of this study was to compare results for the relatively remote Lesser Kuril Islands with those from the near-field zone on the northeastern coast of Honshu Island, Japan, where maximum run-ups of up to 40 m were observed (e.g., MORI et al. 2011). It is especially important to compare the composition of the muds formed under the tsunami with low and high run-ups, in diverse coastal environments and having different sources, and to show the unique properties of tsunami mud deposits within the closed bay coasts.

Another objective of the current paper was to thoroughly study the 2011 Tohoku tsunami deposits and to use the corresponding results to reveal and reconstruct historical tsunami and paleotsunami deposits. The 2011 Tohoku tsunami sediments that were recovered and examined from the coasts of closed bays of the Lesser Kuril Islands were critical to understanding the sedimentological effects of tsunamis in general, and they stimulated further study of paleotsunami muds. While studying the coastal lowlands of closed bays, we have discovered that particular attention has to be paid to thin mud layers alternating with layers of peat. An interesting problem is to compare the Tohoku tsunami deposits in closed bays of the Lesser Kuril Islands with mud layers in the Holocene peat sections, attempting to understand whether they are traces of paleotsunamis and under what conditions they were formed.

2. Regional Setting

Lesser Kuril Ridge is a chain of islands stretching 100 km to the northeast from Hokkaido

Island; it runs parallel to Greater Kuril Ridge and is separated from Kunashir Island by South Kuril Strait, which is 48 km wide and up to 200 m deep (Fig. 1). The Lesser Kuril Islands include six medium-size islands, the largest being Shikotan and Zeleniy, and a number of smaller islands and rocks. Among those, only Shikotan Island has low-mountain topography, which prevented the deep penetration of the tsunami. An asymmetric relief with steeper slopes on the Pacific side of the island in many ways predetermined the configuration and development of the coastal zone. The other islands are flat and low (up to 40 m). The islands are separated by narrow (1.9–22.2-km-wide), shallow straits up to 54 m deep. The strait between Tanfiliev Island and Hokkaido Island is 7 km wide and 89 m deep (ATLAS OF THE KURIL ISLANDS 2009).

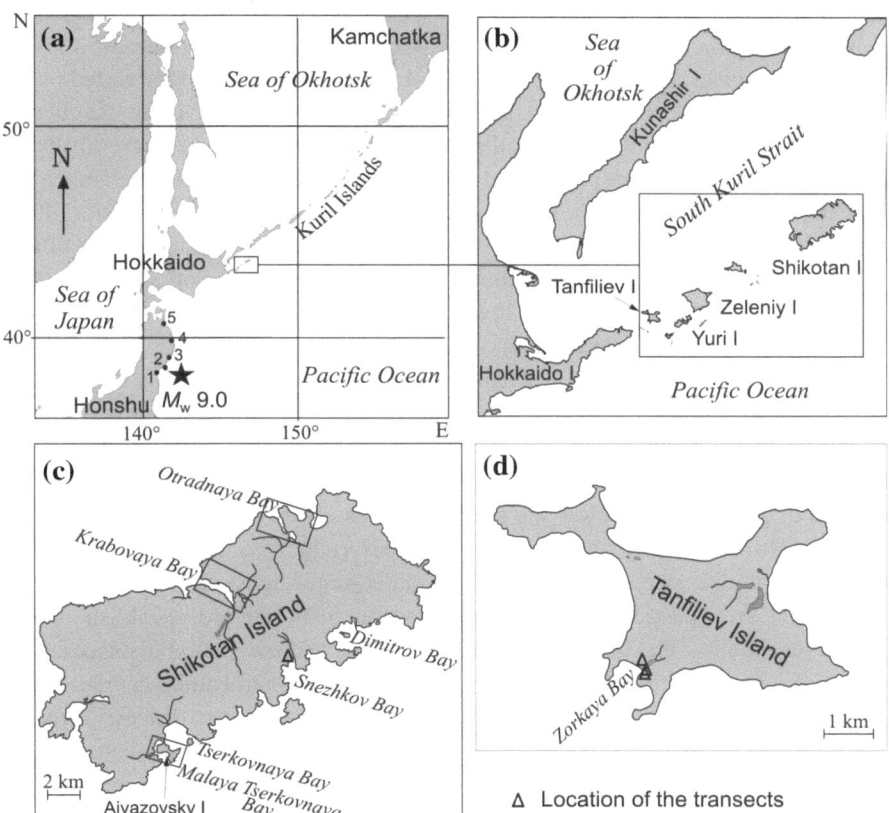

Figure 1
a Map of the study area with the epicenter of the Tohoku-oki earthquake (M_w 9.0) of 11 March 2011 indicated by a *star*. *Small circles* with numbers denote specific sites in Honshu Island (Japan) where tsunami deposits were examined by various authors: *1* Sendai Plain; *2* Matsushima Bay; *3* Hirota Bay; *4* Miyako City; and *5* Misawa coast. **b** The region of South Kuril and Lesser Kuril islands. **c, d** Maps of Shikotan and Tanfiliev islands, respectively, with the location of the transects denoted by *empty triangles*. *Boxes* are shown in Fig. 2

Typical shores of the islands are high, with abrasion and abrasion-denudation coasts with narrow beaches and open bays. In the heads of closed bays there are swampy coastal lowlands with ancient storm ridges directly behind the beaches. Low swampy isthmuses can be found only on Yuri Island. Closed bays can be divided into two types: those formed as a result of selective abrasion, separated from the ocean by protruding capes and small islands composed of more abrasion-resistant rocks. Others were formed due to the penetration of the sea into river valleys with small downstream slopes during the Holocene transgression. After a 0.7-m co-seismic subsidence of the island during the Shikotan Earthquake in 1994 (YEH et al. 1995; IVASHCHENKO et al. 1996; KAIS-TRENKO et al. 1997), the area around the small river mouth in the southern part of Malaya Tserkovnaya Bay was flooded, forming a narrow and shallow inlet. Its bottom, covered by mud and silted sands, is almost fully exposed during low tide.

Deep closed bays are typical for Shikotan Island on the side facing the South Kuril Strait (Fig. 2). The bays are 2.9–3.5 km in length; their depths at the entrance are 10–16 m, gradually decreasing towards the head. There are narrow beaches along the sides of the bays, composed of coarse deposits. A lot of non-rounded debris falls down onto these beaches from cliffs and slopes. Large tidal flats, exposed during low tides, are typical for the bay heads. The amplitudes of the tides are about 1.1–1.9 m (ATLAS OF THE KURIL ISLANDS 2009). The extent of the tidal flats increased drastically after the co-seismic subsidence of the island during the 1994 Shikotan Earthquake. During high tides, large areas of peat-bogs lie below the sea level. Flooded peat in Otradnaya Bay stretches for 300 m. Underwater valleys with steep slopes, cut in loose deposits, are easily seen on tidal flats and in shallow zones near river mouths. The bottoms of the bays are covered by silty sand and mud.

One of the factors for the active development of swamps within the coastal lowlands and river valleys was the tectonic subsidence of the Lesser Kuril Islands in the Holocene, which continues to the present (RAZZHIGAEVA et al. 2008b). The largest swamps are located on the coast of Krabovaya Bay. Thick Holocene marine and lagoon muds are buried beneath layers of peat.

Since the Tohoku tsunami took place in March 2011, it is important to provide data on the ice climatology of the South Kuril region (ATLAS OF THE KURIL ISLANDS 2009). Drifting ice in the straits appears in January, with ice cover reaching its peak in March, when there is heavy consolidated ice in the straits. Ice is also formed locally in shallow waters; open and semi-open bays do not freeze, but sometimes have drifting ice. Closed bays get frozen and the ice sheet breaks in late March–early April. Before the 2011 Tohoku tsunami, Krabovaya and Otradnaya bays were covered by thick ice. Zorkaya Bay on Tanfiliev Island and the strait between Lesser Kuril Ridge and Hokkaido Island were also covered by ice. According to witness reports, the tsunami broke the ice, carrying it out to the strait and back again, and finally depositing it on the shore. The ice blocks were up to 2 m thick.

The last large tsunami in the region of the South Kuril Islands occurred during the 1994 Shikotan Earthquake (IVASHCHENKO et al. 1996). High run-up heights (5–7 m) were measured on the Pacific side of Shikotan Island; maximum heights (up to 15 m) were in Tserkovnaya Bay. In the heads of closed bays of the South Kuril Strait, the maximum heights were 2.3–2.6 m (KAISTRENKO et al. 1997). Deposits from this tsunami were found on top of the Holocene peats on the Pacific side of Shikotan Island (RAZZHIGAEVA et al. 2007).

3. Materials and Methods

The research was based on the data of the 2011 Tohoku tsunami deposit, which was sampled from ice floes right after the tsunami and then half a year later—in August–September 2011, when the coasts were inspected to establish the tsunami run-up heights and inundation distances. Sampling was done along shore-perpendicular transects, using a description of tsunami traces, erosion zones, and identification of the inundation limit zone. Since in most cases the deposits were homogeneous (massive) and thin, one bulk sample was taken for each point. In stratified sections the samples were taken in each layer, depending on material size.

In August–September 2012, another inspection of the shore took place, aiming to find out how well the deposits of the 2011 Tohoku tsunami were preserved 1.5 years after the event. The work was done on Shikotan Island on the shores of the closed bays of Krabovaya, Otradnaya and Malaya Tserkovnaya, and on Tanfiliev Island in Zorkaya Bay. At some locations, the Tohoku tsunami deposits were re-sampled and analyzed to determine whether their composition had changed. Peats were drilled along the profiles in Otradnaya and Krabovaya bays in order to study similar events of the past (Fig. 2). Inter-layers of mud, silty sands, and inclosed peat were also sampled.

Grain-size analysis of mud was made using an "Analysette 22" sedimentograph. Silty sand grain-size composition was studied using sieves with γ step. Large plant detritus was removed through a 1.6-mm sieve. Grain-size statistics were calculated using the logarithmic methods (phi) of moments of GRADI-STAT v.4.0 software (BLOTT and PYE 2001). New data were obtained for Tohoku tsunami deposits at Malaya Tserkovnaya Bay and Zorkaya Bay, and for the paleotsunami muds of Krabovaya Bay and Otradnaya Bay.

Micropaleontological research included diatom and foraminifera analysis. Diatom analysis was made according to the standard techniques (THE DIATOMS OF THE USSR 1974) with a light microscope "Axioskop-Karl-Zeiss" at 1,200× magnification. The taxonomic composition of diatoms and their ecological characteristics were identified following KRAMER and LANGE-BERTALOT (1988) (THE DIATOMS OF THE USSR 1974, 1992). The percentage content was calculated from the total taxa.

Foraminifera that were studied after the deposits were washed through a 0.063-mm sieve under a binocular microscope. The following references were used to identify the ecology of the species (FORAMINIFERA OF FAR EASTERN SEAS OF THE USSR 1979; PREOBRAZHENSKAYA and TROITSKAYA 1996; SEN GUPTA 2002).

A full list of diatom species of the Tohoku tsunami sheets and benthic foraminifera from the tsunami deposits off the ice is given in (RAZJIGAEVA et al. 2013). New foraminifera data were obtained for the tsunami deposits of Shikotan Island and Tanfiliev Island. In an earlier article (RAZJIGAEVA et al. 2013) the data were given only for foraminifera from the Tohoku tsunami deposits sampled from the ice. New diatoms data were obtained for the Tohoku tsunami deposits of Zorkaya Bay and in the paleotsunami deposits of the Otradnaya Bay and Krabovaya Bay coasts.

Age assignments to the paleotsunami are based on radiocarbon dating and tephra stratigraphy. ^{14}C-dating of the samples was obtained on peat (5 cm thick), lying below tsunami deposits. The samples were treated with standard acid and alkali solutions. ^{14}C-dates were produced by liquid scintillation counting at St. Petersburg State University. The ^{14}C-dates were calibrated into calendar ages using CalPal software (www.calpal.de). Identification of the tephra was based on mineral components and glass composition. Volcanic glass was analyzed chemically using a CAMSCAN-4 electron microscope with an AN-10000 semiconductor spectrometer (from the Khlopin Radium Institute, St. Petersburg).

4. Specifics of Tsunami Sedimentation Within Closed Bays

The Tohoku tsunami on the Kuril Islands was seen as a fast sea-level uplift, and up to 3 waves were observed (KAISTRENKO et al. 2013). In spite of the small scale of the Tohoku tsunami in this region, the tsunami led to the accumulation of deposits of different compositions on the coasts of the Lesser Kuril Islands (RAZJIGAEVA et al. 2013). Only spots of sand and cobble were found on the shores of open bays; maximum tsunami run-up heights were marked by marine and anthropogenic waste, wood, grass, and algae heaps. Well-identifiable sediment sheets, presented as organogenic mud, were formed on the coasts of closed bays of Shikotan Island. Thick ice covered these bays before the tsunami. The presence of ice fragments in the water aggravated the eroding capacity of the waves. The material ripped off by the ice fragments was a source of mud deposits, accumulated in Krabovaya and Otradnaya bays (Fig. 3). The ice floes enhanced the bottom erosion on shoals and the destruction of peatland over low-lying coasts, even at modest ranges of run-up heights. Active erosion was found to occur on the peatlands, submerged below sea level or in the high tidal zone.

Figure 2
Location of transects on the coasts of three closed bays on Shikotan Island where the 2011 Tohoku tsunami traces and deposits were studied (*red triangles* the profiles and points) and where paleotsunami deposits were found (*yellow triangles* the profiles). Photo of Otradnaya Bay taken from the head of the bay, photo of Krabovaya Bay from the *right side*, and photo of Malaya Tserkovnaya Bay from the *right side*. *Yellow arrows* show the position of the photographer

The peat bogs were eroded within local areas to a depth of 60 cm and their underlying silts became exposed. At the heads of the bays the tsunami also captured material from the steep sides of the flooded river valleys, cutting into the mud. In Krabovaya Bay, with narrow beaches composed of coarse-grained material, only spots of sands were found. In Otradnaya Bay, the tsunami manifestation was weaker, the ice was broken only at the head of the bay near the shore, and the tsunami flooded only local areas—

far along the river valley and a "tongue" at the head of the bay (Fig. 3a).

Fine silt and clay material were deposited from suspension on the surfaces of ice, peat, and soil. Grain-size characteristics of the deposits were traced along the transects; the mode is uniform, and the proportion of other fractions is redistributed without a distinct pattern. A weakly expressed second mode appears on some curves, which may indicate a greater contribution from another source of material (Fig. 4).

Figure 3
Photos of the coasts **a**, **b** Otradnaya, and **c**, **d** Krabovaya bays on Shikotan Island after the 2011 Tohoku tsunami. The inundation limits are clearly visible. Heavy ice broken by tsunami waves and mud was deposited on the surface of the peatlands. The sea water deposited *dark grey* mud over the snow near the shoreline. In the area of Otradnaya Bay, tsunami run-up heights were insignificant; the tsunami broke thick ice only near the coast of the bay. The maximum inundated area occurred in the southwestern part of the bay. In the head of Krabovaya Bay, the tsunami brought heavy ice into the peatland and river mouth

Sometimes the share of coarser silt or fine sand grows and produces bimodal curves, and sometimes the amount of clay fractions grows in proximity to the water line. Most likely, it is associated with the matter transported by ice, which made the pattern of suspension settling more complex. A mode of 20–30 μm is well-defined in tsunami mud from various bays that suggests a consistency in the settling. Grain-size distributions are unimodal or bimodal, poorly and very poorly sorted (Fig. 5a). The distributions vary between finely skewed and symmetric (skewness from 0.05 to 0.58), to rare and very finely skewed (to 2.30). All distributions are mesokurtis (kurtosis from 2.54 to 3.02).

A pronounced cover of silty sands was formed in the southern part of the Pacific shore of Malaya Tserkovnaya Bay. The wave inundated up to 78 m inland and the sediment cover reached 65 m inland. The cover thickness grows 10–35 m inland, reaching

a maximum depth (up to 14 cm) in small depressions. Farther inland, the thickness of the deposits drops sharply (0.5–1.0 cm). The deposition of silty sands was probably related to the lengthy water presence and sedimentation of material from both the bottom layer and suspension, when the wave had reached maximum inundation and lost its energy. The presence of stratified sections in tsunami deposits in Malaya Tserkovnaya Bay indicates that there were several waves. Grain-size characteristics of the deposits are consistent along transects: the share of silt grows moving inland along the transects, while the size of modal fractions changes very little; the average size of the grains barely changes (Fig. 6).

Shikotan tsunami (1994) deposits were found in this bay (RAZZHIGAEVA *et al.* 2012). These deposits have coarser fractions than the Tohoku tsunami deposits; they have polymodal grain-size curves, probably due to the fact that the tsunami in 1994 was

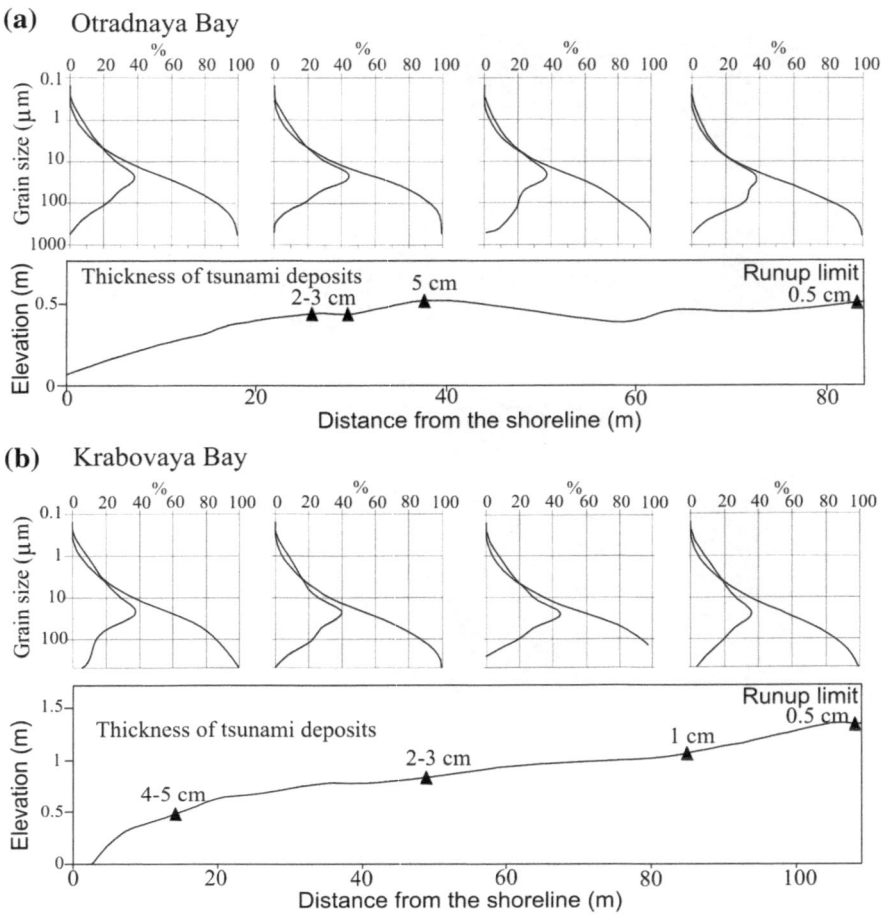

Figure 4
Landward changes in the grain-size distributions of the 2011 Tohoku tsunami deposits in **a** Otradnaya and **b** Krabovaya bays. The presented *curves* are composed from samples collected in the head of Otradnaya Bay (transect 21911) and on *left side* of Krabovaya Bay (transect 20111). *Black triangles* indicate sampling points

stronger, and therefore caused greater erosion with material supplied from a large area and different sources.

Diatom algae in the 2011 Tohoku tsunami deposits were presented by a mixture of marine and freshwater species with different ecologies. In total, 135 marine diatoms were found. Sub-littoral benthic forms prevail, which signals the inshore movement of the material from offshore. The presence of large frustules (up to 30 μm) of such species as *Cocconeis scutellum, Odontella aurita*, and the frequent occurrence of diatom colonies indicate the weak turbulence of the water flow. It is possible that the presence of diatoms in colonies provoked the mass deposition of frustules. Marine benthic species dominate in

stratified deposits of the coarser layers, while fine-grained layers contain increased amounts of plankton species. The composition of the freshwater diatoms depends on the type of terrestrial deposits engaged by the tsunami. As a rule, species from different environments are presented, reflecting the mixture of materials from different sources. The composition of the freshwater diatoms in tsunami deposits can be distinguished from freshwater diatoms of underlying peat, where dominating species are those typical of swamps; various representatives from *Eunotia* genera are noted.

Ten different species of benthic foraminifera inhabiting the littoral and sub-littoral zone of Krabovaya and Otradnaya bays (PREOBRAZHENSKAYA and

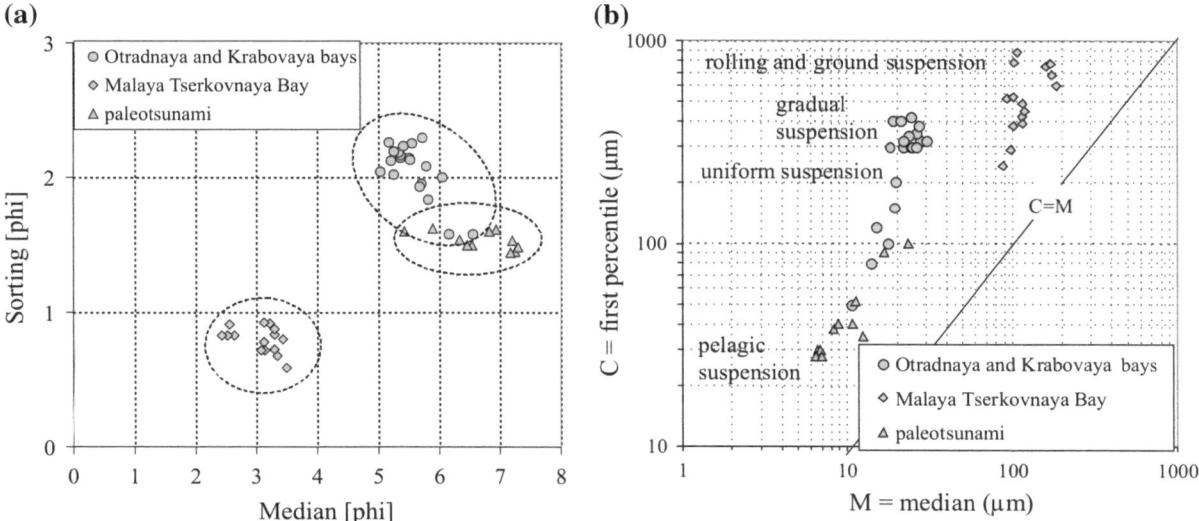

Figure 5
Position of the 2011 Tohoku tsunami deposits on the coasts of closed bays of Shikotan Island (Otradnaya, Krabovaya, Malaya Tserkovnaya bays) and paleotsunami mud in Krabovaya Bay (core 14912) and Otradnaya Bay (core 14512): **a** diagram of sorting and median; **b** Passega C-M diagram

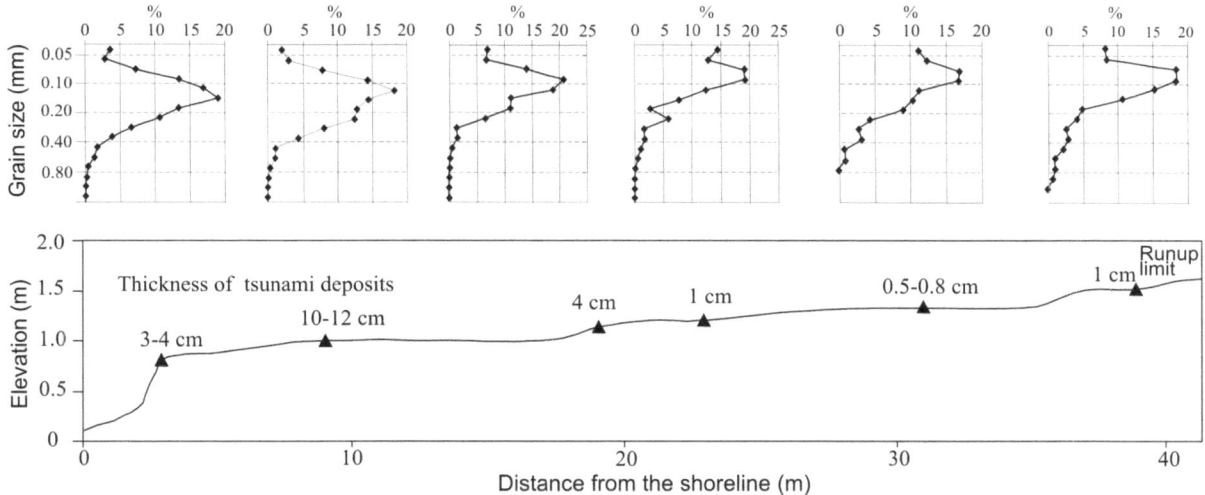

Figure 6
Landward changes in the grain-size distributions of the 2011 Tohoku tsunami deposits for the coast of Malaya Tserkovnaya Bay (the samples were taken in 2012). *Black triangles* indicate sampling points

TROITSKAYA 1996) were found in the tsunami deposits from the ice. *Jadammina macrescens* is dominant. Occasional agglutinating forms of *Trochammina* (*Trochammina japonica, Tr. inflate, Tr. pacifica, Tr. vinogradovi*) and *Miliammina fusca* were noted. Among the secretion species are *Cribroelphidium asterineum, Cr. etigoense*, and representatives of *Buccella* genus. All tests are well-preserved.

In the tsunami mud sheets, a high abundance (fossil concentration 12–27 individuals per 1 g of dry deposit) of benthic foraminifera was found within Krabovaya Bay in the tsunami deposits near the shoreline, where *Jadammina macrescens* prevails. This species is an inhabitant of the littoral zone of this bay (PREOBRAZHENSKAYA and TROITSKAYA 1996). Littoral *Miliammina fusca,* widespread in the intertidal zone (0.5–5 m deep), *Labrospira jeffreysi,* and *Trochammina inflate,* usually living at depths of 5–10 m (FORAMINIFERA OF FAR EASTERN SEAS OF THE USSR 1979), were found in small numbers. The fossil concentration decreased inland (one individual per 1 g of dry deposit).

In Otradnaya Bay, where the tsunami was weak (KAISTRENKO *et al.* 2013), the tsunami mud included low concentrations of foraminifera (less than two individuals per 1 g of dry deposit) such as *Jadammina macrescens, Labrospira jeffreysi, Miliammina fusca,* and *Trochammina* sp., except for mud with abundant *Zostera,* located 26 m from the shoreline, where concentrations of foraminifera reached up to 64 individuals per 1 g of dry deposit. In one sample rare, *Proteonella* sp., and *Rhabdammina* sp. of poor preservation were found. Species of these genera have a wide habitat, from the subtidal zone to the deep sea, and the maximum number usually lives within depths of 70–120 m (FORAMINIFERA OF FAR EASTERN SEAS OF THE USSR 1979). Perhaps the tsunami waves brought these samples from the bottom of the bay.

On the Pacific side of the island (Malaya Tserkovnaya Bay) tsunami deposits had low concentrations of benthic foraminifera (1–3 individuals per 1 g of dry deposit). Rare tests of *Miliammina fusca* were found in the tsunami deposits. The presence of *Labrospira jeffreysi,* living at depths of 5–10 m, and *Eggerella scrippsi* of 15–50 m (FORAMINIFERA OF FAR EASTERN SEAS OF THE USSR 1979) demonstrates the input of material from deeper areas. There were also brown samples of *Trochammina inflate,* redeposited from Holocene marine sediments on the bottom of the bay.

A second study of the tsunami traces, undertaken in 2012, allowed us to evaluate the quality of the preservation of the material. Tsunami deposits on the top of the peat are moist, the redeposition of eolian transfer or washaway does not take place. The deposits are actively colonized by grasses. Technogenic (man-made) constructions, which are free of any tsunami deposits, are the exception. Grain-size analysis of tsunami deposits sampled in 2012 on the coasts of Otradnaya, Krabovaya, and Malaya Tserkovnaya bays shows the composition little changed and similar to the data of 2011. A re-examination of diatoms gave the same result.

In 2012, tsunami deposits were found on the coast of the semi-closed Zorkaya Bay (Tanfiliev Island), where the run-up heights reached 2 m. Before the tsunami, the bay and the strait between Tanfiliev and Hokkaido were covered with ice. The tsunami broke the ice, then the ice was drawn out of the bay into the strait, and finally it was brought back by the tsunami and deposited on the shore. The sedimentological situation was similar to that of open bays. Only small patches of well-sorted, fine sand (up to 4 cm thick) were found (Fig. 7a, b). Grain-size distribution curves are mainly unimodal (modes 0.125–0.16, or 0.2–0.25 mm), moderately well-sorted (σ 0.58–0.67), and symmetric (skewness 0.08–0.10). The main source of the deposits was the beach. Beach sand has almost the same distributions, but is better sorted (σ 0.44–0.49) than the tsunami deposits. In the center of bay behind the storm ridge, the tsunami deposit has a polymodal curve with peaks 0.2–0.25, 0.315–0.4, and 0.5–0.63 mm; the sand is moderately sorted (σ 0.76) and the distribution is symmetric and mesokurtic (Fig. 7c, d). At this place, fine sand from the beach, and medium sand with coarser fractions from the storm ridge were redeposited. The storm ridge sand is moderately sorted (σ 0.81), and has a coarsely skewed distribution.

In the sands of the tsunami in the eastern part of the bay, 36 species of marine and brackish diatoms (up to 58.7 %) were found. The largest content of diatoms was found under a large car tire, which protected the deposits from the rains. Diatoms in the sediment at the top of the bay were rare. In all samples, the majority of the diatom frustules were broken, especially with the larger frustules, which may be due to breakup by ice. Sublittoral benthic species prevail: *Cocconeis scutellum* (14.5 %), *C. costata* (9.6 %), *C. stauroneiformis* (2.6 %), and *Grammatophora hamulifera* (2.6 %); *Navicula directa, Amphora marina, Anaulus maritimus, Arachnoidiscus ehrenbergii, Cocconeis californica, C. decipiens, C. verrucosa, C. scutellum* var. *parva, Cyclotella striata, Grammatophora angulosa, Lyrella pigmaea, L. forcipata, Istmia nervosa, Tabularia tabulate,* and *Achnanthes brevipes* var. *intermedia* were present too. Among plankton species, *Paralia sulcata* (6.1 %) is dominant, and *Hyalodiscus scoticus, H. obsoletus, Trachyneis aspera, Odontella*

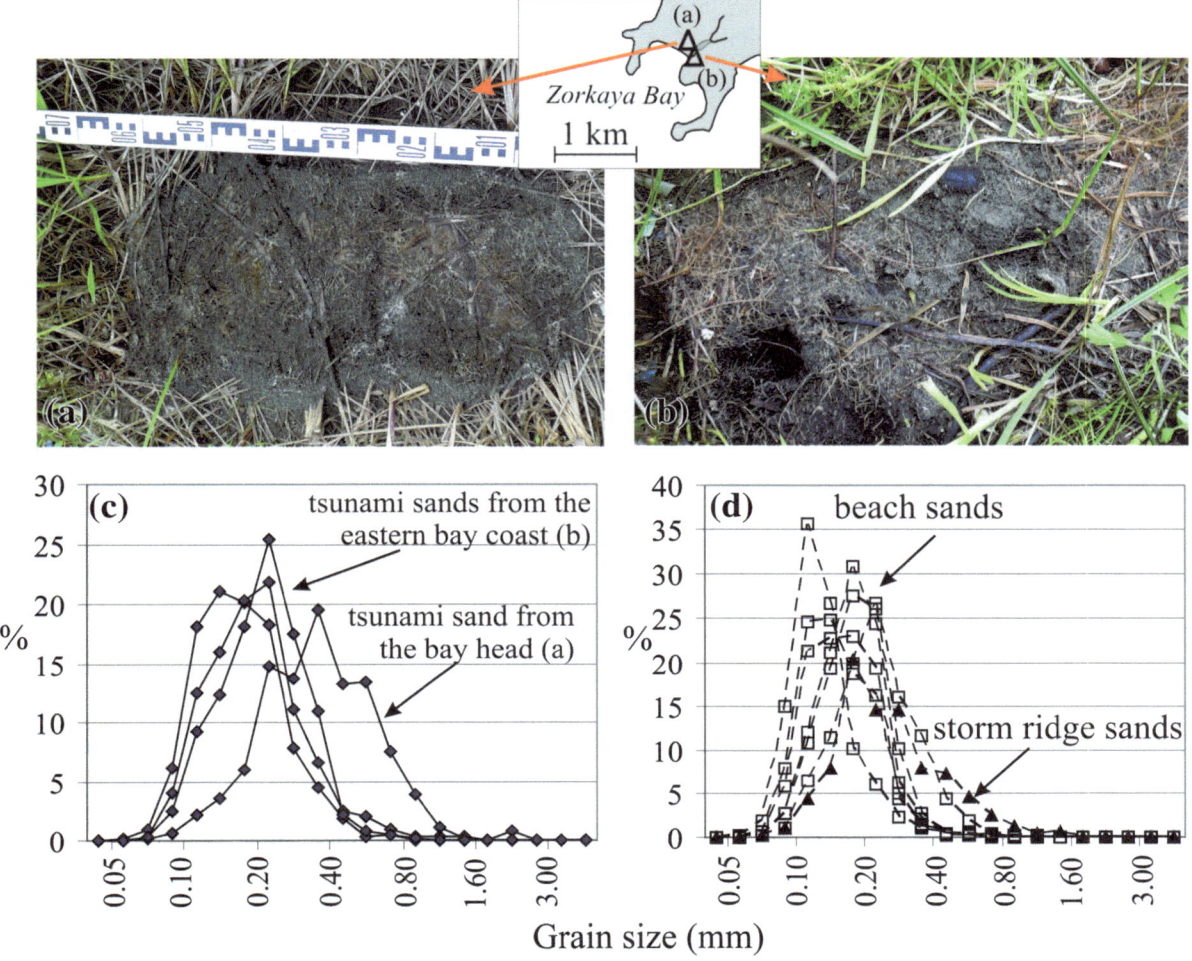

Figure 7
2011 Tohoku tsunami deposits on the coast of Zorkaya Bay, Tanfiliev Island. Patches of well-sorted, fine sand: **a** under a buoy at the head of the bay, distance from shoreline 21 m, height 2.2 m, thickness of deposits 2–2.5 cm; **b** under a big car tire, eastern coast, distance from shoreline 25 m, height 1.4 m, thickness of deposits 4 cm. The grain size of the deposits: **c** Tohoku tsunami sands from the eastern part of the bay and the head of the bay; **d** beach sands (*squares*), and storm ridge sand (*triangles*)

aurita, Actinoptychus senarius, and *Navicula dithmarsica* were found. Neritic and oceanic species are represented by northboreal and arctoboreal *Thalassiosira eccentrica, T. gravida, T. kryophila, T. hyalina, T. decipiens, Thalassiothrix longissima, Coscinodiscus oculus-iridis, Bacterosira fragilis,* and southboreal *Thalassionema nitzschioides, Coscinodiscus asteromphalus,* and *Thalassiosira leptopus.*

Freshwater diatoms include species living in different terrestrial environments. Species typical for weak moisture and weak swamping surfaces dominate: *Pinnularia lagerstedtii* (17.9 %), *P. intermedia* (1.8 %), *Luticola mutica* (up to 8.8 %), *L. ventricosa*

(2.1 %), *Diadesmis contenta* (7.5 %), and *Hantzschia amphioxys* (up to 7.7 %). Species typical for shallow lakes, polls, and little streams were also met: *Achnanthidium minutissimum* (5.1 %), *Navicula perminuta* (8.1 %), *Encyonema silesiacum* (1.3 %), *Nitzschia paleaceae* (2.6 %), *Staurosira venter, Pseudostaurosira brevistriata, Rhoicosphenia abbreviata, Cocconeis placentula,* and plankton *Aulacoseira italica, A. alpigena.* Wetland species *Eunotia bilunaris, E. crista-galli,* and *E. praerupta* are present too.

Only one sample of benthic foraminifera—*Globobulimina* sp. was found in the tsunami sand. A small (1.5 mm), well-preserved shell from the Gastropoda class was also found in one sample.

5. Tsunami Sedimentation of the Periphery Zone in Comparison to the Near-Field Area

For a better understanding of the peculiarities of sedimentation on the 2011 Tohoku tsunami periphery, we tried to compare the data acquired from the Lesser Kuril Islands with those in the near-field zone of maximum run-up heights on the Sendai Plain (northeastern coast of Honshu Island), located in close proximity to the source of the 11 March 2011 earthquake (ABE *et al.* 2012; JAGODZIŃSKI *et al.* 2012; PILARCZYK *et al.* 2012; RICHMOND *et al.* 2012; SZCZUCINSKI *et al.* 2012; TAKASHIMIZU *et al.* 2012) on Sabusawa Island at Matsushima Bay—20 km to the northeast from Sendai (GOTO *et al.* 2012), and on the shore of closed Hirota Bay—130 km from the epicenter (NARUSE *et al.* 2012). The maximum run-up height reached 39.7 m in the area of Miyako City, and the inundation distance was up to 5 km (MORI *et al.* 2011).

5.1. Deposit Distribution and Geometry

Usually, when the run-up height is <5 m and erosion is weak, the tsunami does not leave easily identifiable deposits (DAWSON and SHI 2000). On the periphery of the Tohoku tsunami, on the coast of the South Kuril Islands, the deposit cover had been formed only in the heads of closed bays and the narrow-entrance inlets of Shikotan Island, where the tsunami run-up heights were less than 3 m. In the area adjacent to the epicenter of the tsunami on the coast of Honshu Island, the wave inundated several kilometers inland, while on the Lesser Kuril Islands, the inundation distance was usually only a few tens of meters, whereas along the river valleys it was up to 300 m (KAISTRENKO *et al.* 2013). Continuous deposit sheets have been formed only on the shores of closed bays of Shikotan Island, and they are not extended (maximum is 106 m long); in the heads of the bays, the deposits cover the inundation areas completely. One of the factors that controlled erosion and sedimentation in these bays was the presence of ice in the tsunami waves. In Malaya Tserkovnaya Bay,

the sediment sheet was found up to 65 m inland, which corresponds to 83 % of the inundation distance. Small spots of sand and cobble were found on the edge of closed bays and some open bays, where erosion was weak. Such traces of tsunamis in geological sections are difficult to find.

On the coast of Honshu Island, deposits cover almost the entire inundation area of the Sendai Plain, where the tsunami traveled up to 5.5 km inland (ABE *et al.* 2012; JAGODZIŃSKI *et al.* 2012; PILARCZYK *et al.* 2012; RICHMOND *et al.* 2012; SZCZUCINSKI *et al.* 2012; TAKASHIMIZU *et al.* 2012). If the inundation distance on the Sendai Plain had been shorter than 2.5 km, the maximum limit of the tsunami deposit layer (≥ 0.5 cm thick) extended to over 90 % of the inundation distance, and the sand layer reached 57–76 % of the inundation distance where the tsunami inundated more than 2.5 km inland (ABE *et al.* 2012). Salt contamination down to 15 cm and the formation of salt crusts on the soil after the tsunami were observed up to 4 km inland (CHAGUÉ-GOFF *et al.* 2012). After the water receded, the tsunami sands were actively transported by wind, and 2 months later, the tsunami deposits were covered by eolian sands up to 15–20 cm thick (RICHMOND *et al.* 2012). On the swampy coasts of the Lesser Kuril's closed bays, the Tohoku tsunami deposits are wet, and as a rule, dense sheets were not redeposited.

5.2. Erosion, Basal Contact, and Deposit Thickness

The erosion in the periphery zone of the 2011 Tohoku tsunami was, in general, weak, and its intensity cannot be even remotely compared with the erosion caused by the tsunami on the coasts of Honshu Island. In the closed bays of Shikotan Island, where the tsunami broke the ice, the presence of ice in the water increased the eroding capacity of the waves. The bottoms of shallow bays and coastal peatlands were eroded and large pieces of peat were found in the inundation zones. The most active erosion was observed at the head of Krabovaya Bay where two eroded zones were found (up to 100 m wide) and the soil cover and peat were destroyed (0.6 m thick). The basal contact of the tsunami deposits was sharp; there was no erosion and the deposits lay on the previous year's grass (Fig. 8). The basal contact is clearly visible 1.5 years after the tsunami. The grass and algae are preserved. In the authors' opinion, this is different from deposits in the near-field area close to the tsunami source, i.e., the zones of maximum run-up. On Honshu Island, where the tsunami caused strong erosion, the basal contact of the tsunami deposits on different underlying deposits is sharp (ABE *et al*. 2012; GOTO *et al*. 2012; SZCZUCINSKI *et al*. 2012; TAKASHIMIZU *et al*. 2012).

The thickness of the tsunami deposits strongly depends on the intensity of the erosion processes and the presence of sources of friable material in the coastal zone, and on the relief of the underlying surface. The thickness of the tsunami sheets on Shikotan Island was no more than 5 cm due to the absence of wide, sandy beaches and the deficit of material in the coastal area.

The thickest sheet (up to 14 cm) was found in Malaya Tserkovnaya Bay with a wide tidal flat. This thickness decreases in the inland direction and depends on local topography. In general, the thickness of the tsunami sediment sheets on the periphery is much smaller than in the near-field zone close to the tsunami source area. The thickness of deposits of the 2011 Tohoku tsunami on Sendai Plain on the northeastern coast of Honshu Island is approximately 25–35 cm and, in some places, as much as 50 cm, depending on the proximity of the material source and the relief of the underlying surface (ABE *et al*. 2012; JAGODZIŃSKI *et al*. 2012; SZCZUCINSKI *et al*. 2012; TAKASHIMIZU *et al*. 2012). On Sabusawa Island, the deposit sheet is up to 80 cm thick (GOTO *et al*. 2012).

Figure 8

Tohoku tsunami deposits on Shikotan Island. **a**, **b** The tsunami deposits, Malaya Tserkovnaya Bay (photo 2011), **c**, **d** the tsunami deposits, Malaya Tserkovnaya Bay (photo 2012), **e** the tsunami deposits, Krabovaya Bay (photo 2011)

5.3. Sedimentary Structures and Vertical Variation

As a rule, tsunami deposits on Shikotan Island have a substantial thickness. Sand with mud cups was found on the shore of semi-closed Snezhkov Bay. Stratified deposits were found at only one point on the coast of Malaya Tserkovnaya Bay, where the section had 2–3 distinguishable layers, at the base of which more coarse layers were discovered (up to 1 cm thick) (Fig. 8b). Such deposits indicate that there were 2–3 significant tsunami waves, and that the coarsely-graded material was the first to settle. Eyewitnesses of the event on Shikotan Island observed three tsunami waves, the second and third waves being significantly higher (KAISTRENKO *et al.* 2013). The coarse layers have fewer marine diatom species, and sub-littoral benthic diatoms prevail. The fine deposits have a greater concentration (up to 62 %) and diversity (up to 36 species) of marine species; the number of plankton forms also increases and oceanic diatoms appear. The same pattern was observed in the deposits of the 2004 Indian Ocean tsunami in Thailand (SAWAI *et al.* 2009).

Unlike in the periphery regions, the deposits of the Tohoku tsunami on Sendai Plain in Honshu Island are often well-stratified horizontally, though massive sands can also be found. Deposits on most of this territory were probably brought in by the first major wave; the incoming flow deposits prevail, while deposits of the return flow were found only locally (ABE *et al.* 2012; TAKASHIMIZU *et al.* 2012). In the upper portion of the tsunami deposits, a higher concentration of diatoms was recorded (SZCZUCINSKI *et al.* 2012). There were four units of normal and reversed grading found on the coast of Hirota Bay; this finding is an indicator of several tsunami waves affecting this area and settling the material from incoming and return flows (NARUSE *et al.* 2012).

5.4. Grain-Size Distribution Transportation and Deposition of the Material

The grain-size composition of the Tohoku deposits on the Lesser Kuril Ridge is considerably diverse—from boulders and pebbles on the open bay coasts to mud in the heads of the closed bays. The extended sheets are composed of mud with a prevalence of silt (up to 61.6 %) and contain high concentrations of grains smaller than 10 μm (up to 41.5 %). The deposits are poorly sorted; the grain-size distribution curves are unimodal (mode is 20–30 μm) with a weakly expressed second mode in some cases. The deposits with bimodal curves were found on the coasts where more sandy fractions were brought in during the coast wash-out. The grain-size distribution curves are symmetric and asymmetric—finely skewed due to different proportions of clay fractions. There are no considerable changes in the grain-size composition of mud deposits along the transects. In the area of Malaya Tserkovnaya Bay, where the 2011 tsunami deposited silty sand, inland fining was observed. On the Passega C-M diagram (Fig. 5b), the samples of tsunami mud taken from this closed bay on Shikotan Island form a field of finer material than the Tohoku mud on Sendai Plain (SZCZUCINSKI *et al.* 2012).

Unlike the situation on the periphery, the tsunami deposits of Sendai Plain get finer and their differentiation becomes clearer inland (CHAGUÈ-GOFF *et al.* 2012; JAGODZIŃSKI *et al.* 2012; SZCZUCINSKI *et al.* 2012; TAKASHIMIZU *et al.* 2012). The coarse material, as well as poorly and moderately sorted sands, turn into medium-grained sands, and then into sand with a mud cap; then 2-2.3 km from the shore, mud begins to prevail and the deposits become poorly sorted (SZCZUCINSKI *et al.* 2012). In the Arahama area, located to the north of the Natori River, a layer of mud (up to 4 cm) covering the sand appears in depressions between beach ridges, the mud having settled from stagnant water (TAKASHIMIZU *et al.* 2012). On Sabusawa Island, among the sand left by the tsunami, cobbles and pebbles have been found, together with numerous mollusk shells and their fragments; the mud caps were found only on the eroded areas of rice fields (GOTO *et al.* 2012). On the northeast coast of Honshu Island where the tsunami eroded the landward surface of sand dunes, the composition of deposits is close to that of dune sands and the grain size of the deposits decreases inland (NAKAMURA *et al.* 2012).

5.5. Insights from Diatom Analysis

Rich diatom complexes, represented by the mixture of marine and freshwater species of different ecologies, were found in deposits of the 2011 Tohoku tsunami in closed bays of Shikotan Island and Zorkaya Bay, Tanfiliev Island. Unlike tsunami deposits on Honshu Island (Sendai Plain), where diatom frustules were broken (SZCZUCINSKI *et al.* 2012), the diatom frustules in tsunami deposits on Shikotan Island are well-preserved. Large frustules of sub-littoral benthic species are common; some of the diatom colonies that were found indicate a weak turbulence of the flow and a small transfer distance. Diatoms with small frustules also settled in colonies. On Tanfiliev Island, most of the frustules of marine and brackish diatoms in the tsunami deposits were broken, possibly by ice with sand. The high content (up to 88.3 %) and the diversity of marine diatoms is typical for the deposits. Sub-littoral benthic. as well as sub-littoral plankton species are dominant, and different oceanic and neritic diatoms, typical for different marine water types (from subtropical to Arctic) have also been identified. Freshwater species typical for different environments, such as lakes, rivers, flood-plain lakes, wetlands, swamps, and soils, are present in various proportions.

On Honshu Island, there are no typical marine diatoms in the transect along the Sendai airport (SZCZUCINSKI *et al.* 2012). The content of marine and brackish species in the deposits is <20 %; in general they can be found in the sand cover closer to the sea. Generally, the number of diatoms increases inland. In the mud zone, only freshwater and brackish-water species were found, and their composition was similar to the diatom composition of soils. No more than 2 % of marine diatoms were found in the Arahama area, and their number drops noticeably inland (TAKASHIMIZU *et al.* 2012). The composition of freshwater diatoms is similar to the composition of rice field soils. A lot of plankton species were carried out by the tsunami from the Teizan-bori Canal and Minami Naga-numa Pond located between the beach ridges; the transfer of plankton diatoms occurred over small distances.

5.6. *Insights from Foraminifera Analysis*

The 2011 Tohoku tsunami deposits of closed bays on Shikotan Island have a small number of benthic foraminifera tests. Only sub-littoral species were found

in the Shikotan Island samples collected immediately after the tsunami from the ice surface; these species are extant and widespread in both tidal and littoral zones of the Lesser Kuril Islands (PREOBRAZHENSKAYA and TROITSKAYA 1996). Good shell preservation indicates a short transfer distance and weak turbulence. Sub-littoral benthic foraminifera also prevail in the tsunami mud sheets. The presence of these species, which inhabit both the littoral and sub-littoral zones (5–10 and 15–50 m deep), is evidence that the material seems to have been transported from the seafloor of the bay. Small amounts of re-deposited species were found in deposits on Shikotan Island. This is also related to the poor preservation of foraminifera in the Holocene marine deposits in the South Kuril Islands, where the carbon material dissolves. In contrast to the Tohoku deposits on the Sendai Plain, fossil species were abundant and were eroded from the marine Miocene and Pliocene deposits, and their amount increased inland (PILARCZYK *et al.* 2012). The species composition of the foraminifera is more diverse in the tsunami deposits on the Sendai Plain. Foraminifera tests in locations with high run-up heights have traces of disintegration, destruction, and balling (PILARCZYK *et al.* 2012).

5.7. *Sediment Sources*

The 2011 Tohoku deposits on the coasts of the Lesser Kuril Islands are a mixture of materials from different sources in different proportions. The results of grain-size analysis and studies of biofossil composition allow us to provide the qualitative analysis of the material, though it is difficult to identify the input of individual sources to this material. A simpler picture can be observed in the open bays, where the sources of well-sorted sands were the beaches, tidal flats, and nearshore regions. Pebbles and boulders were transferred from beaches and storm ridges. The patterns of deposit formation were more complex in closed bays. Here, the material came from offshore, where both marine and terrestrial deposits were washed out. The erosion took place in the inundation zone, where various types of terrestrial deposits were eroded.

The unimodal grain-size distribution curves show a re-deposition of the material from one source, while the bimodal curves are typical for those tsunami deposits that have several sources of material. The main source of the mud fractions is the re-deposited marine mud, which is not common for large-scale tsunami deposits, in which the fine fraction usually has a terrestrial origin (RAZZHIGAEVA *et al.* 2006; SAWAI *et al.* 2009; GOTO *et al.* 2011; SZCZUCINSKI *et al.* 2012). The mineral part of the tsunami mud in Otradnaya and Krabovaya bays was brought from the bottom of the bays; the abundant organic deposits indicate that peats were washed from both coastal and submarine areas of the bays.

The fact that the material came from the offshore area is proved by the presence of marine diatoms, among which sub-littoral benthic and plankton species prevail. Neritic and oceanic diatoms, and possibly some of the sub-littoral plankton species, appear to have come in with the marine water, and not because of bottom erosion. Marine diatoms found in debris marking the inundation zone also came with the marine water. The tsunami carried live colonies of diatoms that settled out in the tsunami deposits. The composition of the benthic foraminifera also indicates the contribution of material from the bottom of the bays.

The freshwater diatom complex reflects the supply of material from different terrestrial sources. If diatoms with a similar ecology dominated in the tsunami deposits, then we could assume that a single type of terrestrial deposits had been washed out. If there was a mixture of freshwater diatoms of different ecologies in the deposits, it would mean the active erosion and re-deposition of material from different terrestrial sources. Tsunamis can wash out terrestrial deposits located below the sea level. This is particularly true of the Lesser Kuril Islands, where tectonic subsidence is the main tendency of the Holocene. For example, the composition of freshwater diatoms in the tsunami deposits from Malaya Tserkovnaya Bay indicated the active erosion of river deposits, mainly in the areas near the river mouth flooded by the sea. The composition of freshwater diatoms in Otradnaya and Krabovaya bays proves the re-deposition of the material from various terrestrial sources: peats, soils, alluvial deposits, paleolakes, and ponds of different types. In one instance where freshwater diatoms were the only diatoms in the tsunami deposits—in mud

caps on the shore of the semi-closed Snezhkov Bay—their composition proved the re-deposition of the material from the flood-plain lake deposits.

The sources of the material in the closed bays of Shikotan Island on the periphery of the Tohoku tsunami, and in the near-field area on Honshu Island, adjacent to the epicenter, differ considerably. Severe erosion took place in the flooded area of Sendai Plain (ABE *et al.* 2012; CHAGUÈ-GOFF *et al.* 2012; PILARCZYK *et al.* 2012; SZCZUCINSKI *et al.* 2012). Little material came directly from the sea; the main sources of sandy fractions were the beach and dunes. Because of the beach and dune wash off, the tsunami had formed a cover of sand deposits, stretching up to 1 km inland. Starting from 1 to 2 km inland, the input of these sources dropped and the erosion of the local soil and material coming from the Teizan-bori Canal became the main source of the tsunami deposits. Further inland, the main source of mud is soil erosion, especially the erosion of rice paddy fields (SZCZUCINSKI *et al.* 2012; TAKASHIMIZU *et al.* 2012).

6. Application to Paleotsunami Research

Usually for paleotsunami reconstruction, the layers of marine sands are used; they form sheets, stretching inland beyond the storm surf zone (ATWATER 1987; MINOURA and NAKATA 1994; DAWSON and SHI 2000; PINEGINA and BOURGEOIS 2001; MACINNES *et al.* 2009; etc.). Numerous paleotsunami sands were found on the Pacific side of Shikotan Island that allow reconstruction of large events for the last 6,000 years (RAZZHIGAEVA *et al.* 2008a; NISHIMURA *et al.* 2009). During studies of the last large-scale tsunami deposits, mud caps were often found (ABE *et al.* 2012; CHAGUÉ-GOFF *et al.* 2011; GOTO *et al.* 2011; RAZZHIGAEVA *et al.* 2006; RAZJIGAEVA *et al.* 2013; SAWAI *et al.* 2009), but layers of mud in the sections of coastal lowlands are not commonly used as indicators of paleotsunamis. This relates to the difficulty of their identification. It is hard to prove their marine origin; the research on the Sendai Plain showed that the terrestrial facies are often the sources of mud (SZCZUCINSKI *et al.* 2012; TAKASHIMIZU *et al.* 2012).

The studies of the 2011 Tohoku tsunami deposit distribution on the coasts of the Lesser Kuril Islands showed that even during a weak tsunami, considerably large mud sheets are formed on the coasts of closed bays; this mud is not easily destroyed, and gets actively colonized by grass. Consequently, it can be well-preserved in sections. Peat on the coasts of closed bays may contain traces of a moderate tsunami that occurred in historic times and during the Holocene. Studying paleotsunamis on Shikotan Island, we devoted less time to the closed bays located on the coast of South Kuril Strait than to the open bays of the Pacific side (RAZZHIGAEVA et al. 2012).

Within the Tohoku tsunami inundation zone in Krabovaya and Otradnaya Bays, the peatlands were drilled along the profiles from the shoreline inland. Also, peatland located at the head of Krabovaya Bay beyond the zone of maximum Tohoku tsunami inundation was drilled. Up to 7–9 layers of greenish-gray mud (up to 8 cm thick) and muddy sands (up to 20 cm thick) were found (Fig. 9). The thickness of some layers, which we could correlate, decreases inland. Some thick layers of deposits have vertical stratification and include layers rich with sand and mud. All layers have clear basal contacts, which could be traced from the coastline into the land. We believed that these deposits were of paleotsunami origin, and that was confirmed by the diatom data.

Only tsunamis can deposit mud on the peat surface at such a great distance from the shoreline. Strong storms throw only a lot of algae and Zostera, which lie on the surface of the peat a few meters from the shoreline. There are no deposits within the strong-storm surf zone.

The number of mud layers varies among the cores due to the different preservation of layers in the sections. In some cases, paleotsunami deposits can occur in the form of small lenses, and drilling does not always find these lens. As a result, a layer may be found in one core, but in other cores the deposits of this age are absent. Radiocarbon dating of the underlying peat determined the age of some paleotsunamis (350 ± 110 BP, 390 ± 110 cal. yBP, LU-7058; 390 ± 40 BP, 430 ± 70 cal. yBP, LU-7088; 620 ± 100 BP, 620 ± 60 cal. yBP, LU-7059; $1,010 \pm 100$ BP, 940 ± 110 cal. yBP, LU-7089) (Fig. 9).

The peat sections on the coasts of Otradnaya Bay and Krabovaya Bay recorded that strong tsunamis occurred in this region. Deposits of paleotsunamis of similar age were found and were widespread on the coasts of the Lesser Kurils and Eastern Hokkaido (NANAYAMA et al. 2003, 2007; RAZZHIGAEVA et al. 2012).

More detailed paleotsunami records were found in the coastal lowland sections of Krabovaya Bay. The top of the peatbogs do not have deposits of another historical tsunami that occurred in the early 17th century. So, the uniqueness of the Tohoku tsunami situation was caused by the presence of fragments of ice in the water that increased the erosion effect. Small lenses of mud in the top of the peat bog on the left side of Krabovaya Bay (looking from the head of the bay) can possibly be traced to a tsunami in 1994. The 1994 tsunami strongly manifested on the Pacific side of the island, where its deposits were found on the coasts of several bays (RAZZHIGAEVA et al. 2007, 2012).

At the head of Krabovaya Bay, such layers of mud could be traced more than 500 m inland. The base of the cores exposed the lagoon mud formed during the Late Holocene transgression; the top of this unit is approximately one meter higher than the tidal zone. The volcanic ash Ta-c of the Tarumai Volcano (2.4–2.5 ka) from southwestern Hokkaido Island was found in the upper part of the marine unit. The ash was formed by yellow silt (2–4 cm in thickness), and glass chemistry analysis suggests that this ash is derived from Ta-c eruption of volcano. The glass compositions of the ash are medium-K, similar to the Tarumai Volcano tephra (FURUKAWA and NAKAGAWA 2010). The peat, above Ta-c ash, includes eight layers of mud that could be traces of tsunamis that have happened in the last 2,400 years. Paleotsunami inundation distances have been calculated for the Late Holocene transgression, when the sea level was higher than the present, taking into account the position of the ancient coastline deposits.

Usually, volcanic ashes in the Holocene peat bog sections of Shikotan occur as lenses and have small thicknesses; and it is difficult to find them in the drilling cores. We could not find tephra layers in the section of peatland located in more dynamic environments on the eastern and western sides of Krabovaya Bay and at the head of Otradnaya Bay.

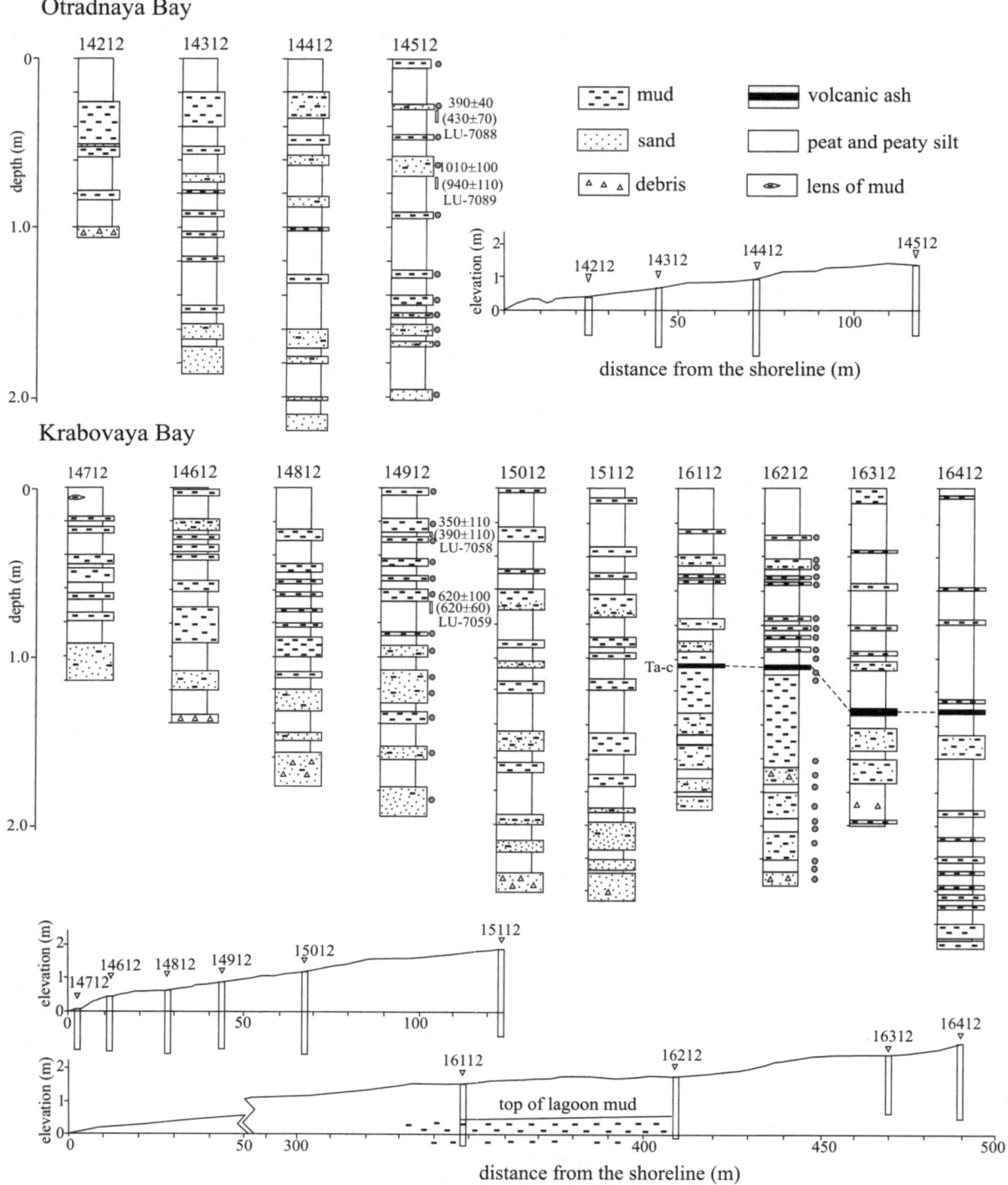

Figure 9
Paleotsunami deposits in peat sections on topographic profiles from **a** Oradnaya and **b** Krabovaya bays, Shikotan Island. *Boxes* indicate the position of the samples for radiocarbon dating, the age is in *brackets*: cal. yBP. *Circles* show the location of samples for the diatom analysis

The grain-size composition of the ancient tsunami mud is similar to the deposits of the 2011 Tohoku tsunami, which proves a likeness of sources and a similarity between their erosion-accumulation processes. If a tsunami occurred at a time of the year when the closed bays were covered by ice, the deposit sheets could have been formed, even with a small run-up, due to increased erosion by water with ice. But, we think that usually, only major tsunamis could lead to the accumulation of long deposit sheets.

Figure 10 shows grain-size curves of the Tohoku tsunami deposits and Late Holocene deposits of different ages, found at the same sites. Grain-size distribution curves in Otradnaya Bay paleotsunami mud is unimodal (mode 20–30 μm, sometimes 10–20 μm), almost symmetrical, with a tail of fine fractions (skewness from 0.65 to 0.84), and mesokurtic. The sand fraction is insignificant (<2.2 %). Grains <10 μm prevail in deposits of Krabovaya Bay (up to 69 %) without sand. The mode is pronounced: 10–20 μm. The distribution curves in more ancient layers are finely skewed (skewness from 0.46 to 0.76), from mesokurtic to platykurtic (kurtosis 2.44–3.03); in the fine pelitic (1–2 μm) area, bimodal properties have been observed. There is a tendency for clay fractions to decrease and the silt amount to increase from ancient to recent deposits. The similarity of grain-size distribution of paleotsunami mud and Tohoku tsunami deposits shows that the sources of the material and the conditions of deposition were very similar.

The marine origin of deposits is confirmed by the diatom data. Marine diatoms are found in all layers. From 14 to 31 marine species are found in deposits of Otradnaya Bay and their content percentage varies from 9 to 39 %. The preservation of frustules differs. As a rule, diatoms are well-preserved only in two layers, probably associated with major tsunamis, and the frustules are broken. Sub-littoral benthic and plankton species prevail in all layers. Plankton species dominate among sub-littoral diatoms in the deposits formed by a stronger tsunami (*Paralia sulcata* up to 13.7 %, *Odontella aurita, Actinocyclus octonarius, Hyalodiscus scoticus, Trachyneis aspera, Thalassiosira bramaputrae*), along with different neritic and oceanic species: northboreal and arctoboreal (*Thalassiosira gravida, T. eccentrica,*

Chaetoceros dyadema, Coscinodiscus granulosus, Actinocyclus curvatulus), and southboreal (*Thalassionema nitzschioides, Actinocyclus divisus, Coscinodiscus*). Freshwater diatoms reflect the redeposition of the material from swamps and shallow ponds with neutral or weakly alkaline pH.

From 15 to 38 marine diatoms are found in the paleotsunami deposits in Krabovaya Bay, which comprise 12.4–66.7 % in section 14912, located on the left side of the bay. The content of marine diatoms decreased (2.4–39.1 %) in the paleotsunami mud in peat section 16212, situated inland. Thicker mud layers include more marine diatoms. The frustules in all layers are well-preserved. Sub-littoral benthic and plankton species are dominant. The deposits with diatoms, typical of shallow semiclosed bays, were most probably formed by the tsunami that provoked the erosion within the shallow water zone and tidal flats. Layers of deposits with marine diatoms, probably indicating deeper bays, were formed during major tsunamis resulting in bottom erosion at greater depths. In most of the layers, neritic and oceanic diatoms are present (*Thalassiosira gravida, T. eccentrica, Porosora glacialis, Coscinodiscus marginatus, Actinocyclus divisus, A. curvatulus, Thalassionema nitzschioides, Coscinodiscus asteromphalus, Thalassiothrix* sp.). Freshwater diatoms indicate that there was material re-deposition from various sources: peats, peaty soils, and deposits of shallow ponds of different types with neutral or weakly alkaline pH. The proportion of material from various terrestrial sources varies in the tsunami deposits of different ages. Among freshwater diatoms in the deposits of ancient tsunamis, we can find different species of *Eunotia and Pinnularia* genera, typical of swamps. These species are almost absent in the 2011 Tohoku tsunami deposits.

7. Conclusions

The comparison of tsunami deposits from the periphery and from the near-field area close to the tsunami showed a considerable dissimilarity in their distribution, grain-size composition, biofossils, and sedimentation. These differences are predetermined by diverse erosion and sedimentological processes,

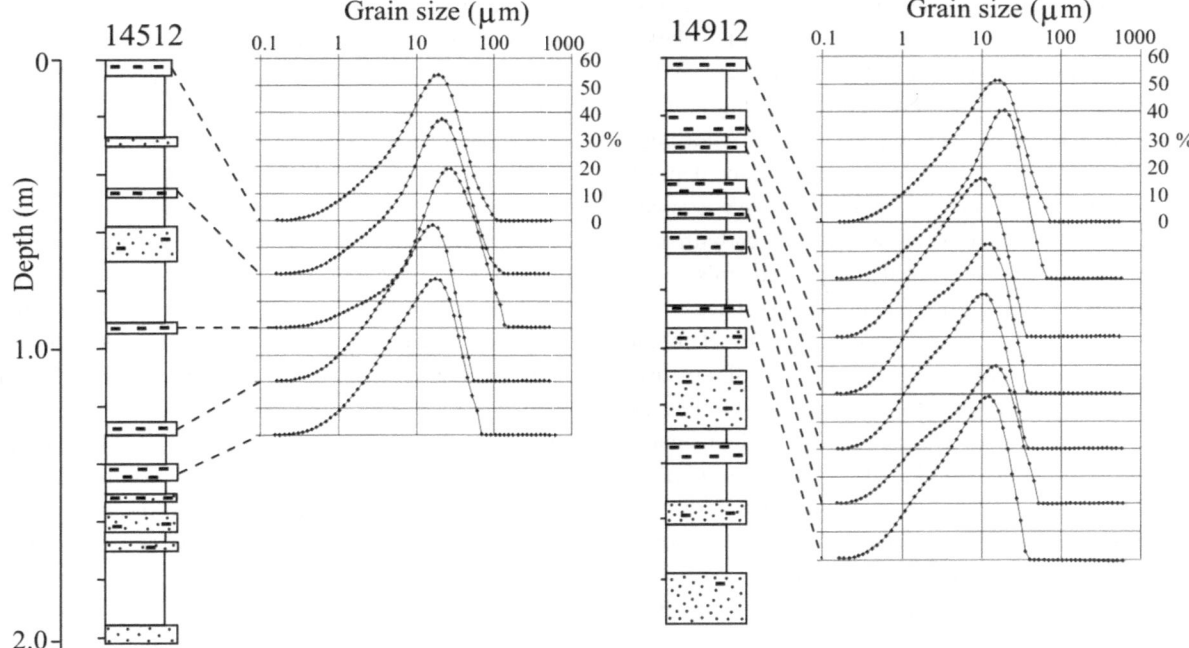

Figure 10
Grain-size distribution curves of the 2011 Tohoku tsunami, and paleotsunami mud samples of different ages from the peatbog of Otradnaya Bay (Core 14512) and the *left side* of Krabovaya Bay (Core 14912). The 2011 Tohoku tsunami and paleotsunami deposits were sampled at the same site

which in turn, indicate the strength of the respective tsunami. Depending on the run-up heights, configuration of the shoreline, and the local topography, the erosion character was different, which reflects the variety of sources of the incoming material. Diverse deposition conditions due to different turbulence of the flow also influenced the deposit composition. In the zones that were remote from the tsunami source, the deposit sheets were formed only on the coasts of closed bays, where the erosion was aggravated by ice fragments in water. The material was mainly transported from the marine source. A lot of fine fractions, good preservation of biofossils, and the presence of diatom colonies indicate the weak turbulence of the flow and short-distance transport. In the near-field zone located in the vicinity of the tsunami source, the most active erosion occurred in flooded areas, which led to the formation of thick deposit sheets. Beaches and dunes were the main sources of sand, while the eroded soil provided mud. In general, terrestrial sources played an important role in the tsunami deposit formation.

The studies of the 2011 Tohoku tsunami on the coasts of the Lesser Kuril Islands showed that muddy layers in the section of coastal lowlands in closed bays could be indicators of paleotsunamis. More detailed records of paleotsunamis could be preserved on the coasts of such bays, in contrast to the coasts of open bays.

Acknowledgments

The authors are thankful to V.N. Boyko, Director of the "Kurilsky" Natural Reserve, for his help in the fieldwork organization. We gratefully acknowledge the contribution from the researchers of the Reserve, I.A. Nevedomskaya, A.G. Savchenko, S.E. Karpenko, and Yu.V. Sinkevich, who sampled the tsunami deposits from the ice floes immediately after the tsunami passage. The work was performed with the financial support of RFBR, grants No. 11-05-00497, 12-05-00757 and FEB RAS grant 12-I-P4-06; the expedition was supported by RFBR grant 11-05-

10027-к, 12-05-10021. We express our sincere gratitude to Mrs. Florence Haiber (New York, USA), who revised the English version. We are very thankful to Fred Stephenson (IOS, Sidney, BC, Canada) for editing the English text. The authors would like to thank Prof. Yuichi Nishimura (Institute of Seismology and Volcanology, Graduate School of Science, Hokkaido University, Japan) and two anonymous reviewers for constructive criticism and productive comments that significantly improved the quality of the paper.

REFERENCES

ABE, T., GOTO, K., and SUGAWARA, D. (2012), *Relationship between the maximum extent of tsunami sand and the inundation limit of the 2011 Tohoku-oki tsunami on the Sendai Plain, Japan*, Sedimentary Geology, *282*, 142–150.

ATLAS OF THE KURIL ISLANDS. (2009), (Design–Information-Cartography, Moscow-Vladivostok).

ATWATER, B.F. (1987), *Evidence for great Holocene earthquakes along the outer coast of Washington State*, Science, *236*, 942–944.

BLOTT, S.J. and PYE, K. (2001), *GRADISTAT: a grain size distribution and statistics package for the analysis of unconsolidated deposits*, Earth Surface and Landforms, *26*, 1237–1248.

CHAGUÉ-GOFF, C., SCHNEIDER, J.-L., GOFF, J.R., DOMINEY-HOWES, D., and STROTZ, L. (2011), *Expanding the proxy toolkit to help identify past events: lessons from the 2004 Indian Ocean Tsunami and the 2009 South Pacific Tsunami*, Earth-Science Reviews, *107*, 107–122.

CHAGUÉ-GOFF, C., NIEDZIELSKI, P., WONG, H.K.Y., SZCZUCIŃSKI, W., SUGAWARA, D., and GOFF, J. (2012), *Environmental impact assessment of the 2011 Tohoku-oki tsunami on the Sendai Plain*, Sedimentary Geology, *282*, 175–187.

DAWSON, A.G. and SHI, S. (2000), *Tsunami deposits*, Pure and Applied Geophysics, *157*, 875–897.

FORAMINIFERA OF FAR EASTERN SEAS OF THE USSR. (1979), (Nauka, Novosibirsk).

FURUKAWA, R., NAKAGAWA, M. (2010), Geological map of Tarumae Volcano 1:30,000. Geological Map of Volcanoes 15. Geological Survey of Japan, AIST.

GOTO, K., CHAGUÉ-GOFF, C., FUJINO, S., GOFF, J., JAFFE B., NISHIMURA, Y., RICHMOND, B., SUGAWARA, D., SZCZUCINSKI, W., TAPPIN, D.R., WITTER, R.C. and YULIANTO, E. (2011), *New insights of tsunami hazard from the 2011 Tohoku-oki event*. Marine Geology. *290*, 46–50.

GOTO, K., SUGAWARA, D., IKEMA, S., and MIYAGI, T. (2012), *Sedimentary processes associated with sand and boulder deposits formed by the 2011 Tohoku-oki tsunami at Sabusawa Island, Japan*, Sedimentary Geology, *282*, 188–198.

INUI, T., YASUTAKA, T., ENDO, K., and KATSUMI, T. (2012), *Geoenvironmental issues induced by the 2011 off the Pacific Coast of Tohoku Earthquake and tsunami*, Soil and Foundations, *52*(5): 856–871.

IVANOV, V.V. (1997), *The investigation of tsunami action on sedimentation by using the traces of October 4, 1994 tsunami*, In

Geodynamics of the Tectonosphere of the Zone of Junction between the Pacific Ocean and Eurasia, v. VIII, (ed. Segreev, K.F.), IMGG FEB RAS, Yuzhno-Sakhalinsk, 1997, pp. 119–128.

IVASHCHENKO, A.I., GUSIAKOV, V.K., DZHUMAGALIEV, V.A., YEH, G., ZHUKOVA L.D., ZOLOTUKHINA, N.D., KAISTRENKO, V.M., KATO, L.N., KLOCHKOV, A.A., KOROLEV, YU.P., KRUGLYAKOV, A.A., KULIKOV, E.A., KURAKIN, V.N., LEVIN, B.V., PELINOVSKY, E.N., POPLAVSKY, A.A., TITOV, V.V., KHARLAMOV, A.A., KHRAMUSHIN, V.N., and SHELTING, E.V. (1996), *The Shikotan Tsunami of October 5, 1994,* Doklady Earth Sciences. 348(4), 693–699.

JAGODZIŃSKI, R., STERNAL, B., SZCZUCIŃSKI, W., CHAGUÉ-GOFF, C., and SUGAWARA, D. (2012), *Heavy minerals in the 2011 Tohoku-oki tsunami deposits—insights into sediment sources and hydrodynamics*, Sedimentary Geology, *282*, 57–64.

KAISTRENKO, V.M., SHEVCHENKO, G.V., and IVELSKAYA, T.N. (2011), *Manifestation of the Tohoku Tsunami of 11 March, 2011 on the Russian Pacific ocean coast*, Questions of Engineering Seismology, *38*(1), 41–64.

KAISTRENKO, V., RAZJIGAEVA, N., KHARLAMOV, A., and SHISHKIN, A. (2013), *Manifestation of the 2011 Great Tohoku tsunami on the coast the Kuril Island: A Tsunami with Ice*, Pure and Applied Geophysics, *170*, 1103–1114.

KAISTRENKO, V. M., GUSYAKOV, V. K., DZHUMAGALIEV, V. A., DYKHAN, G.S., IVASHCHENKO, A.I., YEH, G., KATO, L.N., KLOCHKOV, A.A., PELINOVSKY, E.N., PREDTECHENSKY, G.S., SASOROVA, E.V., TITOV, V.V., KHARLAMOV, A.A. and SHELTING, E.V. (1997), *Manifestation of the Tsunami on October 4, 1994 on Shikotan. Manifestations of Selected Tsunamis.Tsunamis of 1993 and 1994 on the Coasts of Russia,* In Geodynamics of the Tectonosphere of the Zone of Junction between the Pacific Ocean and Eurasia, v. VIII, (ed. Segreev, K.F.), IMGG FEB RAS, Yuzhno-Sakhalinsk, 1997, pp. 55–73.

KRAMMER, K. and LANGE-BERTALOT, H. (1988), Süßwasserflora von Mitteleuropa. Bacillariophyceae, Epithemiaceae, Surirellaceae. VEB Gustav Fisher Verlag, Jena, 596 p.

MACINNES, B.T., PINEGINA, T.K., BOURGEOIS, J., RAZHEGAEVA, N.G., KAISTRENKO, V.M. and KRAVCHUNOVSKAYA, E.A. (2009), *Field survey and geological effects of the 15 November 2006 Kuril tsunami in the middle Kuril Islands*, Pure and Applied Geophysics, *166*(1–2): 3–36.

MINOURA, K., and NAKATA, T. (1994), *Discovery of an ancient tsunami deposit in coastal sequences of southwest Japan: verification of a large historical tsunami*, Island Arc, *3*, 66–72.

MINOURA, K., GUSIAKOV, V.G., KURBATOV, A., TAKEUTI, S., SVENDSEN, J.I., BONDEVIK, S., and ODA, T. (1996), *Tsunami sedimentation associated with the 1923 Kamchatka earthquake*, Sedimentary Geology, *106*, 145–154.

MORI, N., TAKAHASHI, T., YASUDA, T., and YANAGISAWA, H. (2011), *Survey of 2011 Tohoku earthquake tsunami inundation and runup*, Geophysical Research Letters, *38*(7), L00G14, doi:10.1029/2011GL049210, 2011.

NAKAMURA, Y., NISHIMURA, Y., and PUTRA, P.S. (2012), *Local variation of inundation, sedimentary characteristics, and mineral assemblages of the 2011 Tohoku-oki tsunami on the Misawa coast, Aomori, Japan*, Sedimentary Geology, *282*, 216–227.

NANAYAMA, F., SATAKE, K., FURUKAWA, R., SHIMOKAWA, K., ATWATER, B.F., SHIGENO, K., and YAMAKI, S. (2003), *Unusually large earthquakes inferred from tsunami deposits along the Kurile trench*, Nature, *424*, 660–663.

NANAYAMA, F., FURUKAWA, R., SHIGENO, K., MAKINO, A., SOEDA, Y., and IGARASHI, Y. (2007), *Nine unusually large tsunami deposits* from the past 4000 years at Kiritappu march along the southern Kuril Trench, Sedimentary Geology, *200*, 275–294.

NARUSE, H., ARAI, K., MATSUMOTO, D., TAKAHASHI, H., YAMASHITA, S., TANAKA, G., and MURAYAMA, M. (2012), *Sedimentary features observed in the tsunami deposits at Rikuzentakata City,* Sedimentary Geology, *282,* 199–215.

NISHIMURA, Y., NAKAMURA, Y., KAISTRENKO, V.M., and ILIEV, A. YA (2009), *Tsunami deposits and tephras on Kunashir and Shikotan Islands, Southern Kurile Islands,* Chikyu monthly, *31*(6), 311–320.

PILARCZYK, J.E., HORTON, B.P., WITTER, R.C., VANE, C.H., CHAGUÉ-GOFF, C., and GOFF, J. (2012), *Sedimentary and foraminiferal evidence of the 2011 Tohoku-oki tsunami on the Sendai coastal plain, Japan,* Sedimentary Geology, *282,* 78–89.

PINEGINA, T.K., and BOURGEOIS, J. (2001), *Historical and paleo-tsunami deposits on Kamchatka, Russia: long-tern chronologies and long-distance correlations,* Natural Hazards and Earth System Sciences, *1,* 177–185.

PREOBRAZHENSKAYA, T.V. and TROITSKAYA, T.C. (1996), Foraminifera of Far East Seas. Part 1. Foraminifera of littoral of Lesser Kurile Arc. (Dalnauka, Vladivostok 1996).

RAZJIGAEVA, N.G., GANZEY, L.A., GREBENNIKOVA, T.A., IVANOVA, E.D., KHARLAMOV A.A., KAISTRENKO, V.M., and SHISHKIN, A.A. (2013), *Coastal sedimentation associated with the Tohoku tsunami of 11 March 2011 in South Kuril Islands, NW Pacific Ocean,* Pure and Applied Geophysics, *170,* 1081–1102.

RAZZHIGAEVA, N.G., GANZEI, L.A., GREBENNIKOVA, T.A., IVANOVA, E.D. and KAISTRENKO, V.M. (2006), *Sedimentation Particularities during the tsunami of December, 26, 2004, in Northern Indonesia Simelue Island and the Medan coast of Sumatra Island,* Oceanology, *46*(6), 875–890.

RAZZHIGAEVA, N.G., GANZEY, L.A., GREBENNIKOVA, T.A., KHARLAMOV, A.A., IL'EV, A.YA. and KAISTRENKO V.M. (2007), *Deposits of Shikotan earthquake 1994 tsunami,* Oceanology, *47*(4), 579–587.

RAZZHIGAEVA N.G., GANZEY, L.A., GREBENNIKOVA, T.A., KHARLAMOV, A.A., ILYEV, A.YA., KAISTRENKO V.M. (2008a), *Geological record of paleotsumani Striking Shikotan Island, in the Lesser Kurils during Holocene time,* Journal of Volcanology and Seismology, *2*(4), 262–277.

RAZZHIGAEVA N.G., GANZEY, L.A., BELYANINA, N.I., GREBENNIKOVA, T.A., GANZEY, K.S., (2008b) *Paleo-environments and Landscape*

History of Minor Kuril Islands since Late Glacial, Quaternary International, *179.* 83–89.

RAZZHIGAEVA, N.G., GANZEI, L.A., GREBENNIKOVA, T.A., KHARLAMOV, A.A., KAISTRENKO, V.M., ARSLANOV, KH.A., and GORBUNOV, A.O. (2012), *Manifestation of Holocene Tsunamis on the Lesser Kuril Ridge,* Russian Journal of Pacific Geology, *6*(6), 448–456.

RICHMOND, B. SZCZUCIŃSKI, W., CHAGUÉ-GOFF, C., GOTO, K., SUGAWARA, D., WITTER, R., TAPPIN, D.R., JAFFE, B., FUJINO, S., NISHIMURA, Y., and GOFF, J. (2012), *Erosion, deposition and landscape change on the Sendai coastal plain, Japan, resulting from the March 11, 2011 Tohoku-oki tsunami,* Sedimentary Geology, *282,* 27–39.

SAWAI, Y., JANKAEW, K., MARTIN M.E., PREDNERGAST, A., CHOOWONG, M., and CHATOENTITIRAT, T. (2009), *Diatom assemblages in tsunami deposits associated with the 2004 Indian Ocean tsunami at Phra Thong Island, Thailand.* Marine Micropaleontology, *73,* 70–79.

SEN GUPTA, B.K. (2002). Foraminifera in marginal marine environments, In Modern Foraminifera, (ed. Barun K. and Sen Gupta B.K.), Kluwer Academic Publishers, Dortrecht 2002, pp. 141–160.

SHEVCHENKO, G., IVELSKAYA, T., and LOSKUTOV, A. (2014), *Characteristics of the 2011 Great Tohoku Tsunami on the Russian Far East Coast: Deep-water and coastal observations,* Pure Appl. Geophys., *171* (this volume).

SZCZUCIŃSKI, W., KOKOCIŃSKI, M., RZESZEWSKI, M., CHAGUÉ-GOFF, C., CACHĪO, M., GOTO, K., and SUGAWARA, D. (2012), *Sediment sources and sedimentation processes of 2011 Tohoku-oki tsunami deposits on the Sendai Plain, Japan—Insights from diatoms, nannoliths and grain size distribution,* Sedimentary Geology, *282,* 40–56.

TAKASHIMIZU, Y., URABE, A., SUZUKI, K., and SATO, Y. (2012), *Deposition by the 2011 Tohoku-oki tsunami on coastal lowland controlled by beach ridges near Sendai, Japan Original Research Article,* Sedimentary Geology, *282,* 124–141.

THE DIATOMS OF THE USSR. (1974), Fossil and recent (Nauka, Leningrad).

THE DIATOMS OF THE USSR. (1992), Fossil and modern (Nauka, Leningrad).

YEH, H., TITOV, V., GUSIAKOV, V., PELINOVSKY, E., KHRAMUSHIN, V., and KAISTRENKO, V. (1995). *The 1994 Shikotan earthquake tsunamis,* Pure Appl. Geophys. *144*(3–4), 855–874.

(Received June 19, 2013, accepted February 6, 2014, Published online February 22, 2014)

Pure Appl. Geophys. 171 (2014), 3329–3350
© 2013 Springer Basel
DOI 10.1007/s00024-013-0727-1

Characteristics of the 2011 Great Tohoku Tsunami on the Russian Far East Coast: Deep-Water and Coastal Observations

GEORGY SHEVCHENKO,[1] TATIANA IVELSKAYA,[2] and ARTEM LOSKUTOV[1]

Abstract—The source region of the catastrophic Tohoku tsunami of 11 March 2011 was not far from the Russian Far East coast. The Sakhalin Tsunami Warning Center at Yuzhno-Sakhalinsk issued an Alert for threatened coasts of the Kuril Islands and Kamchatka; the observed tsunami heights were up to 2–2.5 m along the Pacific coast of the Kuril Islands. The tsunami was clearly recorded by a number of coastal tide gauges and by deep-ocean bottom pressure stations located in the vicinity of the Kuril Islands, as well as by the Russian open-ocean DART station 21401, located eastward from the South Kuril Islands. The data from other DART stations, in particular those located near Japan and the Aleutian Islands, were used for comparison. The records from these instruments were used to estimate the major characteristics of the tsunami waves, including arrival times, maximum heights, durations of the signal and main wavelength periods, as well as for comparison with the results of numerical modelling. In contrast to deep-sea stations where the first waves were the highest, at Russian coastal sites the highest waves occurred several hours after the arrival of the first tsunami wave. Further analysis indicated significant differences in spectral characteristics of tsunami waves propagating eastward toward North America and those directed in the northeast direction towards the Russian Far East. The main peaks of the eastward propagating tsunami waves were relatively high-frequency (periods ranging from 6–8 to 15–20 min), while those propagating in the northeast direction were mainly low-frequency (ranging from 25–40 to 60–80 min). Additionally, pronounced spectral peaks with similar long periods were found in the "middle-field" records at Hanasaki (on the northeastern coast of Hokkaido Island), at Yuzhno-Kurilsk (Kunashir Island) and at Malokurilsk (Shikotan Island). At remote stations, the resonant periods associated with local topographic features were predominant in the spectra.

Key words: Tsunami measurements, tsunami warning service, Kuril Islands, long waves, resonant mode, tide gauges, bottom pressure stations, DART, spectral analysis, numerical modelling, tsunami source.

1. Introduction

On 11 March 2011, at 05:46 UTC, a mega-thrust earthquake with moment magnitude $M_w = 9.0$ (the largest instrumentally recorded earthquake in Japan) with a focal depth of 32 km occurred off the northeastern coast of Honshu (cf. SONG et al. 2012). This earthquake generated a catastrophic tsunami that killed about 20,000 people, and caused enormous damage. The tsunami devastatingly affected the Japanese Pacific coast from Hokkaido to Kyushu (about 2,000 km); the maximum inundation was observed on the northeastern coast of Honshu Island (MORI et al. 2011). Tsunami waves were clearly recorded by a number of deep-ocean DART tsunameters (TANG et al. 2012; MUNGOV et al. 2013; HEIDARZADEH and SATAKE 2013) and by the NEPTUNE-Canada geophysical network (FINE et al. 2013). These data were actively used for comparison and verification of numerical models. The Tohoku tsunami was also recorded by numerous coastal tsunami gauges in far-field areas; for example, on the coasts of California (CA), French Polynesia, Australia and New Zealand (BORRERO and GREER 2013; BORRERO et al. 2013; REYMOND et al. 2013; HINWOOD and McLEAN 2013). The extreme distant wave heights (with amplitudes of more than 2.5 m) were observed at Crescent City, CA; the 2011 Tohoku tsunami caused great damage here (WILSON et al. 2013).

The source of the 2011 Tohoku tsunami was located close to the Russian Far East; the tsunami presented a serious threat for the coasts of the Kuril Islands and Kamchatka. The Russian Tsunami Warning Service (TWS) declared a Tsunami Alert for these regions; the people were evacuated from low-

¹ Institute of Marine Geology and Geophysics, FEB RAS, Yuzhno-Sakhalinsk, Russia. E-mail: shevchenko@imgg.ru
² Sakhalin Tsunami Warning Center, Yuzhno-Sakhalinsk, Russia.

Reprinted from the journal

lying areas and the ships were directed into the open ocean. This tsunami was clearly recorded by a number of Russian coastal tide gauges and bottom pressure stations (SHEVCHENKO and IVELSKAYA 2012). A network of precise coastal telemetric sea level gauges, deployed by Russia during the last two years, has effectively measured the Tohoku tsunami waves at 17 sites: Primorye (Box A in Fig. 1; three stations); Sakhalin Island (Box B; seven), the Kuril Islands (Box C; four), Kamchatka (Box D; two), and the Commander Islands (Box E; one). This shows recent rapid development of the Russian TWS network. For comparison, the 2009 Samoan tsunami was recorded by only one Russian TWS station, and the 2010 Chilean tsunami was recorded by four stations (SHEVCHENKO et al. 2013). The progress of the Russian tsunami monitoring system tsunami gauges improves the TWS efficiency and enables us to provide a more detailed study of the tsunami features in various parts of the Russian Far East coast.

This tsunami was also recorded by the Russian open-ocean DART station 21401, located east of the South Kuril Islands, and by several temporary autonomous bottom pressure stations near the Kuril Islands. The data from DARTs 21418 (located near Japan), 21419 (the Middle Kuril Islands), 21416 (southeastern Kamchatka) and 21415 (western part of the Aleutian Islands) were used for comparison (locations of DARTS are shown in Fig. 1). These instruments enabled us to estimate major characteristics of the observed tsunami waves, including their arrival times, maximum wave heights, durations of the signal and main periods. The FFT and wavelet analysis were used to describe the spectral content of the tsunami signal. This study provides a comprehensive analysis of the middle-field (the South Kuril Islands and adjacent northeastern part of Hokkaido Island) and far-field (Kamchatka, the Commander Islands, Sakhalin Island and Primorye) tsunami characteristics on the Russian Far East coast, which has not experienced a major tsunami in the Pacific Ocean in nearly 50 years. The main goal of the present study was to examine tsunami parameters and the effects of local bottom topography and coastline geometry on the spatial variability of tsunami heights in this particular region.

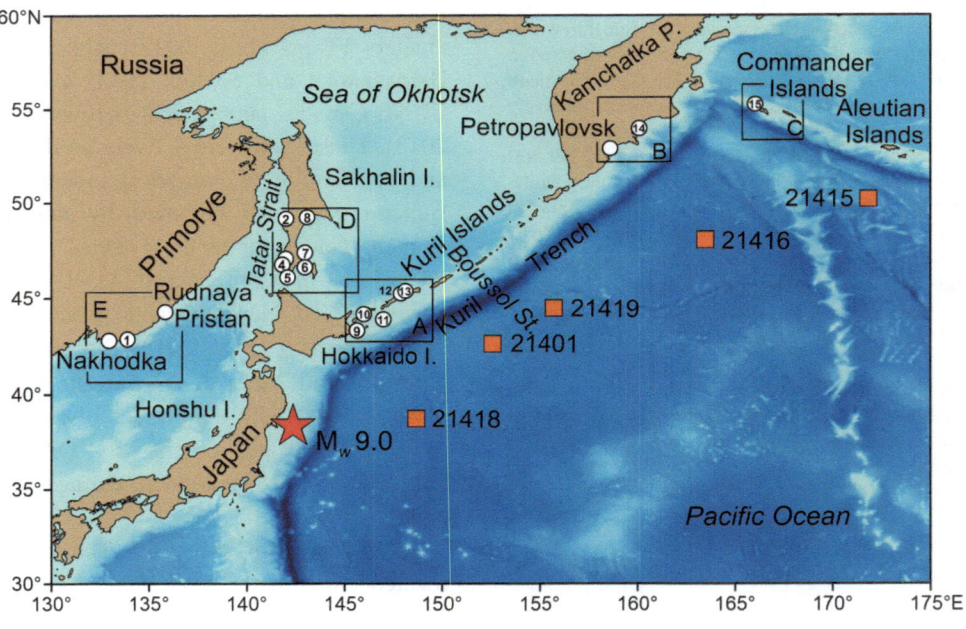

Figure 1

The locations of the open-ocean bottom pressure DART stations (*orange squares*) in the northwestern Pacific Ocean and the TWS tide gauges (*white circles, 1* Preobazhenie, *2* Uglegorsk, *3* Kholmsk, *4* Nevelsk, *5* Krilion, *6* Korsakov, *7* Starodubsk, *8* Poronaisk, *9* Hanasaki (Hokkaido, Japan), *10* Yuzhno-Kurilsk, *11* Malokurilsk, *12* Kurilsk, *13* Reidovo, *14* Semyachik, *15* Nikolskoe) on the Russian Far East coast. The position of earthquake epicenter is shown by a *red star*. The *boxes A, B, C, D, E* are shown below (Figs. 6, 9, 10, 15)

2. Data and Methods

The 2011 Tohoku tsunami was recorded by the above deep-ocean DART stations in the vicinity of the Kuril Islands. The tsunami records from DART stations 21418 (off northeastern Honshu) and 21415 (near the western Aleutian Islands) were also used for comparison. Unfortunately, the record of DART 21416 (southeastern Kamchatka) was of poor quality; so we could use only fragments of this record.

DART stations are located in the open ocean where the bottom is approximately flat (see Fig. 1). The respective records contain a relatively clean tsunami signal, not distorted by the impact of steep depth variations, typical for the area of the continental shelf and slope. The records could therefore be used to study subtle effects, such as the wave dispersion. The calculation of frequency–time (f–t) diagrams (e.g. KULIKOV and GONZÁLEZ 1996) is an effective approach for the assessment of the dispersion effect on tsunami wave propagation. This method allows us to determine the arrival time of wave components with different frequencies (DZIEWONSKI et al. 1969), which is an efficient way to detect the tsunami dispersion. This method was used to analyze Tohoku tsunami records in several papers (HEIDARZADEH and SATAKE 2013; BORRERO and GREER 2013).

Tsunami waves, which travel for long distances over a relatively uniform bottom, have to exhibit frequency dispersion. For slow bottom variation along the tsunami ray, the known dispersion equation may be written as:

$$\omega^2 = gk \tanh kH(s), \qquad (1)$$

where ω is the angular frequency of a harmonic constituent of the wave packet, k is the wave number, g is the gravity acceleration, s is the ray length and $H(s)$ is the along-ray depth profile.

Using dispersion Eq. (1), we can estimate and plot the theoretical dispersion curves of tsunami component arrival times t_A at specific site as function of frequency ω:

$$t_A(\omega) = \frac{L}{c_g(\omega)}, \qquad (2)$$

where the group speed c_g is determined by

$$c_g = \frac{1}{2}c\left[1 + \frac{2kH}{\sinh(2kH)}\right], \qquad (3)$$

and c is the phase speed of the surface waves:

$$c = \sqrt{gH\frac{\tanh(kH)}{kH}}. \qquad (4)$$

The theoretical arrival time curves were then compared with the observed f–t (wavelet) diagrams which enable us to verify front wave dispersion effects as discussed later.

The 2011 Tohoku tsunami was also clearly recorded by a number of coastal tide gauges and bottom pressure stations on the Pacific coast of Russia (Fig. 1). Unfortunately, because of technical problems, some of the TWS stations were not in operation during the event. In the area of the South Kuril Islands, the Tohoku tsunami was recorded by two TWS gauges: Yuzhno-Kurilsk (Yuzhno-Kurilsk Bay on the east coast of Kunashir Island, overlooking Yuzhno-Kurilskiy Strait) and Kurilsk (Kitoviy Bay on the Okhotsk Sea coast of Iturup Island). There was also another station, an autonomous bottom pressure gauge deployed by the Institute of Marine Geology and Geophysics (IMGG) on this coast (Reidovo Village, in the Gulf of Prostor, Iturup Island), which recorded the 2011 tsunami. This bay is separated from Kitoviy Bay by the Chirip Peninsula, which juts out into the sea.

A number of autonomous IMGG stations in bays of Shikotan Island, which recorded the 2009 Samoan and 2010 Chilean tsunamis (SHEVCHENKO et al. 2013), have since been lost due to strong currents caused by the Tohoku tsunami.

The Tokoku tsunami was also clearly recorded by seven TWS stations on both the east and west coasts of Sakhalin Island and by three stations that are located on the Primorye coast (Fig. 1). The 2011 tsunami data collected at coastal stations enabled us to estimate the tsunami arrival times and wave heights (Tables 1, 2, 3, and 4). Additionally, the dominating oscillation periods have been estimated, which are mainly determined by the frequency-selective properties of the local topography.

In order to examine the spectral properties of recorded tsunami oscillations at each site and compare them with the properties of background

Table 1

Statistical characteristics of the 11 March 2011 Tohoku tsunami estimated from open-ocean DART stations in the Northwestern Pacific

DART station	ETA (TAT) (UTC)	Wave	First wave		Maximum wave	
			Height (cm)	Arrival time	Height (cm)	Arrival time
21418	6:11 (6:12)	Crest	183.0	6:19	183.0	6:19
		Trough	−97.0	6:26	−97.0	6:26
21401	6:42 (6:43)	Crest	65.0	6:53	65.0	6:53
		Trough	−27.0	7:04	−27.0	7:04
21419	7:04 (7:03)	Crest	48.0	7:17	48.0	7:17
		Trough	−24.0	7:27	−24.0	7:27
21416	7:51 (7:53)	Crest	33.0	8:09	33.0	8:09
		Trough	−24.0	8:21	−24.0	8:21
21415	8:42 8:45)	Crest	26.6	8:58	26.6	8:58
		Trough	−16.6	9:10	−16.6	9:10

ETA estimated time of tsunami arrival, *TAT* actual (observed) tsunami arrival time (UTC)

Table 2

Statistical characteristics of the 11 March 2011 Tohoku tsunami estimated from tide gauges located on the Pacific coast of Russia (Kuril Islands and Kamchatka) and at Hanasaki (Hokkaido Island, Japan)

Station (region)	ETA (TAT) (UTC)	Wave	First wave		Maximum wave	
			Height (cm)	Arrival time	Height (cm)	Arrival time
Hanasaki (Hokkaido I.)	6:26 (6:43)	Crest	256.8	6:58	256.8	6:58
		Trough	<−110[a]	7:18	<−110[a]	7:18
Malokurilsk (Shikotan I.)	6:53 (6:59)	Crest	83.1	7:10	149.9	9:09
		Trough	−65.2	7:25	−79.0	9:04
Yuzhno-Kurilsk (Kunashir I.)	7:29 (7:25)	Crest	91.9	7:47	92.9	10:59
		Trough	−73.8	8:17	−102.3	10:40
Kurilsk (Iturup I.)	7:23 (7:07)	Crest	10.4	7:18	27.1	9:00
		Trough	−8.1	7:43	−29.3	8:48
Reidovo (Iturup I.)	7:24 (7:44)	Crest	19.2	8:04	78.1	17:51
		Trough	−14.0	8:15	−67.8	17:40
Nikolskoe (Bering I.)	8:43 (8:47)	Crest	10.3	9:19	8.3	23:22
		Trough	−10.3	9:38	−15.6	23:06
Semyachik (Kamchatka)	8:30 (8:49)	Crest	30.7	9:11	90.3	16:21
		Trough	−46.3	9:39	−47.2	15:45
Petropavlovsk (Kamchatka)	8:38 (8:44)	Crest	27.2	9:06	20.7	17:51
		Trough	10.6	9:29	−2.9	17:25

[a] The maximum wave trough was chopped off at a level of −110 cm (relative to zero tide) because the instrument was not designed to measure such strong oscillations

oscillations, we used one-day time intervals of sea level records just before the tsunami arrival (on March 10) and during the tsunami (from March 11, 6:00 UTC to March 12, 6:00 UTC). It is well known that tsunami waves observed at coastal stations are strongly influenced by local topographic effects. To improve the spectral estimates, we used a Kaiser-Bessel spectral window with half-window overlapping prior to the Fourier transform (cf. EMERY and THOMSON 2003). The length of the window was chosen as 6 h, yielding a degree of freedom equal to 14. According to the length of the window, we did not analyze the spectral characteristics for periods of more than 3 h. To suppress these effects and to reconstruct the spectral characteristics of the source, RABINOVICH (1997) suggested a method based on the estimation of tsunami to background spectral ratios. We used this method to eliminate local resonant

Table 3

Statistical characteristics of the 11 March 2011 Tohoku tsunami estimated from tide gauges located on the southeastern coast of Sakhalin Island

Station	ETA (TAT) (UTC)		First wave		Maximum wave	
			Height (cm)	Arrival time	Height (cm)	Arrival time
Krilion	8:46 (9:04)	Crest	7.6	9:19	12.8	13:45
		Trough	−0.7	9:34	−9.3	14:11
Korsakov	8:54 (8:00)	Crest	22.4	9:52	48.9	18:30
		Trough	7.0	10:36	−12.9	18:05
Starodubsk	8:25 (7:35)	Crest	22.9	8:57	33.1	15:14
		Trough	6.2	9:11	−19.7	14:44
Poronaisk	10:10 (9:40)	Crest	27.1	10:44	39.2	8:53[a]
		Trough	−6.0	11:03	−37.8	9:28[a]

[a] Observed on 12 March 2011

Table 4

Statistical characteristics of the March 2011 Tohoku tsunami estimated from tide gauges located on the coasts of southwestern Sakhalin Island and Primorye

Station (Region)	ETA (TAT) (UTC)		First wave		Maximum wave	
			Height (cm)	Arrival time	Height (cm)	Arrival time
Uglegorsk (Sakhalin)	? (8:58)	Crest	?	?	12.2	3:49[a]
		Trough	?	?	−5.9	4:04[a]
Kholmsk (Sakhalin)	? (7:22)	Crest	?	?	16.0	2:41[a]
		Trough	?	?	−9.2	3:01[a]
Nevelsk (Sakhalin)	7:32 (7:10)	Crest	2.2	10:33	13.7	14:25
		Trough	−4.8	10:18	−12.7	14:40
Preobrazhenie (Primorye)	6:36 (6:34)	Crest	0.8	6:37	21.5	4:56[a]
		Trough	−4.4	6:40	−9.9	5:13[a]
Rudnaya Pristan (Primorye)	6:51 (6:48)	Crest	8.6	6:56	18.2	0:14[a]
		Trough	−7.5	7:00	−18.3	23:39
Nakhodka (Primorye)	6:58 (7:25)	Crest	2.8	7:33	15.4	20:40
		Trough	1/4	7:36	−13.3	21:26

? indicates that the height and arrival time of first wave could not be determined

[a] Observed on 12 March 2011

effects and estimate the source spectral properties. In this case, to get more consistent estimates, we used two-day intervals to calculate background spectra.

3. Numerical Modelling

To study the propagation of the 2011 Tohoku tsunami waves to the Russian Far East, we used the same numerical tsunami model used by RABINOVICH *et al.* (2008) to examine the 2006 and 2007 Kuril Islands tsunamis. The model is based on the shallow-water finite difference formulation, which is similar to that described by IMAMURA (1996). The 1-arc min seafloor topographic grid was derived from the website http://www.shipdesign.ru/Invent/index.html, created by V. N. Khramushin for the northwestern Pacific Ocean, the Sea of Japan and the Sea of Okhotsk. Tsunami waveforms simulated by this model were compared with those from the open-ocean DARTs and coastal tide gauges, thus verifying the model formulation and the initial tsunami source parameters (see Figs. 4, 7, 11, and 13 in the following text).

In this simulation, we applied the finite-fault source model constructed by HAYES (2011). The slip distribution was then converted into vertical displacements of the ocean bed using OKADA (1985) formulas for the semi-infinite elastic media. The same model of tsunami source was used by FINE *et al.* (2013) to study the propagation and decay of the 2011 Tohoku tsunami waves in the Pacific Ocean. The source model provides detailed information about the spatial displacements in the tsunami generation area and minimizes the uncertainties in the size of the source.

As will be shown below, the tsunami waveforms observed at deep-water stations are in good agreement with those numerically simulated. This is also true for most of the coastal stations, except some on the Primorye coast and for some located deep in bays on the Pacific Ocean coast. This suggests that the model correctly describes the propagation of the 2011 Tohoku tsunami and can be used to explain the principal features of the tsunami waves along the Pacific coast of Russia. The difference between the estimated time of tsunami arrival (ETA), and the observed tsunami arrival time (TAT, see Table 1) recorded at DARTs 21418, 21401 and 21419, is only about 1 min. The difference between ETA and TAT increases with distance and reaches 2 min at DART 21416 and 3 min at DART 21415 near the western Aleutian Islands.

Several obvious features, possibly small-scale bottom irregularities, wave dispersion, coupling with elastic bottom and gravity potential change injecting deviation to dispersion law, may produce observed ETA–TAT difference (WATADA *et al.* 2013). The numerical shallow water model has its own numerical dispersion, which depends on both spatial and temporal steps. Numerical dispersion, according to IMAMURA *et al.* (1988), can match the physical dispersion when grid size satisfies a certain condition. For practical purposes, the optimal grid size for the Pacific Ocean should be 4–5 arc min according to the Imamura relation.

For coastal stations, the difference between TAT and ETA is even bigger (see Table 2); it reaches 2–6 min for deep coast areas and 12–19 min for shallow areas. The difference between ETA and TAT increases even further on the coast of Sakhalin Island (Table 3). It appears that the inaccuracies and insufficient resolution in the approximation of bottom topography in the shallow-water areas increase errors in calculated tsunami arrivals.

Figure 2 shows the results of numerical modelling of the 2011 Tohoku tsunami 70 min after the earthquake when the waves reached the South Kuril Islands and 190 min after the earthquake when the waves reached the southern part of Sakhalin Island. It can be seen in Fig. 2 that for 70 min the tsunami waves propagated relatively slowly through Yuzhno-Kurilsk Strait and the relatively small straits of the Lesser Kuril Islands. At the deeper shelf of Iturup Island, the tsunami propagated more rapidly and reached Ekaterina Strait (separating Kunashir and I-turup islands) from the northeast. At about the same time, the tsunami approached Friz Strait (between I-turup and Urup islands), which is wider and deeper. Through Friz Strait, the tsunami entered the Sea of

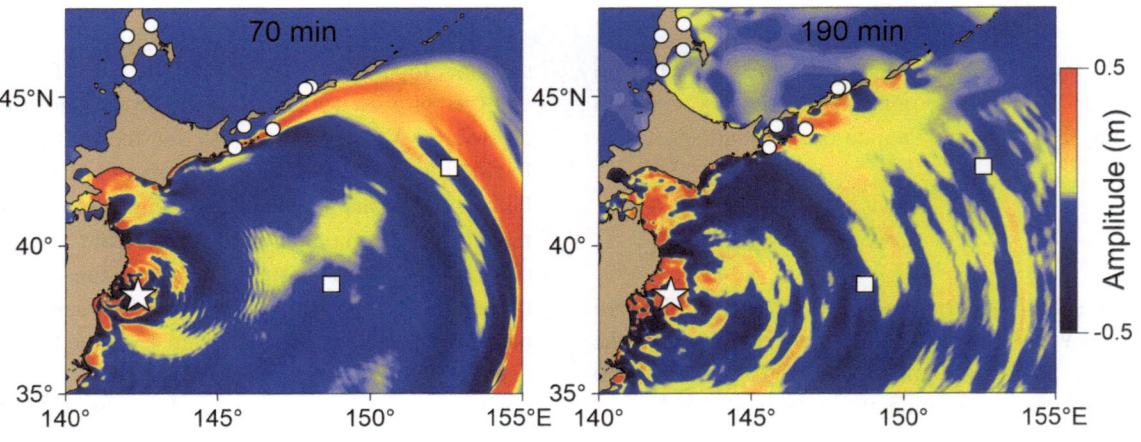

Figure 2

Snapshots of numerically simulated waves of the 2011 Tohoku tsunami for 70 and 190 min after the main shock. The position of earthquake epicenter is shown by a *white star*

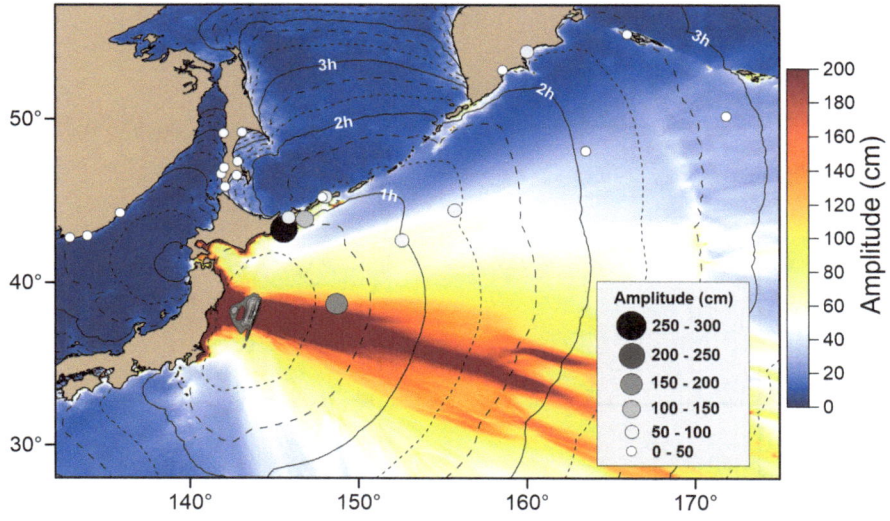

Figure 3
The spatial distribution of the maximum simulated Tohoku tsunami wave amplitudes. The maximum observed wave amplitudes at various stations are shown by different *color* and size *circles* (TWS and DART)

Okhotsk and reached the Gulf of Prostor, where the Reidovo settlement is located (Fig. 6).

The 190 min plot shows very complex oscillations in the area of the South Kuril Islands, and particularly in Yuzhno-Kurilsk Strait. Even visually, it can be noticed that expressed lower frequency oscillations prevail in this region in contrast to the shorter and deeper shelf of Iturup Island.

The spatial distribution of the maximum Tohoku tsunami simulated wave heights is shown in Fig. 3. From this figure, it follows that the Russian Far East coast is located in a shadow from the main sector of the tsunami energy propagation. Similar numerical results were obtained by Tang *et al.* (2012). However, in the area adjacent to the South Kuril Islands (and to some extent on the shallow wide shelf of the Northern Kuril Islands and Kamchatka), we also see substantial wave heights. On the coasts of Sakhalin Island, the maximum simulated wave heights were found on the northeastern shelf, specifically in the area of the offshore oil and gas development. The tsunami-caused sea-ice motions (Ikonnikova, USSR Hydromet, reported about such motions in the case of the Chilean tsunami of 22 May, 1960) were probably most dangerous for oil and natural gas production facilities (drilling platforms, pipelines, etc.).

4. Deep-Ocean DART Stations

Let's consider tsunami records from deep-ocean stations located in the Northwest Pacific near Honshu and the Kuril Islands (Fig. 1). Eight-hour segments (from 4:00 to 12:00) were selected and the preliminary calculated tides were subtracted from the observed series. A similar variation pattern was recorded at all stations: first, a strong solitary wave and then a long train of much weaker oscillations with amplitudes <10 cm (Fig. 4). In all cases, high-frequency variations were recorded shortly after the main earthquake shock (i.e. prior to the main wave arrival); this is a typical feature of near-bottom pressure records associated with seismic Rayleigh waves (e.g. Rabinovich *et al.* 2013b). The DART station 21418, which is the closest to the earthquake epicenter, recorded the tsunami arrival at 06:11 UTC, with the main peak at 06:19 (the elevation from the normal tide level was 187 cm).

The positive elevation is most important: this parameter determines the distance of the tsunami inundation and the character of its impact on shores and coastal localities. The negative elevation is also important (maximum trough was 94 cm at 06:26), since it is related to the dynamic loading on the coast (many buildings and coastal structures

161

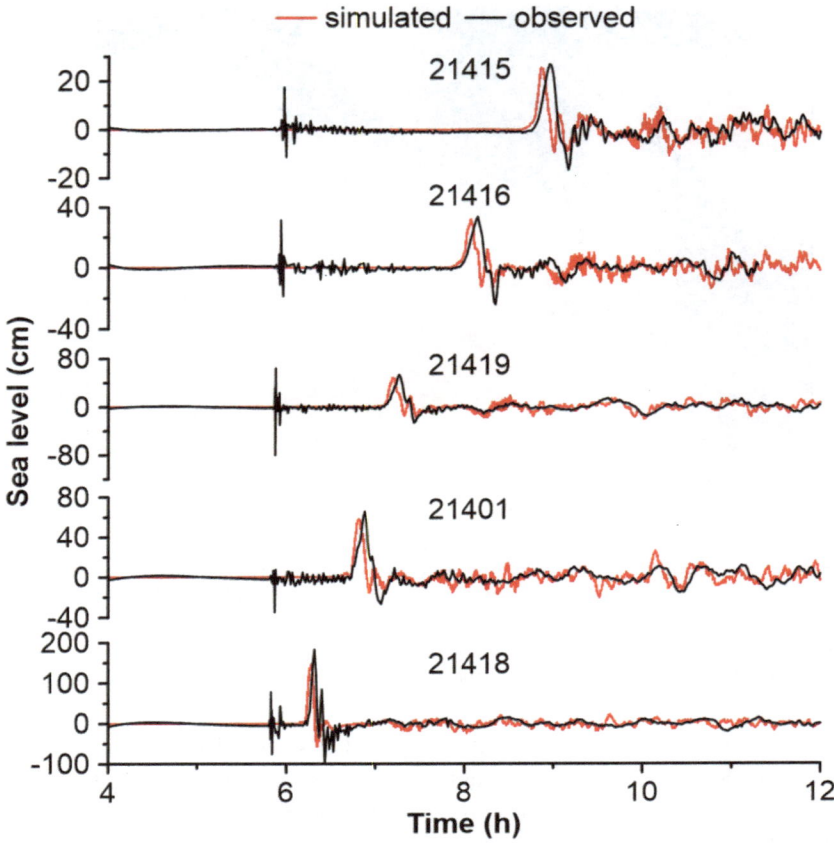

Figure 4

The 8-h simulated and observed segments of the Tohoku tsunami records (4:00–12:00 UTC on 11 March 2011) for DART stations installed near the Kuril Islands, Honshu Island and the western Aleutian Islands

are damaged by tsunami waves rolling back to the sea), but this parameter is usually considered separately.

At DART 21401, located beyond the deep-ocean trench in the area of the South Kuril Islands, the tsunami arrived at 06:43 UTC. Figure 4 indicates a gradual decrease in wave amplitudes for tsunami waves spreading from the source to various parts of the Kuril Islands and the western Aleutian Islands (it is related to the increase in the wave frontal line with distance). Observed and simulated tsunami waveforms are in good agreement for all stations (although the difference in arrival times slowly but progressively increases with the distance from the source).

To investigate the main periods of tsunami-induced sea level oscillations and their evolution in time, we estimated frequency–time (*f–t*) diagrams

(DZIEWONSKI *et al.* 1969; EMERY and THOMSON 2003) which are similar to wavelet analysis. The calculation was performed for periods from 2 to 100 min (frequencies from 0.5 to 0.01 cpm); the matrix of spectral amplitudes was scaled to 30 cm for DART 21418 and to 10 cm for the other four DARTs. The calculation results are presented in Fig. 5. Certain differences in tsunami behavior were derived for various stations. In particular, at DART 21418 (nearest to the source), the signal had higher frequencies, with the main maxima at periods of 6–8 min and 15–20 min. At other DART stations, the tsunami signal was different from those at 21418 but similar to each other; in general, the respective records have more energy at lower frequencies with the main peak at periods of 20–30 min and considerable energy at periods of 50–80 min.

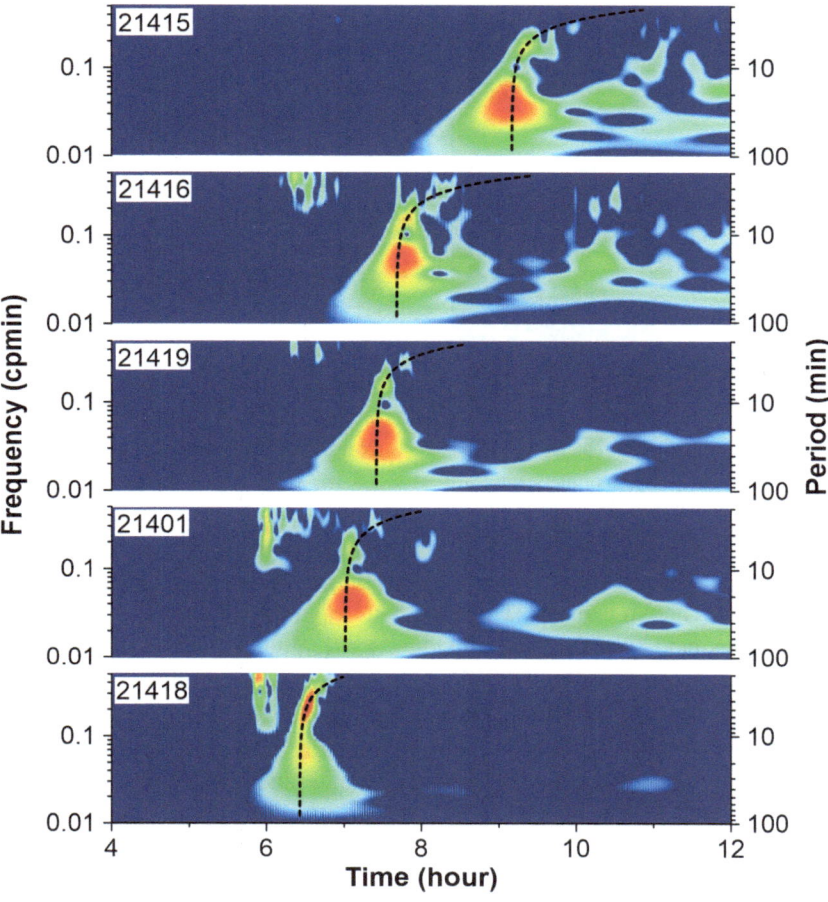

Figure 5
Frequency–time (*f–t*) diagrams of the 2011 tsunami waves observed at DART stations. The maximum amplitude is 30 cm for DART 21418 and 10 cm for other DARTs. The isolines are drawn with 1 dB step. *Black dashed curves* show the theoretical dispersion curves

Such substantial spatial differences in the open-ocean energy distribution are typical for tsunamis and are probably caused by the spatial extent of the source: longer waves propagate in the direction of the long source axis, while shorter waves in the direction of the short axis. This feature of non-isotropic tsunami radiation is clearly visible in the results of the numerical simulation (Fig. 2). The previously discussed example (the observed DART signals) provides obvious evidence of this property.

An important result revealed during this analysis is the clearly expressed wave dispersion: high-frequency oscillations delayed in comparison with lower-frequency components. This effect is hardly identifiable at coastal stations due to the strong influence of reflected and refracted waves on the shelf

and in the near-coast zone. In the open ocean, the effect of wave dispersion on propagating tsunami waves was first identified by GONZÁLEZ and KULIKOV (1993) and KULIKOV and GONZÁLEZ (1996). The theoretical estimation of wave dispersion for different distances from the source was found to be in good agreement with observations (Fig. 5).

5. The 2011 Tohoku Tsunami on the Coasts of the Kuril Islands and Kamchatka

5.1. The South Kuril Islands

The South Kuril Islands, the nearest islands to the earthquake epicenter in Russian territory, were at the

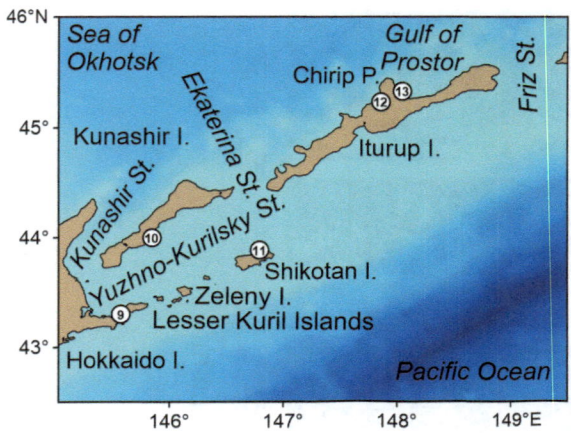

Figure 6
The locations of TWS tide gauges on the coasts of the South Kuril Islands (*white circles*, *10* Yuzhno-Kurilsk, *11* Malokurilsk, *12* Kurilsk, *13* Reidovo) and Hanasaki station on Hokkaido Island, Japan (*9*) (corresponds to box "A" in Fig. 1)

highest risk during the 2011 Tohoku tsunami. Therefore, the main attention was paid specifically to this area (KAISTRENKO *et al.* 2013; RAZJIGAEVA *et al.* 2013). Tsunami waves that propagated towards the South Kuril Islands were first detected at Hanasaki on the northeastern coast of Hokkaido Island, Japan (Figs. 2, 6). The tsunami data recorded at this station were very important due to the proximity of this site to the South Kuril Islands. The tsunami was not instrumentally recorded on the unoccupied Pacific coast of Shikotan Island; only maximum runup heights were estimated on this coast during the field survey (KAISTRENKO *et al.* 2013). It is not easy to compare tsunami wave heights, calculated for some offshore distance at a depth of 10 m, to the onshore runup height.

At Hanasaki, the first wave was the highest (Fig. 7). Its amplitude exceeded 2.5 m above the normal tide level. The maximum wave trough was clipped at a level of −110 cm (relative to zero tide) because the instrument was not designed to measure such strong oscillations. Intense oscillations with amplitudes of about 1 m continued until the end of the day.

It is important to note the complexity of the oscillations observed at coastal stations in comparison with the open ocean: in coastal areas prolonged intense oscillations were observed, while in the deep ocean, a single high impulse followed by relatively weak oscillations was typical.

The northeastern tip of the Nemuro Peninsula, where Hanasaki station is located, is directed toward a group of small islands (the Russian name is the Lesser Kuril Islands; the Japanese name is the Habomai Islands). The spatial grid with 1-arc min resolution is not good enough to produce high-quality simulation results for small islands; that is why we considered only the largest in the group, Zeleny Island. The tsunami arrived at the island at 06:33 UTC, 7 min later than it reached Hanasaki (ETA). The simulated amplitude of the first crest wave was about 1.5 m, quite similar maximum was observed in the *f–t* diagram for DART 21401. The maximum trough-to-crest height of modelled wave was over 3.7 m, with a positive elevation of 1.8 m. The field observations on Zeleny Island demonstrated that this value is in good agreement with runup heights ranging from 1 to 4 m (KAISTRENKO *et al.* 2013).

Approximately 6 min later, the waves arrived at the southwestern end of Shikotan Island. The information about runup heights in this region was available for Tserkovnaya Bay (ocean side) and Delphin Bay (the coast, facing the Lesser Kuril Islands and the Nemuro Peninsula). The estimated wave amplitudes on the approach to Delphin Bay (0.8 m) were significantly lower than those near Tserkovnaya Bay (1.2 m), in agreement with field observations (KAISTRENKO *et al.* 2013). The people fishing on ice in Delphin Bay also noticed some weak oscillations associated with this tsunami. On the oceanic side of Shikotan Island the first simulated wave was the highest, although it was followed by a long train of intense oscillations with periods of about 25–30 min. There were also low-frequency oscillations with periods of 70–90 min, according to observations.

KAISTRENKO *et al.* (2013) noted that on the inner (west) coast of Shikotan Island, from the side of Yuzhno-Kurilskiy Strait (where villages Malokurilsk and Krabozavodsk are located), the observed wave heights were about the same as on the oceanic side. This could be due to the fact that the survey was conducted mainly on the coasts of bays with well-defined resonance properties. The fundamental period of Krabovaya Bay (a long fjord-like inlet) is 29 min (RABINOVICH and MONSERRAT 1996); this period is close to the dominant periods of the tsunami.

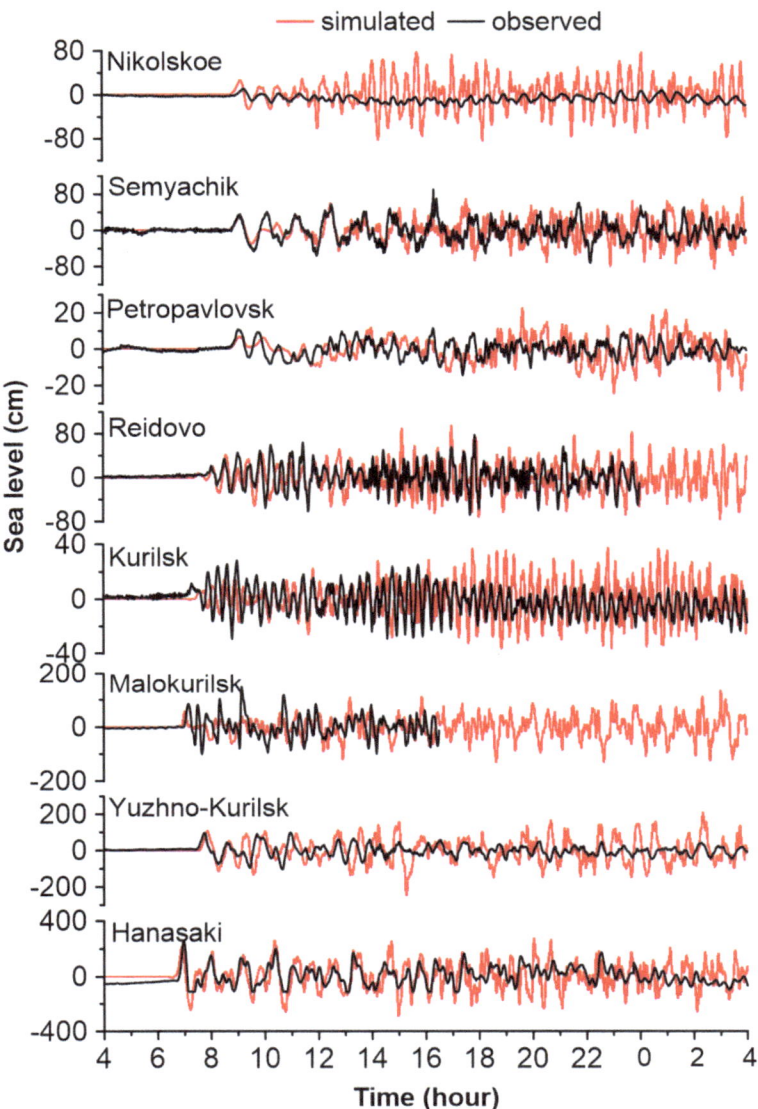

Figure 7
The 24-h simulated and observed segments of the Tohoku tsunami records of 11 March 2011 for stations installed on the Pacific coast of Russia, and for the JMA station Hanasaki. *Time scale* is given in UTC hours

Probably this is the reason for tsunami amplification in this bay (KAISTRENKO *et al.* 2013). However, in Malokurilskaya Bay (a bottle-like bay with a narrow neck), which is normally characterized by strong dominance of the fundamental (Helmholtz) mode with period of about 19 min (RABINOVICH and LEVYANT 1992; RABINOVICH and MONSERRAT 1996), the tsunami record (Fig. 7) was atypical and dominated by lower-frequency oscillations with periods of about 80 min. Calculated waveforms for this site are in poor agreement with the observations, apparently because the grid size did not allow the bottom topography of the bay to be resolved with sufficient accuracy. For this reason, the calculation point for the model was chosen outside of the bay. The duration of the oscillations is difficult to assess since the record was too short. For this reason, the tsunami spectrum at this station had insufficient resolution.

The tsunami behavior in the region of the Yuzhno-Kurilsk station, on the eastern coast of Kunashir Island, needs more detailed consideration (Fig. 7). The station is located on the opposite side of Yuzhno-Kurilskiy Strait relative to Malokurilskaya and Krabovaya bays (Fig. 6). At this station, despite the complex topography of the area, the agreement between the calculated and the recorded data, especially for the first 5–6 h after the tsunami arrival, is quite good. Yuzhno-Kurilsk Bay is an open wide bay that does not have pronounced resonant properties (RABINOVICH and LEVYANT 1992).

At stations Kurilsk and Reidovo, on the Okhotsk coast of Iturup Island (Fig. 6), the tsunami was significantly weaker than on the coast of Yuzhno-Kurilskiy Strait, and the periods of the oscillations were very different (Fig. 7). The amplitudes of the oscillations observed at Reidovo (40–50 cm) were about two times higher than at Kurilsk. Also, the agreement between the calculated and observed waveforms for Reidovo was much better than for Kurilsk. This is due to the fact that the Kurilsk tide gauge is located in the port area, which is sheltered from the arriving tsunami waves. It is surprising that the records at Kurilsk were strongly dominated by persistent oscillations with a period of about 20 min. Such monochromatic oscillations are typical for bays of the closed type (as, for example, Malokurilskaya Bay), but not for wide bays with a long open boundary, like Kitoviy Bay, adjacent to Kurilsk. The oscillations at Reidovo had similar periods. Low-frequency oscillations, like those observed at Hanasaki, Yuzhno-Kurilsk and Malokurilsk, were not detected at Kurilsk and Reidovo.

Figure 8 shows spectra of the tsunami and background signal and their ratios for the oceanic stations. The Hanasaki tsunami spectrum has weak maxima at periods of about 75, 36 and 22 min. The maximum of the spectral ratio was found at low frequencies (at periods of 70–100 min), gradually decreasing towards the higher frequencies. In the direction of the lower frequencies, the decrease is steeper. These estimates match well the open-ocean estimates of spectral ratios based on analysis of the 2011 Tohoku DART records presented in (RABINOVICH et al. 2013a).

There is quite good agreement between the spectra of the observed and modelled low-frequency

tsunami oscillations. In contrast, for high-frequency oscillations there are some differences, we found the very low values of the modelled spectral density at periods <5 min. Such small values of the modelled high-frequency oscillations are typical for all stations. This indicates the small contribution of the high-frequency component in the model source.

The pattern of the tsunami spectrum and the tsunami/background spectral ratio at Yuzhno-Kurilsk is similar to those for Hanasaki, as discussed above. Even the periods of the main peaks (∼75 and 35 min) are the same. It is important to note that for the high-frequency band (for periods <5 min), the observed tsunami/background ratio decreases faster than for the open Pacific coast.

In contrast to the stations described above, the spectrum at Kurilsk has a major, well-defined peak at a period of about 20 min, corresponding to the fundamental eigen mode of Kitoviy Bay. The tsunami/background spectral ratio has a maximum at this period and decreases rapidly both in the direction of lower and higher frequencies. There were also several weaker peaks, probably associated with other bay or harbour modes; the most significant peak with a period of 5 min corresponds to the main resonant mode of the internal port area. Most likely, the pronounced suppression of the high frequency components was due to the influence of Ekaterina Strait (separating Kunashir and Iturup islands), the main gate for the 2011 Tohoku tsunami to penetrate into the southern part of the Sea of Okhotsk and arrive at Kurilsk. We also found a well-expressed peak in the Kurilsk spectral ratio at 70 min, but it was much weaker than similar peaks observed at Hanasaki and Yuzhno-Kurilsk.

In the region of Iturup Island, the ocean shelf is narrow and steep; therefore, the low-frequency tsunami component was substantially weaker. This impeded the penetration of these components through Ekaterina Strait, as well as through Friz Strait, which separates Iturup and Urup islands. It was through Friz Strait that the waves reached Reidovo, where the observed low-frequency oscillations were quite weak. The main spectral maximum at this station was at 25 min (the primary resonant mode in the Gulf of Prostor). Another well-defined spectral peak was at about 7 min, corresponding to the resonant mode of

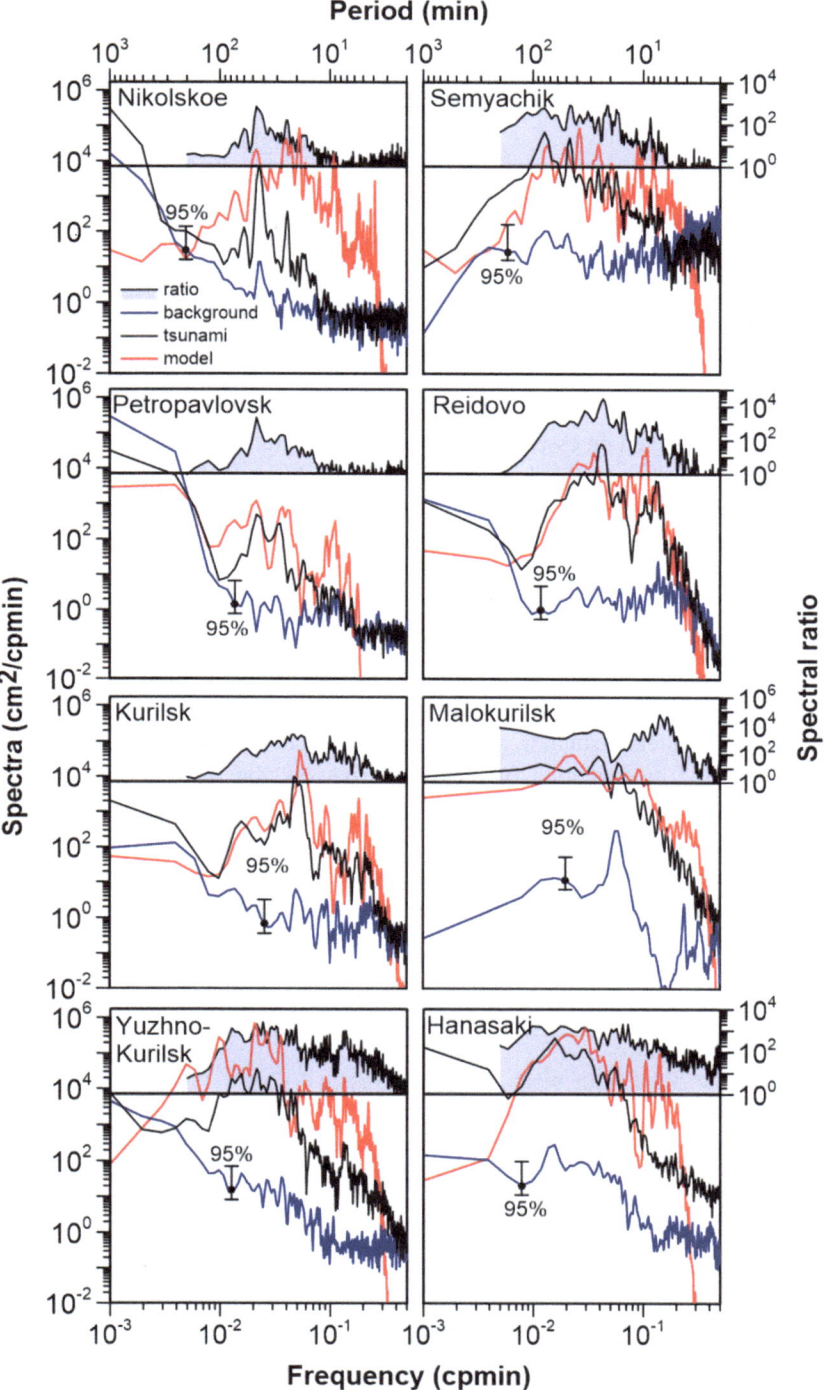

Figure 8
Spectra of background and the 2011 Tohoku tsunami sea level oscillations (observed and modeled), and their spectral ratios at stations located on the Pacific coast of Russia. Spectral ratios for periods longer than 180 min are not shown

167

Olya Bay. Significant attenuation of low-frequency tsunami components along the west coast of Iturup Island was probably also attributed to the fact that the shelf there was even narrower than at the oceanic side of the island.

5.2. Kamchatka and the Commander Islands

On the east coast of Kamchatka, there are two TWS stations. The first is in the port of Petro-pavlovsk, located in Avacha Bay and connected to the ocean via a narrow channel. This bay is characterized by a system of well-defined resonant modes (KOROLEV et al. 2003), which played an essential role during the tsunami. However, due to the long and narrow channel, the amplitudes of these modes rarely reach high magnitudes. The second gauge is at the meteorological station, Semyachik, located on the coast of Kronotsky Gulf (Fig. 9), in the proximity of the Semyachik Volcano. One more station is located in the port of Nikolskoe, Bering Island, the Com-mander Islands. The unexpectedly weak manifestation of the 2010 Chilean tsunami at this station led to a more detailed investigation of long wave properties in the area. It was found that an underwater ridge, which protects the approaches to the port, significantly weakens arriving tsunami waves at this site (SHEVCHENKO et al. 2013).

The tsunami waves reached the eastern coast of Kamchatka and the Commander Islands about an hour later than the South Kuril Islands (see Table 2 and Fig. 3). In the ports of Petropavlovsk and Nikolskoe, the Tohoku tsunami caused weak sea level oscillations with the maximum trough-to-crest wave height of <25 cm (Fig. 7). Initially, it was assumed that the northeastern flank of the Pacific coast of Russia was not in danger of the 2011 Tohoku tsunami. However, the record from the Semyachik station showed that the trough-to-crest wave height at this station was about 1.4 m. It has already been noted that the Petropavlovsk station is located in Avacha Bay connected to the ocean by a narrow channel, while the port of Nikolskoe (Bering Island) is protected from arriving waves by an underwater ridge. Nevertheless, the difference in tsunami wave height at Semyachik from that at Petropavlovsk and Nikolskoe looked too large.

However, the results of numerical modelling supported these findings. The simulated tsunami waveforms for the first six waves for Semyachik station were in good agreement with the observed waveforms. The numerical simulation could not reproduce high-frequency oscillations at Petro-pavlovsk, but low-frequency oscillations associated with the fundamental resonant mode of Avacha Bay, with a period of 4.8 h, were reproduced quite well after approximately 15:00 UTC (but not during the initial interval). Only at Nikolskoe were there significant differences between the computed and observed tsunami waveforms.

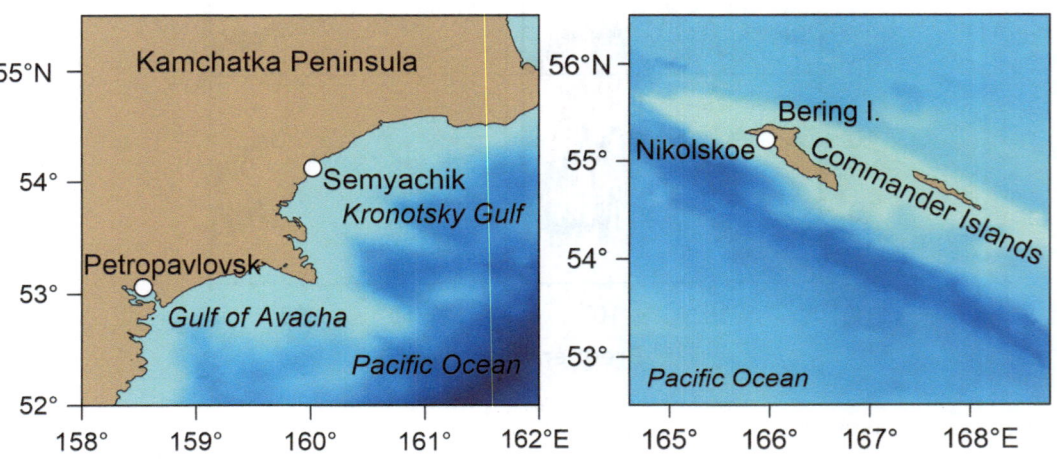

Figure 9
The location of TWS tide gauges on the coasts of Kamchatka Peninsula (*left figure*) and Bering Island (*right*) (corresponds to *boxes "B"* and *"C"* in Fig. 1)

According to the results of numerical simulations for Kronotskiy Gulf, the tsunami waves in this gulf were much higher than in the Gulf of Avacha (the lowest wave heights for the latter gulf were on the approach to Avacha Bay); this is due to the fact that the Kronotskiy Gulf is much shallower (Fig. 9).

The spectral ratio for Semyachik (Fig. 8) was generally similar to that described for Hanasaki: there was a weakly pronounced maximum at a period of about 80 min and smooth decay in the direction of both higher and lower frequencies. At periods less than 5 min, the spectral ratio was less than 1.0. This was probably due to the sea activity and high level of the associated background oscillations in days preceding the tsunami event.

At Petropavlovsk and Nikolskoe, the tsunami frequency band, i.e. the band where the spectral ratios were greater than 1.0, was narrower due to the low-frequency band of the spectra. In contrast to Semyachik, there were well-defined peaks in the tsunami spectra at these stations. At Nikolskoe, the oscillations with periods of 45 and 25 min had particularly high Q factors. The Q factor characterizes the sharpness of the resonant peak and commonly becomes a measure of energy dissipation at corresponding frequency (EMERY and THOMSON 2003). The Q factor of resonant system is given by the relation:

$$Q = \frac{f_r}{\Delta f_{1/2}} \qquad (5)$$

where f_r resonant frequency, and $\Delta f_{1/2}$ frequency width of the resonant half-peak.

At Petropavlovsk, the pronounced peaks were observed at periods of 45, 30 and 19 min, corresponding to the resonant modes of Avacha Bay (KOROLEV et al. 2003).

In general, the comparison of the spectral ratios at the different stations on the Pacific coast of Russia revealed certain differences (Fig. 8). For instance, at Hanasaki, Yuzhno-Kurilsk and Semyachik the main maximum was at periods of 75–80 min. The decays of this parameter in the directions of both lower and higher frequencies were relatively smooth. At Nikolskoe and Petropavlovsk stations, the main peak was observed at a period of 45 min. The decay of ratios was steep on both sides (especially, at lower

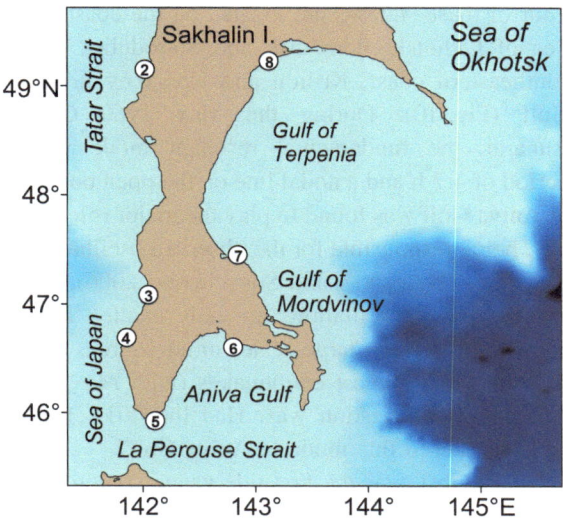

Figure 10
The locations of the TWS tide gauges on the coasts of Sakhalin Island (*white circles*, 2 Uglegorsk, 3 Kholmsk, 4 Nevelsk, 5 Krilion, 6 Korsakov, 7 Starodubsk, 8 Poronaisk); corresponds to *box "D"* in Fig. 1

frequencies). On the Okhotsk coast of Iturup Island (Reidovo and Kurilsk), the main maxima were at periods of 20 and 25 min, and the general characteristics of the decay spectral ratio functions were approximately between the two cases described above.

6. Sakhalin Island and Primorye

6.1. Southeastern Coast of Sakhalin Island (the Sea of Okhotsk)

Through Ekaterina and Friz straits, the 2011 tsunami waves penetrated into the southern part of the Sea of Okhotsk and reached the southeastern coast of Sakhalin Island (Fig. 10). As can be seen in Figs. 2 and 3, the tsunami propagation through Kunashirskiy Strait, between Hokkaido and Kunashir islands, was insignificant. The widest and deepest Boussol Strait is the main entrance for tsunami waves arriving from the Pacific Ocean (see, for example, RABINOVICH et al. 2008); the 2011 tsunami waves mostly penetrated through this strait into the northern and northeastern parts of the Sea of Okhotsk.

The 2011 Tohoku tsunami was recorded by seven TWS stations in the southern part of Sakhalin Island.

169

Four of these stations are located on the coast of the Sea of Okhotsk: Poronaisk and Starodubsk on the southeastern coast; Krilion and Korsakov in Aniva Gulf (Fig. 10). During the May 1960 Chilean tsunami, the fundamental resonant mode with a period of 4.7 h and a nodal line on the open boundary of Aniva Gulf was found to play the major role in this bay, being responsible for the observed oscillations at Korsakov (IVELSKAYA and SHEVCHENKO 2006). These low-frequency oscillations are only excited by tsunamis with a very large-scale initial source, caused only by the strongest earthquakes (with $M_w \geq 8.5$). An important question was: Had the 2011 Tohoku tsunami caused this mode in Aniva Gulf?

Starodubsk was the first tide gauge station on the eastern coast of Sakhalin Island reached by the Tohoku tsunami. It is located on the southeastern coast of Sakhalin Island; the adjacent shelf is narrow and deep. The maximum wave recorded at this station had a trough-to-crest height of about 0.5 m (Fig. 11) and approximately 0.8 m at Poronaisk (it was observed almost a day after the first wave). The agreement between the calculated and observed

waveforms was much better for Poronaisk than for Starodubsk. On the southeastern coast of Sakhalin Island, the Tohoku tsunami did not produce any noticeable damage, although extensive evacuation took place there: about 10,000 fishermen were evacuated from the ice in the Gulf of Mordvinov (south of Starodubsk).

There was an evident maximum with a period of about 75 min in the tsunami spectra at these stations (Fig. 12). A similar peak was also observed in the spectra of Hanasaki and Yuzhno-Kurilsk. This peak was better defined at Poronaisk, probably because this period is close to the main resonant period of the Gulf of Terpenia (~ 80 min).

Surprisingly, at the station in Korsakov, located at the head of the bay, we found distinctive low-frequency oscillations with a period of about 4.7 h. These oscillations are poorly defined at Cape Krilion, apparently because of the closeness of this station to the nodal line of this mode. The 4.7-h oscillations had an amplitude of about 30 cm. As was mentioned above, similar low-frequency oscillations were observed in the Avacha Bay and were found to be

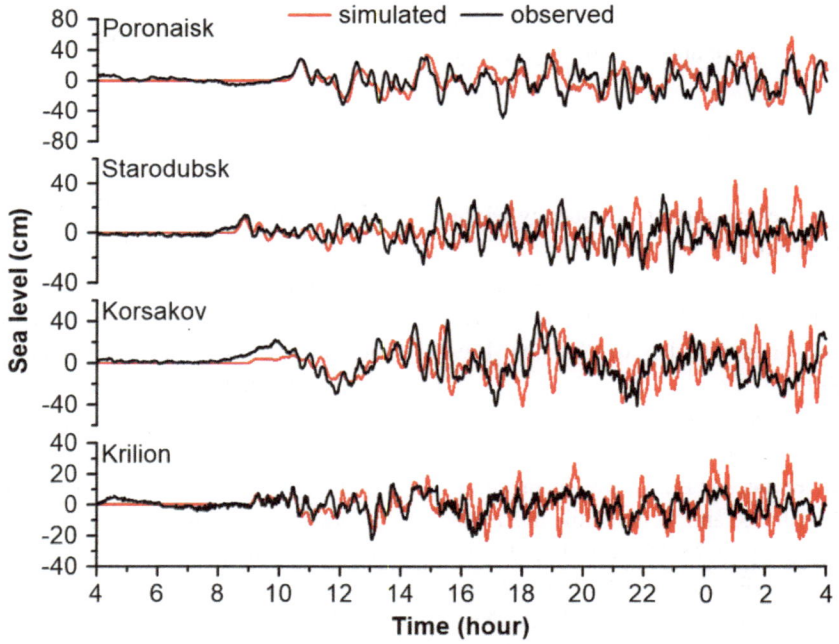

Figure 11
The 24-h simulated and observed segments of the Tohoku tsunami records of 11 March 2011 from stations installed on the eastern coasts of Sakhalin Island. *Time scale* is given in UTC hours

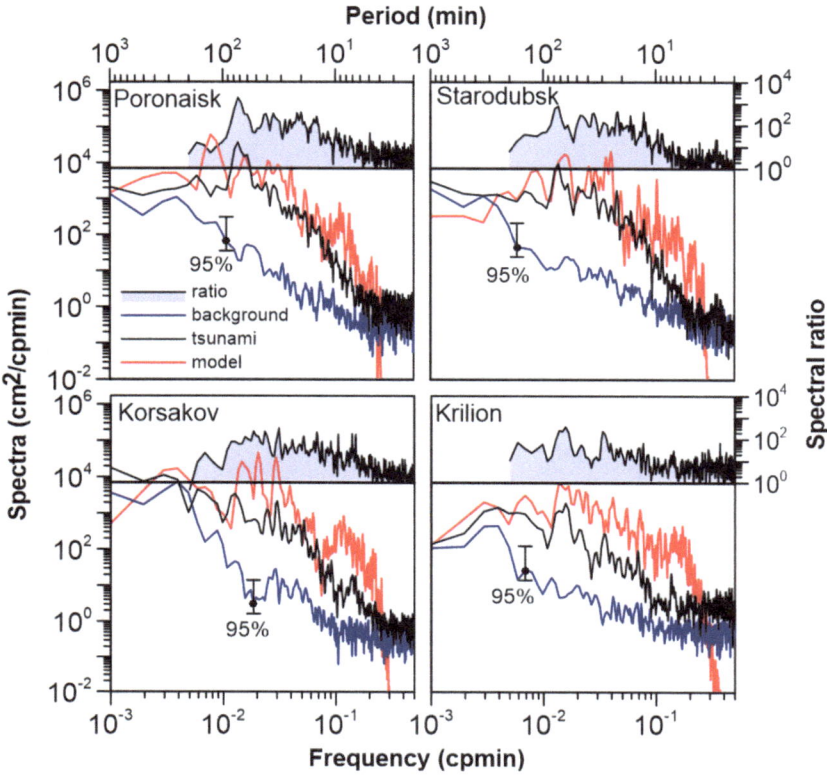

Figure 12
Spectra of background and the 2011 Tohoku tsunami sea level oscillations (observed and modeled), and their spectral ratios at stations located on the southeastern coast of Sakhalin Island. Spectral ratios for periods longer than 180 min are not shown

related to the zeroth mode of the bay. The fact that the 2011 Tohoku tsunami induced such low-frequency oscillations is surprising, taking into account that the tsunami source did not have components with long spatial scales. The tsunami spectra at Korsakov and Petropavlovsk do not have similar peaks, since the spectral estimates were based on 1-day records with 6-h windows, which are too short to resolve such long periods.

6.2. Southwestern Coast of Sakhalin Island and Primorye (the Sea of Japan)

Three stations are located on the western coast of Sakhalin Island, in Tatar Strait, Sea of Japan: Nevelsk, Kholmsk and Uglegorsk. Hazardous tsunamis on this coast are produced only by earthquakes occurring in the adjacent region. These gauges were installed in port harbours of relatively small size;

consequently, the resonant oscillations at these sites have relatively high frequencies. For comparison, three tide gauges located on the opposite coast of Tatar Strait, in the ports of Primorye: Nakhodka, Preobrazhenie and Rudnaya Pristan, are in much larger bays, with much lower eigen frequencies. Also, resonant oscillations of Tatar Strait itself could be expected to be excited.

The 2011 Tohoku tsunami penetrated into the Sea of Japan earlier than into the Sea of Okhotsk. This is because the waves propagated into the former basin through Tsugaru Strait separating Honshu and Hokkaido islands (see Fig. 2). That is why the tsunami waves reached the southwestern coast of Sakhalin Island before they arrived at the southeastern coast of this island (at Nevelsk in the Sea of Japan the waves arrived 25 min earlier than at Starodubsk in the Sea of Okhotsk, see Tables 3, 4; Fig. 13). The waves that passed through the straits of the South Kuril Islands

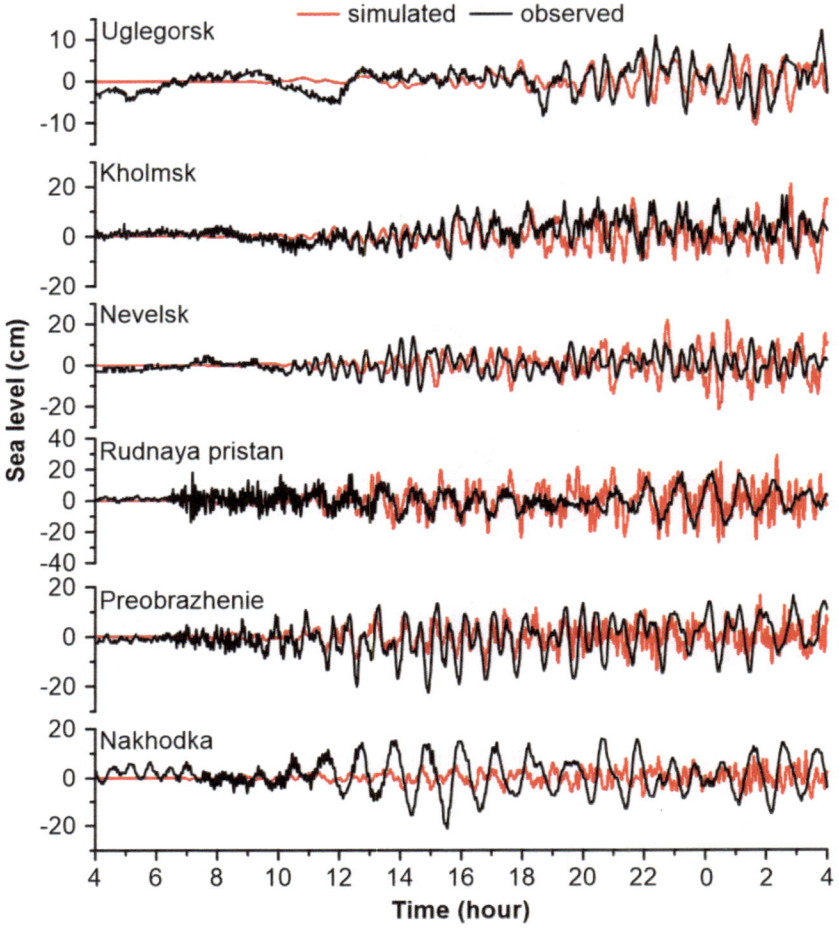

Figure 13
The 24-h simulated and observed segments of the Tohoku tsunami records of 11 March 2011 for stations installed on the coasts of southwestern Sakhalin Island and Primorye. *Time scale* is given in UTC hours

and La Perouse Strait (between Hokkaido and Sakhalin) had been significantly weakened, and did not play any noticeable role in the formation of the tsunami wave field in the northern part of the Sea of Japan. There is good agreement between the simulated and observed tsunami waveforms at Kholmsk; the differences between the modelled and observed waves at Nevelsk and Uglegorsk are more significant.

The tsunami/background spectral ratios were similar for all stations on the west coast of Sakhalin Island (Fig. 14); the maximum values were observed at periods of 20–60 min, and decayed in the directions of both higher and lower frequencies (the steepest decrease was observed at Uglegorsk). In the tsunami spectra there were some similarities but also

evident differences. The common spectral feature at all stations was the spectral peak at a period of about 1 h, corresponding to the transverse seiche mode of Tatar Strait (KOVALEV *et al.* 2009).

The 2011 Tohoku tsunami waves propagating through Tsugaru Strait arrived at the coast of Primorye even earlier than at the South Kuril Islands (Table 4); in particular, at Preobrazhenie Bay (Fig. 15), the tsunami was recorded 25 min earlier than at Malokurilsk (Shikotan Island). However, satisfactory agreement between the observed and simulated waves was found only for Rudnaya Pristan (see Fig. 13). This is probably due to the fact that the respective tide gauge was located in Tetyukhe Bay, which has a long open boundary. The

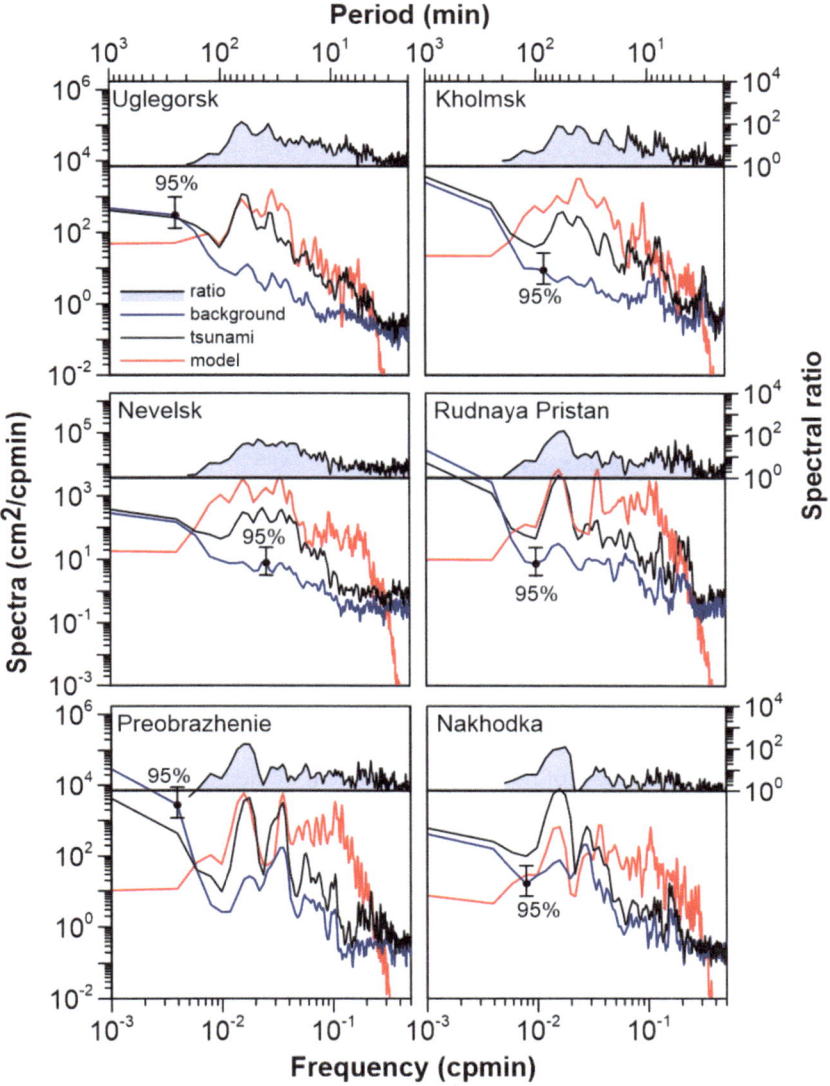

Figure 14
Spectra of background and 2011 Tohoku tsunami sea level oscillations (observed and modeled), and their spectral ratios at stations located on the coasts of southwestern Sakhalin Island and Primorye. Spectral ratios for periods longer than 180 min are not shown

topography of the coastline in the regions of Preobrazhenie Bay and Nakhodka Harbour (located inside the larger Gulf of Nakhodka) is much more complicated and probably is not reproduced well by the numerical model.

The tsunami spectra at the stations of Primorye include well-defined peaks at periods of about 60 and 30 min (Fig. 14). The latter peak is very pronounced both in the tsunami and background spectra of Preobrazhenie Bay, but is not defined in the background spectrum of Rudnaya Pristan. We have insufficient information about the resonant oscillations in the bays of Primorye, so we can only make some assumptions. The 60-min period is likely associated with the resonant oscillations in the southern part of Tatar Strait. The period of 30 min is apparently associated with the fundamental eigen period of Preobrazhenie Bay, while the period of 7 min corresponds to the resonant mode of Nakhodka Harbour, which has a relatively small size.

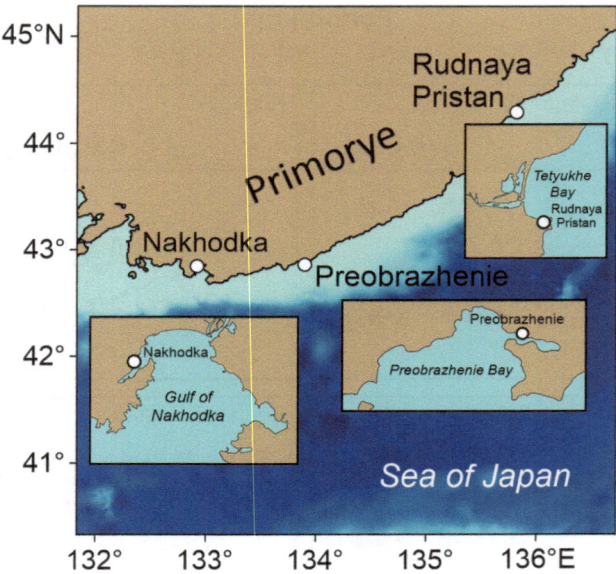

Figure 15
The location of TWS tide gauges on the coasts of Primorye (corresponds to *box "E"* in Fig. 1)

In general, the frequency-selective properties of the local topography on the Russian coast of the Sea of Japan (both Sakhalin and Primorye), have a stronger effect than on the Okhotsk coast of Sakhalin Island.

7. Conclusions

We have analyzed a great number of records from the Tohoku tsunami of 11 March 2011 obtained at deep-ocean and shore-based gauges in the Russian Far East. A significant difference was found between the open-ocean and coastal tsunami records: in the former case, a large solitary impulse-like wave followed by relatively weak oscillations were observed; while in the latter case, long-term intensive oscillations were recorded (only at Hanasaki, Japan did the first wave have the maximum height). The tsunami oscillations near the shore were mainly determined by local topographic effects; the general properties of the records associated with the processes at the tsunami source were less evident.

Analysis of the deep-ocean data showed there was anisotropic spreading of the tsunami energy: shorter waves radiated eastward, to the open ocean, while longer waves propagated toward the Kuril Islands.

Significant energy was also identified in the low-frequency band of the spectrum at periods of 50–80 min. The wave dispersion effect was clearly evident in the high-frequency band of the spectrum at periods <10 min.

The low-frequency component significantly intensified on the extended shelf of the South Kuril Islands and played an important role in the formation of tsunami-caused oscillations at Hanasaki, Malokurilsk, and Yuzhno-Kurilsk. In the far-field, the major role in the formation of tsunami-caused oscillations was played by the resonant bay modes.

In the most populated settlements of the South Kuril Islands (Yuzhno-Kurilsk, Krabozavodskoe and Malokurilsk) tsunami heights were up to 2–2.5 m. The highest waves were observed in Krabovaya Bay, Shikotan Island (KAISTRENKO *et al.* 2013). Most likely, this was caused by the resonance of arriving tsunami waves with dominant period of about 30 min and the eigen period of the fundamental (Helmholtz) mode for this bay (~ 29 min). Very low-frequency oscillations with periods about 4.7–4.8 h were produced by the Tohoku tsunami in Aniva Gulf and Avacha Bay. Such oscillations can be excited only by the most powerful earthquakes with very large spatial scales of the source (another example of this kind is the 1960 Great Chile tsunami).

The results of numerical modelling of the 2011 Tohoku tsunami waves for the Russian Far East were mainly in good agreement with the observations. Simulated and observed tsunami waveforms were in agreement for all deep-ocean stations and for most of the coastal stations. The results of numerical modelling show that the Russian Far East coast was in the "shadow" of the main energy flux of the 2011 tsunami. The maximum wave heights in this region were simulated to occur in areas with adjacent wide shallow shelves: on the oceanic side of the South Kuril Islands, near the North Kuril Islands, on the southeastern coast of the Kamchatka Peninsula and on the northeastern coast of Sakhalin Island.

Acknowledgments

We are grateful to Isaac Fine (IOS, Sidney, BC, Canada) for his help with the source model and productive comments, and to Fred Stephenson (IOS, Sidney, BC, Canada) for editing the text. We are also quite thankful to our reviewers Paul Whitmore (WC/ATWC, Palmer, AK, USA) and Kenji Satake (ERI, University of Tokyo, Japan) for their thorough evaluation and helpful comments and suggestions that significantly improved this manuscript. This study for was partly supported by the RFBR grants 13-05-00936-a, 12-05-00757-a (last one for TI only) and by a grant of the Russian Academy of Sciences (Far East Branch–Siberian Branch) No. 12-II-0-08-003.

REFERENCES

BORRERO, J.S., and GREER, S.D. (2013). *Comparison of the 2010 Chile and 2011 Japan tsunamis in the far-field.* Pure Appl. Geophys., *170*, 1249–1274.

BORRERO, J.S., BELL R., CSATO, C., GORING, D., GREER, S.D., PICKETT, V., and POWER, W. (2013). *Observations, effects and real time assessment of the March 11, 2011 Tohoku-oki tsunami in New Zealand.* Pure Appl. Geophys., *170*, 1229–1248.

DZIEWONSKI, A., BLOCH, S., and LANDISMAN, M. (1969), *Technique for the analysis of transient seismic signals*, Bull. Seism. Soc. Am., *59*, 427–444.

EMERY, W.J., and THOMSON, R.E. (2003). *Data Analysis Methods in Physical Oceanography*, Second and revised edition, Elsevier, New York, 638 p.

FINE, I.V., KULIKOV, E.A., and CHERNIAWSKY, J.Y. (2013). *Japan's 2011 tsunami: characteristics of wave propagation from observations and numerical modelling.* Pure Appl. Geophys., *170*, 1297–1307.

GONZÁLEZ, F.I., and KULIKOV, E.A. (1993). *Tsunami dispersion observed in the deep ocean.* In: *Tsunamis in the World*, S. Tinti (ed.), Kluwer Pub., Dordrecht, pp. 7–16.

HAYES, G. (2011). *Finite fault model. Updated result of the March 11, 2011 Mw 9.0 earthquake offshore Honshu, Japan,* http://earthquake.usgs.gov/earthquakes/eqinthenews/2011/%20usc0001xgp/finite_fault.php.

HEIDARZADEH, M., and SATAKE, K. (2013). *Waveform and spectral analyses of the 2011 Japan tsunami records on tide gauge and DART stations across the Pacific Ocean.* Pure Appl. Geophys., *170*, 1275–1293.

HINWOOD, J.B., and MCLEAN, E.J. (2013). *Effects of the March 2011 Japanese tsunami in bays and estuaries of SE Australia.* Pure Appl. Geophys., *170*, 1207–1227.

IMAMURA, F., SHUTO, N., and GOTO, C. (1988). *Numerical simulation of the transoceanic propagation of tsunamis*, The Sixth Congress of the Asian and Pacific Regional Division, Int. Assoc. Hydraul. Res., Kyoto, Japan, pp. 265–272.

IMAMURA, F. (1996). *Review of tsunami simulation with a finite difference method.* In: Long-Wave Run-up Models, (eds. by H. Yeh, P. Liu and C. Synolakis), World Scientific, London, pp. 25–42.

IVELSKAYA, T.N., and SHEVCHENKO, G.V. (2006). *Amplification of low-frequency component of the Chilean tsunami (May 1960) on the northwestern shelf of the Pacific Ocean,* Russian Meteorology and Hydrology, 2, 69–81.

KAISTRENKO, V., RAZJIGAEVA, N., KHARLAMOV, A., and SHISHKIN, A. (2013). *Manifestation of the 2011 Great Tohoku tsunami on the coast of the Kuril Islands: A tsunami with ice.* Pure Appl. Geophys., *170*, 1103–1114.

KOROLEV, YU.P., and SHEVCHENKO, G.V. (2003) *The propagation of tsunami waves in the Petropavlovsk-Kamchatski area,* Volcanology and Seismology, 6, 62–70.

KOVALEV, P.D., SHEVCHENKO, G.V., KOVALEV, D.P., CHERNOV, A.G., and ZOLOTUKHIN, D.E. (2009). *Simushir and Nevelsk tsunami recording in the Port of Kholmsk,* Pacific Geology, 28(5), 36–43.

KULIKOV, E.A., and GONZÁLEZ, F.I. (1996). *Recovery of the shape of a tsunami signal at the source from measurements of oscillations in the ocean level by a remote hydrostatic pressure sensor,* Trans. (Doklady) Russian Acad. Sci., *345A*, 585–591.

MORI, N., TAKAHASHI, T., YASUDA, T., and YANAGISAWA H. (2011). *Survey of 2011 Tohoku earthquake tsunami inundation and run-up,* Geophys. Res. Lett., *38*, L00G14, 2011. doi:10.1029/2011GL049210.

MUNGOV, G., EBLE, M., and BOUCHARD, R. (2013). *DART tsunameter retrospective and real-time data: A reflection on 10 years of processing in support of tsunami research and operations.* Pure Appl. Geophys., *170*, 1369–1384.

OKADA, Y. (1985). *Surface deformation due to shear and tensile faults in a half-space,* Bull. Seism. Soc. Am. *75*, 1135–1154.

RABINOVICH, A.B. (1997). *Spectral analysis of tsunami waves: Separation of source and topography effects,* J. Geophys. Res., *102*(C6), 12663–12676.

RABINOVICH, A.B., and LEVYANT, A.S. (1992). *Influence of seiche oscillations on the formation of the long-wave spectrum near the coast of the Southern Kuriles,* Oceanology, *32*(1), 17–23.

RABINOVICH, A.B., and MONSERRAT, S. (1996). *Meteorological tsunamis near the Balearic and Kuril Islands: Descriptive and statistical analysis,* Natural Hazards, *13*(1), 55–90.

RABINOVICH, A. B., LOBKOVSKY, L.I., FINE, I.V., THOMSON, R.E., IVELSKAYA, T.N., and KULIKOV, E.A. (2008). *Near-source observationsand modeling of the Kuril Islands tsunamis of 15 November 2006 and 13 January 2007*, Adv. Geosc., *14*, 105–116.

RABINOVICH, A.B., CANDELLA, R.N., and THOMSON, R.E., (2013a), *The open ocean energy decay of three recent trans-Pacific tsunamis*. Geophys. Res. Lett., *40*. doi:10.1002/grl.50625.

RABINOVICH, A.B., THOMSON, R. E. and FINE I. V. (2013b). *The 2010 Chilean tsunami off the west coast of Canada and the northwest coast of the United States*. Pure Appl. Geophys., *170*, 1529–1565.

RAZJIGAEVA, N.G., GANZEY, L.A., GREBENNIKOVA, T.A., IVANOVA, E.D., KHARLAMOV, A.A., KAISTRENKO, V.M., and SHISHKIN, A.A. (2013). *Coastal sedimentation associated with the Tohoku tsunami of 11 March 2011 in South Kuril Islands, NW Pacific Ocean*. Pure Appl. Geophys., *170*, 1081–1102.

REYMOND, D., HYVERNAUD, O., and OKAL, E.A. (2013). *The 2010 and 2011 tsunamis in French Polynesia: Operational aspects and field surveys*. Pure Appl. Geophys., *170*, 1169–1187.

SHEVCHENKO, G.V., and IVELSKAYA, T.N. (2012). *The Tohoku tsunami of 11 March 2011 as recorded on the Russian Far East*, Science of Tsunami Hazards, *31*(4), 268–262.

SHEVCHENKO, G., IVELSKAYA, T., LOSKUTOV, A., and SHISHKIN, A. (2013), *The 2009 Samoan and 2010 Chilean tsunamis recorded on the Pacific coast of Russia*. Pure Appl. Geophys., *170*, 1511–1527. doi:10.1007/s00024-012-0562-9.

SONG, Y.T., FUKUMORI, I., SHUM, C.K., and YI, Y. (2012). *Merging tsunamis of the 2011 Tohoku-Oki earthquake detected over the open ocean*, Geophys. Res. Lett. *39*, L05606. doi:10.1029/2011GL050767.

TANG., L., TITOV, V.V., BERNARD, E.N., WEI, Y., CHAMBERLIN, C.D., NEWMAN, J.C., MOFJELD, H.O., ARCAS, D., EBLE, M.C., MOORE, C., USLU, B., PELLS, C., SPILLANE, M.C., WRIGHT, L., and GICA, E. (2012). *Direct energy estimation of the 2011 Japan tsunami using deep-ocean pressure measurements*, J. Geophys. Res., *117*, C08008, doi:10.1007/s00024-012-0492-6.

WATADA, S., KUSUMOTO, S. and SATAKE, K. (2013). *Cause of traveltime difference between observed and synthetic tsunami waveforms at distant locations*, In: Abstracts of Knowledge for the Future (Joint Assembly IAHS-IAPSO-IASPEI), 22–26 July 2013, Gothenburg, Sweden.

WILSON, R.I., ADMIRE, A.R., BORRERO, J.C., DENGLER, L.A., LEGG, M.R., LYNETT, P., MCCRINK, T.P., MILLER, K.M., RITCHIE, A., STERLING, K., and WHITMORE, P.M. (2013). *Observations and impacts from the 2010 Chilean and 2011 Japanese tsunamis in California (USA)*, Pure Appl. Geophys., *170*, 1207–1227.

(Received July 24, 2013, revised October 17, 2013, accepted October 19, 2013, Published online November 22, 2013)

Pure Appl. Geophys. 171 (2014), 3351–3363
© 2013 Springer Basel
DOI 10.1007/s00024-013-0757-8

⌐ Pure and Applied Geophysics

Marshall Islands Fringing Reef and Atoll Lagoon Observations of the Tohoku Tsunami

Murray Ford,[1] Janet M. Becker,[2] Mark A. Merrifield,[3] and Y. Tony Song[4]

Abstract—The magnitude 9.0 Tohoku earthquake on 11 March 2011 generated a tsunami which caused significant impacts throughout the Pacific Ocean. A description of the tsunami within the lagoons and on the surrounding fringing reefs of two mid-ocean atoll islands is presented using bottom pressure observations from the Majuro and Kwajalein atolls in the Marshall Islands, supplemented by tide gauge data in the lagoons and by numerical model simulations in the deep ocean. Although the initial wave arrival was not captured by the pressure sensors, subsequent oscillations on the reef face resemble the deep ocean tsunami signal simulated by two numerical models, suggesting that the tsunami amplitudes over the atoll outer reefs are similar to that in deep water. In contrast, tsunami oscillations in the lagoon are more energetic and long lasting than observed on the reefs or modelled in the deep ocean. The tsunami energy in the Majuro lagoon exhibits persistent peaks in the 30 and 60 min period bands that suggest the excitation of closed and open basin normal modes, while energy in the Kwajalein lagoon spans a broader range of frequencies with weaker, multiple peaks than observed at Majuro, which may be associated with the tsunami behavior within the more irregular geometry of the Kwajalein lagoon. The propagation of the tsunami across the reef flats is shown to be tidally dependent, with amplitudes increasing/decreasing shoreward at high/low tide. The impact of the tsunami on the Marshall Islands was reduced due to the coincidence of peak wave amplitudes with low tide; however, the observed wave amplitudes, particularly in the atoll lagoon, would have led to inundation at different tidal phases.

Key words: 2011 Tohoku tsunami, tide gauge records, Marshall Islands, coral reef, atoll.

[1] School of Environment, The University of Auckland, Private Bag 92019, Auckland 1142, New Zealand. E-mail: m.ford@auckland.ac.nz
[2] Department of Geology and Geophysics, The University of Hawaii at Manoa, Honolulu, HI 96822, USA.
[3] Department of Oceanography, The University of Hawaii at Manoa, Honolulu, HI 96822, USA.
[4] Jet Propulsion Laboratory, California Institute of Technology, Pasadena, CA 91109, USA.

1. Introduction

The magnitude 9.0 earthquake centered off the coast of Sendai, Japan (38.297°N, 142.372°E) on 11 March 2011 at 05:46:24 UTC triggered a tsunami with a highest recorded runup of 38.9 m, impacting much of the east coast of Honshu (Mori *et al.* 2011). The tsunami propagated across the Pacific Ocean (Fig. 1) with damage reported in Hawaii (Yamazaki *et al.* 2012), along the west coast of North America (Allan *et al.* 2012; Wilson *et al.* 2012), New Zealand (Borrero *et al.* 2012) and within parts of South America [National Geophysical Data Center/(NGDC/WDS) Global Historical Tsunami Database, Boulder, CO, USA (Available at http://www.ngdc.noaa.gov/hazard/tsu_db.shtml) 2011]. The impacts at Pacific Ocean atolls were relatively minor with no reports of significant injuries or damage, although wave overtopping at Midway Atoll in the Northwestern Hawaiian Islands resulted in the deaths of thousands of seabirds (US Fish and Wildlife Service 2011).

Historically, mid-ocean atolls have been vulnerable to coastal flooding and overwash during tsunami events. The 26 December 2004 magnitude 9.1 earthquake off the west coast of Sumatra generated a destructive tsunami which impacted numerous atolls in the Indian Ocean (Merrifield *et al.* 2005; Rabinovich and Thomson 2007). The Maldives, an atoll nation in the central Indian Ocean, was heavily impacted with an estimated death toll of 108 and US $295 million damage (NGDC/WDS). Tide gauge measurements of the Sumatran tsunami within the Maldives indicate maximum trough-to-crest wave heights of 2.15 and 2.17 m at Male and Hanimaadhoo atolls (Rabinovich and Thomson 2007). Geomorphic investigations showed that the tsunami was high enough to wash over many low-lying islands, depositing significant volumes of sand and gravel to the islands (KENCH *et al.* 2006). The highest tsunami-associated water level recorded by a tide gauge on a Pacific atoll was 1.90 m at Midway atoll following the magnitude 8.2 Kamchatka earthquake in 1952 (NGDC/WDS). A tide gauge operating since 1946 on Kwajalein Atoll ($8.732°N$, $167.735°E$) recorded 26 unique tsunami events prior to the Tohoku tsunami, the highest of which was 0.38 m (NGDC/WDS).

Local topographic effects on tsunamis have been noted at mid-ocean atolls. VAN DORN (1984) summarized attempts to account for the modification of a mid-ocean tsunami by Wake Island, and noted the excitation of high frequency oscillations (31 cph) in tide gauge records located within a small boat harbor. HEIDARZADEH and SATAKE (2013) characterized the Tohoku tsunami at the Kwajalein and Wake atolls based on tide gauge records, and noted a broad frequency response in the first hour of arrival followed by more narrow-banded oscillations at both sites. In general, tide gauge observations of tsunamis at atolls and island stations tend to show a more rapid decay than stations along continental margins (MERRIFIELD *et al.* 2005; HEIDARZADEH and SATAKE 2013), likely due to the lack of well-defined shelves at most islands (HEIDARZADEH and SATAKE 2013). Atolls encompass a range of topographically complex sub-environments within enclosed or semi-enclosed lagoons and along ocean-facing reefs at the atoll rim. Typically a single lagoon tide gauge record may be available, with few direct observations to assess the response in the various atoll sub-regions.

Fringing coral reefs are highly efficient attenuators of wind wave energy (e.g. LUGO-FERNÁNDEZ *et al.* 1998; PEQUIGNET *et al.* 2011), affording the protection necessary for human occupation of low-lying reef islands. Impact assessments following the 2004 Sumatran tsunami found no clear association between the presence of coral reefs and reduced runup and inundation (ADGER *et al.* 2005; BAIRD *et al.* 2005; GOFF *et al.* 2006). In an analysis of 62 sites impacted by the Sumatran tsunami, CHATENOUX and PEDUZZI (2007) noted a positive relationship between the presence of a coral reef and the extent of inundation. We find few applications of numerical models of tsunami propagation within reef settings (KUNKEL *et al.* 2006; LYNETT 2007; GELFENBAUM *et al.* 2011) and none within mid-ocean atoll settings. In a

numerical study of tsunami propagation around an idealised circular island KUNKEL (2006) found that wide, shallow lagoons were effective at reducing tsunami impact. Similarly, in a detailed numerical study of tsunami propagation across fringing coral reefs GELFENBAUM *et al.* (2011) found that wide and rough reef flats provide the greatest attenuation of tsunami energy. However, GELFENBAUM *et al.* (2011) report shoaling of tsunami exceeded dissipation on narrow reefs (<200 m) resulting in higher amplitudes at the shoreline than incident at the reef edge.

Here we describe high-frequency water level measurements of the 2011 Tohoku tsunami recorded within a range of environments on two mid-ocean atolls in the Marshall Islands (Fig. 1). Sixteen pressure gauges and two permanent tide gauges at Kwajalein and Majuro atolls provide concurrent measurements of tsunami waves in the inner and outer waters of a mid-ocean atoll. The atoll water level observations are complemented by information of the tsunami in deep-water from pressure measurements at a nearby Deep-Ocean Assessment and Reporting of Tsunamis (DART) buoy and from numerical model simulations of the event (Fig. 1). The observations provide the first high-frequency

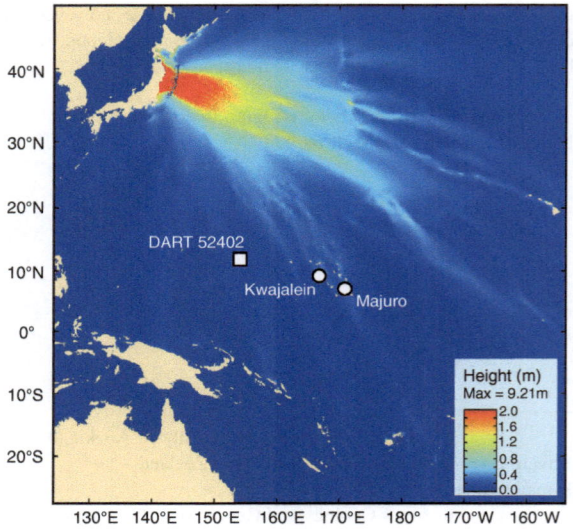

Figure 1
Modelled maximum predicted height of the Tohoku tsunami (SONG 2012), DART buoy #52402 and study sites within the Marshall Islands

record of tsunami propagation throughout a reef/atoll setting and afford a unique opportunity to assess the tsunami response at different geomorphic locations throughout an atoll. In addition, the study provides further context of the tsunami signal detected by atoll lagoon tide gauges, which are an important component of Pacific and Indian Ocean tsunami warning networks.

2. Description of Measurements and Model Output

As part of a study to investigate wind wave processes on fringing coral reefs, bottom mounted pressure sensors were deployed within the lagoon and on the ocean-facing fringing reef at Majuro and Kwajalein atolls within the Republic of the Marshall Islands (Fig. 2). Instruments at Majuro were

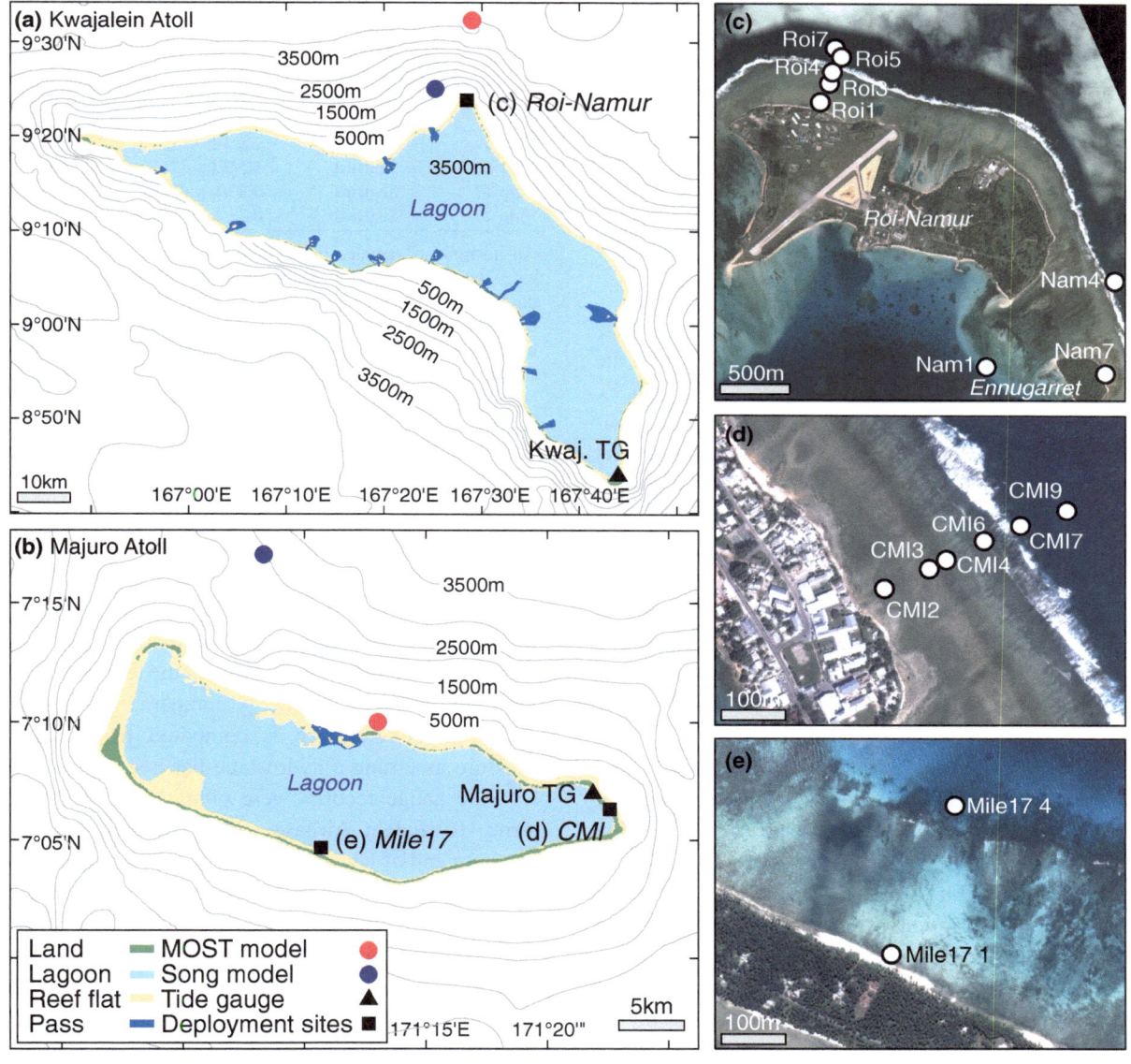

Figure 2

Locations of numerical simulation of tsunami propagation, tide gauge and deployment sites on Kwajalein (**a**) and Majuro (**b**) atolls. Pressure sensor locations at Roi-Namur (**c**), College of the Marshall Islands (CMI) (**d**) and Mile 17 (**e**). Imagery copyright DigitalGlobe, all rights reserved

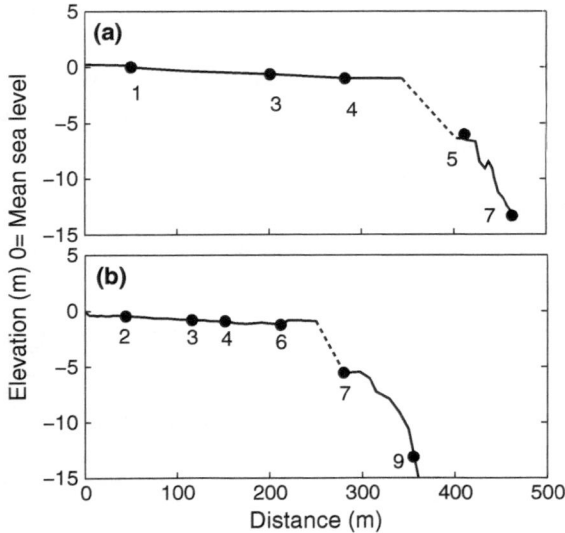

Table 1

Instrument type and sampling scheme for the pressure data used in this analysis

Instrument	Type	Sampling	Depth (m)
Roi 7	AWAC	0.21/1 h	13.8
Roi 5	SeaBird	1.5/3 h	6.2
Roi 4	Aquadopp	1.5/3 h	0.8
Roi 3	SeaBird	1.5/3 h	0.7
Roi 1	SeaBird	1.5/3 h	0.4
Nam 4	SeaBird	1.5/3 h	7.7
Nam 7	SeaBird	1.5/3 h	0.4
Nam 1	SeaBird	1.5/3 h	3.1
CMI 9	AWAC	0.21/1 h	13.8
CMI 7	SeaBird	1.5/3 h	5.3
CMI 6	Aquadopp	1.5/3 h	1.0
CMI 4	SeaBird	1.5/3 h	0.8
CMI 3	Aquadopp	1.5/3 h	0.8
CMI 2	SeaBird	1.5/3 h	0.5
Mile 17 4	SeaBird	0.33/1 h (2 Hz)	5.1
Mile 17 1	SeaBird	0.33/1 h (2 Hz)	0.9

Figure 3
Cross-reef bathymetry at Roi (**a**) and CMI (**b**) showing instrument locations

All instruments sampled at 1 s unless indicated. "AWAC" indicates the Nortek Acoustic Wave and Current Profiler, "Seabird" the Seabird SBE26 plus Wave and Tide Record, and "Aquadopp" the Nortek Aquadopp acoustic Doppler current meter. Sampling indicates the duration of the burst sample/interval between burst samples

deployed on the reef adjacent to the College of the Marshall Islands (CMI, Figs. 2b, d, 3) and within the lagoon at a site known locally as Mile 17 (Fig. 2b, e). Instruments at Kwajalein were deployed at Roi-Namur, two islands connected by reclaimed land at the north of Kwajalein atoll (Fig. 2a, c). Five instruments were deployed along a shore perpendicular transect extending from the shoreline at Roi (Fig. 3). At Namur instruments were deployed across the reef flat between Namur and Ennugarret Island, including instruments at the toe of beach at Ennugarret Island (Nam7) (Fig. 2c). The fringing reef flats at each site are near-horizontal, intertidal surfaces with steep (5°–8°) reef faces (Fig. 3).

The pressure sensors were configured and deployed to investigate wind waves (periods of 3–30 s), infragravity waves (30–600 s), and breaking wave setup on the shallow reef flats. As such, 1-s samples over 1.5 h records every 3 h were collected. Instrument descriptions and sampling schemes are presented in Table 1. Infragravity waves have complicated the detection of tsunami signals within nearshore records and are typically low-pass filtered prior to analysis to isolate the tsunami signal (RABINOVICH *et al.* 2006, 2011). While infragravity waves at Kwajalein and Majuro have been observed with periods exceeding 6 min during large swell events in

the complete data set, the moderate incident wave heights during the tsunami suggest that the observed low frequency energy may be attributed to the tsunami. As a result, the predicted tide was removed from each pressure time series and a low-pass filter was applied with a cut-off period of 6 min. For the Mile 17 pressure time series in the Majuro lagoon, a 30 s cut-off period was used due to the reduced wind-wave and infragravity energy compared to the outer reef. Surface elevation is computed from bottom pressure assuming a hydrostatic balance.

Tide gauge records were obtained from the US Army Kwajalein Airbase within the lagoon at the southern end of Kwajalein Atoll (1 min averages) and at Uliga dock, on the lagoon shoreline in the east of Majuro Atoll (6 min averages). The tide gauges sampled continuously throughout the event with the exception of a 24 min gap in the Kwajalein record (at 10:23 in Fig. 4a). Time series from the nearest DART buoy that was operational during the tsunami (buoy 52402) provides deep water measurements (1 min averages) of the tsunami wave approximately 1,500 km WNW of Kwajalein (Fig. 1).

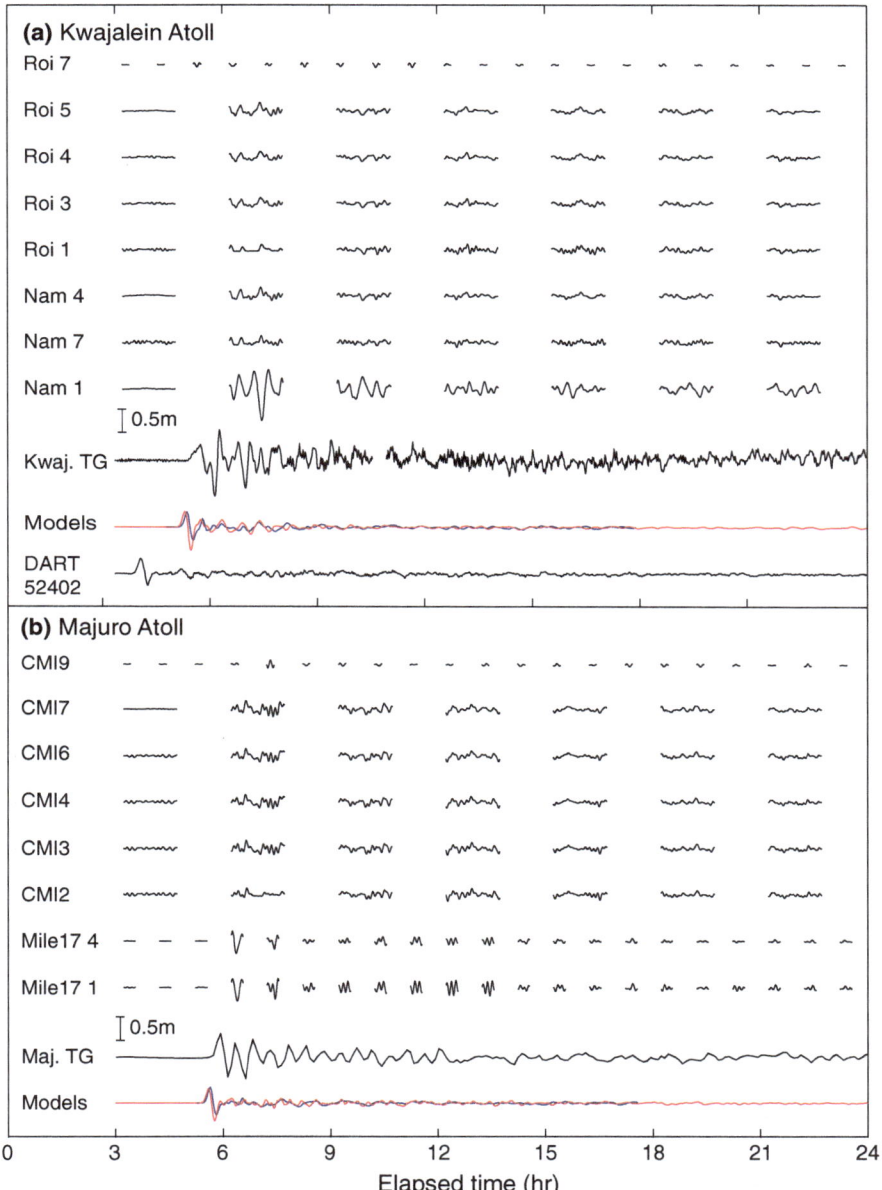

Figure 4
a De-tided surface elevation time series showing tsunami waves at **a** Kwajalein Atoll, and **b** Majuro Atoll. The model results are from deep water grid points near the respective atolls (*red line* = MOST, *blue line* = Song) (Fig. 2)

As a result of the significant distance between both Kwajalein and Majuro atolls and the DART buoy, outputs of two previously published tsunami propagation models are presented to provide deep water estimates of tsunami amplitude close to each atoll rim. The numerical models are referred to as the Song model (SONG *et al.* 2012) and the MOST model

(TITOV and GONZALEZ 1997). The Song tsunami model is a simplified version of the three-dimensional ocean circulation model of SONG and HOU (2006). The tsunami is generated by the earthquake-induced seafloor motions, as formulated in SONG *et al.* (2008). The model grid is configured to cover the entire Pacific Ocean with a 1/12° horizontal resolution and 30 terrain-following vertical levels. The time step is 3 s and the results are saved every 40 steps. Similarly, outputs of the MOST model cover the Pacific Ocean and are stored spatially at 16 arcmin with a 60 s time step (TANG et al. 2012). Time series were extracted from both models at the closest model points northwest of each atoll (Fig. 2). The Marshall Islands and the DART buoy are located south of the main beam of tsunami energy radiating from the source region in the model simulations (Fig. 1).

3. Observations of the Tohoku Tsunami in the Marshall Islands

The Kwajalein and Majuro tide gauge records indicate a leading positive elevation wave at 10:51 and 11:24 UTC respectively (or 05:06 and 05:39 elapsed time in Fig. 4a, b), with the first crest highest at Majuro (0.50 m) and the second crest highest at Kwajalein (0.61 m). Applying a 6 min running mean filter to the Kwajalein record, to match the Majuro record, reduced the peak amplitude to 0.45 m. The 1-min sampling at Kwajalein shows an initial minima following the first crest, and then a substantial trough of −0.81 m. The initial wave height at Kwajalein is, therefore, somewhat uncertain, but the height measured from the crest to the larger trough corresponds to 1.42 m. The initial crest-to-trough wave height at the Majuro tide gauge is 0.93 m, which is greater than the height measured and estimated in the deep ocean, with the DART buoy and the models showing a prominent initial crest-to-trough height of 0.58–0.69 m (Fig. 4a, b).

Most of the Roi-Namur and Majuro pressure sensors began a burst record at 12:00 UTC, which missed the initial wave arrival but captured the wave form soon after the second crest as indicated by the tide gauge records (Fig. 4a, b). The water level oscillations in the lagoons (Nam1, Mile17) resemble those captured by the tide gauges, with a peak amplitude at Nam1 of 0.41 m. The tsunami amplitudes on the ocean-facing fringing reefs at both study sites are significantly weaker than in the lagoons (Fig. 4). The maximum recorded water level amplitudes are 0.18 m on the reef faces (Nam4, Roi5, CMI7), reducing to 0.13–0.14 m on

the reef flats. During the same burst sample records, the peak amplitudes in the lagoons are 0.35 and 0.38 m at the Kwajalein and Majuro tide gauges respectively.

Wave model simulations of the tsunami elevation at grid points in deep water (Fig. 2) compare favorably with the reef face pressure time series at Majuro and Kwajalein (CMI7, Roi5, Nam4) during the first two measured sampling bursts, low-pass filtered with a cut-off period of 6 min (Fig. 5). We emphasize that the model time series and the reef face observations are not sampled at the same location. The agreement is better during the more energetic initial burst, and in general the MOST model matches the observations more closely than the Song model, although both models provide a reasonable approximation of the observations. The first wave of elevation was not measured on the reef flats; however, if the first waves scale with the deep water signal in a similar manner as shown in Fig. 5, then we would estimate the first wave arrival to be ∼0.3 m in amplitude. When added to the tidal level, this would lead to a total water level that is ∼0.5 m below the previous high tide and >1 m below maximum spring tide levels. The relatively low total water level is consistent with the lack of coastal inundation reported during the tsunami on the outer shores of both atolls.

Hours after the initial wave arrival, the tsunami energy on the outer reefs is less energetic than in the atoll lagoons, but more energetic than in the deep ocean as indicated by the models or the DART buoy observations. We note that cross-shore standing mode periods for the shallow reef flats are on the order of minutes, compared to the much longer periods of modal oscillations in the deeper, broader atoll lagoons (see Sect. 4). The spectral characteristics of the deep water tsunami are discussed in Sect. 4 below. Deep-water properties of the Tohoku tsunami also have been discussed by HEIDARZADEH and SATAKE (2013), BORRERO and GREER (2013), RABINOVICH et al. (2013) and others.

Instruments in the Majuro lagoon at Mile 17 were deployed across a shallow subtidal reef, with Mile17 1 located near the toe of the island beach and Mile17 4 in ∼5.1 m depth. During the first post-arrival burst sample, 20 min water level records at Mile17 show similar wave forms at the shallow

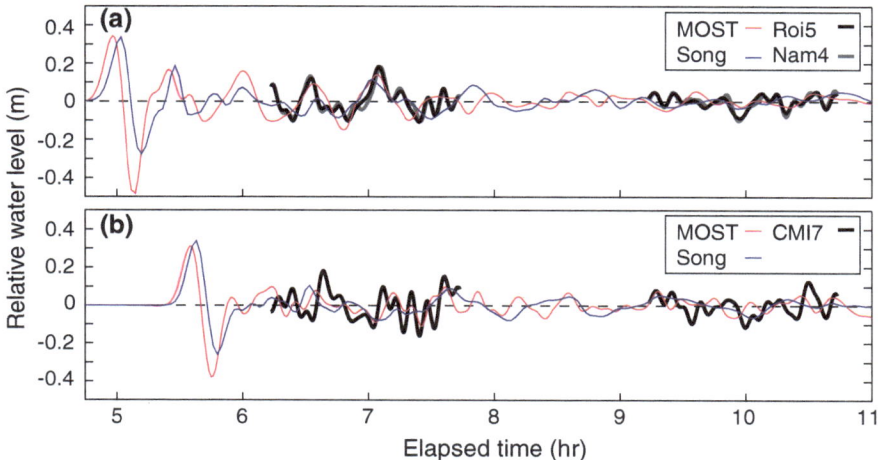

Figure 5
Surface elevation time series showing tsunami waves on the reef face at Kwajalein Atoll (**a**, **b**), and Majuro Atoll (**c**). The model results are from deep water grid points shown in Fig. 2

and deep sensors, with amplitudes of 0.31 and 0.29 m at Mile17 1 and Mile17 4, respectively (Fig. 6a). Phase and amplitude differences between the two locations begin to appear in the next sample (Fig. 6b). During subsequent bursts, the deep and shallow oscillations are approximately 180° out of phase, the dominant period falls below 6 min, and the maximum amplitudes at the shoreline are between 0.05 and 0.17 m higher than at 5.1 m depth (Fig. 6c–e). The high frequency behavior suggests the presence of edge waves in the lagoon, that would not be well captured in the Majuro tide gauge record, with a 6 min sample period.

On the outer fringing reefs, the effects of tidal level on tsunami propagation across the reef flats is evident (Fig. 7). Focussing on the first post-arrival burst sample measured by the pressure sensors at Roi, we see that wave crests propagate on to the reef flat at low tide. Cross-reef time lags and amplitude decay are observed as the wave crests separated by 20–30 min propagate over the tidally exposed reef flat (Fig. 7a). The low water level causes slow wave speeds and enhanced dissipation. In contrast, at high tide (Fig. 7b), wave crests and troughs are observed on the reef, the time lags are short between sensors, and short period (<30 min) wave amplitudes increase toward shore, presumably due to enhanced shoaling relative to dissipation with higher water levels over the reef flat than at low tide.

4. Tide Gauge and Lagoon Oscillations

As is evident from Fig. 4, the lagoon response to the tsunami is amplified compared to the outer reefs of the two atolls. To characterize the spectral character of the tsunami in the atoll lagoons further, a wavelet analysis of the continuous tide gauge records is performed (Fig. 8). We use a Morlet wavelet following the procedures described by TORRENCE and COMPO (1998). The 24-min gap in the Kwajalein time series is filled with linear interpolation and all time series have the predicted tide removed and are high-passed filtered (8-h cut-off period). The wavelet spectra are normalized by the variance at each station and plotted in Fig. 8 as decibels below the peak value.

As discussed by HEIDARZADEH and SATAKE (2013), the tsunami signal at Kwajalein initially (first 2 h after arrival) exhibits energy over a broad range of periods, 15–80 min with a peak ~30 min, followed by sustained energy in the 30–80 min period range (Fig. 8a). In contrast, Majuro shows energy in two prominent bands, 20–40 min and 50–80 min, with peak periods of ~30 and 70 min (Fig. 8b). Ringing in these bands persists for days at Majuro, particularly in the 20–40 min band. A notable energy maxima approximately 40–50 h after the tsunami event likely represents the arrival of reflected wave energy, as noted by HEIDARZADEH and SATAKE (2013).

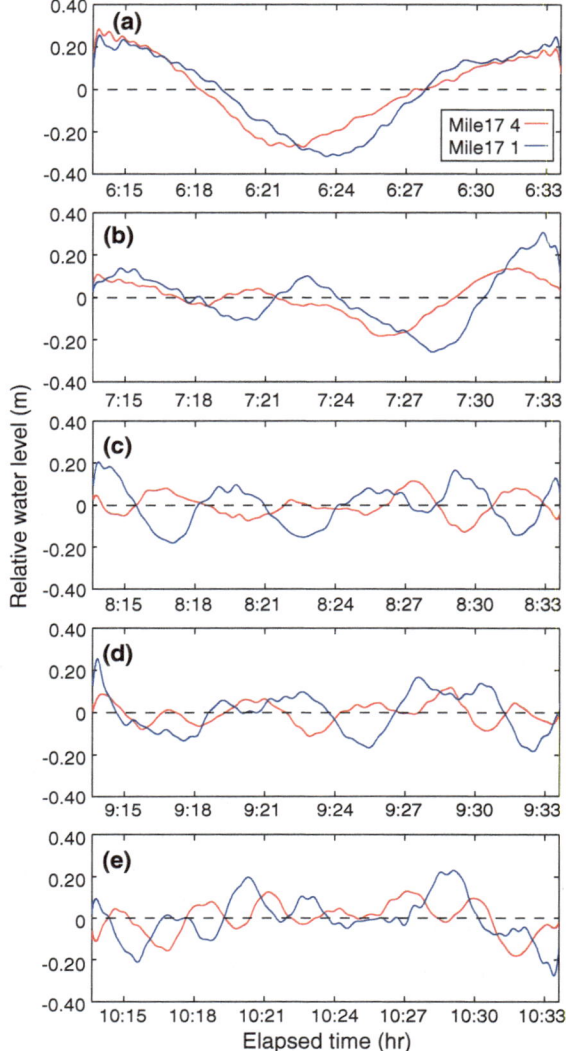

Figure 6
Spectrally filtered (30 s cut-off) water level records from Mile 17 during the first five post-tsunami bursts

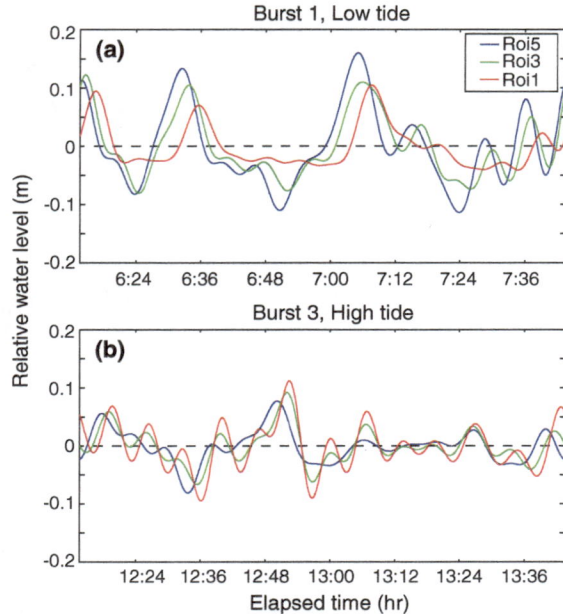

Figure 7
Effects of tidal level on tsunami propagation across the reef at Roi. The *top panel* shows the decay in amplitude and propagation speed of the first measured burst of the tsunami across the reef at low tide. The *bottom panel* shows an increase in the amplitude of the tsunami shoreward during high tide as measured in burst 3. The location of sensors are indicated in the *legend*

We next use autospectra to characterize the frequency content of the tsunami signals at each tide gauge in comparison to deep ocean estimates from the model simulations. Based on the wavelet results (Fig. 8), we focus on the 15-h time span following the initial arrival of the tsunami. At the Kwajalein tide gauge, the tsunami energy is broad-banded with four spectral peaks that are just barely significant at the 95 % level (Fig. 9a). In the deep ocean, the models show energy over a similar broad band without the noted spectral peaks, except for the MOST model,

which indicates prominent deep ocean energy near 30 min period. The difference in lagoon and deep ocean spectra is even more pronounced at Majuro, where the two spectral peaks seen in the wavelet spectra (Fig. 8b) are prominent in the tide gauge autospectrum (Fig. 9b), and both peaks are absent in the deep ocean model spectra. It appears that the tsunami generates basin modes specific to each lagoon.

We further describe the spectral properties of the tsunami from the lagoon tide gauge records following RABINOVICH *et al.* (2013). In Fig. 10, we compare the background spectra (S_b) calculated from the tide gauge records for ~ 10.5 days prior to the onset of the tsunami, to that estimated from the tide gauge records for 15 h during the tsunami (S_t). The background spectra have been smoothed over 112 frequency bands and the tsunami spectra over seven frequency bands. For the Kwajalein background spectral estimates, gaps ranging from 2 to 19 min were filled using linear interpolation. While the

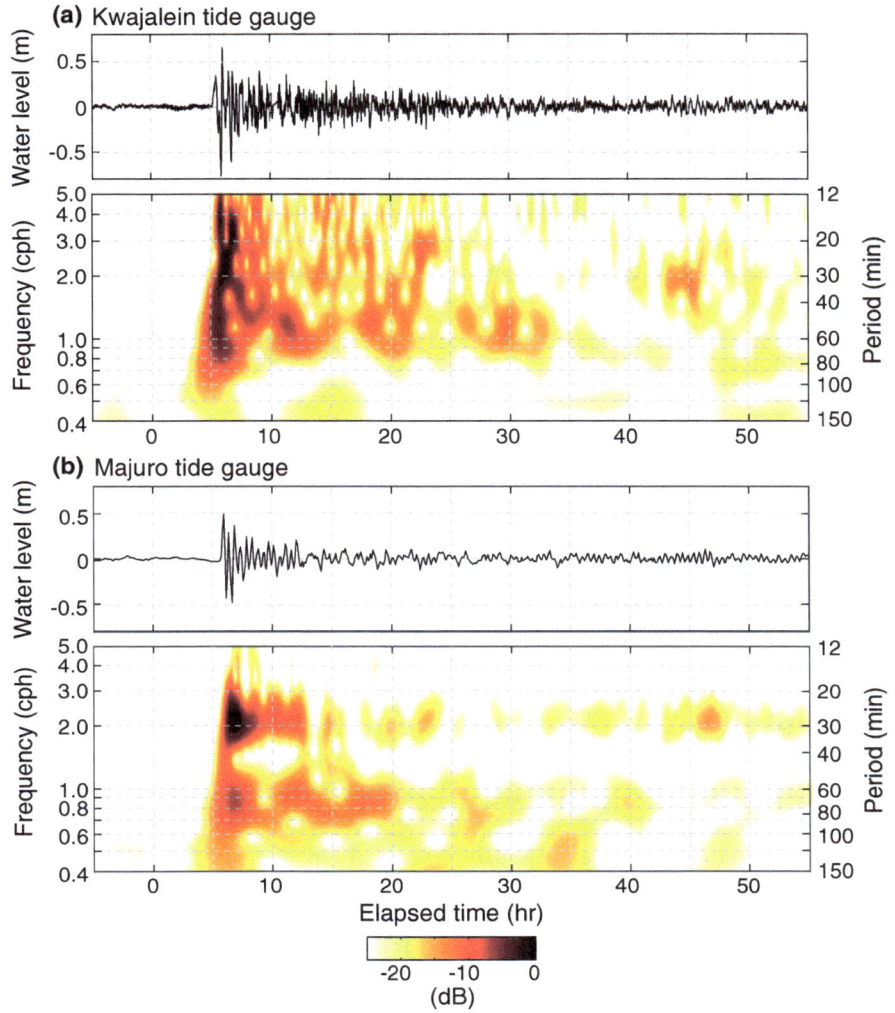

Figure 8

Time series and wave spectra for the **a** Kwajalein and **b** Majuro tide gauge records. Elapsed time is referenced to the start of the Tohoku tsunami event at 3/11/2011 5:46:24 UTC. The wavelet spectra are computed after removing the predicted tide and high-pass filtering (8-h cut-off period) the time series. The spectra are normalized by the variance of each time series, and the contour plots are presented in decibels (dB) below the maximum value of each spectral plot

Majuro tide gauge record was continuous, the 6-min sample period set the high frequency cut-off for the Majuro spectral estimates. At both locations, S_b and S_t exhibit similar energy levels at periods <150 min (Fig. 10a, b). Energy during the tsunami also is higher than the background at Kwajalein out to the highest frequencies measured, which agrees with the high frequency energy observed in the Majuro lagoon pressure records. The ratio of the spectra during the tsunami to the background, S_t/S_b, is used to minimize the effects of the local topography (RABINOVICH *et al.*

2013). The broad spectral peak between the 20 and 60 min period is fairly consistent at the two sites when correcting for the background spectra (Fig. 10c). The distinct spectral peaks at each site are well above background levels. The Marshall Islands spectral ratios (Fig. 10c) also resemble ratios obtained from an average of all available open-ocean DART records (RABINOVICH *et al.* 2013) and from tide gauge records along the Pacific coast of Russia (SHEVCHENKO *et al.* 2014). The spectral character of the Tohoku tsunami, normalized by background

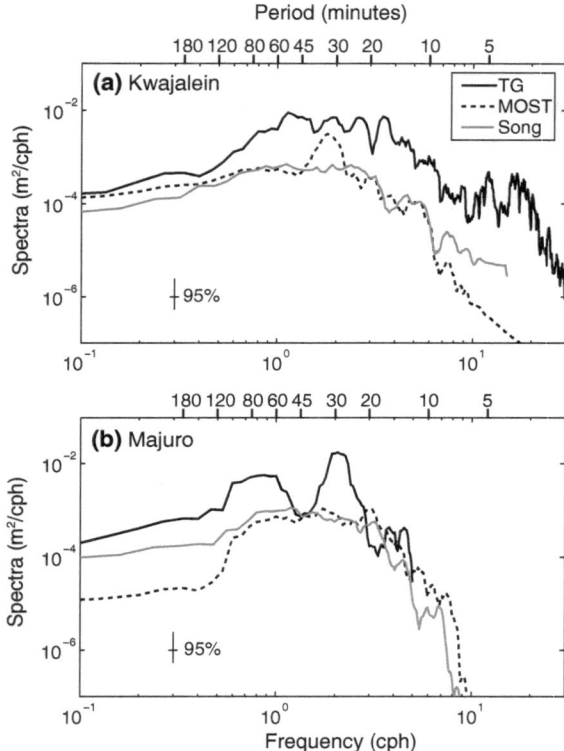

Figure 9
Autospectra of water level from the tide gauge observations (TG), MOST model output (MOST), and Song model output (Song) for **a** Kwajalein and **b** Majuro. Spectra are computed for the time period 3/11/2011 10:49 UTC to 3/12/2011 1:49. The predicted tide has been removed from the TG time series. 95 % confidence intervals are obtained by smoothing over 7 adjacent frequencies

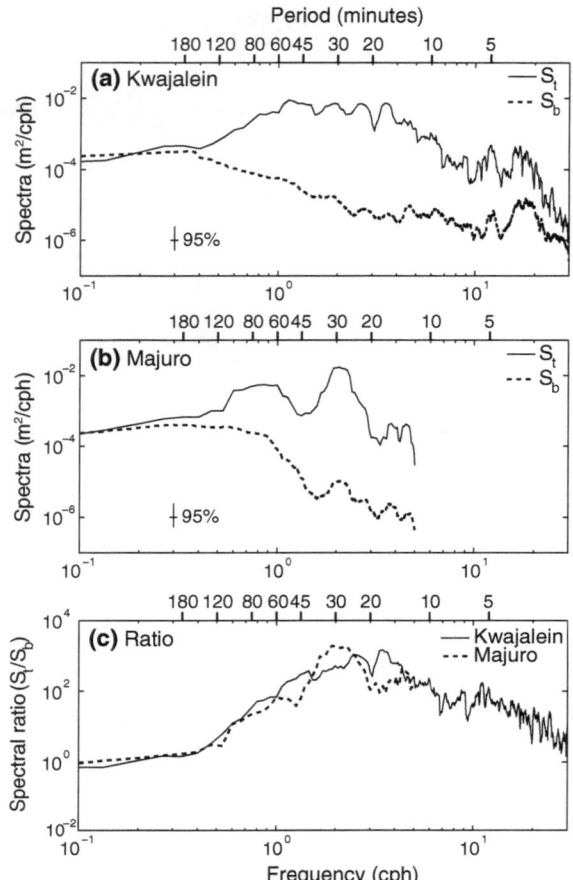

Figure 10
Autospectra computed before, S_b (3/1/2011 0:00 UTC to 3/11/2011 10:49), and after, S_t (3/11/2011 10:49 to 3/12/2011 1:49), the arrival of tsunami waves observed at the **a** Kwajalein and **b** Majuro tide gauges. 95 % confidence intervals are computed based on the S_t estimate obtained by smoothing over seven adjacent frequencies. **c** The ratio S_t/S_b for the Kwajalein and Majuro tide gauges

spectra, appears to have been remarkably consistent, even at sites well north (Russia) and south (Marshall Islands) of the main beam of wave energy.

The presence of spectral peaks and persistent "ringing" of the tsunami signal at the Majuro atoll lagoon suggests that natural modes of oscillation have been excited by the deep water tsunami, with dominant periods illustrated in Figs. 8, 9, 10. At Majuro, the observed spectral peaks agree with qualitative estimates of the dominant modal periods. Treating the Majuro lagoon as rectangular in shape ($L = 37$ km, $W = 8$ km) and predominantly enclosed with an average depth of $h = 40$ m, we estimate that the period of a one-dimensional 1/2 wavelength mode at Majuro is $2L(gh)^{-1/2} = 62$ min, and 1 wavelength mode is $L(gh)^{-1/2} = 31$ min, roughly matching the periods of the observed spectral peaks. While the

Majuro lagoon is predominantly enclosed, with a dry or shallow water reef enclosing most of the lagoon, one ~3.5 km wide pass exists on the north-facing rim of the atoll and a narrow, artificial boat channel is found on the southern rim. Hence, the possibility exists that a 1/4 wavelength mode in width, $4W(gh)^{-1/2}$, may be excited in the lagoon. For Majuro, the period of the 1/4 wavelength mode is 27 min. Comparing the tide gauge time series with the pressure records at Mile17 indicates an approximate 180° phase shift between the sites, which appears to favor the one-wavelength (closed basin) major axis mode rather than the quarter-wavelength (open basin) minor axis mode. At Kwajalein, the complicated geometry of the lagoon precludes simple estimates of the closed and open basin normal modes. A more detailed model analysis using the actual bathymetry and boundary conditions is required to determine the character of the normal modes at both atolls, which at present is limited by the lack of high resolution topography.

5. Discussion and Implications for Atoll Islands

The amplitude of the Tohoku tsunami was the largest recorded at Kwajalein atoll since the tide gauge began operation in 1946. However, the Tohoku tsunami caused no reported inundation within the Marshall Islands. This is a fortuitous function of the initial and potentially highest amplitude motions coinciding with low-tide. The wave amplitude along the ocean-facing reef flat was considerably lower than within the lagoon, similar to deep ocean values, at least after the initial crest arrival. The initial peak of the tsunami was unrecorded by the fringing reef instruments. The total water level fluctuations recorded at all sites, including the tidal component, did not exceed the level of the preceding high tide. Although direct observations are lacking, we speculate that the same would hold for reef-protected shorelines around the perimeter of the atoll. Assuming that the amplitude of the initial crest on the reef flats scales with the modeled deep-ocean tsunami in a similar manner as observed for subsequent lower amplitude crests (Fig. 5), then we would expect the initial tsunami crest amplitudes on the reef flat to be lower than what was observed by the lagoon tide gauges.

Observations of tsunami behavior on the ocean-facing reef flats are broadly consistent with results of numerical models (GELFENBAUM et al. 2011). With the exception of the first post-tsunami burst, during which time the tsunami propagated across a dry reef flat and subsequent measurements at low tide, we find that the tsunami increases in height while propagating across the reef flat. GELFENBAUM et al. (2011) show that on narrow reef flats (<200 m) shoaling exceeds dissipation, driving higher shoreline runup levels. Reef flats at the Roi-Namur and CMI sites range between 240

and 350 m wide. Elsewhere in Kwajalein and Majuro ocean-facing reef flats range in width; in places, intertidal reef flats are narrower than 100 m. Similarly, the width and depth of sub- and intertidal reefs within the lagoon vary considerably, likely to result in considerable spatial variability in shoreline runup levels. At present, assessing potential tsunami-driven inundation, particularly at Majuro is problematic as no high-quality topographic datasets exist.

Atoll islands are particularly low-lying, with areas neighboring the lagoon typically the lowest coastal relief (WOODROFFE 2008). Relatively small positive sea level anomalies (<0.30 m) associated with ENSO cycles have been documented to drive inundation of lagoon-facing sections of Majuro during spring tides (FORD et al. 2012). As a result, a wave sharing similar characteristics as the Tohoku tsunami would have been high enough to exceed the topographic threshold had the peak coincided with higher tidal levels. Lagoon shorelines are typically sheltered from ocean swells and have lower levels of exposure to storm-driven inundation. Since WWII Majuro atoll has undergone rapid population growth and intensification of development (SPENNEMANN 1996). Significant key infrastructure including the airport, port and power station are located along the lagoon shore, along with dense housing, which occupies nearly all available land within the urban sections of Majuro.

Relative to storm-driven inundation, tsunamis have to date caused no documented impact in the Marshall Islands. However, the impact of the 2004 Sumatran tsunami in the Maldives does illustrate the potential for high impact tsunami events on atolls (NGDC/WDS). As a result, hazard mitigation and preparedness plans for atolls require careful consideration of the risks associated with tsunamis. Within atoll settings ocean-facing sections of the islands are most exposed to storm-associated hazards, particularly wave overwash. Our results show the likelihood that settings considered sheltered from storm-driven inundation, namely the inner lagoon shoreline, are not immune to inundation associated with the long-wave forcing caused by tsunamis, particularly if the resonant frequencies of the lagoon match energetic portions of the tsunami spectrum.

Acknowledgments

This study was supported by a grant from the National Science Foundation (OCE-0927407). DART buoy data were obtained from the NOAA National Data Buoy Center (http://www.ndbc.noaa.gov/dar.shtml), Kwajalein tide gauge data from the NOAA Center for Operational Oceanographic Products and Services (http://tidesandcurrents.noaa.gov), and Majuro tide gauge data from the National Tidal Centre, Australia. Comments by Alexander Rabinovich and two anonymous reviewers greatly improved the manuscript. We thank Rachel Tang for the MOST model data. Christopher Kontoes, Carly Fetherolf, and Derek Young were responsible for the Marshall Islands field observations. Hyang Yoon provided graphics and data processing support.

REFERENCES

ADGER, W.N., T.P. HUGHES, C. FOLKE, S.R. CARPENTER, and J. ROCKSTRÖM (2005). *Social-ecological resilience to coastal disasters*, Science, 309, 1036–1039. doi:10.1126/science.1112122.

ALLAN, J.C., P.D. KOMAR, P. RUGGIERO, and R. WITTER (2012). *The March 2011 Tōhoku Tsunami and Its Impacts along the US West Coast*, J. Coast. Res. 28, 1142–1153. doi:10.2112/JCOASTRES-D-11-00115.1.

BAIRD, A.H., S.J. CAMPBELL, A.W. ANGGORO, R.L. ARDIWIJAYA, N. FADLI, Y. HERDIANA, T. KARTAWIJAYA, D. MAHYIDDIN, A. MUKMININ, and S.T. PARDEDE (2005). *Acehnese reefs in the wake of the Asian tsunami*, Curr. Biol., 15, 1926–1930. doi:10.1016/j.cub.2005.09.036.

BORRERO, J.C., BELL, R., CSATO, C., DELANGE, W., GORING, D., GREER, S.D., PICKETT, V. and POWER, W. (2012). *Observations, effects and real time assessment of the March 11, 2011 Tohoku-oki tsunami in New Zealand*, Pure and Appl. Geophys., 170, 1229–1248. doi:10.1007/s00024-012-0492-6.

BORRERO, J.C. and S.D. GREER (2013). *Comparison of the 2010 Chile and 2011 Japan tsunamis in the far field*, Pure Appl. Geophys. 170, 1249–1274. doi:10.1007/s00024-012-0559-4.

CHATENOUX, B. and P. PEDUZZI (2007). *Impacts from the 2004 Indian Ocean Tsunami: analysing the potential protecting role of environmental features*, Nat. Hazards, 40, 289–304. doi:10.1007/s11069-006-0015-9.

FORD, M.R., J.M. BECKER, and M.A. MERRIFIELD (2012). *Spatial and temporal controls of atoll island inundation: implications for urbanized atolls in the Marshall Islands*. (Invited), 2012 AGU Fall Meeting.

GELFENBAUM, G., A. APOTSOS, A.W. STEVENS, and B. JAFFE (2011). *Effects of fringing reefs on tsunami inundation: American Samoa*, Earth-Sci. Rev., 107, 12–22. doi:10.1016/j.earscirev.2010.12.005.

GOFF, J., P.L. LIU, B. HIGMAN, R. MORTON, B.E. JAFFE, H. FERNANDO, P. LYNETT, H. FRITZ, C. SYNOLAKIS, and S. FERNANDO (2006). *Sri Lanka field survey after the December 2004 Indian Ocean tsunami*, Earthquake Spectra, 22, 155–172. doi:10.1193/1.2205897.

HEIDARZADEH, M. and K. SATAKE (2013). *Waveform and spectral analyses of the 2011 Japan tsunami records on tide gauge and DART stations across the Pacific Ocean*, Pure Appl. Geophys., 170, 1275–1293. doi:10.1007/s00024-012-0558-5.

Kench, P.S., R.F. McLean, R.W. Brander, S.L. Nichol, S.G. Smithers, M.R. Ford, K.E. PARNELL, and M. ASLAM (2006). *Geological effects of tsunami on mid-ocean atoll islands: The Maldives before and after the Sumatran tsunami*, Geology, 34, 177–180. doi:10.1130/G21907.1.

KUNKEL, C.M., R.W. HALLBERG, and M. OPPENHEIMER (2006). *Coral reefs reduce tsunami impact in model simulations*, Geophys. Res. Lett., 33, L23612. doi:10.1029/2006GL027892.

LUGO-FERNÁNDEZ, A., H.H. ROBERTS, and J.N. SUHAYDA (1998). *Wave transformations across a Caribbean fringing-barrier coral reef*, Cont. Shelf Res., 18, 1099–1124.

LYNETT, P.J. (2007). *Effect of a shallow water obstruction on long wave runup and overland flow velocity*, J. of Waterway, Port, Coastal, and Ocean Eng., 133, 455–462. doi:10.1061/(ASCE)0733-950X(2007)133:6(455).

MERRIFIELD, M., Y. FIRING, T. AARUP, W. AGRICOLE, G. BRUNDRIT, D. CHANG-SENG, R. FARRE, B. KILONSKY, W. KNIGHT, and L. KONG (2005). *Tide gauge observations of the Indian Ocean tsunami, December 26, 2004*, Geophys. Res. Lett., 32, L09603. doi:10.1029/2005GL022610.

MORI, N., T. TAKAHASHI, T. YASUDA, and H. YANAGISAWA (2011). *Survey of 2011 Tohoku earthquake tsunami inundation and runup*, Geophys. Res. Lett., 38, L00G14. doi:10.1029/2011GL049210.

NATIONAL GEOPHYSICAL DATA CENTER (NGDC) (2012). *March 11, 2011 Japan Earthquake and Tsunami*. http://www.ngdc.noaa.gov/hazard/honshu_11mar2011.shtml. Accessed 09/28/2013.

PÉQUIGNET, A., J. BECKER, M. MERRIFIELD, and S. BOC (2011). *The dissipation of wind wave energy across a fringing reef at Ipan, Guam*, Coral Reefs, 30(1), 71–82. doi:10.1007/s00338-011-0719-5.

RABINOVICH, A.B., R.E. THOMSON, and F.E. STEPHENSON (2006). *The Sumatra tsunami of 26 December 2004 as observed in the North Pacific and North Atlantic oceans*, Surv. Geophys., 27(6), 647–677. doi:10.1007/s10712-006-9000-9.

RABINOVICH, A.B. and R.E. THOMSON (2007). *The 26 December 2004 Sumatra tsunami: Analysis of tide gauge data from the world ocean Part 1. Indian Ocean and South Africa.* Pure Appl. Geophys. 164, 261–308. doi:10.1007/s00024-006-0164-5.

RABINOVICH, A.B., K. STROKER, R. THOMSON, and E. DAVIS (2011). *DARTs and CORK in Cascadia Basin: High-resolution observations of the 2004 Sumatra tsunami in the northeast Pacific*, Geophys. Res. Lett., 38(8), L08607. doi:10.1029/2011GL047026.

RABINOVICH, A.B., R.N. CANDELLA, and R.E. THOMSON (2013). *The open ocean energy decay of three recent trans-Pacific tsunamis*, Geophys. Res. Lett. 40, 3157–3162. doi:10.1002/grl.50625.

SHEVCHENKO, G., T. IVELSKAYA, and A. LOSKUTOV (2014). *Characteristics of the 2011 Great Tohoku Tsunami on the Russian Far East Coast: Deep-Water and Coastal Observations*, Pure Appl. Geophys., 1–22. doi:10.1007/s00024-013-0727-1.

SONG, Y.T. and T.Y. HOU (2006). *Parametric vertical coordinate formulation for multiscale, Boussinesq, and non-Boussinesq ocean modeling*, Ocean Modelling, 11(3), 298–332. doi:10.1016/j.ocemod.2005.01.001.

SONG, Y.T., L. FU, V. ZLOTNICKI, C. JI, V. HJORLEIFSDOTTIR, C. SHUM, and Y. YI (2008). *The role of horizontal impulses of the faulting* continental slope in generating the 26 December 2004 tsunami, Ocean Modelling, 20(4), 362–379. doi:10.1016/j.ocemod.2007.10.007.

SONG, Y.T., I. FUKUMORI, C. SHUM, and Y. YI (2012). Merging tsunamis of the 2011 Tohoku-Oki earthquake detected over the open ocean, Geophys. Res. Lett., 39(5). doi:10.1029/2011GL050767.

SPENNEMANN, D.H. (1996). Nontraditional settlement patterns and typhoon hazard on contemporary Majuro atoll, Republic of the Marshall Islands, Environ. Manage., 20, 337–348.

Tang, L., V.V. Titov, E.N. Bernard, Y. Wei, C.D. Chamberlin, J.C. Newman, H.O. MOFJELD, D. ARCAS, M.C. EBLE, and C. MOORE (2012). Direct energy estimation of the 2011 Japan tsunami using deep-ocean pressure measurements, J. Geophys. Res., 117(C8). doi:10.1029/2011JC007635.

TITOV, V. and F. GONZALEZ (1997). Implementation and testing of the Method of Splitting Tsunami (MOST) model, NOAA Tech. Memo. ERL PMEL-112 (PB98-122773), NOAA/Pacific Marine Environmental Laboratory, Seattle, WA, 11 pp.

TORRENCE, C. and G.P. COMPO (1998). A practical guide to wavelet analysis, Bull. Am. Meteorol. Soc., 79, 61–78.

U.S. FISH and WILDLIFE SERVICE (2011). Seabird loses at Midway Atoll Wildlife Refuge greatly exceed early estimates. Press Release, 18th March 2011.

VAN DORN, W. (1984). Some tsunami characteristics deducible from tide records, J. of Phys.Ocean., 14, 353–363.

WILSON, R.I., A.R. ADMIRE, J.C. BORRERO, L.A. DENGLER, M.R. LEGG, P. LYNETT, T.P. MCCRINK, K.M. MILLER, A. RITCHIE, and K. STERLING (2012). Observations and impacts from the 2010 Chilean and 2011 Japanese tsunamis in California (USA), Pure Appl. Geophys., 170, 1127–1147. doi:10.1007/s00024-012-0527-z.

WOODROFFE, C.D. (2008). Reef-island topography and the vulnerability of atolls to sea-level rise, Global Planet. Change, 62, 77–96. doi:10.1016/j.gloplacha.2007.11.001.

YAMAZAKI, Y., K.F. CHEUNG, G. PAWLAK, and T. LAY (2012). Surges along the Honolulu coast from the 2011 Tohoku tsunami, Geophys. Res. Lett., 39: L09604. doi:10.1029/2012GL051624.

(Received October 1, 2013, revised December 1, 2013, accepted December 3, 2013, Published online December 22, 2013)

Pure Appl. Geophys. 171 (2014), 3365–3384
© 2014 Springer Basel
DOI 10.1007/s00024-014-0781-3

Waves and Currents in Hawaiian Waters Induced by the Dispersive 2011 Tohoku Tsunami

HONGQIANG ZHOU,[1,2] YONG WEI,[1,2] LINDSEY WRIGHT,[1,2] and VASILY V. TITOV[1]

Abstract—This study focuses on the effects of frequency dispersion on tsunami-induced coastal water waves and currents, exemplified by the 2011 Tohoku tsunami event. The investigation relies on numerical simulations. We start from a tsunami source constrained through the inversion algorithm of NOAA's tsunami inundation forecast system. The trans-Pacific propagation and the hydrodynamic processes in the Hawaiian Islands region are simulated with a weakly dispersive Boussinesq model and a shallow-water model that neglects dispersion effects. From these modeling results, boundary conditions are derived to force the high-resolution simulations in the coastal waters in the Hawaiian Islands region through MOST, a tsunami simulating code based on the shallow-water theory. We note that the dispersion effects generally lower the amplitudes of leading waves. Trailing waves of short wavelengths and high amplitudes can develop in coastal waters. A model neglecting dispersion effects could under-predict the wave heights and current speeds at the trailing waves.

Key words: Water waves, Currents, Coastal waters, Dispersion, 2011 Tohoku tsunami.

1. Introduction

Frequency dispersion may affect tsunamis propagating in the ocean. This process can be interpreted as follows. Since the dynamic evolution of tsunamis is a linear process in open ocean (KÂNOGLU and SYNOLAKIS, 2006), the wave train can be seen as being composed of numerous independent components, each with a period that mainly depends on the horizontal size and water depth of the source area (RABINOVICH, 1997). Dispersion relation states that wave celerity decreases for shorter wave periods. Propagating in the ocean, longer components will be

subsequently separated from shorter ones, leaving the latter in the trailing group. In general, this process results in lower and smoother leading waves (LØVHOLT et al., 2012). The behavior of dispersion effects is more complicated in the trailing wave groups. In some cases, by neglecting dispersion, the numerical model could under-predict the amplitudes of some trailing waves. This may be the reason for the observations that non-dispersive models predict lower maximum water surface elevations in some areas in the ocean (e.g., KIRBY et al., 2013). Since the separation happens faster in waves of higher frequencies, in general, dispersion effects are more significant if there is more wave energy concentrated in high-frequency modes. This can be exemplified by the observations in previous events. For example, the 2009 Samoa tsunami, with predominantly high energy concentration in periods of 2–30 min, exhibits stronger dispersion effects than the 2011 Tohoku event, which has wave energy more evenly distributed in periods of 2–180 min (RABINOVICH et al., 2013; ZHOU et al., 2012; LØVHOLT et al., 2012). In order to quantify the significance of dispersion effects, GLIMSDAL et al. (2013) defined a non-dimensional parameter, "dispersion time",

$$\tau = \frac{6h^2 L}{\lambda^3}, \tag{1}$$

where h is still water depth, L is propagation distance, and λ is the length of the wave front. Based on the examination of several seismic and landslide-generated tsunamis, they suggested that dispersion effects can be neglected for $\tau < 0.01$ but must be considered when $\tau > 0.1$.

Nearshore wave height is a parameter commonly employed to evaluate the severity of tsunami hazards. In addition to water waves, damage induced by a tsunami may also be a result of strong water currents

───────────
[1] Pacific Marine Environmental Laboratory, National Oceanic and Atmospheric Administration, Seattle, WA 98115, USA. E-mail: hongqiang.zhou@noaa.gov
[2] Joint Institute for the Study of the Atmosphere and Ocean, University of Washington, Seattle, WA 98105, USA.

(LYNETT *et al.*, 2012). Because of the lack of field measurements, studies on tsunami-induced currents, especially in the events with significant dispersion effects, are rare.

The 2011 Tohoku tsunami, which was triggered by a moment magnitude 9.0 earthquake off the northeastern coast of Honshu Island (Tohoku region), Japan on 11 March 2011, is reported to have considerable dispersion effects (LØVHOLT *et al.*, 2012; GRILLI *et al.* 2013; KIRBY *et al.*, 2013; GLIMSDAL *et al.*, 2013). GLIMSDAL *et al.* (2013) estimate $\tau \simeq 0.45$ at DART (Deep-Ocean Assessment and Reporting of Tsunamis) buoy 43413 west of Guatemala. At DART 51407 south of the Hawaiian Islands, dispersion effects are less but still clear (GLIMSDAL *et al.*, 2013). This event resulted in severe devastation to coastal cities near the epicenter. Strong waves and currents, as well as localized damage, were also reported in many other areas in the Pacific basin. For example, wave amplitudes of ~ 1 m and current speeds of ~ 3–4 knots (1.5–2.1 m/s) were observed at the University of Hawaii Marine Center near the entrance of Honolulu Harbor (YAMAZAKI *et al.*, 2012). In New Zealand, the tsunami induced maximum wave heights ranging from 0.4 to 2.0 m, and peak current speeds of approximately 2 knots (~ 1.0 m/s) (BORRERO *et al.*, 2013). In Crescent City, California, the damage caused by this tsunami was greatly reduced as it arrived during low tide. Even so, a maximum wave amplitude of 2.47 m was recorded by the tide gauge. A small-boat basin was nearly completely destroyed by the strong currents, which were estimated to be as high as 4.5 m/s (WILSON *et al.*, 2013). During this event, time series of water surface elevations were widely recorded by the DART tsunameters in deep ocean and tide gauges in coastal waters. In the Hawaiian Islands region, the University of Hawaii has deployed an Acoustic Doppler Current Profiler (ADCP) at the Kilo Nalu Nearshore Observatory near the entrance of Honolulu Harbor (PAWLAK *et al.*, 2009). Water surface elevations and current velocities at this station were recorded during this event. Water current velocities at multiple sites in the Hawaiian Islands region were also measured by the ADCPs deployed by the National Ocean Service (NOS). These data provide an opportunity to validate the numerical models and

to investigate the dispersion effects in tsunami-induced water waves and currents.

In this study, we visit the 2011 Tohoku tsunami event with an interest focused on the dispersion effects in tsunami-induced hydrodynamic processes in Hawaii waters. This study mainly relies on numerical simulations. We first simulate the trans-Pacific propagation with a Boussinesq model that considers weakly dispersive effects and a non-dispersive shallow-water (SW) model. The hydrodynamic processes in the Hawaiian Islands region are then computed with the same models in a nested grid at a higher resolution. Water waves and currents in the coastal waters of the Hawaiian Islands are simulated with a non-dispersive model, MOST (Method of Splitting Tsunamis), in three layers of nested grids at successively increased resolutions.

2. Numerical Approach

Most conventional tsunami simulating models are based on the SW theory, which neglects dispersion effects. Also neglected in these models are the variation of water velocities along water depth and the hydrodynamic pressure due to vertical acceleration of water particles. The availability of high-performance computers has made the SW models applicable to real-time forecast simulations.

In recent years, the Boussinesq models have been applied to tsunamis (e.g., GRILLI *et al.*, 2007, 2010, 2013; HORILLO *et al.*, 2006; IOUALALEN *et al.*, 2007; LØVHOLT *et al.*, 2008, 2012; ZHOU *et al.*, 2011, 2012; KIRBY *et al.*, 2013). Dispersion effects are considered in the Boussinesq models through perturbation approximation. Compared with the SW models, the Boussinesq models exhibit great improvement in simulating dispersive water waves. Because of the high-order partial derivative equations they employ, the Boussinesq models usually consume more computer resources. Parallel computing techniques have been applied to some Boussinesq models to increase the computation speed (e.g., SITANGGANG and LYNETT, 2005; KIRBY *et al.*, 2013). The high-order derivative terms may become singular at grid points with abrupt variation of bathymetry, and cause instability of the numerical model. Numerical filtering can be applied to remove the numerical noises due to singularity (e.g., KIRBY et al., 1998). Stability of the Boussinesq models may also be improved through advanced numerical algorithms (e.g., SHI et al., 2001; ROEBER et al., 2010; KIRBY et al., 2013).

Dispersion effects are usually important in long-distance propagation in the open ocean, and may be less significant in regional propagation and coastal inundation. A combined modeling approach is suggested by LØVHOLT et al. (2010), which applies a Boussinesq model to dispersive oceanic propagations and an SW model to the dynamic processes in coastal waters. This approach is also adopted in the present study.

2.1. The Boussinesq Model

Early versions of the Boussinesq models were developed in the Cartesian coordinates for water waves in nearshore shallow and intermediate-depth water (e.g., NWOGU, 1993; WEI and KIRBY, 1995). For basin-scale simulations, LØVHOLT et al. (2008) developed a Boussinesq model in the geographic coordinate system. KIRBY et al. (2013) introduced a more comprehensive scaling system and derived a set of the Boussinesq equations in a close form of LØVHOLT et al.'s (2008) model. A slightly different form of these equations was obtained by ZHOU et al. (2011), which transformed NWOGU's (1993) original version from the Cartesian coordinates into the geographic reference frame. In the Appendix, we re-derive the Boussinesq equations by adopting the scaling system of KIRBY et al. (2013), and following the steps of LYNETT and LIU (2002). The convective terms, i.e., $(u_\alpha v_\alpha)/R \tan\theta$ in Eq. 41 and $u_\alpha^2/R \tan\theta$ in Eq. 42, have a magnitude of $(\epsilon\delta)/\mu$. Here, u_α and v_α are longitudinal and meridional water velocities at the reference water depth, R is the radius of the Earth, θ is latitude, ϵ and μ are parameters of nonlinearity and dispersion, respectively, and δ is the ratio of characteristic water depth to the Earth's radius. For weakly dispersive water waves in the ocean basin, KIRBY et al. (2013) assumed $O(\delta) = O(\mu^3)$ and recommended that the convective terms be dropped from these equations. In this study, this assumption is not followed and the convective terms are retained. This makes the model applicable to a broader scope of problems, such as when $O(\delta) \sim O(\mu)$. In this

situation, the convective terms have a magnitude of $O(\epsilon)$ and may play an important role in wave evolution. Another advantage of Eqs. 39, 41, and 42 is that they can be readily converted into the nonlinear SW equations by turning off the dispersion terms. A numerical solution to the present Boussinesq equations is obtained through the finite difference scheme of WEI and KIRBY (1995). In general, the numerical model has a second-order accuracy in space and fourth-order accuracy in time. A one-way grid-nesting scheme is also included in this model. In this study, the trans-Pacific propagation of the 2011 Tohoku tsunami, as well as the waves and currents in the Hawaiian Islands region, is simulated with the present Boussinesq model.

2.2. MOST

The present study employs the numerical code MOST to compute the hydrodynamic processes in selected coastal areas in the Hawaiian Islands region. MOST is the numerical core of SIFT (Short-Term Inundation Forecast of Tsunamis), a real-time tsunami forecast system developed by the NOAA Center for Tsunami Research (NCTR). The code employs a characteristic formulation of the SW theory and solves it numerically through a finite difference scheme that is accurate to the second order in space and first order in time. A one-way grid nesting approach is employed to zoom simulations from large-scale to high-resolution nested grids. A moving boundary scheme is employed to track the advance and retreat of shoreline (TITOV and SYNOLAKIS, 1995; 1998; TITOV and GONZÁLEZ, 1997). MOST has numerical dispersion, whose magnitude is mainly determined by the spatial resolution of a computational grid. The numerical dispersion can be manipulated to compensate for some neglected physical dispersion effects (BURWELL et al., 2007). On the other hand, high resolutions need to be applied if the numerical dispersion is to be minimized.

3. Tsunami Source and Trans-Pacific Propagation

Constraint of the tsunami source (initial water wave profiles) is the first step in tsunami modeling and a determinant factor for the accuracy of modeling results. Generated water wave structure is primarily dictated by the features of an earthquake, the most important of which may be seabed displacement. Time history of seabed deformation may also play a role in tsunami generation. HAMMACK (1973) indicated that when the bottom disturbance occurs in a very short time, the time-displacement history has a negligible effect on tsunami generation; the generated water waves generally resemble the final shape of a deformed seabed. For the 2011 Tohoku event, some studies assumed impulsive seabed deformation in tsunami source constraint (e.g., LØVHOLT et al., 2012; Tang et al., 2012). On the other hand, some studies suggested that the seabed deformation in this event may have lasted for hundreds of seconds, and this timing aspect needs to be considered (e.g., GRILLI et al., 2013).

The SIFT system aims to estimate the possible tsunami impact quickly after an earthquake happens. Instead of the details of an earthquake, which may be difficult to determine shortly after an event, SIFT relies on real-time measurements of water surface displacements at DART stations near the epicenter and a pre-computed tsunami propagation database, and constrains tsunami sources through an inversion algorithm (PERCIVAL et al., 2011). The propagation database is composed of tsunami source functions, i.e., time series of water surface elevations and water velocities in the ocean due to a unit earthquake source with predefined seismic parameters. The inversion algorithm looks for the best fit of water surface displacements between measurements and the linear combination of selected tsunami source functions. The accuracy of the inversion algorithm has been demonstrated in numerous studies (e.g., TANG et al., 2008; 2009; WEI et al., 2008; 2013). Although the propagation database is computed with MOST, in which frequency dispersion is not considered, a recent study (ZHOU et al., 2012) indicated that this approach can also be applied to events with considerable dispersion effects. The main reason is that the inversion algorithm employs measurements near the epicenter, where dispersion effects have not become significant.

Immediately after the Tohoku earthquake on 11 March 2011, NCTR conducted an experimental tsunami forecast for the U.S. Pacific coast with the SIFT system. A tsunami source was constrained based on the measurements at DART stations 21401 and 21418 (TANG et al., 2012; WEI et al., 2013). In this study, we refine the inversed result employing the time series recorded at DART stations 21413 and 21419, and obtain a source that provides better agreement for the Boussinesq model. Figure 1 shows the locations of involved unit earthquake sources and the resulting tsunami source. The parameters of unit earthquake sources are presented in Table 1. This tsunami source is input into the Boussinesq model as initial conditions to initiate the simulation of tsunami propagation in the northern Pacific basin at a resolution of 4′ (Fig. 2). Bathymetry in this grid is derived from the ETOPO2 2-min global relief model. Time series of computed water surface elevations are output at selected DART stations and compared with those recorded in the field. In order to demonstrate the effects of dispersion, we turn off the dispersion terms in the Boussinesq model and rerun the same process. For convenience, we refer to the model without dispersion terms as the "SW model". As observed in Fig. 3, both models have reasonable agreement with the measurements in leading waves. At DART stations 51407 and 43413, the SW model predicts a higher leading wave, similar to that

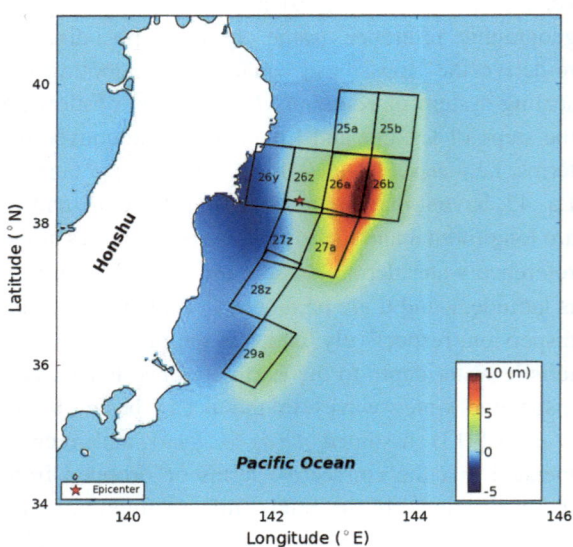

Figure 1
Tsunami source of the 2011 Tohoku event

Table 1

Source parameters of the 2011 Tohoku tsunami

Unit source	Latitude (°N)	Longitude (°E)	Dip (°)	Strike (°)	Slip (m)	Depth (km)
25b	39.418	143.425	19.0	185.0	2.7	5.00
25a	39.454	142.884	21.0	185.0	5.0	21.28
26b	38.525	143.293	19.0	188.0	3.4	5.00
26a	38.584	142.762	21.0	188.0	27.3	21.28
26z	38.642	142.231	21.0	188.0	22.6	39.20
26y	38.700	141.699	21.0	188.0	18.0	57.12
27a	37.783	142.532	21.0	198.0	15.9	21.28
27z	37.913	142.027	21.0	198.0	11.2	39.20
28z	37.223	141.667	21.0	208.0	3.7	39.20
29a	36.264	141.597	21.0	211.0	8.6	21.28

Each unit source has an area of 100×50 km^2 and a rake of 90°. The location of each unit source represents the center of the deformation rectangle side parallel to the foot wall (on the subsidence side of the rectangle). Details of the acronym for all unit sources are provided by GICA *et al.* (2008).

reported by LØVHOLT *et al.* (2012) and GLIMSDAL *et al.* (2013).

In coastal waters and open ocean, tsunami waves may be influenced by local bathymetry. For example, amplitudes may be greatly increased at specific frequencies due to resonance. RABINOVICH (1997) recommended the employment of "spectral ratios", i.e., the ratios of tsunami spectra (S_{tsu}) to background

spectra (S_b) at the same water level stations, and indicated that this approach can separate the tsunami characteristics from bathymetric influences. Figure 4 presents the spectral ratios for the present event at four DART stations near the epicenter. The background spectra are calculated based on the 10-day wave records preceding this event. The tsunami spectra are calculated employing the 15-h measurements after the tsunami arrived. All the raw data were taken at intervals of 15 s, and de-tided through the least-squares method. To smooth the spectra, the time series are first separated into 6-h segments with 3-h overlaps, and then are passed through a Kaiser-Bessel window. Spectra are computed for each segment and block averaging is performed over all segments (EMERY and THOMSON, 1998). The spectra of simulated waves are computed over the 15-h time series output from both models. In the spectral ratios of measurements, we note considerable wave amplitudes in the periods between 2 and 100 min, in good agreement with RABINOVICH *et al.*'s (2013) findings. If we take 4,500 m as the mean water depth and 2 min as the wave period, the phase speed is estimated to be 600 km/hr and the wavelength is approximately 20.0 km or 4.4 times the water depth, which suggests considerable dispersion effects in the high-frequency

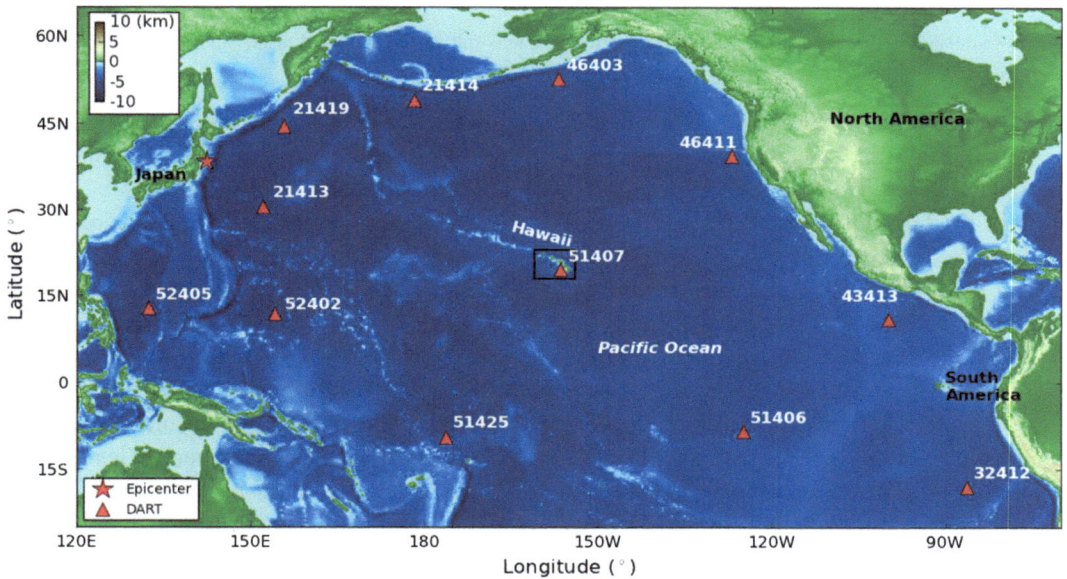

Figure 2
Northern Pacific basin. The *black rectangle* indicates the boundaries of the nested grid covering the Hawaiian Islands

195

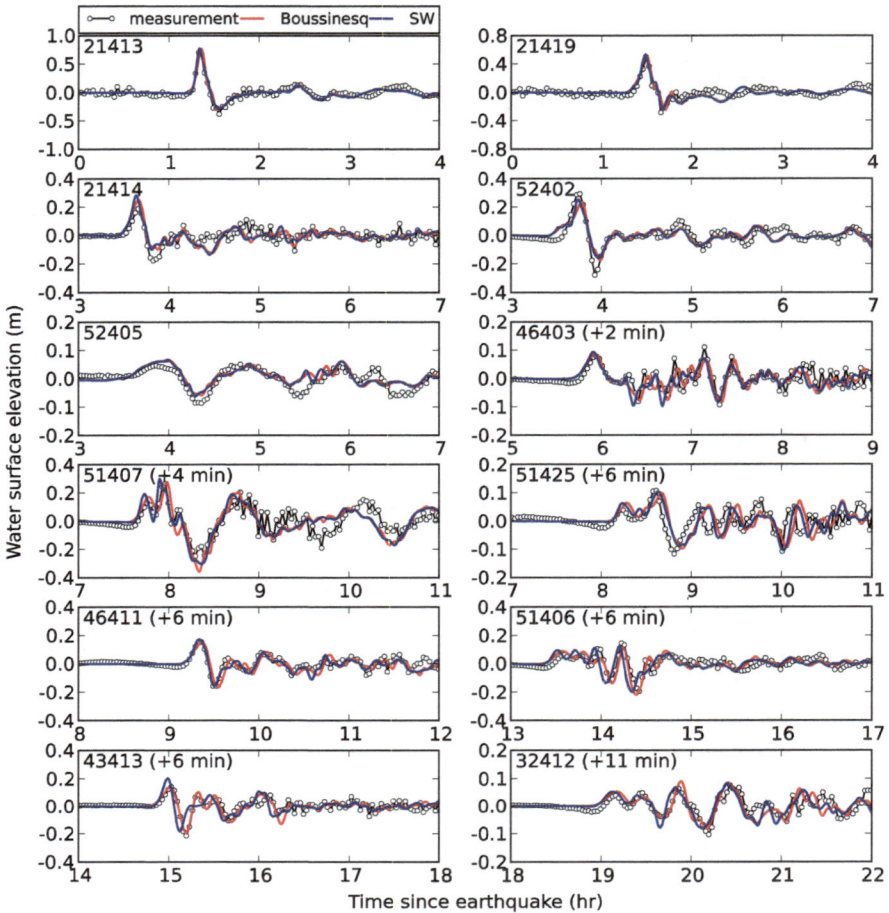

Figure 3
Time series of water surface elevations at selected DART stations. Time is shifted for numerical results at some stations as indicated in the bracelets

trailing waves in the ocean basin. Comparison between the modeled and measured results shows good agreement in the periods above 10 min and big differences in the higher frequency band. The mismatch at high frequencies is a reason for the poor agreement in the trailing waves at some DART stations. The errors in high-frequency waves may be due to the way that we constrain the tsunami source. In the propagation database, all unit earthquake sources have a length of 100 km and a width of 50 km, and the generated water waves are estimated by directly copying the seabed deformation. Short wave components are strongly suppressed in this method. Employing smaller-sized faulting segments may improve the accuracy of tsunami sources, but this approach requires longer operational time for source

constraint. The present configuration of unit earthquake sources comes from a balance between efficiency and accuracy.

Maximum water surface elevations are usually employed as an indicator of the severity of tsunami hazard. In Fig. 5a, we present the maximum water surface elevations computed with the Boussinesq model in the northern Pacific grid. A great amount of wave energy is projected along a few major trajectories in the ocean as a result of tsunami source alignment and ocean bathymetry (GRILLI et al., 2013). Noticeable differences between the Boussinesq and the SW models are shown in Fig. 5b. These differences are presented as normalized with respect to the Boussinesq model result. The positive sign indicates that the SW model predicts higher values, and vice

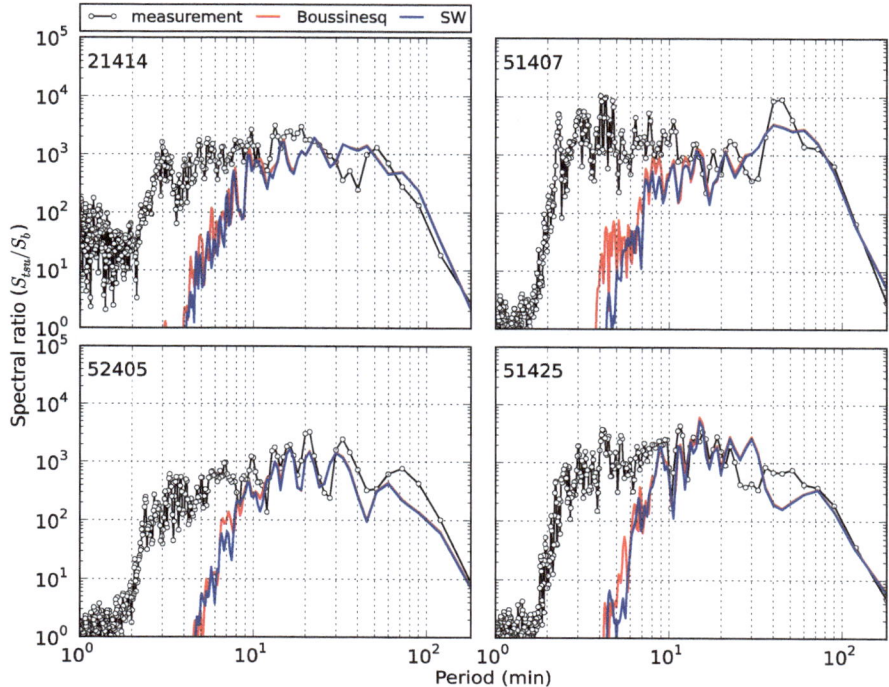

Figure 4
Spectral ratios (S_{tsu}/S_b) for the 2011 Tohoku tsunami at selected DART stations

versa. The Boussinesq model yields smaller water surface elevations in most areas. We note a beam of high wave energy sweeps across the Hawaiian Islands and moderate dispersion effects are present there.

4. Waves and Currents in Hawaiian Waters

Hydrodynamic processes in the Hawaiian Islands region are simulated with the Boussinesq and the SW models at a resolution of $1'$ in a grid nested to the Pacific basin (Fig. 2). The bathymetric and topographic data are interpolated from the ETOPO2 dataset. Boundary conditions of this grid are derived from the respective models in the Pacific basin. We further simulate the waves and currents in four selected harbors, i.e., Nawiliwili, Honolulu, Kahului, and Hilo, and neighboring coastal waters (Fig. 6). These simulations are performed with MOST in three telescoped grids, denoted as A, B, and C, at successively increased resolutions of $12''$, $3''$, and $2/3''$, respectively. Bathymetric and topographic data in these grids are derived from the high-resolution digital

elevation models that were employed to develop the tsunami forecast models in this region (TANG et al., 2009). The boundary conditions of A-grids are interpolated from the Boussinesq and SW model results in the Hawaiian Islands grid. For convenience, we refer to the model employing MOST with input boundary conditions derived from the Boussinesq results as "Boussinesq/MOST" and that with boundary conditions from the SW results as "SW/MOST". The differences between the two models in the same grids are mainly a result of the frequency dispersion effects in the oceanic propagation.

Numerical results are first validated against the data recorded during this event. The tide gauges sampled water surface elevations at an interval of 15 s. Water surface elevations and current velocities were measured at the Kilo Nalu station every 2 s. The sampling frequency of the NOS ADCPs was lower, i.e., once every 6 min. These data first go through a 120-min high-pass Butterworth filter to remove the signals due to astronomical tides. The measured time series may also be affected by the infragravity waves with typical periods ranging from 30 s to a few

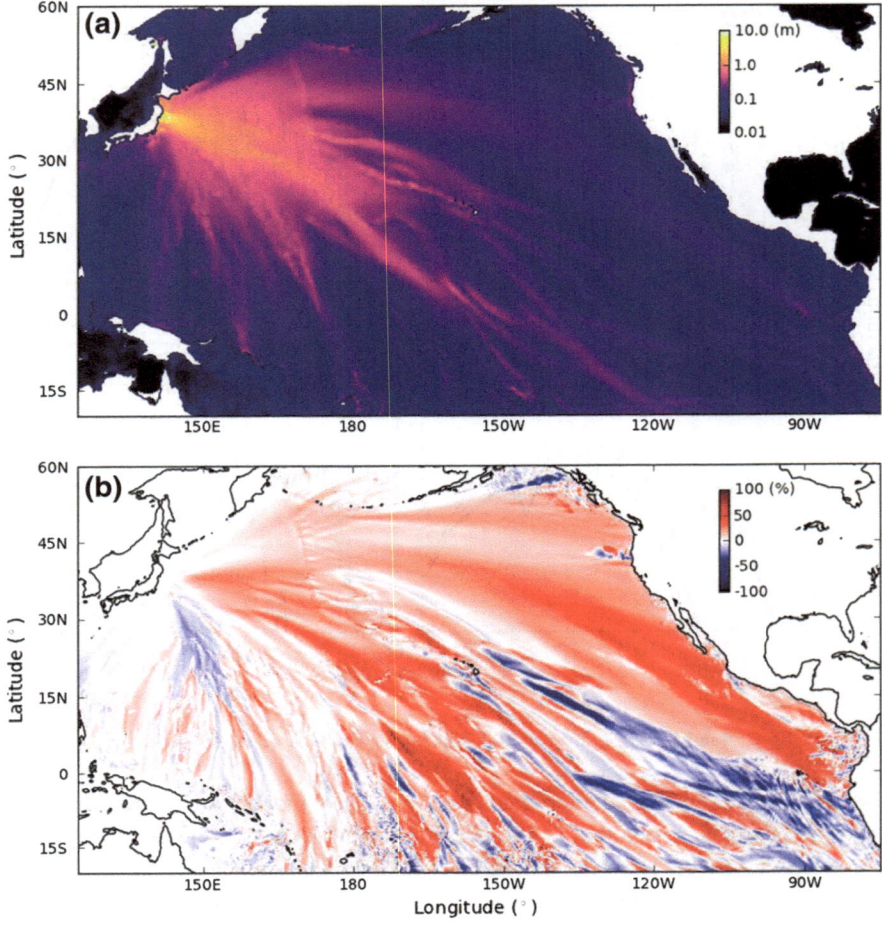

Figure 5
Modeling results of the trans-Pacific propagation: **a** maximum water surface elevations computed with the Boussinesq model; **b** difference between the Boussinesq and SW model results

minutes (RABINOVICH, 2009). To eliminate this effect, the data recorded at the tide gauges and Kilo Nalu station are further smoothed through a 2-min low-pass Butterworth filter. We note that it is difficult to derive the depth-averaged water velocities from the measured profiles. The ADCP sensors do not cover the entire water column. In addition, measurements near the sea surface and bottom may be contaminated due to side-lobe interference. In this study, we calculate the arithmetic mean of water velocities over sensors around the midpoint of the still water column, and consider it equivalent to the depth-averaged velocity. This approximation is indeed quite accurate as we note very high uniformity in the measured water velocity profiles in this event.

Figure 7 compares the simulated water surface elevations with those measured at the tide stations. Both models agree well with the gauges at the leading wave peaks. Following the leading peaks are trailing waves characterized by shorter wavelengths and high amplitudes. In these trailing waves, the discrepancy between numerical and measured time series becomes larger. This may be due to several factors, including inaccurate tsunami source, errors in local bathymetry, and the 4′ resolution in the Pacific basin, which may introduce conceivable computational errors in the shorter wave components. We note that some data points are missing in the measured time series at the Kahului tide station, possibly caused by an instrumental error with the tide gauge. In Fig. 8,

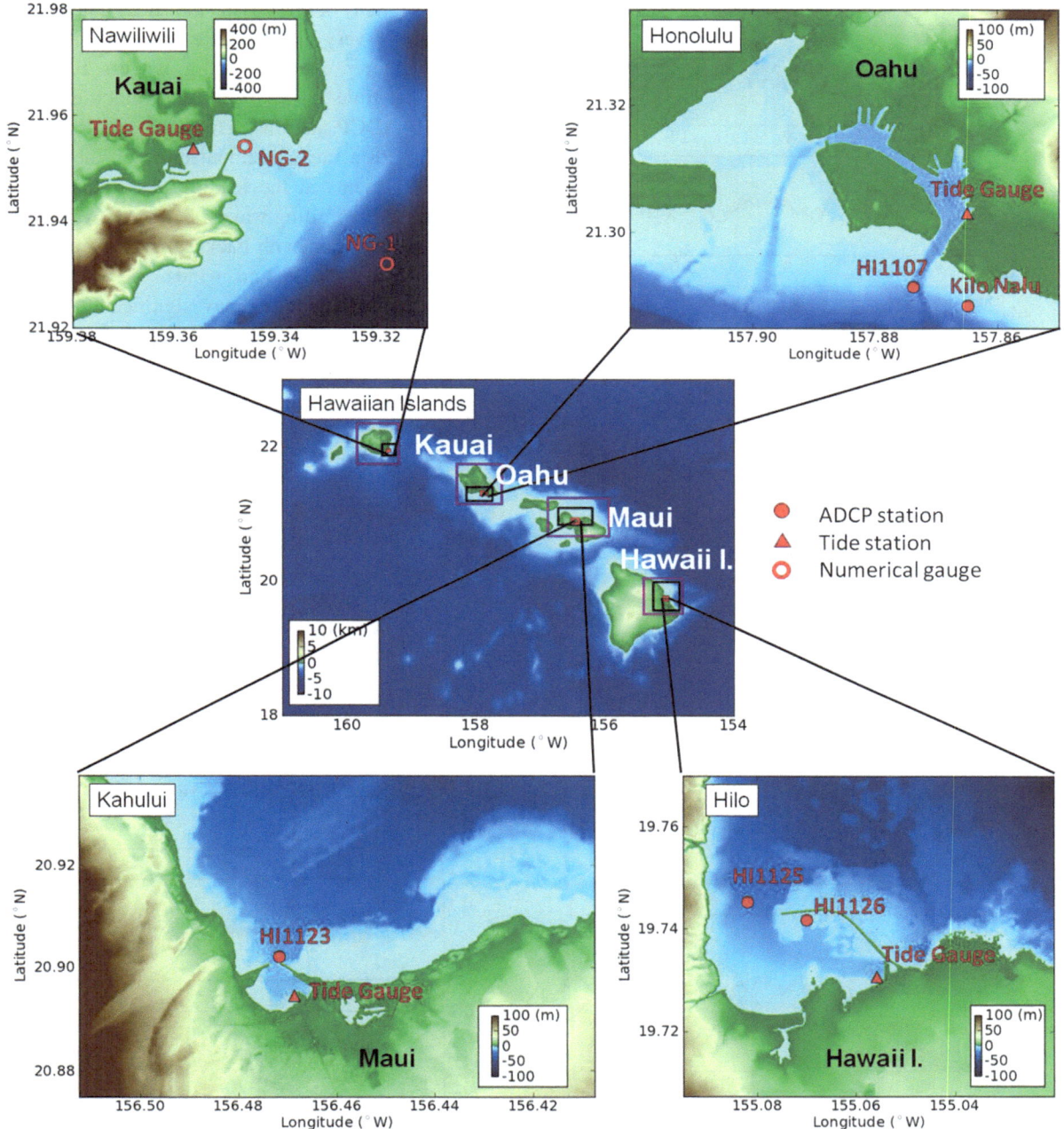

Figure 6

Nested grids in the Hawaiian Islands region: the A-, B-, and C-grid boundaries are depicted with *magenta*, *black*, and *red rectangles*, respectively

we also compute the spectral ratios of tsunami waves at the tide stations in Nawiliwili, Honolulu, and Hilo harbors. Compared with the DART tsunameters, where wave energy is concentrated in low-frequency modes, at the tide stations wave energy is more evenly distributed in all frequencies. This may suggest that some of the shorter wave components are generated in the local coastal water. In Fig. 9, we compare the models with measurements for both water waves and currents at the Kilo Nalu station.

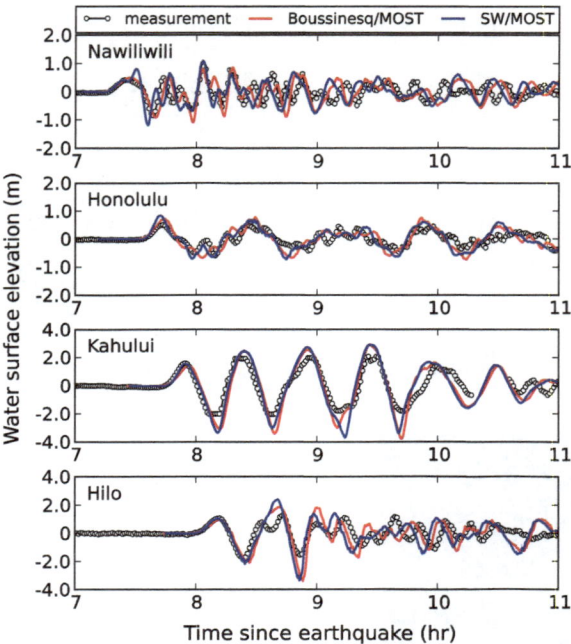

Figure 7
Water surface elevations at tide stations in the Hawaiian Islands region. Computed results are shifted +4 min in time

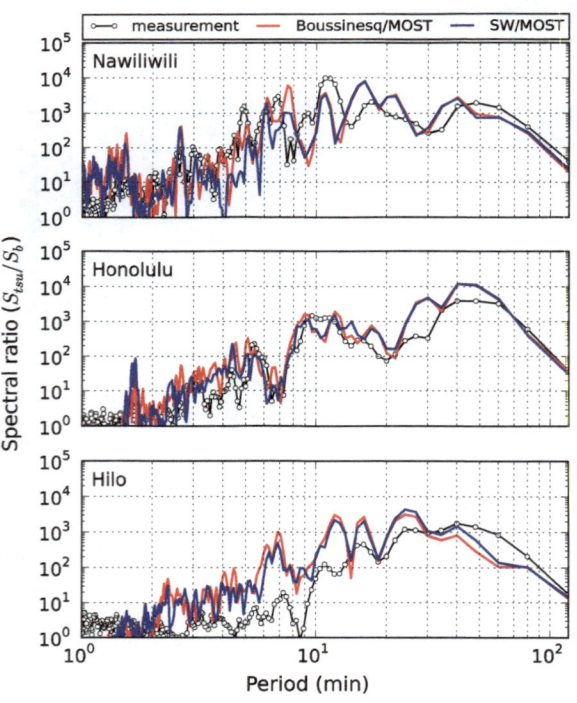

Figure 8
Spectral ratios (S_{tsu}/S_b) for the 2011 Tohoku tsunami at tide stations in the Hawaiian Islands region

Similar to that observed at the tide gauges, the agreement is very good at the leading waves but becomes poorer at the trailing waves. Figure 10 presents the time series of current velocities at the four NOS ADCP stations. The comparison is characterized with a fair agreement between the models and measurements at the leading peaks and a poor agreement at the trailing waves. In coastal waters, the nonlinear interaction with background currents, such as those due to tides and winds, as well as the influence of stratification, may significantly change the amplitudes and directions of the tsunami currents. These processes are not included in the numerical models and may cause considerable errors in the modeled tsunami currents. In addition, the sampling frequency of the ADCPs may be too low for the shorter trailing waves, and, therefore, some high velocity values may be missed in the measurements. In most cases, the Boussinesq/MOST model has a better agreement with the measurements. Compared with the Boussinesq/MOST model, SW/MOST generally predicts higher amplitudes of water waves and current velocities at the leading waves.

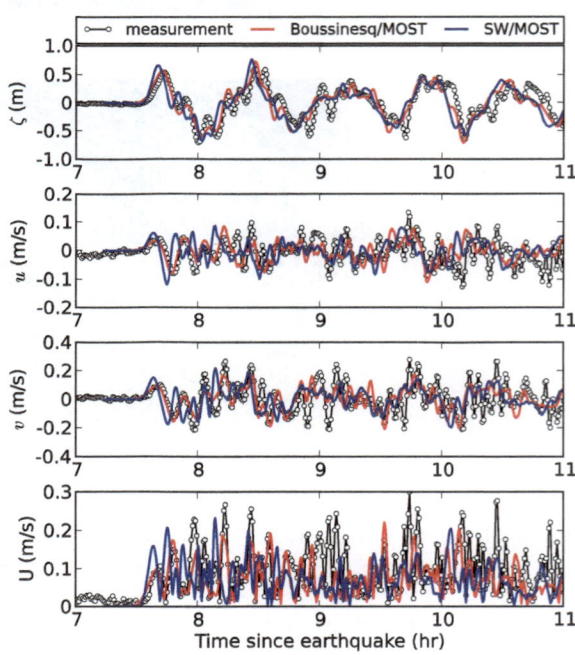

Figure 9
Water surface elevations (ζ), eastward (u) and northward (v) current velocities and current speeds ($U = \sqrt{u^2 + v^2}$) at the Kilo Nalu station. Computed results are shifted +4 min in time

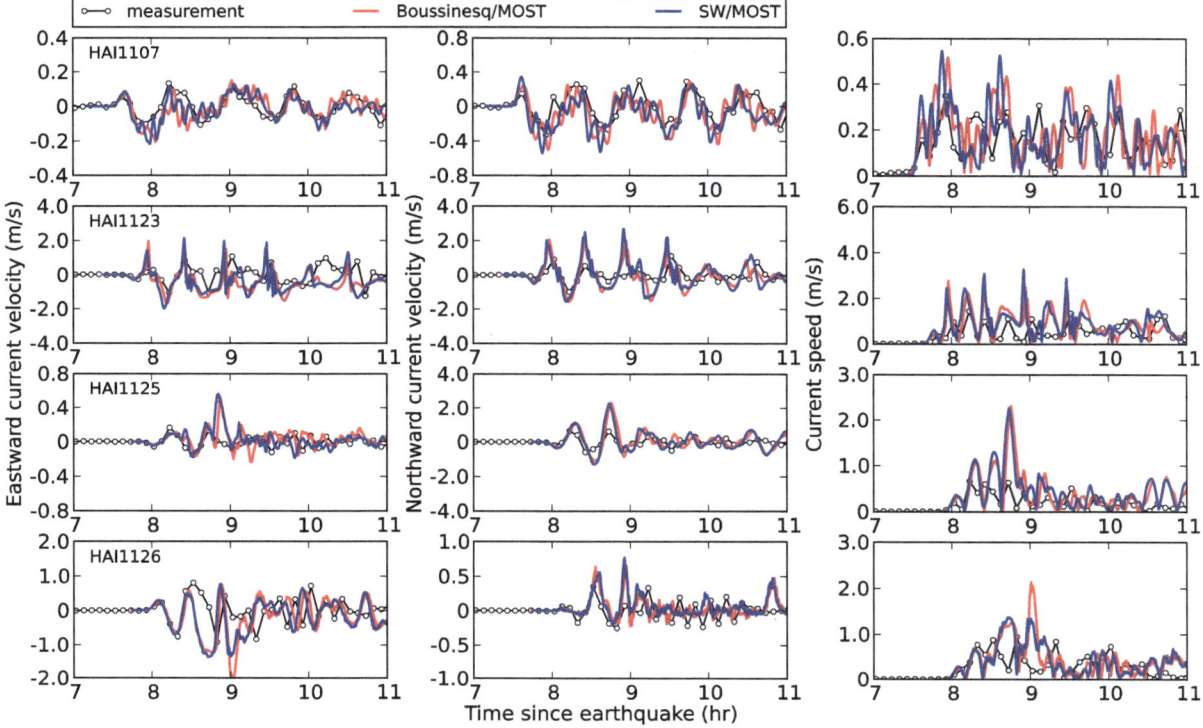

Figure 10
Eastward and northward current velocities and current speeds at NOS ADCP stations. Computed results are shifted +4 min in time

In Fig. 11, we compare between the Boussinesq and the SW models for the computed maximum water surface elevations and current speeds in the Hawaiian Islands region. In the deep ocean west and north of the islands, before the waves pass through the archipelago, the SW model generally predicts higher maximum wave heights and current speeds. The waves then undergo processes of reflection, refraction, and diffraction while propagating between the islands. High-frequency and high-amplitude waves develop following the leading waves and become dominant in the wave trains in some areas near the coasts and in the deep ocean behind the islands. By neglecting the dispersion effects, the SW model may under-predict the amplitudes of the trailing waves, as well as the maximum water surface elevations and current speeds in these areas.

In order to demonstrate the dispersion effects in nearshore water, we output the water surface elevations and current velocities at two "numerical gauges", NG-1 at a depth of 345 m and NG-2 at a depth of 14 m near the entrance of Nawiliwili Harbor (Fig. 6) from the high-resolution simulations. The time series are presented in Fig. 12. At NG-1, the waves are characterized by a dominant leading peak. The height of the this peak is over-predicted by SW/MOST, so as the current speed under this peak. At NG-2, the height of the leading peak increases due to shoaling effect. High-amplitude trailing waves of shorter wavelengths develop following the leading peak. The SW/MOST model significantly under-predicts the heights of some trailing waves.

Figs. 13, 14, 15, 16 show the numerical results from the high-resolution simulations. The differences are normalized with regard to the Boussinesq/MOST results. The positive sign indicates that the SW/MOST model computes higher values, and vice versa. Offshore of Nawiliwili and Hilo (Figs. 13 and 16), SW/MOST first over-predicts the maximum water surface elevations in deeper water, and then under-predicts these parameters outside the harbors. We note the spatial variation of current speeds is greater and more complicated. In some areas, for example, in the entrance channel of Honolulu Harbor

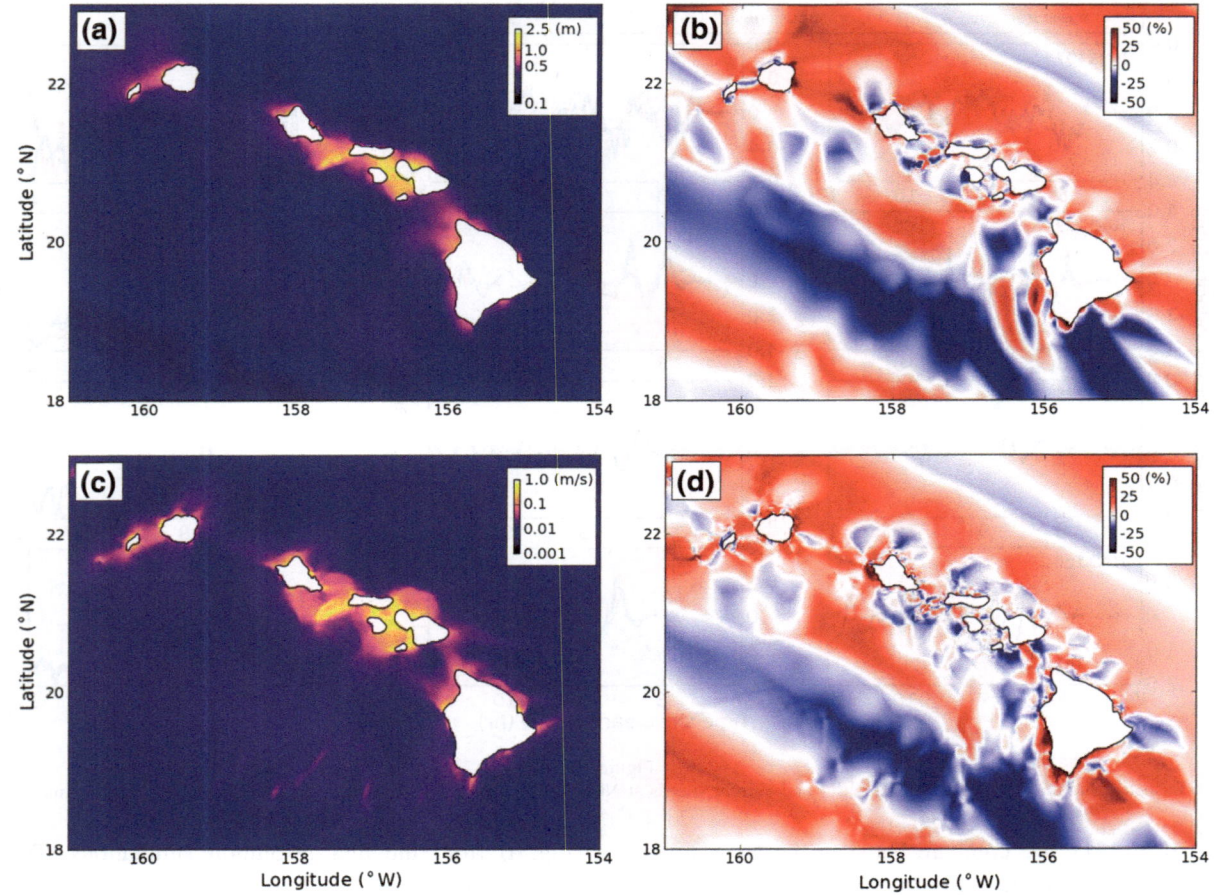

Figure 11
Water waves and currents in the Hawaiian Islands region: **a** maximum water surface elevations computed with the Boussinesq model; **b** differences in maximum water surface elevations between the Boussinesq and SW models; **c** maximum current speeds computed with the Boussinesq model; **d** differences in maximum current speeds between the Boussinesq and the SW models

(Fig. 14), the SW/MOST model over-predicts the maximum water surface elevations but meanwhile greatly under-predicts the maximum current speeds, compared with Boussinesq/MOST.

5. Conclusions

In this study, we investigate the waves and currents in the coastal waters of the Hawaiian Islands induced by the 2011 Tohoku tsunami. The propagations of this tsunami in the Pacific basin and the Hawaiian Islands region are simulated with a Boussinesq model and a non-dispersive SW model for comparison purposes. Boundary

conditions are derived from these models, respectively, and employed in the high-resolution simulations in the coastal waters. Computations in this stage are performed with MOST in three layers of nested grids.

We note that by neglecting the dispersion effects, a numerical model generally over-predicts the height of the leading waves. The effects of frequency dispersion are more complicated in the trailing waves of shorter wavelengths. Trailing waves of high amplitudes may develop in coastal waters and can be significantly under-predicted by a model neglecting the dispersion effects. This is widely observed along the coasts of the Hawaiian Islands in the present event.

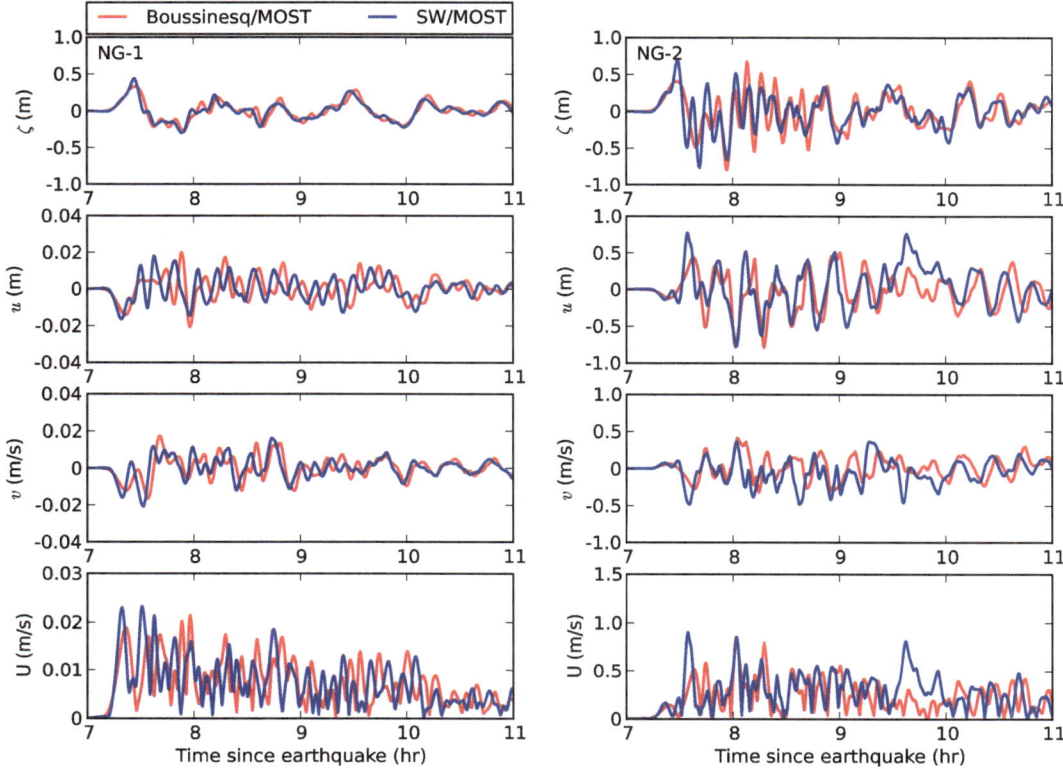

Figure 12

Water surface elevations (ζ), eastward (u) and northward (v) current velocities and current speeds ($U = \sqrt{u^2 + v^2}$) at numerical gauges near Nawiliwili Harbor. Computed results are shifted +4 min in time

Acknowledgments

This study is partially funded by the Joint Institute for the Study of the Atmosphere and Ocean, Contribution No. 2171, the Pacific Marine Environmental Laboratory, Contribution No. 4046, under NOAA Cooperative Agreement No. NA10OAR4320148, and the US Nuclear Regulatory Commission (NRC), Office of Nuclear Regulatory Research under Interagency Agreement RES-07-004 Project N6401. We thank G. Mungov at the National Geophysical Data Center for the wave data at DART stations, P. Burke at NOS for the current data at the NOS ADCP stations, and E. Pawlak at the University of California, San Diego for the wave and current data at the Kilo Nalu station. We are grateful to R. Anoosheh-poor at NRC, A. Rabinovich at the Russian Academy of Sciences and Institute of Ocean Sciences of Canada, and two anonymous reviewers for helpful revision advice. This report was prepared as an account of work sponsored by an agency of the US Government. Neither the US Government nor any agency thereof, nor any of their employees, makes any warranty, expressed or implied, or assumes any legal liability or responsibility for any third party's use, or the results of such use, of any information, apparatus, product, or process disclosed in this report, or represents that its use by such third party would not infringe privately owned rights. The views expressed in this paper are not necessarily those of the US Nuclear Regulatory Commission.

Appendix A: Derivation of the Boussinesq Equations in Geographic Coordinates

To derive the Boussinesq equations in geographic coordinates, we adopt the scaling system of KIRBY *et al.* (2013), and follow the steps presented by LYNETT and LIU (2002). We start from the Euler

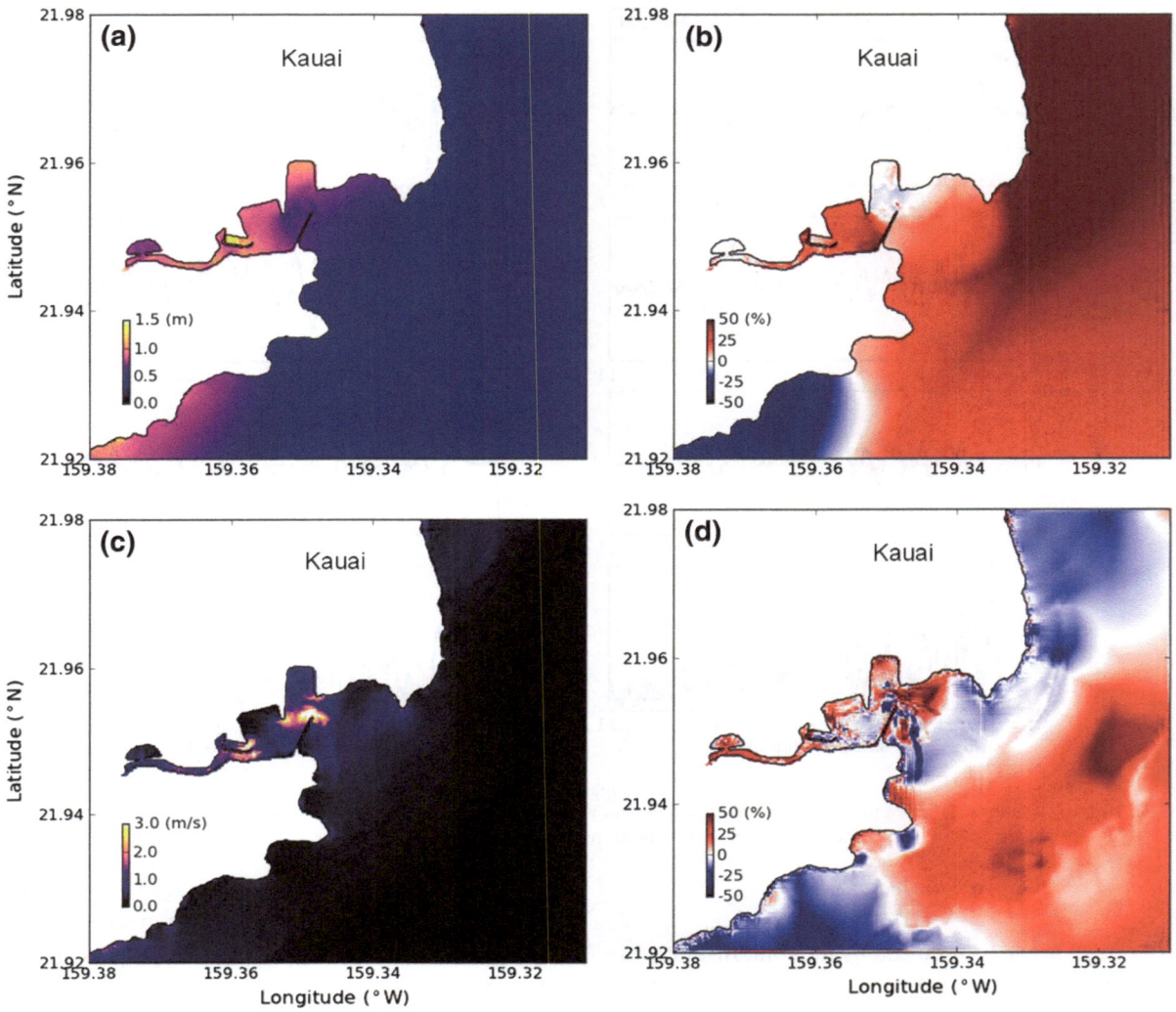

Figure 13

Water waves and currents in the Nawiliwili area: **a** maximum water surface elevations computed with Boussinesq/MOST; **b** differences in maximum water surface elevations due to frequency dispersion; **c** maximum current speeds computed with Boussinesq/MOST; **d** differences in maximum current speeds due to frequency dispersion

equations in geographic coordinates (see, e.g., HOL-TON, 1992),

$$\frac{1}{R\cos\theta}\left[\frac{\partial u}{\partial \phi} + \frac{\partial}{\partial \theta}(v\cos\theta)\right] + \frac{\partial w}{\partial z} + \frac{2w}{R} = 0, \quad (2)$$

$$\frac{\partial u}{\partial t} + \frac{u}{R\cos\theta}\frac{\partial u}{\partial \phi} + \frac{v}{R}\frac{\partial u}{\partial \theta} + w\frac{\partial u}{\partial z} + \frac{uw}{R} - \frac{uv}{R}\tan\theta$$
$$+ 2\Omega(w\cos\theta - v\sin\theta) = -\frac{1}{\rho R\cos\theta}\frac{\partial p}{\partial \phi}, \quad (3)$$

$$\frac{\partial v}{\partial t} + \frac{u}{R\cos\theta}\frac{\partial v}{\partial \phi} + \frac{v}{R}\frac{\partial v}{\partial \theta} + w\frac{\partial v}{\partial z} + \frac{u^2}{R}\tan\theta + \frac{vw}{R}$$
$$+ 2\Omega u\sin\theta = -\frac{1}{\rho R}\frac{\partial p}{\partial \theta}, \quad (4)$$

$$\frac{\partial w}{\partial t} + \frac{u}{R\cos\theta}\frac{\partial w}{\partial \phi} + \frac{v}{R}\frac{\partial w}{\partial \theta} + w\frac{\partial w}{\partial z} - \frac{u^2 + v^2}{R}$$
$$- 2\Omega u\cos\theta = -\frac{1}{\rho}\frac{\partial p}{\partial z} - g. \quad (5)$$

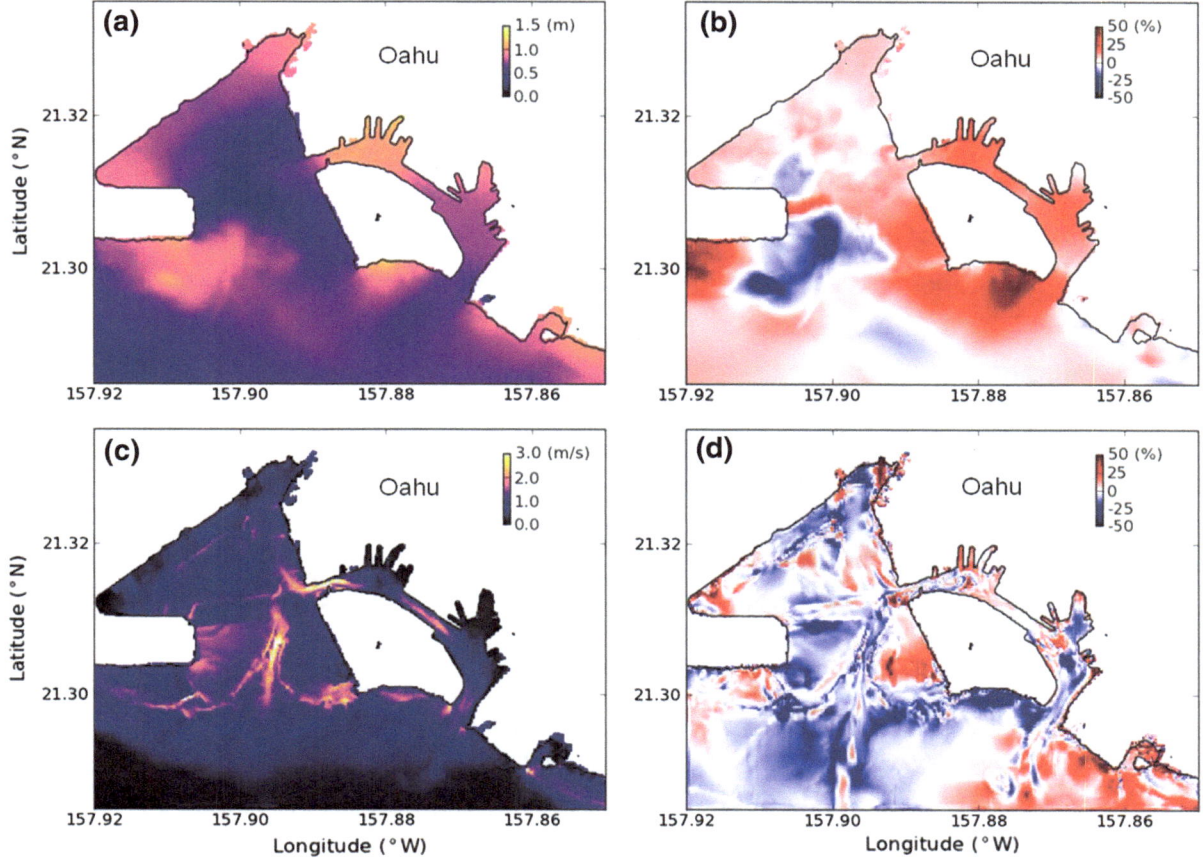

Figure 14
Water waves and currents in the Honolulu area: **a** maximum water surface elevations computed with Boussinesq/MOST; **b** differences in maximum water surface elevations due to frequency dispersion; **c** maximum current speeds computed with Boussinesq/MOST; **d** differences in maximum current speeds due to frequency dispersion

In the above equations, R is the radius of the Earth, Ω is angular speed of Earth's rotation, (u, v, w) is water velocity, g is gravitational acceleration, ϕ is longitude, θ is latitude, z is the vertical coordinate, ρ is density of water, p is pressure, and t is time. The boundary conditions on the free surface and seabed are described as

$$w = \frac{\partial \zeta}{\partial t} + \frac{u}{R \cos \theta} \frac{\partial \zeta}{\partial \phi} + \frac{v}{R} \frac{\partial \zeta}{\partial \theta}, \quad z = \zeta, \quad (6)$$

$$p = 0, \quad z = \zeta, \quad (7)$$

$$w = -\left(\frac{u}{R \cos \theta} \frac{\partial h}{\partial \phi} + \frac{v}{R} \frac{\partial h}{\partial \theta} \right), \quad z = -h, \quad (8)$$

where ζ is water surface elevation and h is still water depth. KIRBY *et al.*'s (2013) scaling system employs

the length scales of characteristic wave amplitude, wavelength, and water depth, which are denoted as a_0, l_0, and h_0, respectively. The mean radius of the Earth R_0 is scaled to the water depth through a parameter $\delta = h_0/R_0$. The non-dimensional variables in this system read

$$R' = \frac{\delta}{\mu l_0} R, \quad z' = \frac{z}{h_0}, \quad (u', v') = \frac{(u, v)}{\epsilon \sqrt{gh_0}},$$

$$w' = \frac{w}{\epsilon/\mu \sqrt{gh_0}}, \quad p' = \frac{p}{\epsilon \rho gh_0}, \quad t' = \frac{t\sqrt{gh_0}}{l_0}, \quad \zeta' = \frac{\zeta}{a_0}. \quad (9)$$

In the study of KIRBY *et al.* (2013), longitudes and latitudes are scaled as

$$(\phi^*, \theta^*) = \frac{\mu}{\delta} (\phi, \theta), \quad (10)$$

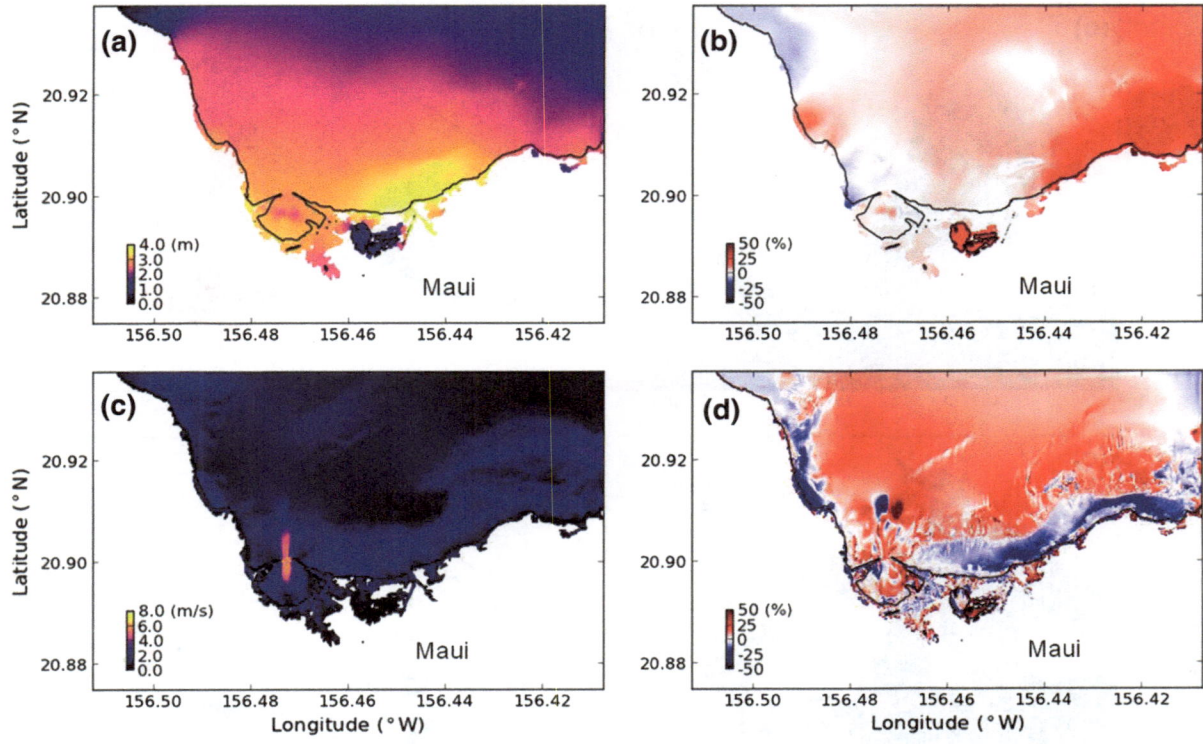

Figure 15
Water waves and currents in the Kahului area: **a** maximum water surface elevations computed with Boussinesq/MOST; **b** differences in maximum water surface elevations due to frequency dispersion; **c** maximum current speeds computed with Boussinesq/MOST; **d** differences in maximum current speeds due to frequency dispersion

such that horizontal coordinates vary with a magnitude of $O(1)$ in the range of $O(l_0)$. The angular frequency of the Earth's rotation is scaled as

$$\Omega' = \frac{\mu}{\delta}\frac{l_0}{\sqrt{gh_0}}\Omega. \tag{11}$$

The Euler equations can then be described in non-dimensional forms as follows, after the primes are dropped for convenience:

$$\mu^2 \frac{1}{R\cos\theta}\left[\frac{\partial u}{\partial \phi^*} + \frac{\partial}{\partial \theta^*}(v\cos\theta)\right] + \frac{\partial w}{\partial z} + \delta\frac{2w}{R} = 0, \tag{12}$$

$$\frac{\partial u}{\partial t} + \epsilon\left\{\frac{u}{R\cos\theta}\frac{\partial u}{\partial \phi^*} + \frac{v}{R}\frac{\partial u}{\partial \theta^*}\right\} + \frac{\epsilon}{\mu^2}w\frac{\partial u}{\partial z} - \frac{\epsilon\delta}{\mu}\frac{uv}{R}\tan\theta$$
$$+ \frac{\epsilon\delta}{\mu^2}\frac{uw}{R} + 2\frac{\delta}{\mu^2}\Omega(w\cos\theta - \mu v\sin\theta) = -\frac{1}{R\cos\theta}\frac{\partial p}{\partial \phi^*}, \tag{13}$$

$$\frac{\partial v}{\partial t} + \epsilon\left\{\frac{u}{R\cos\theta}\frac{\partial v}{\partial \phi^*} + \frac{v}{R}\frac{\partial v}{\partial \theta^*}\right\} + \frac{\epsilon}{\mu^2}w\frac{\partial v}{\partial z} + \frac{\epsilon\delta}{\mu}\frac{u^2}{R}\tan\theta$$
$$+ \frac{\epsilon\delta}{\mu^2}\frac{vw}{R} + \frac{2\delta}{\mu}\Omega u\sin\theta = -\frac{1}{R}\frac{\partial p}{\partial \theta^*}, \tag{14}$$

$$\frac{\partial w}{\partial t} + \epsilon\left(\frac{u}{R\cos\theta}\frac{\partial w}{\partial \phi^*} + \frac{v}{R}\frac{\partial w}{\partial \theta^*}\right) + \frac{\epsilon}{\mu^2}w\frac{\partial w}{\partial z} - \epsilon\delta\frac{u^2+v^2}{R}$$
$$- 2\delta\Omega u\cos\theta = -\frac{\partial p}{\partial z} - \frac{1}{\epsilon}. \tag{15}$$

The boundary conditions read in non-dimensional forms as

$$w = \mu^2\frac{\partial \zeta}{\partial t} + \epsilon\mu^2\left(\frac{u}{R\cos\theta}\frac{\partial \zeta}{\partial \phi^*} + \frac{v}{R}\frac{\partial \zeta}{\partial \theta^*}\right), \quad z = \epsilon\zeta, \tag{16}$$

$$p = 0, \quad z = \epsilon\zeta, \tag{17}$$

$$w = -\mu^2\left(\frac{u}{R\cos\theta}\frac{\partial h}{\partial \phi^*} + \frac{v}{R}\frac{\partial h}{\partial \theta^*}\right), \quad z = -h. \tag{18}$$

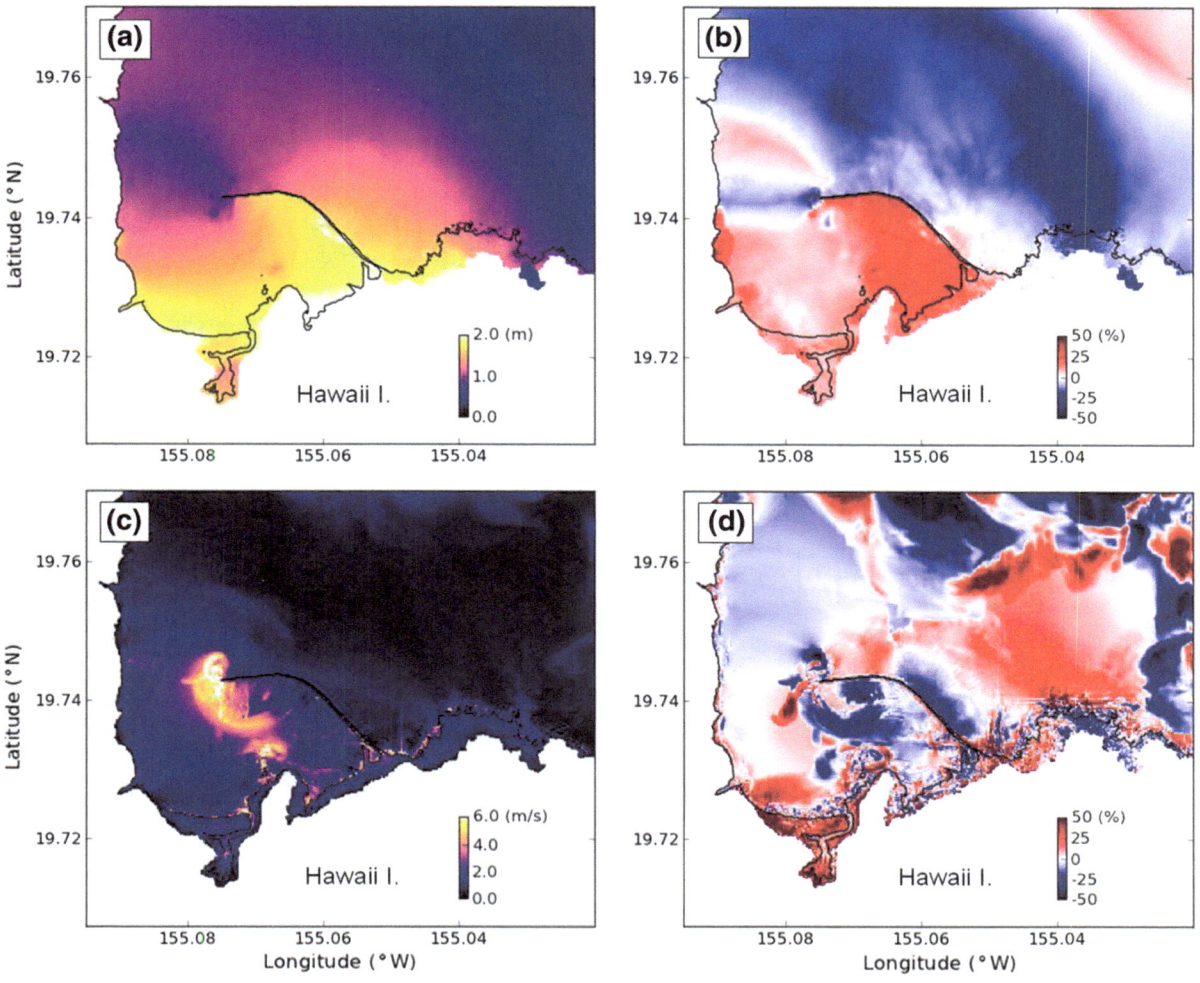

Figure 16

Water waves and currents in the Hilo area: **a** maximum water surface elevations computed with Boussinesq/MOST; **b** differences in maximum water surface elevations due to frequency dispersion; **c** maximum current speeds computed with Boussinesq/MOST; **d** differences in maximum current speeds due to frequency dispersion

Integrating Eq. 12 with respect to z, and applying boundary conditions (16) and (18), we obtain the following equation:

$$\frac{\partial \zeta}{\partial t} + \frac{1}{R\cos\theta}\left[\frac{\partial}{\partial\phi^*}\int_{-h}^{\epsilon\zeta} u\,dz + \frac{\partial}{\partial\theta^*}\int_{-h}^{\epsilon\zeta}(v\cos\theta)dz\right]$$

$$+ \frac{\delta}{\mu^2}\int_{-h}^{\epsilon\zeta}\frac{2w}{R}dz = 0. \tag{19}$$

We apply perturbation expansion to water velocities

$$u = u_0 + \mu^2 u_1 + O(\mu^4), \tag{20}$$

$$v = v_0 + \mu^2 v_1 + O(\mu^4), \tag{21}$$

$$w = \mu^2 w_1 + O(\mu^4), \tag{22}$$

where, u_0 and v_0 are uniform along water depth. Under the assumption of zero horizontal vorticity, we have

$$\frac{\partial u_1}{\partial z} = \frac{1}{R\cos\theta}\frac{\partial w_1}{\partial\phi^*}, \tag{23}$$

$$\frac{\partial v_1}{\partial z} = \frac{1}{R}\frac{\partial w_1}{\partial \theta^*}. \tag{24}$$

Substituting Eqs. 20–22 into Eqs. 12, 16, and 18, and collecting leading order terms, we have

$$\frac{1}{R\cos\theta}\left[\frac{\partial u_0}{\partial \phi^*} + \frac{\partial}{\partial \theta^*}(v_0\cos\theta)\right] + \frac{\partial w_1}{\partial z} = 0, \tag{25}$$
$$-h < z < \epsilon\zeta,$$

$$w_1 = \frac{\partial \zeta}{\partial t} + \epsilon\left(\frac{u_0}{R\cos\theta}\frac{\partial \zeta}{\partial \phi^*} + \frac{v_0}{R}\frac{\partial \zeta}{\partial \theta^*}\right), \quad z = \epsilon\zeta, \tag{26}$$

$$w_1 = -\left(\frac{u_0}{R\cos\theta}\frac{\partial h}{\partial \phi^*} + \frac{v_0}{R}\frac{\partial h}{\partial \theta^*}\right), \quad z = -h. \tag{27}$$

Integrating Eq. 25 along the z-axis and applying the bottom boundary condition in Eq. 27, we have

$$w_1 = -\frac{z+h}{R\cos\theta}\left[\frac{\partial u_0}{\partial \phi^*} + \frac{\partial}{\partial \theta^*}(v_0\cos\theta)\right]$$
$$-\left(\frac{u_0}{R\cos\theta}\frac{\partial h}{\partial \phi^*} + \frac{v_0}{R}\frac{\partial h}{\partial \theta^*}\right). \tag{28}$$

Let

$$S_0 = \frac{1}{R\cos\theta}\left[\frac{\partial u_0}{\partial \phi^*} + \frac{\partial}{\partial \theta^*}(v_0\cos\theta)\right], \tag{29}$$

$$T_0 = \frac{1}{R\cos\theta}\left[\frac{\partial}{\partial \phi^*}(hu_0) + \frac{\partial}{\partial \theta^*}(hv_0\cos\theta)\right], \tag{30}$$

Eq. 28 can then be written as

$$w_1 = -zS_0 - T_0. \tag{31}$$

Integrating Eqs. 23 along the z-axis yields

$$u_1 = \frac{1}{R\cos\theta}\left\{-\frac{z^2}{2}\frac{\partial S_0}{\partial \phi^*} - z\frac{\partial T_0}{\partial \phi^*}\right\} + C, \tag{32}$$

where C is a function of (ϕ^*, θ^*, t). Substituting Eq. 32 into 20, we have

$$u = u_0 - \frac{\mu^2}{R\cos\theta}\left(\frac{z^2}{2}\frac{\partial S_0}{\partial \phi^*} + z\frac{\partial T_0}{\partial \phi^*} + C\right) + O(\mu^4), \tag{33}$$

We define u_α as the longitudinal component of horizontal velocity at the reference level $z = z_\alpha$,

$$u_\alpha = u_0 - \frac{\mu^2}{R\cos\theta}\left(\frac{z_\alpha^2}{2}\frac{\partial S_0}{\partial \phi^*} + z_\alpha\frac{\partial T_0}{\partial \phi^*} + C\right) + O(\mu^4), \tag{34}$$

By subtracting Eq. 34 from 33, we have

$$u = u_\alpha - \frac{\mu^2}{R\cos\theta}\left\{\frac{z^2 - z_\alpha^2}{2}\frac{\partial S_\alpha}{\partial \phi^*} + (z - z_\alpha)\frac{\partial T_\alpha}{\partial \phi^*}\right\} + O(\mu^4), \tag{35}$$

where

$$S_\alpha = \frac{1}{R\cos\theta}\left[\frac{\partial u_\alpha}{\partial \phi^*} + \frac{\partial}{\partial \theta^*}(v_\alpha\cos\theta)\right], \tag{36}$$

$$T_\alpha = \frac{1}{R\cos\theta}\left[\frac{\partial}{\partial \phi^*}(hu_\alpha) + \frac{\partial}{\partial \theta^*}(hv_\alpha\cos\theta)\right]. \tag{37}$$

Repeat the above steps to the meridional component of velocity and we obtain

$$v = v_\alpha - \frac{\mu^2}{R}\left\{\frac{z^2 - z_\alpha^2}{2}\frac{\partial S_\alpha}{\partial \theta^*} + (z - z_\alpha)\frac{\partial T_\alpha}{\partial \theta^*}\right\} + O(\mu^4). \tag{38}$$

Substituting Eqs. 35 and 38 into 19, and neglecting higher-order terms, we have the continuity equation as

$$\frac{\partial \zeta}{\partial t} + \frac{1}{R\cos\theta}\left[\frac{\partial}{\partial \phi^*}(Hu_\alpha) + \frac{\partial}{\partial \theta^*}(Hv_\alpha\cos\theta)\right]$$
$$+ \mu^2\left\{\frac{1}{R^2\cos^2\theta}\frac{\partial}{\partial \phi^*}\left[h\left(\frac{z_\alpha^2}{2} - \frac{h^2}{6}\right)\frac{\partial S_\alpha}{\partial \phi^*} + h\left(z_\alpha + \frac{h}{2}\right)\frac{\partial T_\alpha}{\partial \phi^*}\right]\right.$$
$$+ \frac{1}{R^2\cos\theta}\frac{\partial}{\partial \theta^*}\left[h\left(\frac{z_\alpha^2}{2} - \frac{h^2}{6}\right)\frac{\partial S_\alpha}{\partial \theta^*}\cos\theta\right.$$
$$\left.\left. + h\left(z_\alpha + \frac{h}{2}\right)\frac{\partial T_\alpha}{\partial \theta^*}\cos\theta\right]\right\} = O(\epsilon\mu^2, \mu^4, \delta), \tag{39}$$

where $H = \epsilon\zeta + h$. Substituting Eqs. 18–22, 31, 35, and 38 into 15, integrating this equation along the z-axis, and neglecting higher-order terms, we have

$$p(z) = \zeta - \frac{z}{\epsilon} + \mu^2\left(\frac{z^2}{2}\frac{\partial S_\alpha}{\partial t} + z\frac{\partial T_\alpha}{\partial t}\right) + O(\epsilon\mu^2, \mu^4, \delta). \tag{40}$$

Substituting Eqs. 35, 38 and 40 into 13, and evaluating at $z = z_\alpha$, we have the momentum equation along longitude

$$\frac{\partial u_\alpha}{\partial t} + \frac{\epsilon}{R\cos\theta}\left(u_\alpha\frac{\partial u_\alpha}{\partial \phi^*} + v_\alpha\frac{\partial u_\alpha}{\partial \theta^*}\cos\theta\right) - \frac{\epsilon\delta}{\mu}\frac{u_\alpha v_\alpha}{R}\tan\theta$$
$$+ \frac{1}{R\cos\theta}\frac{\partial \zeta}{\partial \phi^*} + \frac{\mu^2}{R\cos\theta}\left(\frac{z_\alpha^2}{2}\frac{\partial^2 S_\alpha}{\partial \phi^*\partial t} + z_\alpha\frac{\partial^2 T_\alpha}{\partial \phi^*\partial t}\right)$$
$$- \frac{2\delta}{\mu}\Omega v_\alpha\sin\theta = O(\epsilon\mu^2, \mu^4, \delta). \tag{41}$$

Repeat this step for Eq. 14, we obtain the meridional momentum equation,

$$\frac{\partial v_\alpha}{\partial t} + \frac{\epsilon}{R\cos\theta}\left(u_\alpha\frac{\partial v_\alpha}{\partial \phi^*} + v_\alpha\frac{\partial v_\alpha}{\partial \theta^*}\cos\theta\right) + \frac{\epsilon\delta}{\mu}\frac{u_\alpha^2}{R}\tan\theta$$
$$+ \frac{1}{R}\frac{\partial \zeta}{\partial \theta^*} + \frac{\mu^2}{R}\left(\frac{z_\alpha^2}{2}\frac{\partial^2 S_\alpha}{\partial \theta^*\partial t} + z_\alpha\frac{\partial^2 T_\alpha}{\partial \theta^*\partial t}\right)$$
$$+ \frac{2\delta}{\mu}\Omega u_\alpha\sin\theta = O(\epsilon\mu^2,\mu^4,\delta). \tag{42}$$

Eqs. 39, 41, and 42 compose the Boussinesq equations in the geographic coordinates. KIRBY et al. (2013) assumes $O(\delta) = O(\mu)$ for problems of very weak dispersion effects. Under this condition, the equations become

$$\frac{\partial \zeta}{\partial t} + \frac{1}{R\cos\theta}\left[\frac{\partial}{\partial \phi^*}(Hu_\alpha) + \frac{\partial}{\partial \theta^*}(Hv_\alpha\cos\theta)\right]$$
$$= O(\mu,\delta), \tag{43}$$

$$\frac{\partial u_\alpha}{\partial t} + \frac{\epsilon}{R\cos\theta}\left(u_\alpha\frac{\partial u_\alpha}{\partial \phi^*} + v_\alpha\frac{\partial u_\alpha}{\partial \theta^*}\cos\theta\right) - \epsilon\frac{u_\alpha v_\alpha}{R}\tan\theta$$
$$+ \frac{1}{R\cos\theta}\frac{\partial \zeta}{\partial \phi^*} - 2\Omega v_\alpha\sin\theta = O(\mu,\delta), \tag{44}$$

$$\frac{\partial v_\alpha}{\partial t} + \frac{\epsilon}{R\cos\theta}\left(u_\alpha\frac{\partial v_\alpha}{\partial \phi^*} + v_\alpha\frac{\partial v_\alpha}{\partial \theta^*}\cos\theta\right) + \epsilon\frac{u_\alpha^2}{R}\tan\theta$$
$$+ \frac{1}{R}\frac{\partial \zeta}{\partial \theta^*} + 2\Omega u_\alpha\sin\theta = O(\mu,\delta), \tag{45}$$

which are equivalent to the non-dispersive shallow water equations. The velocity components u_α and v_α become equal to the depth-uniform variables u_0 and v_0 after the dispersive terms are removed, and in turn become equivalent to the depth-averaged velocities $\bar{u}$ and $\bar{v}$ that are commonly employed in the shallow-water equations.

REFERENCES

BORRERO, J.C., BELL, R., CSATO, C., DELANGE, W., GORING, D., GREER, S.D., PICKETT, V., and POWER, W. (2013). *Observations, effects and real time assessment of the March 11, 2011 Tohoku-oki tsunami in New Zealand*, Pure Appl. Geophys., *170*, 1229–1248, doi:10.1007/s00024-012-0492-6.

BURWELL, D., TOLKOVA, E., and CHAWLA, A. (2007). *Diffusion and dispersion characterization of a numerical tsunami model*, Ocean Modell., *19*, 10–30.

EMERY, W.J., and THOMSON, R.E. (1998). *Data Analysis Methods in Physical Oceanography*, first edition, Elsevier, New York, NY.

GICA, E., SPILLANE, M., TITOV, V.V., CHAMBERLIN, C., and NEWMAN, J.C. (2008). Development of the forecast propagation database for NOAA's Short-term Inundation Forecast for Tsunamis (SIFT), NOAA Tech. Memo. OAR PMEL-139, 89 pp.

GLIMSDAL, S., PEDERSEN, G.K., HARBITZ, C.B., and LØVHOLT, F. (2013). *Dispersion of the tsunamis: does it really matter?*, Nat. Hazards Earth Syst. Sci., *13*, 1507–1526, doi:10.5194/nhess-13-1507-2013.

GRILLI, S.T., IOUALALEN, M., ASAVANANT, J., SHI, F., KIRBY, J.T., and WATTS, P. (2007). *Source constraints and model simulation of the December 26, 2004, Indian Ocean tsunami*, J. Waterwy. Port, Coastal Ocean Eng., *133*, 414–428.

GRILLI, S.T., DUBOSQ, S., POPHET, N., PÉRIGNON, Y., KIRBY, J.T., and SHI, F. (2010). *Numerical simulation and first-order hazard analysis of large co-seismic tsunamis generated in the Puerto Rico trench: Near-field impact on the North shore of Puerto Rico and far-field impact on the US East Coast*, Nat. Hazards Earth Syst. Sci., *10*, 2109–2125, doi:10.5194/nhess-10-2109-2010.

GRILLI, S.T., HARRIS, J.C., TAJALLI BAKHSH, T.S., MASTERLARK, T.L., KYRIAKOPOULOS, C., KIRBY, J.T., and SHI, F. (2013). *Numerical simulation of the 2011 Tohoku tsunami based on a new transient FEM co-seismic source: Comparison to far- and near-field observations*, Pure Appl. Geophys., *170*, 1333–1359, doi:10.1007/s00024-012-0528-y.

HAMMACK, J.L. (1973). *A note on tsunamis: Their generation and propagation in an ocean of uniform depth*, J. Fluid Mech., *60*(4), 769–799.

HOLTON, J.R. (1992). *An Introduction to Dynamic Meteorology*, Third Edition, Academic Press, San Diego, CA.

HORILLO, J., KOWALIK, Z., and SHIGIHARA, Y. (2006). *Wave dispersion study in the Indian Ocean tsunami of December 26, 2004*, Marine Geodesy, *29*, 149–166.

IOUALALEN, M., ASAVANANT, J., KAEWBANJAK, N., GRILLI, S.T., KIRBY, J.T., and WATTS, P. (2007). *Modeling the 26 December 2004 Indian Ocean tsunami: Case study of impact in Thailand*, J. Geophys. Res., *112*, C07024, doi:10.1029/2006JC003850.

KÂNOGLU, U., and SYNOLAKIS, C.E. (2006). *Initial value problem solution of nonlinear shallow water wave equations*, Phys. Rev. Lett., *97*(14), 148501, doi:10.1103/PhysRevLett.97.148501.

KIRBY, J.T., WEI, G., CHEN, Q., KENNEDY, A.B., and DALRYMPLE, R.A. (1998). FUNWAVE 1.0 Fully Nonlinear Boussinesq Wave Model. Documentation and User's Manual, Research Report No. CACR-98-06, Center for Applied Coastal Research, Department of Civil Engineering, University of Delaware, 70 pp.

KIRBY, J.T., SHI, F., TEHRANIRAD, B., HARRIS, J.C., and GRILLI, S.T. (2013). *Dispersive tsunami waves in the ocean: Model equations and sensitivity to dispersion and Coriolis effects*, Ocean Modell., *62*, 30–55.

LØVHOLT, F., PEDERSEN, G., and GISLER, G. (2008). *Oceanic propagation of a potential tsunami from the La Palma Island*, J. Geophys. Res., *113*, C09026, doi:10.1029/2007JC004603.

LØVHOLT, F., PERDERSEN, G., and GLIMSDAL, S. (2010). *Coupling of dispersive tsunami propagation and shallow water coastal response*, in Proceedings of the "First Caribbean Waves workshop in Guadeloupe", Dec. 2008, The Open Oceanography Journal, *4*, edited by: ZAHIBO, N., PELINOVSKY, E., YALÇINER, A., and TITOV, V., 71–82.

LØVHOLT, F., KAISER, G., GLIMSDAL, S., SCHEELE, L., HARBITZ, C.B., and PEDERSEN, G. (2012). *Modeling propagation and inundation* of the 11 March 2011 Tohoku tsunami, Nat. Hazards Earth Syst. Sci., 12, doi:10.5194/nhess-12-1017-2012.

LYNETT, P., and LIU, P.L.-F. (2002). A numerical study of submarine-landslide-generated waves and run-up, Proc. R. Soc. Lond. A, 458, 2885–2910.

LYNETT, P., BORRERO, J.C., WEISS, R., SON, S., GREER, D., and RENTERIA, W. (2012). Observations and modeling of tsunami-induced currents in ports and harbors, Earth Planet. Sci. Lett., 327-328, 68–74, doi:10.1016/j.epsl.2012.02.002.

NWOGU, O. (1993). Alternative form of Boussinesq equations for nearshore wave propagation, J. Waterwy. Port Coastal Ocean Eng., 119, 618–638.

PAWLAK, G., DE CARLO, E., FRAM, J., HEBERT, A., JONES, C., MCLAUGHLIN, B., MCMANUS, M., MILLIKAN, K., SANSONE, F., STANTON, T., and WELLS, J. (2009). Development, deployment, and operation of Kilo Nalu Nearshore Cabled Observatory, IEEE OCEANS 2009 Conference, Bremen, Germany.

PERCIVAL, D.B., DENBO, D.W., EBLE, M.C., GICA, E., MOFJELD, H.O., SPILLANE, M.C., TANG, L., and TITOV, V.V. (2011). Extraction of tsunami source coefficients via inversion of DART buoy data, Nat. Hazards, 58(1), doi:10.1007/s11069-010-9688-1, 567–590.

RABINOVICH, A.B. (1997). Spectral analysis of tsunami waves: Separation of source and topography effects, J. Geophys. Res., 102(C6), 12663–12676.

RABINOVICH, A.B. (2009). Seiches and harbour oscillations. In: Handbook of Coastal and Ocean Engineering (ed. Y.C. Kim), World Scientific, Singapore, 193–236.

RABINOVICH, A.B., CANDELLA, R.N., and THOMSON, R.E. (2013). The open ocean energy decay of three recent trans-Pacific tsunamis, Geophys. Res. Lett., 40, doi:10.1002/grl.50625.

ROEBER, V., CHEUNG, K.F., KOBAYASHI, M.H. (2010). Shock-capturing Boussinesq-type model for nearshore wave processes, Coastal Eng., 57, 407–423, doi:10.1016/j.coastaleng.2009.11.007.

SHI, F., DALRYMPLE, R.A., KIRBY, J.T., CHEN, Q., and KENNEDY, A. (2001). A fully nonlinear Boussinesq model in generalized curvilinear coordinates, Coastal Eng., 42, 337–358.

SITANGGANG, K. and LYNETT, P. (2005). Parallel computation of a highly nonlinear Boussinesq equation model through domain decomposition. Int. J. Numer. Methods Fluids, 49, 57–74.

TANG, L., TITOV, V.V., WEI, Y., MOFJELD, H.O., SPILLANE, M., ARCAS, D., BERNARD, E.N., CHAMBERLIN, C.D., GICA, E., and NEWMAN, J. (2008). Tsunami forecast analysis for the May 2006 Tonga tsunami, J. Geophys. Res., 113, C12015, doi:10.1029/2008JC004922.

TANG L., TITOV, V.V., and CHAMBERLIN, C.D. (2009). Development, testing, and applications of site-specific tsunami inundation

models for real-time forecasting, J. Geophys. Res., 114, C12025, doi:10.1029/2009JC005476.

TANG, L., TITOV, V.V., BERNARD, E.N., WEI, Y., CHAMBERLIN, C.D., NEWMAN, J.C., MOFJELD, H.O., ARCAS, D., EBLE, M.C., MOORE, C., USLU, B., PELLS, C., SPILLANE, M., WRIGHT, L., and GICA, E. (2012). Direct energy estimation of the 2011 Japan tsunami using deep-ocean pressure measurements, J. Geophys. Res., 117, C08008, doi:10.1029/2011JC007635.

TITOV, V.V., and SYNOLAKIS, C.E. (1995). Modeling of breaking and nonbreaking long-wave evolution and runup using VTCS-2, J. Waterwy. Port Coast. Ocean Eng., 121(6), 308–316.

TITOV, V. V. and GONZÁLEZ, F.I. (1997). Implementation and testing of the Method of Splitting Tsunami (MOST) model, NOAA Tech. Memo. ERL PMEL-112 (PB98-122773), NOAA/Pacific Marine Environmental Laboratory, Seattle, WA.

TITOV, V.V., and SYNOLAKIS, C.E. (1998). Numerical modeling of tidal wave runup, J. Waterwy. Port Coast. Ocean Eng., 124(4), 157–171.

WEI, G., and KIRBY, J.T. (1995). A time-dependent numerical code for extended Boussinesq equations, J. Waterwy. Port Coastal Ocean Eng., 120, 251–261.

WEI, Y., BERNARD, E.N., TANG, L., WEISS, R., TITOV, V.V., MOORE, C., SPILLANE, M., HOPKINS, M., and KÂNOGLU, U. (2008). Real-time experimental forecast of the Peruvian tsunami of August 2007 for U.S. coastlines, Geophys. Res. Lett., 35, L04609, doi:10.1029/2007GL032250.

WEI, Y., CHAMBERLIN, C., TITOV, V.V., TANG, L., and BERNARD, E. N. (2013). Modeling of the 2011 Japan tsunami: Lessons for near-field forecast, Pure Appl. Geophys., 170(6-8), doi:10.1007/s00024-012-0519-z, 1309–1331.

WILSON, R.I., ADMIRE, A.R., BORRERO, J.C., DENGLER, L.A., LEGG, M.R., LYNETT, P., MCCRINK, T.P., MILLER, K.M., RITCHIE, A., STERLING, K., and WHITMORE, P.M. (2013). Observations and impacts from the 2010 Chilean and 2011 Japanese Tsunamis in California (USA), Pure Appl. Geophys., 170, 1127–1147, doi:10.1007/s00024-012-0527-z.

YAMAZAKI, Y., CHEUNG, K.F, PAWLAK, G., and LAY, T. (2012). Surges along the Honolulu coast from the 2011 Tohoku tsunami, Geophy. Res. Lett., 39, L09604, doi:10.1029/2012GL051624.

ZHOU, H., MOORE, C.W., WEI, Y., and TITOV, V.V. (2011). A nested-grid Boussinesq-type approach to modelling dispersive propagation and runup of landslide-generated tsunamis, Nat. Hazards Earth Syst. Sci., 11, 2677–2697, doi:10.5194/nhess-11-2677-2011.

ZHOU, H., WEI, Y., and TITOV, V.V. (2012). Dispersive modeling of the 2009 Samoa tsunami, Geophy. Res. Lett., 39, L16603, doi:10.1029/2012GL053068.

(Received September 20, 2013, revised January 17, 2014, accepted January 20, 2014, Published online February 12, 2014)

Pure Appl. Geophys. 171 (2014), 3385–3403
© 2014 Springer Basel
DOI 10.1007/s00024-014-0797-8

Observed and Modeled Currents from the Tohoku-oki, Japan and other Recent Tsunamis in Northern California

AMANDA R. ADMIRE,[1] LORI A. DENGLER,[1] GREGORY B. CRAWFORD,[2] BURAK U. USLU,[3] JOSE C. BORRERO,[4,5] S. DOUGAL GREER,[5] and RICK I. WILSON[6]

Abstract—We investigate the currents produced by recent tsunamis in Humboldt Bay and Crescent City, California. The region is susceptible to both near-field and far-field tsunamis and has a historic record of damaging events. Crescent City Harbor, located approximately 100 kms north of Humboldt Bay, suffered US $28 million in damages from strong currents produced by the 2006 Kuril Islands tsunami and an additional US $26 million from the 2011 Japan tsunami. In order to better evaluate these currents in northern California, we deployed a Nortek Aquadopp 600 kHz 2D acoustic Doppler current profiler (ADCP) with a 1-min sampling interval in Humboldt Bay, near the existing National Oceanic and Atmospheric Administration (NOAA) National Ocean Service (NOS) tide gauge station. The instrument recorded the tsunamis produced by the M_w 8.8 Chile earthquake on February 27, 2010 and the M_w 9.0 Japan earthquake on March 11, 2011. One other tsunami was recorded on the Humboldt Bay tide gauge during the period of ADCP operation, but was not visible on the ADCP, suggesting a threshold water level value of about 0.2 m to produce an observable ADCP record. The 2010 tsunami currents persisted in Humboldt Bay for approximately 30 h with peak amplitudes of about 0.35 m/s. The 2011 tsunami signal lasted for over 40 h with peak amplitude of 0.84 m/s. The strongest currents corresponded to the maximum change in water level approximately 67 min after the initial wave arrival. No damage was observed in Humboldt Bay for either event. In Crescent City, currents for the first three and one-half hours of the 2011 Japan tsunami were estimated using security camera video footage from the Harbor Master, approximately 70 m away from the NOAA–NOS tide gauge station. The largest amplitude tide gauge water-level oscillations and most of the damage occurred within this time window. The currents reached a velocity of approximately 4.5 m/s and six cycles exceeded 3 m/s during this period. Measured current velocities both in Humboldt Bay and in Crescent City were compared to calculated velocities from the Method of Splitting Tsunamis (MOST) numerical model. The frequency and pattern of current amplification and decay at both locations are replicated by the MOST model for the first several hours after the tsunami onset. MOST generally underestimates 2011 peak current velocities by about 10–30 %, with a few peaks by as much as 50 %. At Humboldt Bay, MOST predicted attenuation of the signal after 4 h but the actual signal persisted at a nearly constant level for at least twice as long. The results from this project demonstrate that ADCPs can effectively record tsunami currents for small to moderate events and can be used to calibrate and validate models (i.e., MOST) in order to better understand hazardous tsunami conditions within harbors.

Key words: Tsunamis, ADCP, Tsunami current, 2010 Chile tsunami, 2011 Japan tsunami, Natural hazards.

[1] Department of Geology, Humboldt State University, #1 Harpst St., Arcata, CA 95521, USA. E-mail: amanda.admire@humboldt.edu; lori.dengler@humboldt.edu

[2] Vancouver Island University, 900 Fifth St., Nanaimo, BC V9T 1J7, Canada. E-mail: greg.crawford@viu.ca

[3] NOAA, Pacific Marine Environmental Laboratory, 7600 Sand Point Way NE, Seattle, WA 98115, USA. E-mail: burak.Uslu@noaa.gov

[4] Department of Civil and Environmental Engineering, University of Southern California, 3620 S. Vermont Ave., KAP 210, Los Angeles, CA 90089, USA. E-mail: jborrero@usc.edu

[5] eCoast Ltd., Box 151, Raglan, New Zealand. E-mail: d.greer@ecoast.co.nz; jose@ecoast.co.nz

[6] California Geological Survey, 801 K St., MS12-30, Sacramento, CA 95814, USA. E-mail: rick.wilson@conservation.ca.gov

1. Introduction

The north coast of California is susceptible to both near-field and far-field tsunamis. Thirty-seven tsunamis have been observed since a tide gauge was installed in Crescent City in 1933, including twelve with maximum peak-to-trough wave heights exceeding 1 m and five that caused damage (DENGLER and USLU, 2011; NATIONAL GEOPHYSICAL DATA CENTER (NGDC) 2013). Crescent City, located in Del Norte County just south of the Oregon border (Fig. 1), is particularly vulnerable to tsunamis. The M_w 9.2 March 28, 1964 Alaska earthquake triggered a tsunami that produced significant damage and 11 deaths in Del Norte County, more than any other location outside of Alaska (DENGLER and MAGOON, 2005). The most recent

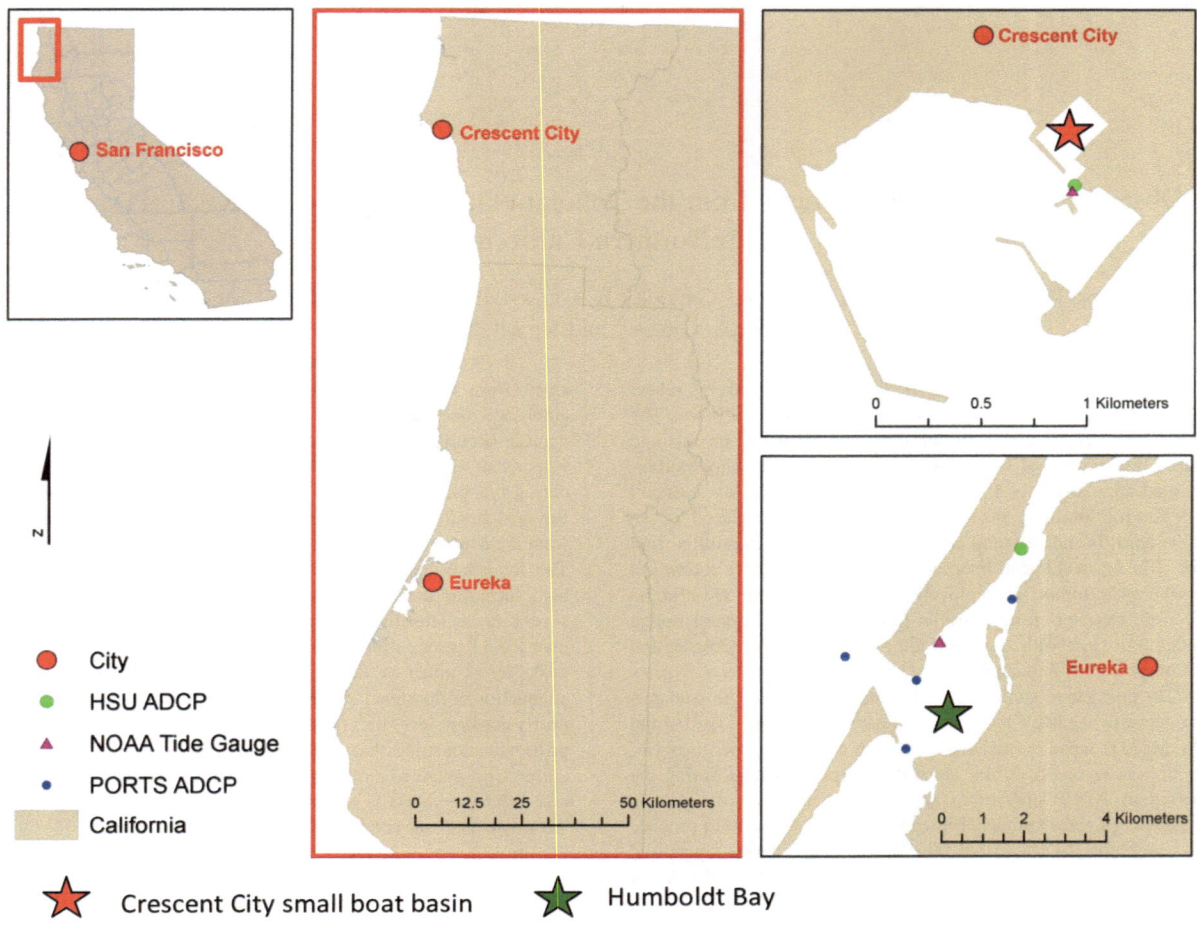

Figure 1
Study area map where tsunami current measurements were collected. *Left panel* California map indicating location of the north coast (*red box*). *Middle panel* location of Eureka and Crescent City. *Right panel* the Crescent City small boat basin near Crescent City (*upper right*) and Humboldt Bay near Eureka (*lower right*) with the NOAA NOS tide gauge sites indicated (*pink triangles*). The Humboldt Bay ADCP and Crescent City ADCP sites are noted by the *green dots*. The *blue dots* show the four ADCPs deployed in Humboldt Bay as a part of the NOAA PORTS project

tsunamis to cause damage in Crescent City were triggered by the 2006 M_w 8.3 Kuril Islands and the 2011 M_w 9.0 Japan earthquakes. Damage from the 2006 tsunami was estimated at US $28 million dollars (DENGLER et al., 2008, 2009; YOUNG, 2013). The 2006 damage had yet to be repaired when the 2011 tsunami demolished almost all remaining structures in the harbor and caused an additional US $26 million in damage (WILSON et al., 2012a; YOUNG, 2013).

All damage from the 2006 and 2011 tsunamis at Crescent City was the result of strong currents in the harbor. There was no on-land flooding by the 2006 tsunami, and only minimal flooding during the 2011

event. Currents produced by the Japan tsunami also caused significant damage at over two dozen other California locations (WILSON et al., 2012a). These events illustrate the vulnerability of ports and harbors to strong tsunami currents and the importance of understanding their impacts.

Although currents have long been recognized as a significant tsunami hazard, few direct measurements of tsunami currents exist. This project began in 2009 in Humboldt Bay (Fig. 1), about 100 km south of Crescent City, to study the feasibility of using a commercially available acoustic Doppler current profiler (ADCP) to directly measure the currents produced by tsunamis. Since deployment, five tsunamis have been recorded on tide gauges on the west coast of North America. Only three were recorded on the NOS Humboldt Bay tide gauge. The ADCP detected two of these events. This paper summarizes the preliminary analysis of the ADCP data for the 2010 Chile and 2011 Japan tsunamis. We also examine three and one-half hours of security video camera footage taken from the entrance to the Crescent City small boat basin and present a method to produce a time series of the current oscillations from the video footage. We compare the observed Humboldt Bay ADCP current values and the Crescent City video velocity estimates to the expected currents from numerical models to test the validity of the model output.

2. *Previous Estimates of Tsunami Current Velocity*

Tsunami currents are routinely predicted in numerical modeling studies; however, few studies have validated those estimates by direct measurements. Peak current estimates for the 1993 Hokkaido tsunami based on displaced objects were on the order of 10–18 m/s (SHIMAMOTO *et al.*, 1995; TSUTSUMI *et al.*, 2000). Several studies have inferred current velocities from analysis of tsunami deposits. CHOO-WONG *et al.* (2008) estimated a flow velocity between 7 m/s and 21 m/s from the thickness and grain size of sediment deposited by the 2004 Indian Ocean tsunami in Phuket, Thailand, and JAFFE *et al.* (2011) estimated tsunami flow velocities of 3–8 m/s based on boulder transport calculations and inverse sediment transport modeling of sediment sheets deposited by the 2009 Samoa tsunami. Tsunami deposit studies can infer peak currents but not the complete time series of flow.

To analyze a continuous time sequence of currents, instrumental recording is necessary. The 2006 Kuril Islands tsunami was observed with ADCP and pressure data from offshore Oahu, Hawaii (BRICKER *et al.*, 2007). An ADCP on the inner shelf of Monterey Bay in California recorded the 2010 Chile tsunami (LACEY *et al.*, 2012) that had a period of approximately 16 min for the first couple hours of the event and persisted several days due to resonance.

The maximum depth averaged velocity was 0.36 m/s during the largest tsunami wave (LACEY *et al.*, 2012). Tide gauge and current data were analyzed in New Zealand following the 2011 Japan event (BORRERO *et al.*, 2012) at the Port of Tauranga and Lyttelton Harbor. In Tauranga, maximum tsunami speeds of 1 m/s were observed at the harbor entrance. Drogues deployed at the entrance to Lyttelton Harbor were tracked (via GPS and boat) every 30 min, and the data were used to produce velocity field snapshots (visual representations of the flow over a certain time span) with currents on the order of 0.6 m/s.

Video analysis can provide surface current information in areas where in situ instrumentation is lacking. FRITZ *et al.* (2012) determined tsunami current velocities from videos recorded by two disaster survivors, ranging in length from 5 to 55 min, taken from rooftops at Kesennuma Bay along Japan's Sanriku coast after the 2011 Japan tsunami. The method (initially developed after the 2004 Indian Ocean tsunami) involved four steps: (1) using LiDAR point clouds to calibrate the camera field of view in real world coordinates; (2) determining the motion of the camera during the recordings; (3) rectifying the video images using direct linear transformation; and (4) analyzing the rectified images to determine the instantaneous tsunami flow velocity fields. The maximum tsunami outflow current at Kesennuma Bay was 11 m/s and arrived <10 min after the measured maximum tsunami height of 9 m in the Kesennuma Bay narrows (FRITZ *et al.*, 2012). A problem with many video recordings is that the recording time is often short and the perspective and view may frequently vary making quantitative analysis difficult, and observations can only be made from the surface which limits the velocities within the flow.

Previous studies in northern California have estimated tsunami current velocities by interpreting tsunami deposits or through numerical modeling. ABRAMSON'S (1998) study at Lagoon Creek (90 km north of Eureka), determined a likely tsunami flow velocity of 2–5 m/s to deposit the observed sand layers within the marsh. A finite-difference tidal model by LAMBERSON *et al.* (1998) predicted peak tsunami current velocities of 5 m/s at the entrance to Humboldt Bay and 2 m/s immediately within the bay.

WHITMORE (1993) modeled a M_w 8.8 Cascadia subduction zone rupture scenario that predicted a maximum 3.5 m/s tsunami velocity along the U.S. West Coast and British Columbia. Tsunami current velocities modeled by DENGLER and USLU (2011) ranged between 1.5 and 2 m/s at the entrance to the small boat basin in Crescent City produced by the 1933 Sanriku and 2006 Kuril Islands tsunamis.

Previous studies, both locally and distant, demonstrate peak current values and the applicability of ADCPs and videos to tsunami current studies. However, no ADCPs with 1 min resolution have been in place continuously for as long a time as the Humboldt Bay instrument described in this paper, have captured more than one tsunami after installation, or have been used to help calibrate numerical modeling.

3. Measuring Tsunamis on the Coast of Northern California

3.1. Humboldt Bay

In 2009, a Nortek Aquadopp 600 kHz 2D ADCP was deployed from the Fairhaven Terminal in Humboldt Bay as part of a pilot project to directly measure tsunami current velocities. The instrument site was chosen because of its close proximity to NOAA's NOS North Spit tide gauge, and easy access from Humboldt State University to make technical adjustments in the pilot phase of the project (Fig. 2). The ADCP emits two acoustic beams that calculate the current velocity along the beams using the Doppler Effect from the change in pitch or frequency of the echo to measure the velocity components of the channel. The navigational channel east of Fairhaven Terminal trends approximately 25° east of true North, and the ADCP is oriented so that the beams extend across the channel (Fig. 2). The velocity measurements are recorded in the local coordinate system relative to magnetic North. That coordinate system was then rotated so that the velocity measurements presented in this paper are relative to true North. Net currents in Humboldt Bay are almost entirely parallel to the channel. We, therefore, chose to convert the velocity data to a scalar representation, with the magnitude reflecting the net current speed and the sign of the value reflecting flood (positive) or ebbing

(negative) conditions, as determined from the sign of the northward component of the flow.

The ADCP is at a fixed depth of 5 m below the mean lower low water (MLLW) tidal datum, and is mounted to the piling at 5 m above the channel bed. The difference between MLLW and MHHW (mean higher high water) at the site is approximately 2 m, which corresponds to a tidal variation above the instrument of 5–7 m. While boundary conditions affect the velocity near the channel bed, the ADCP is mounted within a region of the vertical velocity profile representative of the average flow velocity. The measured velocity is approximately 3–6 % higher than the average velocity; therefore, the velocity data presented in this study are more likely representative of the depth-averaged current.

The instrument is configured to sort the acoustic reflections into bins that represent sampling intervals along the beams (ADMIRE, 2013). In this study we report on results from the most distant bin (bin 5) extending 33–41 m from the dock, in a region near the middle of the channel where water depth is approximately 12 m. Measurements are made every 1 s and averaged over 1 min intervals with a manufacturer's estimated uncertainty of ±0.02 m/s (NORTEK, 2008). This gives a continuous, high resolution velocity measurement near the Fairhaven Terminal.

Since the deployment of the ACDP in Humboldt Bay, five tsunamis have been recorded on the Crescent City tide gauge which is one of the most sensitive locations on the U.S. west coast: 2009 Samoa, 2010 Chile, 2011 Japan, 2012 Haida Gwaii, and 2013 Solomon Islands. Both the 2009 Samoa tsunami (WILSON et al., 2011) and the 2013 Solomon Islands tsunami were recorded at a number of California locations, but neither was detected by either the tide gauge or ADCP in Humboldt Bay. The three that were recorded on the North Spit tide gauge, 2010 Chile (0.21 m), 2011 Tohoku (0.97 m), and 2012 Haida Gwaii (0.12 m), typically had amplitudes of about 30–35 % of the Crescent City value. The 2012 Haida Gwaii tsunami was observed on the tide gauge in Humboldt Bay but is not apparent in the ADCP data. This suggests that at Humboldt Bay, a threshold value of about 0.2 m on the tide gauge is required to produce a detectable signal on the ADCP.

Figure 2
Location of the pilot ADCP project in Humboldt Bay, near Eureka, California. **a** View of the North Spit and Humboldt Bay entrance (image obtained from Google Earth). The NOAA NOS North Spit tide gauge is 3 km SSW of the ADCP site on the Fairhaven Terminal. **b** Detailed view of the Fairhaven Terminal (image obtained from Google Earth). **c** The ADCP site located at the north end of the Terminal (image provided by J. Powell), indicating the orientation of the local coordinate system

To separate the tsunami signal from tidal and high-frequency noise in both the water level and velocity time series, we followed a methodology similar to LYNETT *et al.* (2012) and BORRERO and GREER (2012), which involved using harmonic analysis to identify the tidal variations and de-noising using a wavelet methodology. Spectral analysis, performed by BORRERO and GREER (2012), on the Humboldt Bay tide gauge data for the 2010 and 2011 events evaluated the duration and the dominant periods of the tsunami waves.

In classical harmonic analysis, the tide is modeled as a combination of sinusoids at very specific frequencies determined by astronomical motions (see, for example, EMERY and THOMSON, 2004). We used the freely-available MATLAB toolbox called

T-tide (PAWLOWICZ *et al.*, 2002) to analyze the water level and velocity data to determine the amplitudes and phases of the dominant tidal frequencies in Humboldt Bay. Water level data were compiled from NOAA TIDES and CURRENTS (2013). For the 2010 tsunami, we used a 3 month period (November 8, 2009–February 26, 2010) to determine tidal constituents for water level and a 1 month period (March 3–April 7, 2010) in a similar way for velocity. The shorter time range used for determining the velocity tidal constituents is a consequence of issues in maintaining continuous velocity observations during the early stages of this project. The 2011 tsunami was processed in a similar fashion, using the water level between December 9, 2010 and March 10, 2011, and velocity between January 26 and February 28, 2011.

In general, when computing tidal constituents, the longer the record the better in order to accurately separate frequencies; however, the tsunami periods of interest are significantly shorter than those that require extended record lengths.

These amplitudes, phases and frequencies can then be used to predict tidal heights and currents for any time. The predicted tides can be subtracted from the observations to provide detided time series. In the present case, however, after the data were detided, the residual time series outside the tsunami period exhibited fairly regular periodicity in the tidal height at periods of 10–12 h and in the velocity at periods of 4–6 h. The residual may be due to an insufficiently long period of data to adequately separate two closely-tuned tidal constituents, S2 (period 12.00 h) and K2 (period 11.97 h); in the case of the velocity estimates, there may also be higher order harmonics sometimes observed in shallow water conditions (e.g., M4, period 6.21 h, caused by nonlinear self-interaction of the M2 constituent). In order to remove these residual oscillations, we found it necessary to apply a 4th order, phase-preserving, high-pass Butterworth filter to the detided water level and velocity data with cutoff periods of 8 and 4 h, respectively. Following the approach of BORRERO and GREER (2012) and LYNETT et al. (2012) again, the signals were subsequently passed through a one-dimensional wavelet de-noising function (DONOHO and JOHNSTONE, 1994).

3.2. Crescent City

After the 2006 Kuril Islands tsunami, a security camera was mounted in a fixed location to the roof of the two-story Harbor Master's building just opposite the entrance to the small boat basin in Crescent City Harbor (Fig. 3) and oriented to optimize viewing the mouth of the innermost harbor (small boat basin). The camera faces westward with a view of the 55 m wide entrance to the small boat basin. The Crescent City harbor area and small boat basin were evacuated an hour before the expected arrival of the first tsunami surges in 2011, and the power was shut off in anticipation of potential flooding. The camera filmed the onset of the tsunami and the first three and one-half hours of

the event, beginning at 1508 UTC and ending at 1845 UTC, until the battery power ran out.

To estimate velocities from the time stamped video footage, we measured the distance objects within the flow moved in 10 s windows (Fig. 4). In most cases, only one object was present in the flow or easily identified for observation for each 10 s frame. The start and end locations of flotsam and debris were identified in the video footage and then transferred to a Google Earth image (with the same angle of view as the video camera) looking towards the west with an eye altitude of 5 m at 41.746452°N, 124.183704°W. The ruler tool was used to approximate the distance between the start and end locations for every 10 s pair of observations. The 10 s estimates were averaged over 1 min intervals to generate a surface velocity time series for Crescent City. The velocity time series is essentially continuous during the period when the tsunami is present, but it is difficult to determine the pre-tsunami currents because of the lack of surface flow.

The standard deviation of velocity was computed for each minute by calculating the standard deviation over the six estimates per minute using Excel. Throughout the time series, the standard deviation ranged between 0.02 and 1.84 m/s, with a mean overall of 0.29 m/s during the period of observation. Over 60 % of the standard errors were <0.20 m/s; 95 % of the standard errors were <1.00 m/s. In general, standard errors were significantly larger near slack flow conditions. During these periods, it was clear from the video footage that there was substantial variability in the flow across the entrance region. Due to the distribution of the standard deviation values and the majority falling below 1.00 m/s, we take 0.5 m/s as a rough, if generally conservative, estimate of the uncertainty in the current speed.

The 2011 event was videotaped by several eye witnesses in Crescent City in addition to the security camera footage used in this study. WILSON et al. (2012b) combined bathymetry and sediment analysis with observations from various ground-level and aerial video (including the findings in this study as well as 13 additional video clips) to develop a tsunami peak current in-flow regime map identifying areas of strong currents, sediment erosion and

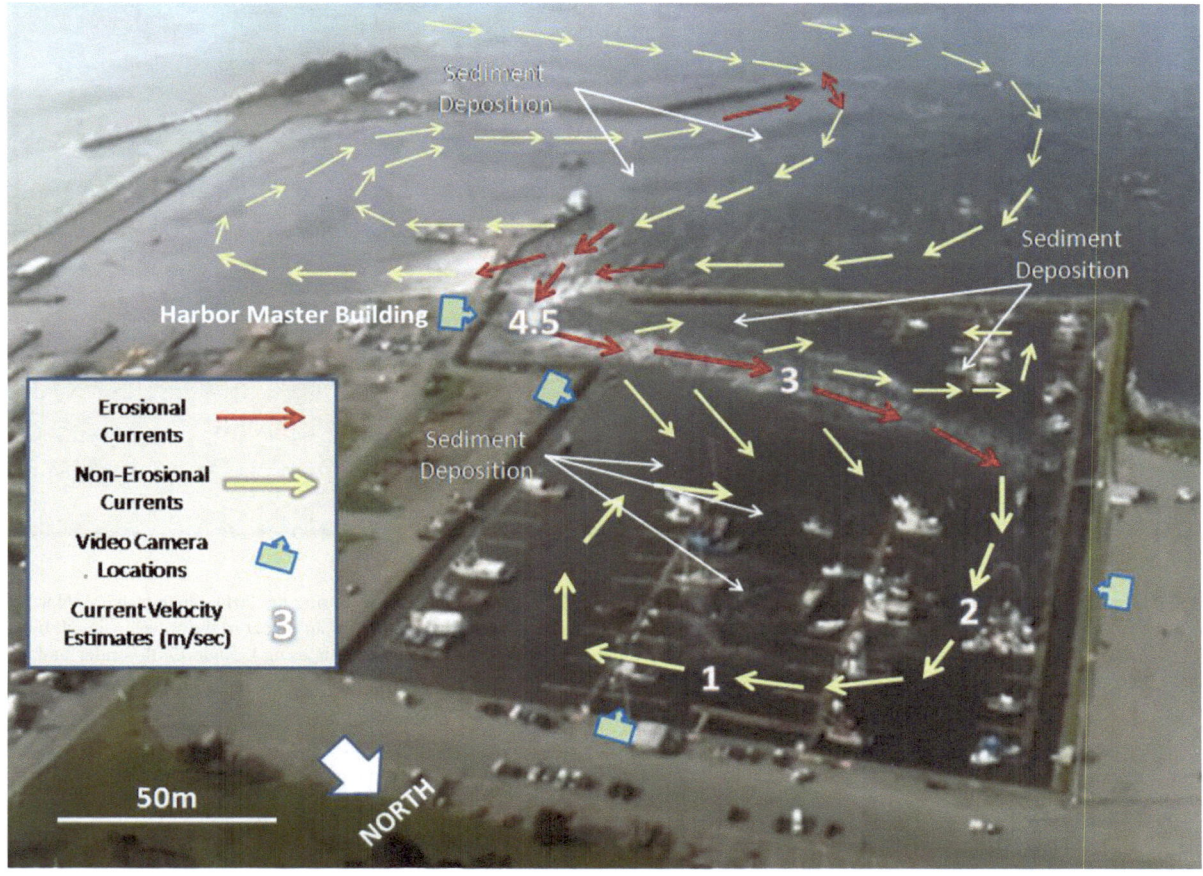

Figure 3
Tsunami in-flow regime map for Crescent City Harbor showing the maximum observed tsunami current velocity values over the span of the first 3.5 h of the 2011 tsunami. *Numbers* show estimated surface velocities (m/s), and *arrows* indicate flow direction based on 14 ground-level and aerial videos taken of the event

sedimentation (Fig. 3). Velocity estimates for the eyewitness videos (ranging in length from 33 s to 49 min) were calculated by comparing distance moved to travel time for the first several hours (between 1700 and 1900 UTC) of the event from several vantage points around the harbor.

The water level record for the Crescent City tide gauge during the time of the tsunami was analyzed in the same way as Humboldt Bay using data from December 9, 2010 to March 10, 2011 to compute the amplitudes, phases and frequencies. Due to the short record obtained from the video footage, it was not possible to detide the current velocity data; however, during this period it was clear from the tide gauge record that the tsunami dominated the tidal flows. The maximum predicted tidal change during the entire

video record is on the order of 0.45 m with slack tide conditions (~ 0 m/s) near the end of the video recording window.

3.3. MOST Model

The tsunami currents and water levels produced by the 2010 Chile and 2011 Japan tsunamis were modeled using the NOAA's Tsunami Propagation Database (GICA *et al.*, 2008; WEI *et al.*, 2008; TANG *et al.*, 2009) and the MOST model (TITOV and GONZALEZ, 1997; TITOV and SYNOLAKIS, 1998). TANG *et al.* (2012) used the Deep-ocean Assessment and Reporting of Tsunamis (DART®) to compute the energy transmitted by the tsunami. The MOST model complies with the standards and procedures outlined

Figure 4
Method for determining tsunami current surface velocities from Crescent City video footage taken during the 2011 Japan tsunami. Note the location of the Battery Point Lighthouse (BPL) in the background. **a** At the start time, the location of an object in the water is noted. **b** The location of the same object 10 s later is noted. **c** The distance between the start and end locations using Google Earth's ruler tool

Table 1

Source parameters from the propagation database for 2010 Chile and 2011 Japan tsunamis. For additional details, see Gica et al. (2008) and the NOAA Center for Tsunami Research webpage (http://nctr.pmel.noaa.gov)

Date	Time (UTC)	Location	Epicenter	M_w	Tsunami Source
27/02/10	0634	Chile	35.909°S 72.733°W	8.8	$17.24 \times a88 + 8.82 \times a90 + 11.86 \times b88 + 18.39 \times b89 + 16.75 \times b90 + 20.78 \times z88 + 7.06 \times z90$
11/03/11	0546	Japan	38.322°N 142.369°E	9.0	$4.66 \times kiszb24 + 12.23 \times kiszb25 + 26.31 \times kisza26 + 21.27 \times kiszb26 + 22.75 \times kisza27 + 4.98 \times kiszb27$

in SYNOLAKIS *et al.* (2008) and have shown to provide reliable forecast and modeling capabilities. The bathy-topo digital elevation model resolution in this study is 1/3 arc-s (10 m) and assumes mean high water as the vertical datum, WGS 1984 for the horizontal datum, and 0.03 for the Manning roughness coefficient of friction. Detailed earthquake source parameters from NOAA's Tsunami Propagation Database are listed in Table 1 and the details of the nested grids used are listed in Table 2. In this paper, we present the tsunami current velocity results from the MOST model. The velocity values have been orientated so that the coordinate system agrees with that of the ADCP and video footage measurements.

4. 2010 and 2011 Observations of the Tsunamis in Northern California

4.1. February 2010 Chile Tsunami

On February 27, 2010 at 0634 UTC, a M_w 8.8 earthquake occurred in the Maule region of Chile, which generated a tsunami that caused severe damage to coastal towns and ports in the source area and was recorded throughout the Pacific (FRITZ *et al.*, 2011). Although inundation was not observed in California, a dozen maritime locations in the southern and central part of the state were affected by strong currents that caused US $2–3 million in damage (WILSON *et al.*, 2010, 2012a; DENGLER *et al.*, 2011).

Table 2

Extents of the three nested grids used for Crescent City and Humboldt Bay

Grid	Region	Coverage		Cell size (″)	Grid size (nx × ny)
		Lat. (°N)	Lon. (°E)		
A	Northern California	39.03500–41.98836	234.25917–236.37251	24	318 × 444
B	Del Norte County	41.65262–41.92650	235.65381–235.91575	1	987 × 944
C-1	Crescent City	41.72800–41.75800	235.782000–235.835981	1/3	584 × 325
Friction coefficient (Manning roughness)		0.03			
B	Humboldt County	40.52418–41.95419	235.61168–236.11835	12	153 × 430
C-2	Humboldt Bay	40.67005–40.87097	235.67042–235.88060	1/3	2,271 × 2,171
Friction coefficient (Manning roughness)		0.03			

A and B grids are common, whereas C-1 covers Crescent City harbor and C-2 covers the north spit of Humboldt Bay

The first tsunami surges arrived close to low tide in Humboldt Bay and Crescent City. Both the NOAA NOS tide gauge and the ADCP in Humboldt Bay recorded the event (Fig. 5).

Figure 5a shows the measured velocity magnitude beginning 12 h prior to arrival of the 2010 Chile tsunami and continuing for the following 45 h in Humboldt Bay as well as the tidal prediction base on harmonic analysis. During this event, the currents are dominated by the tidal flux. The negative values indicate ebb velocities (to the south), and the positive values indicate flood velocities (to the north). Figure 5b is the de-noised tsunami signal plotted after the tide was removed and the phase-preserving, high-pass filter was applied. The tsunami signal was very small relative to the inherent background noise, which can be seen in the pre-tsunami portion of the plot. The tsunami current peak amplitude was 0.35 m/s, arriving 16 h after the tsunami arrival, and the amplitude slowly decays over time to 0.15 m/s approximately 25 h after the tsunami arrival.

The water level data from the NOAA NOS tide gauge beginning 12 h prior to the tsunami arrival and the following 45 h are shown in Fig. 5c plotted with the tidal prediction determined using harmonic analysis. In this plot, the tsunami can be seen superimposed on the tidal signal. Figure 5d presents the detided, high-pass filtered and de-noised water level data for the same time period. The tsunami signal is more observable in the water level data than the velocity data likely due to the velocity measurements being noisier than the water level data. The arrival of the peak water levels and peak velocities

coincide at approximately 16 h after the initial arrival with water level amplitudes on the order of 0.23 m.

4.2. March 2011 Japan Tsunami

The March 11, 2011 Japan tsunami was generated by a M_w 9.0 earthquake off the Tohoku-oki coast of Honshu, Japan at 0546 UTC. A very large tsunami was generated locally that caused catastrophic damage and loss of life in Japan. It also produced a Pacific-wide teletsunami that damaged over two dozen harbors and ports along California's coast, causing over US $50 million in damage (WILSON et al., 2011, 2012a). The greatest damage in California occurred in Crescent City and Santa Cruz harbors. The tsunami arrived on the north coast of California at approximately 1530 UTC on March 11, 2011 during low tide. The tsunami destroyed nearly all of the docks and infrastructure in the Crescent City small boat basin that had survived the 2006 tsunami, sinking 17 boats, transporting volumes of sediment into the basin and causing about $26 million worth of damage (WILSON et al., 2012a; YOUNG, 2013). The tsunami was recorded by the NOAA NOS tide gauges in Humboldt Bay and Crescent City Harbor (Figs. 6c, d, 7b, respectively), as well as by the ADCP deployed in Humboldt Bay (Fig. 6a, b). It was also recorded by a video camera in Crescent City Harbor (Fig. 7a).

Figure 6a shows the observed total current velocities and the tidal prediction based on harmonic analysis, 12 h prior to the tsunami arrival and the following 48 h for Humboldt Bay. The tsunami signal is clearly superimposed on top of the tidal

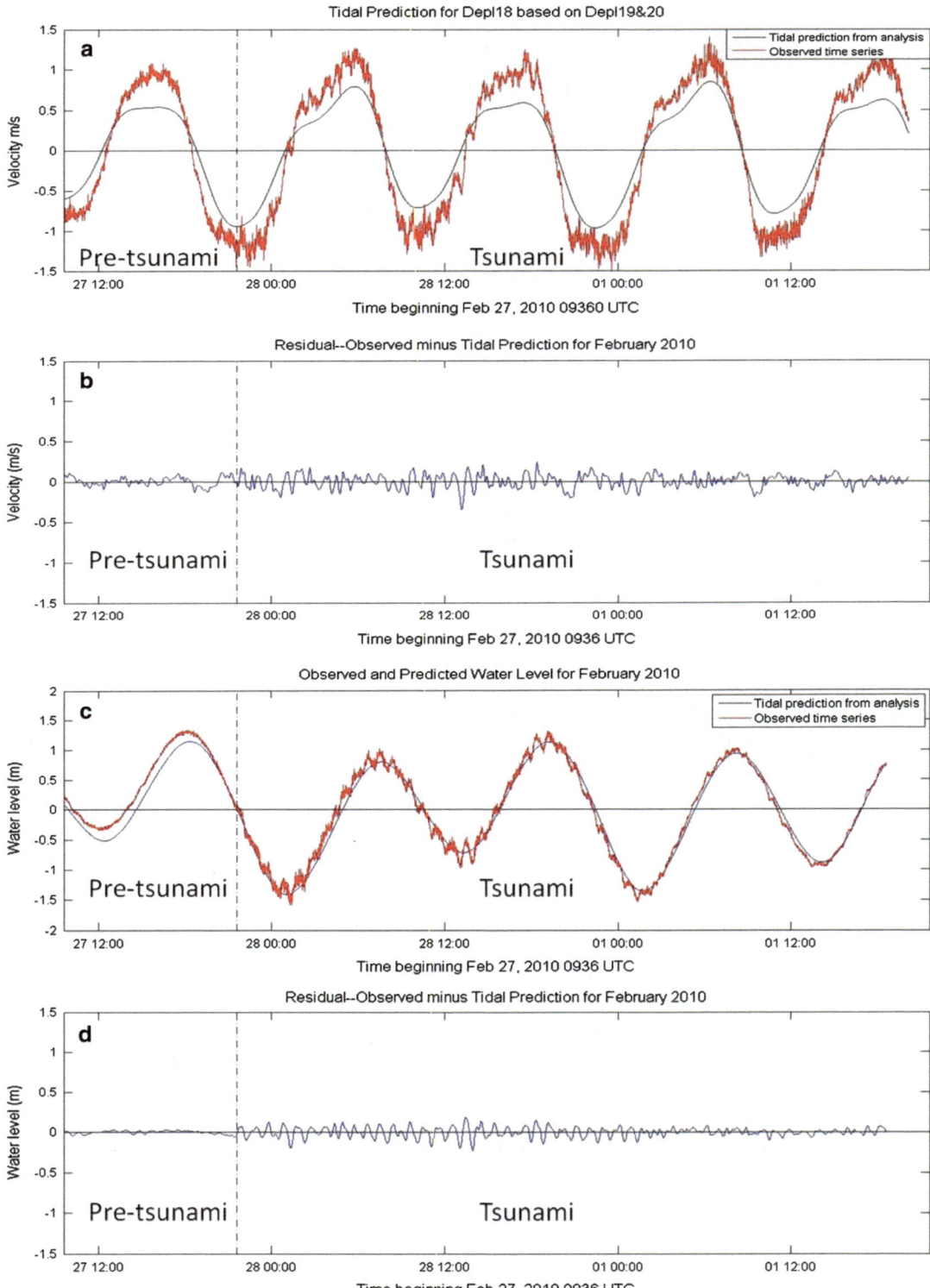

◀ Figure 5
ADCP velocity and NOAA NOS tide gauge water level for February 27, 2010 Chile tsunami in Humboldt Bay. **a** Observed current (*red line*) and the tidal prediction based on harmonic analysis (*blue line*) for data. **b** Observed tsunami velocity signal after detiding, high-pass filtering and de-noising (details provided in the text). **c** Observed water level (*red line*) and the tidal prediction based on harmonic analysis (*blue line*) for the North Spit tide gauge. **d** Observed tsunami water level signal after detiding, high-pass filtering and de-noising. The *graphs* show the 12 h prior to the tsunami arrival and the following 45 h. *Vertical, dashed black line* indicates the observed arrival of the tsunami at 2136 UTC

signal. Figure 6 presents the first 48 h of the event, but it's important to note that the tsunami signal appears to persist longer than the time window. The tsunami current velocity is shown in Fig. 6b after the tide and noise had been removed using the method described previously. The tsunami signal is substantially larger than the inherent background noise (pre-tsunami), with a peak tsunami current of approximately 0.84 m/s.

Figure 6c presents the water level data from the NOAA–NOS tide gauge in Humboldt Bay 12 h prior to the tsunami arrival and the following 48 h. The tsunami is clearly evident in the water level plot. The detided, phase-preserving, high-pass filtered and de-noised water-level data during the tsunami are shown in Fig. 6d. The peak velocities occurred during the largest water level transitions, and near-zero velocities were observed during slack tide. The strongest current (∼0.84 m/s) was observed 67 min after the arrival of the tsunami during the third surge into Humboldt Bay corresponding to a peak water level of 0.81 m.

The tsunami currents were much stronger in Crescent City. The surface currents at the entrance to the small boat basin in Crescent City as determined by the video analysis are presented in Fig. 7. The peak velocity measured during the video was approximately 4.5 m/s, with an average period of 20–30 min. The NOAA–NOS tide gauge recorded the tsunami for over six days in Crescent City. Figure 7b shows the detided, filtered and de-noised water level data corresponding to the time period of the measured surface currents from the video footage. There is a 25 min gap in Fig. 7b that is a result of missing 1-min water level data. The third cycle had the highest water level at 2 m and a current velocity of approximately 2 m/s. The fifth wave had the

strongest current at approximately 4.5 m/s with a water level of approximately 0.6 m. Half of the measured currents met or exceeded 3 m/s. All, with the exception of the first wave, met or exceeded 2 m/s (Fig. 7a). The peak velocity estimates from the additional eye-witness videos and the peak velocity observed in the security footage ranged from 4.5 m/s at the entrance to the small boat basin to approximately 1 m/s towards the back of the basin (Fig. 3) providing a representative in-flow current of the peak velocities observed during the first three and one-half hours of the tsunami in Crescent City.

5. MOST Model Comparison

Figures 8 and 9 show the maximum computed amplitudes and currents in Humboldt Bay and Crescent City Harbor, respectively, from the 2010 Chile and 2011 Japan tsunamis. The MOST model predicted wave amplitudes up to 0.4 m along the South Spit with 0.4 m/s maximum speed within the entrance for the 2010 Chile tsunami, whereas the 2011 Japan tsunami amplitudes were predicted to exceed 2 m along the North Spit with a maximum speed of 2 m/s at the entrance to the bay (Fig. 8). The model predicted larger wave amplitudes in Crescent City than in Humboldt Bay for both the 2010 and 2011 tsunamis. The 2010 Chile tsunami was predicted to reach 0.8 m in the Crescent City small boat basin with a speed of 1.5 m/s near the basin entrance, whereas the model predicted the 2011 Japan tsunami reaching up to 4 m within the small boat basin and a maximum speed of 9 m/s near the entrance (Fig. 9).

The comparisons between the observed and modeled tsunami current velocities for the Humboldt Bay ADCP and the Crescent City small boat basin entrance are shown in Figs. 10 and 11. The observations and the model outputs for the 2010 and 2011 tsunami currents in Humboldt Bay and the 2011 currents in Crescent City had an arrival discrepancy of 10 min. The model results indicated that the tsunamis arrived 10 min prior to the observed arrival for both events (2010 and 2011) in Humboldt Bay and for the 2011 event in Crescent City. This difference has been observed in many real-time forecasting

Reprinted from the journal

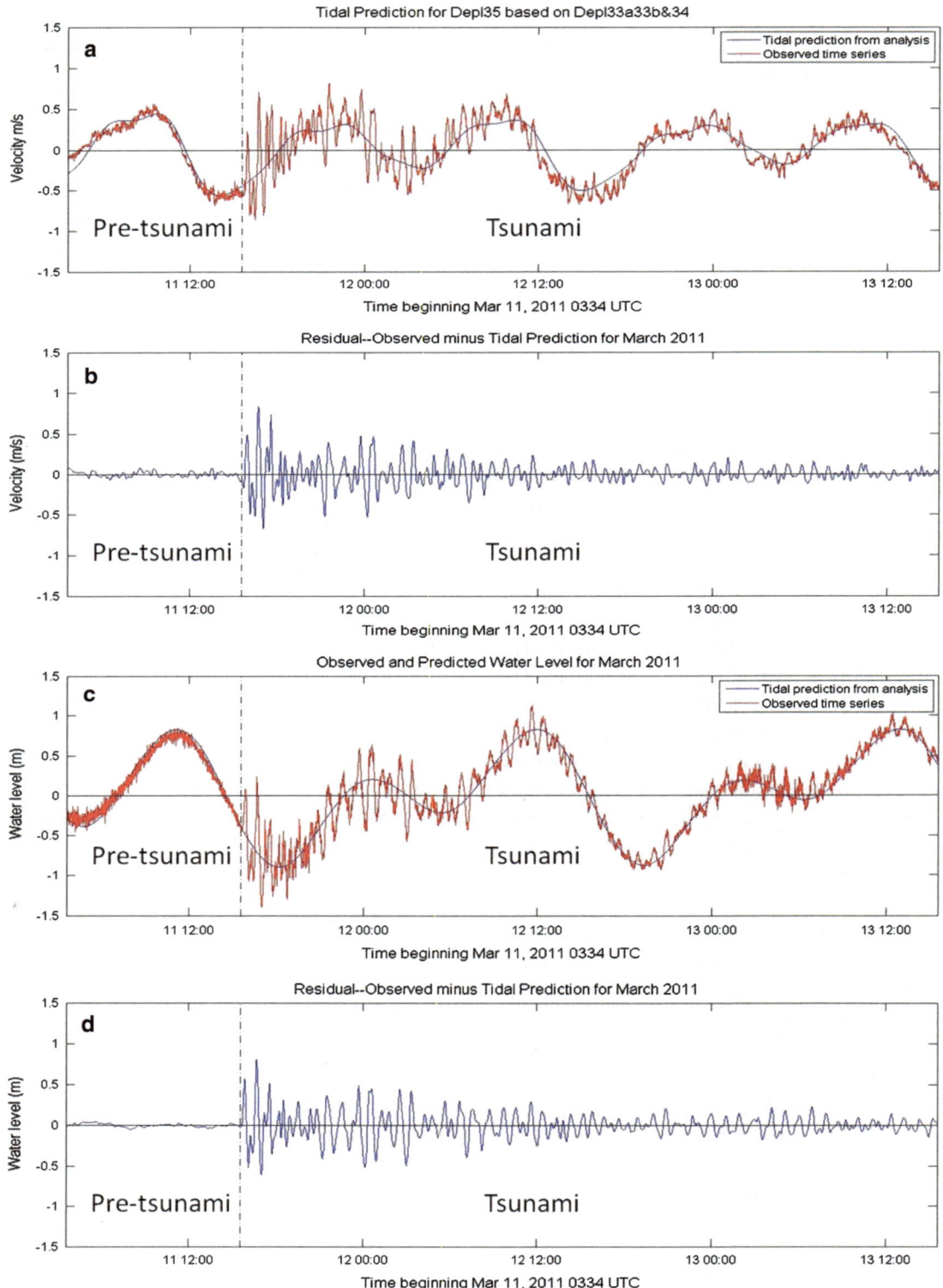

◄ Figure 6
ADCP velocity and NOAA NOS tide gauge water level for March 11, 2011, Japan tsunami in Humboldt Bay. **a** Observed current (*red line*) and the tidal prediction based on harmonic analysis (*blue line*) for data. **b** Observed tsunami velocity signal after detiding, high-pass filtering and de-noising (details provided in the text). **c** Observed water level (*red line*) and the tidal prediction based on harmonic analysis (*blue line*) for the North Spit tide gauge. **d** Observed tsunami water level signal after detiding, high-pass filtering and de-noising. The *graph* is plotted with the same time scale as in Fig. 5, showing the 12 h prior to the tsunami arrival and the following 48 h. *Vertical, dashed black line* indicates the observed arrival of the tsunami at 1534 UTC

observations at NOAA's Center for Tsunami Research, and it is believed to be a combination of deep ocean propagation results and inversion of the source (TANG *et al.*, 2012). Difference in the arrival has been observed in earlier studies at Crescent City (USLU *et al.*, 2007), and some of the discrepancy is attributed to the instantaneous-rupture assumption numerical models. The model outputs are shifted 10 min compared to the observations to align the initial onset of the predicted and actual signals.

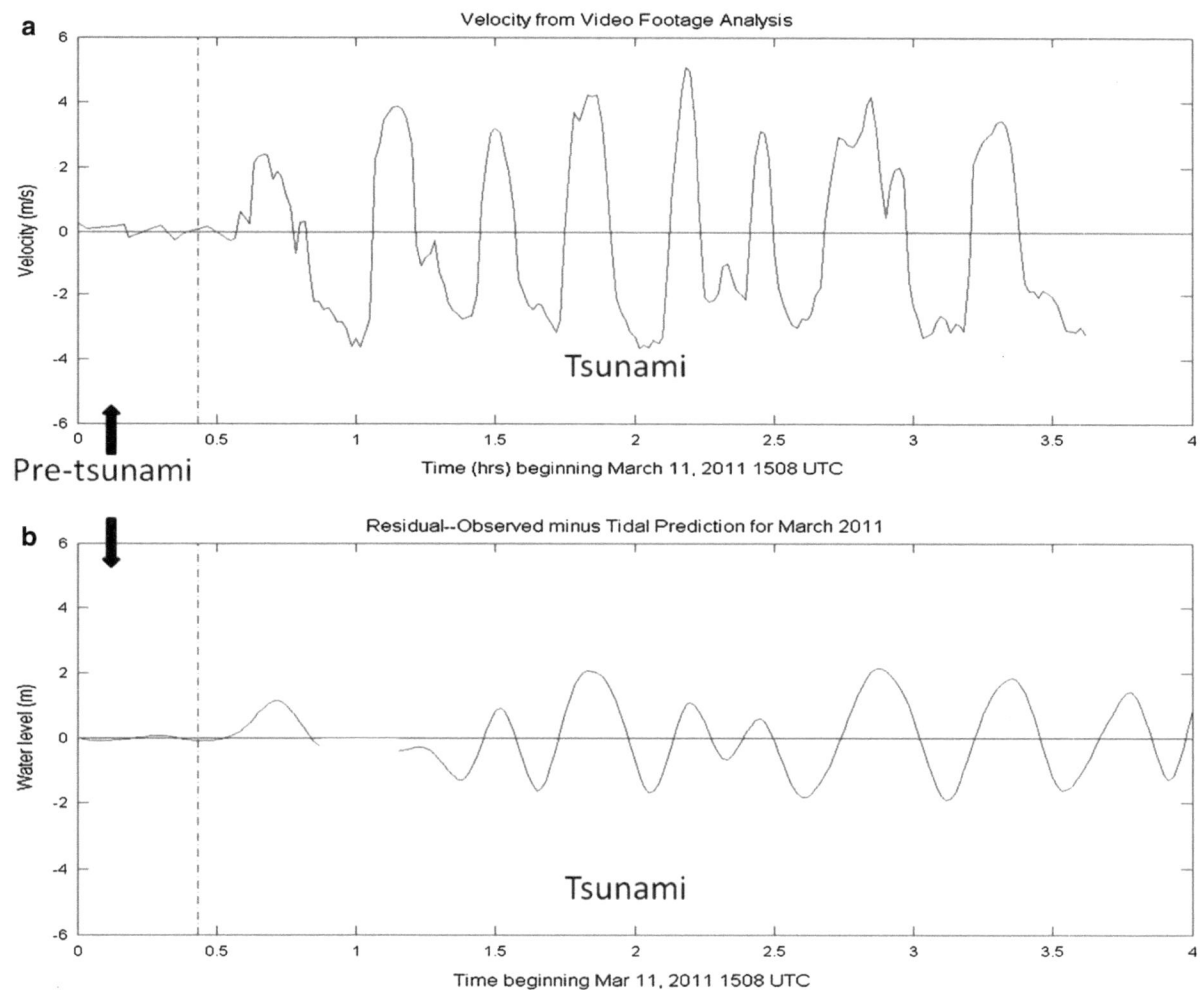

Figure 7
a Surface currents (derived from video footage) at Crescent City for the March 11, 2011 Japan tsunami for the first three and one-half hours of the event. The tide has not been removed due to the brevity of the camera observations. **b** The water level time series after detiding, high-pass filtering and de-noising (details provided in the text) for the time corresponding to the video footage. The gap in (**b**) accounts for 1 min data not collected on the tide gauge. Each plot shows the first 3 h of the event, and the *vertical, dashed black line* indicates the arrival of the tsunami at 1534 UTC. Note that the time scale on this figure is different than Figs. 5 and 6

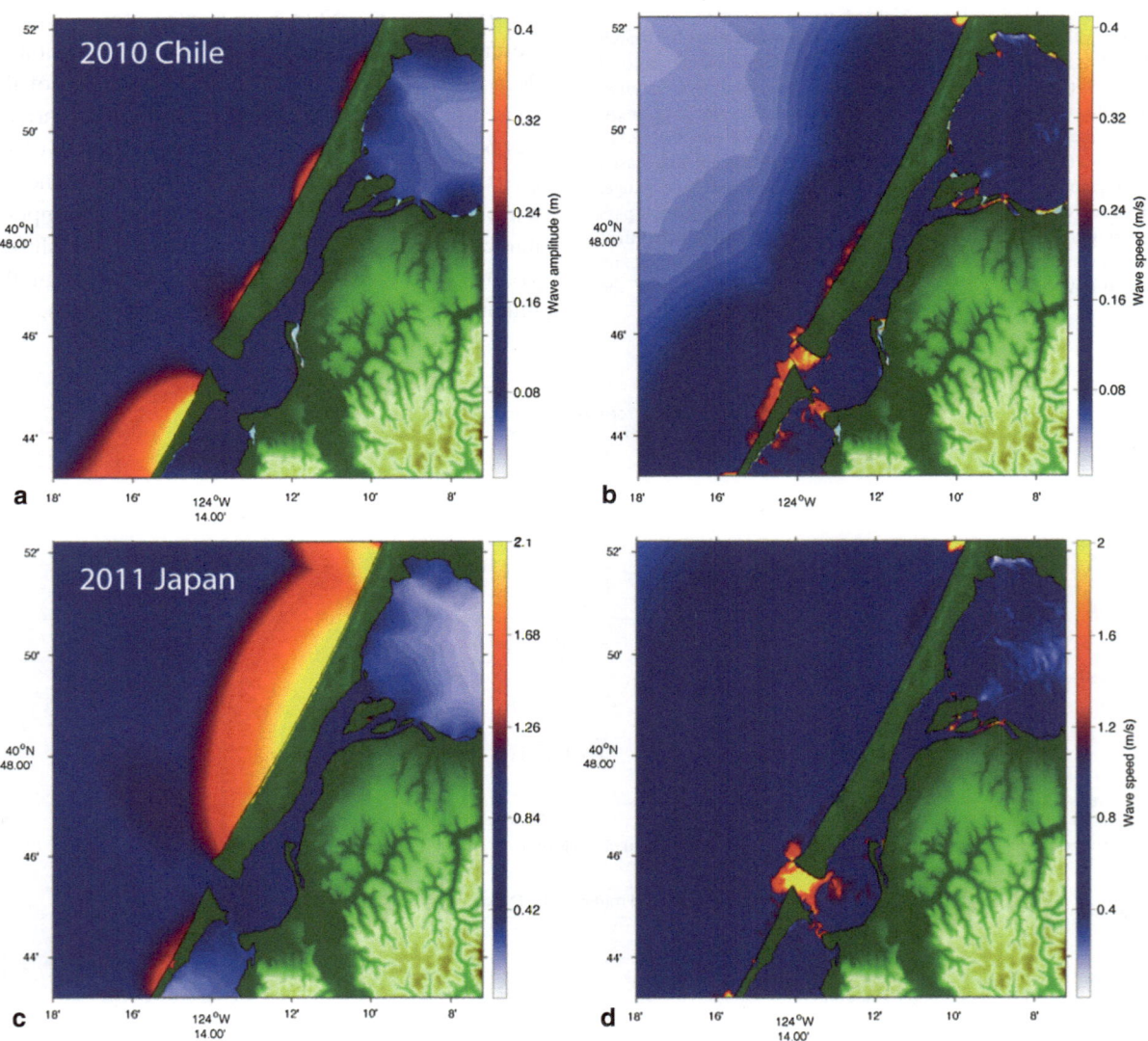

Figure 8
Peak tsunami wave amplitudes (**a**, **c**) and peak current speeds (**b**, **d**) computed by the MOST model for the 2010 Chile and 2011 Japan tsunamis for the coast and entrance to Humboldt Bay. Amplitudes were predicted to be largest on the coastal side of the North Spit with increased currents focused at the harbor entrance

The 2010 Chile tsunami measured in Humboldt Bay is plotted with the MOST model prediction in Fig. 10. Although the tsunami velocity signal is small, the model predicts the frequency and the pattern of amplification and decay of the signal quite well for the first 2 h after the arrival. However, the amplitudes are underestimated on average by 10–20 % and for several of the largest velocities by as much as 50 %. After the first 2 h, the model predicts attenuation of the signal, and consistently underestimates peak positive and negative velocities. This discrepancy may be due to the similarity between the inherent background noise and tsunami signal.

The 2011 Japan tsunami in Humboldt Bay is compared to the MOST model prediction in Fig. 11a. Like the 2010 event, the frequency and pattern of the currents are well replicated by MOST, especially for the first 4 h of the event. The peak flood velocities are underestimated by MOST, with the first, third and

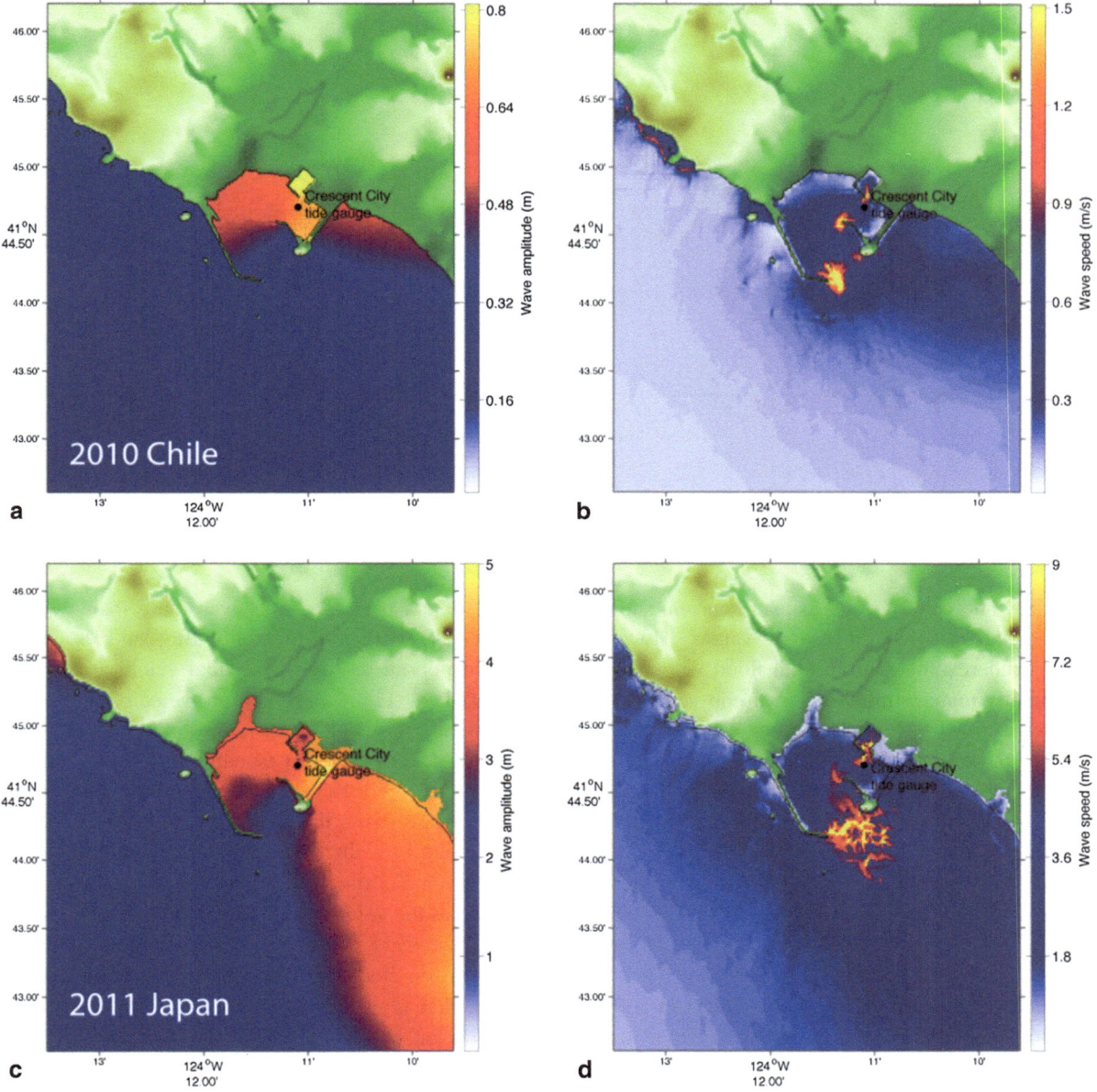

Figure 9

Peak wave amplitudes (**a**, **c**) and peak current speeds (**b**, **d**) computed by the MOST model for the 2010 Chile and 2011 Japan tsunamis for Crescent City Harbor and the small boat basin. Amplitudes were predicted to be largest within the bay and along the coast with current speeds focused at the bay entrance and small boat basin entrance

fifth modeled waves smaller than the observed values by 30–50 %. The maximum change in velocity corresponded to the ebb flow after the third wave of ~ 1.5 m/s (from ~ 0.84 to -0.67 m/s) while the model estimated a change 0.5 m/s less than the observed. After 4 h, the model predicted diminished oscillations, but observations suggest continued

motions in the bay on the time scale of 20–60 min for two days. Significant oscillations of the observed velocity signal persist for more than twice as long as the MOST prediction.

Figure 11b compares the velocity data and the MOST model prediction for the 2011 tsunami in Crescent City. As in Humboldt Bay, the predicted

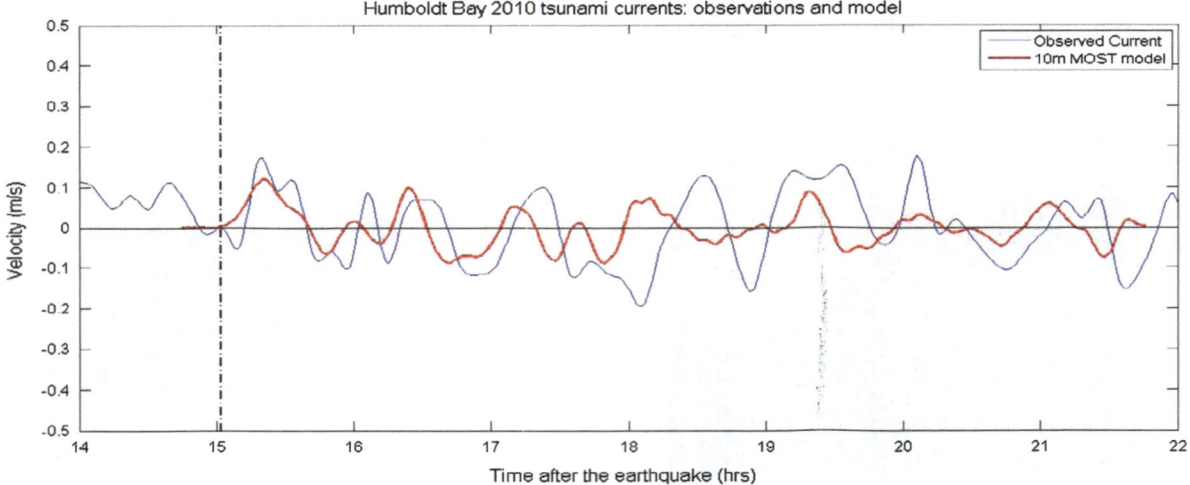

Figure 10
Comparison of the observed tsunami current velocity (*blue line*) and the MOST model prediction (*red line*) for the February 27, 2010 Chile tsunami in Humboldt Bay, California. The *vertical dashed line* indicates the tsunami arrival time of 2136 UTC. MOST predicts signal attenuation after 4 h that is not observed on the ADCP

frequency and pattern of the tsunami current shows a good match with the measured values. However, MOST underestimates the initial flood velocity, and consistently underestimates several ebb velocities by as much as 50–70 % during the first one and one-half hours of the event. After one and one-half hours, the model overestimates flood velocities by 30–60 % and begins to predict ebb velocities more consistent with those measured in the video footage.

6. Discussion

Since the initiation of the pilot project in 2009 in Humboldt Bay, the ADCP has recorded two tsunamis (the 2010 Chile and the 2011 Japan), and has demonstrated the utility of ADCPs to directly measure currents for small to moderately-sized tsunami events.

The 2010 Chile tsunami had a small velocity signal in Humboldt Bay with maximum velocities reaching 0.35 m/s and maximum water levels of 0.23 m. Figure 5b and d show the comparison of the observed velocity and water level signals during this event, relevant to the pre-tsunami background noise. The water level signal stands out more clearly from the background than the velocity signal does. The

2011 Japan tsunami signal was more than twice as large as 2010 in Humboldt Bay and is clearly superimposed on top of the tidal velocity and water level data (Fig. 6). Maximum velocity amplitudes between 0.6 m/s and 0.84 m/s were measured within the first 2 h of the 2011 tsunami arrival in Humboldt Bay. It is interesting to note that while no previous studies have been done near the deployed ADCP in Humboldt Bay, the maximum velocity amplitudes for Fairhaven Terminal are on the same order of magnitude as the values predicted in the LAMBERSON et al. (1998) study (2 m/s immediately within the bay). The Lamberson study was based on an arbitrary design wave that is not easily compared with actual water heights on March 11. It also predicted velocities in the open channel near the entrance of the bay which is about 5 km south of the Fairhaven Terminal where the ADCP is deployed in a more protected area of the bay.

At Crescent City, the observed velocities at the entrance ranged from 2 to 4.5 m/s during the first three and one-half hours of the event with strongest currents occurring at approximately 1730 UTC with a corresponding water level of 0.6 m. The tide gauge data show that the tsunami persisted for over six days in Crescent City Harbor. WHITMORE'S (1993) 3.5 m/s estimated tsunami current along the U.S. West Coast

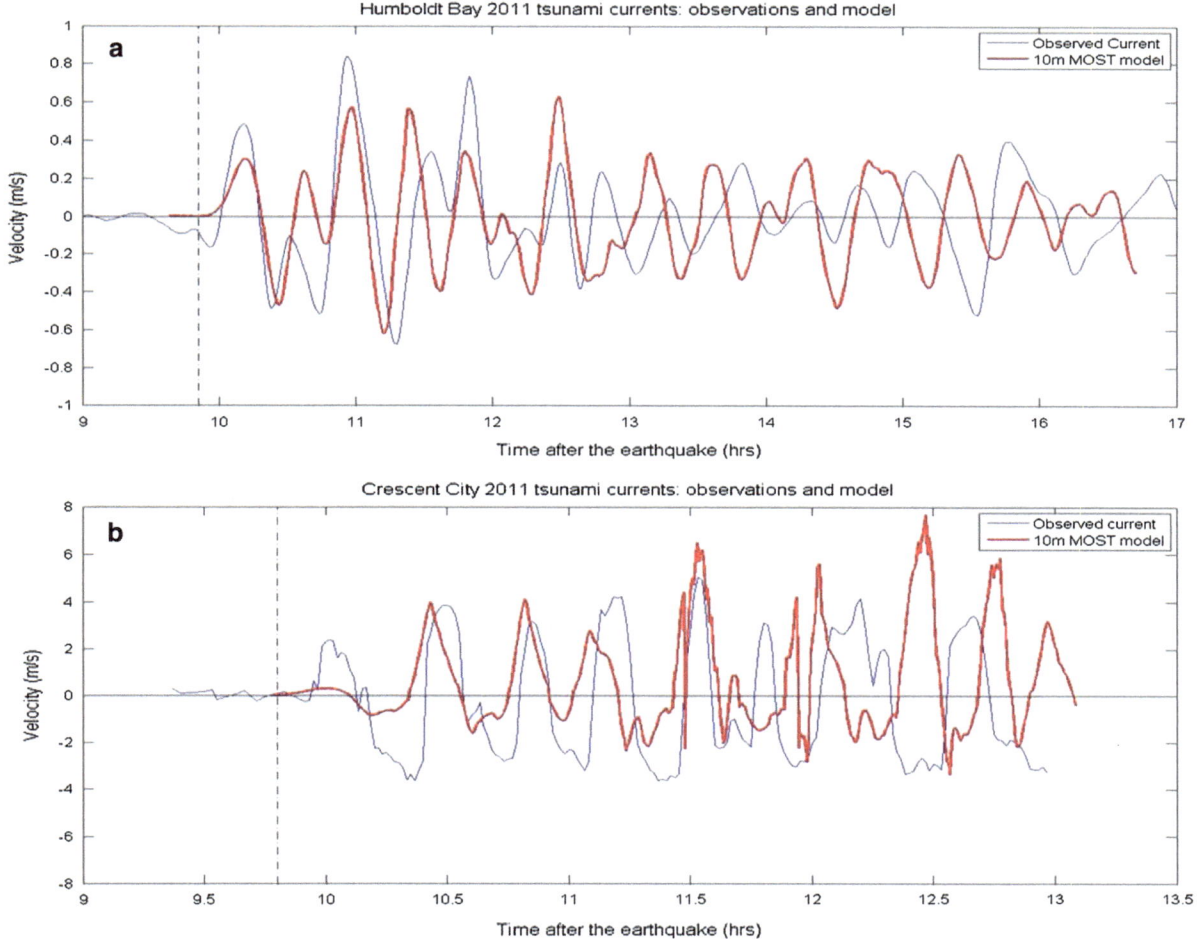

Figure 11

Comparison of the observed tsunami current velocity and MOST prediction for March 11, 2011 Japan tsunami. **a** Graph of the observed tsunami currents (*blue line*) compared with the MOST model prediction (*red line*) for Humboldt Bay. **b** Graph of the surface currents (*blue line*) compared with the MOST model prediction (*red line*) for Crescent City during the first 3.5 h of the event. Note the different scales used in **a** and **b** with velocity (m/s) along the *Y*-axis and time (hours after the earthquake) along the *X*-axis. The *vertical dashed line* indicates the observed tsunami arrival time of 1534 UTC

from a M_w 8.8 Cascadia rupture scenario, and DENGLER and USLU'S (2011) prediction of velocities exceeding 2 m/s at the entrance to the small boat basin produced by the 1993 Sanriku and 2006 Kuril Islands tsunamis were exceeded by the velocities from the March 11, 2011 Japan tsunami in Crescent City. Several eye witnesses in Crescent City estimated currents on the order of 10–13 m/s (WILSON *et al.*, 2011), which is more than double the measured velocity from the security camera video footage. The video camera estimates of velocity illustrate how dubious eyewitness current estimates can be. They also illustrate the real value in using security cameras to estimate velocity due to their fixed location as well as the ability to capture footage in places where tsunami currents are known to be high and cause damage, increasing the risk of losing in situ instruments. However, in many security monitoring situations, the video footage is routinely reused within a day or two of the recording. It is important to contact all sources of such footage as soon as possible in order to preserve the data. Also, security camera managers should be encouraged to use high-speed capture rates and time-stamping.

The frequency signature and the general pattern of current amplification and decay in Humboldt Bay (2010 and 2011) and Crescent City (2011) are reasonably replicated by the MOST model during the first several hours after the tsunami arrival. At both locations, MOST underestimates peak current velocities by about 10–30 %, with a few peaks by as much as 50 %. Crescent City saw overestimations one and one-half hours after the tsunami arrival. In Humboldt Bay, MOST predicts attenuation of the signal after 4 h, when the observed currents continue to fluctuate within a nearly constant range for considerably longer.

Since the initiation of the pilot project in 2009 in Humboldt Bay, three tsunamis have been recorded on the Humboldt Bay tide gauge, two of which were recorded by the ADCP. The 2012 Haida Gwaii tsunami was observed in the tide gauge record (with a peak water level of 0.12 m) but was not detectable in the ADCP data. The 2010 Chile tsunami (0.23 m peak water level), produced a small velocity signal in Humboldt Bay with maximum velocities reaching 0.35 m/s. This suggests a tsunami-based change in water level on the order of 0.2 m is required in order to obtain a detectable tsunami current for this ADCP and this location.

This project is the first phase of a larger one. Four additional ADCPs have been installed in Humboldt Bay and offshore of the bay mouth as part of the Physical Oceanographic Real-Time System (PORTS) project (location shown in Fig. 1). An ADCP was installed near the NOAA NOS tide gauge at Crescent City in August 2013 to measure currents within the main channel near the entrance to the small boat basin. Finally, as a part of a statewide maritime hazard analysis and mapping project, the video analysis technique used in this paper is being applied to other coastal areas in California where security camera or amateur video footage exists. Better estimates of the currents generated by tsunamis can be used to improve numerical modeling and to provide better understanding of the hazards in ports and harbors caused by currents.

Acknowledgments

We would like to thank Pacific Gas & Electric Company for providing the support to install the ADCP in Humboldt Bay. Thanks to David Hull, Adam Wagschal, Alan Bobillot and John Powell (Humboldt Bay, Conservation and Recreation District) for help with the mounting design, development and deployment of the ADCP as well as images taken of the deployment location on Fairhaven Terminal, Katrina Sigler and Mike Willcutt (Security Nation Properties) for assistance in establishing a deployment site, Jose Montoya (Humboldt State University) for instrument maintenance, Richard Alvarez and Steve Monk (Humboldt State University) for diving support, Alan Winogradov and Richard Young in Crescent City for the security camera video footage, and Karen Earwaker (NOAA NOS Center for Operational Oceanographic Products and Services) for discussions regarding tidal harmonics in Humboldt Bay.

REFERENCES

ABRAMSON, H.F., (1998), Evidence for tsunamis and earthquakes during the last 3500 years from Lagoon Creek, a coastal freshwater marsh, northern California, M.S. thesis, Humboldt State University, Arcata, California.

ADMIRE, A.R., (2013), Observed and modeled tsunami current velocities on California's north coast, M.S. thesis, Humboldt State University, Arcata, California.

BORRERO, J.C., BELL, R., CSATO, C., DELANGE, W., GORING, D., GREER, S.D., PICKETT, V., and POWER, W., (2012), Observations, effects and real time assessment of the March 11, 2011 Tohoku-oki tsunami in New Zealand, Pure Appl. Geophys. doi:10.1007/s00024-012-0492-6.

BORRERO, J.C., and GREER, S. D., (2012), Comparison of the 2010 Chile and 2011 Japan Tsunamis in the far field, Pure Appl. Geophy. doi:10.1007/s00024-012-0559-4.

BRICKER, J. D., MUNGER, S., PEQUIGNET, C., WELLS, J.R., PAWLAK, G., CHEUNG, K.F., (2007), ADCP observations of edge waves off Oahu in the wake of the November 2006 Kuril Islands tsunami, Geophys. Res. Lett. 34(23), L23617.

CHOOWONG, M., MURAKOSHI, N., HISADA, K., CHARUSIRI, P., CHAROENTITIRAT, T., CHUTAKOSITKANON, V., JANKAEW, K., KANJANAPAYONT, P., and PHANTUWONGRAJ, S., (2008), 2004 Indian Ocean tsunami inflow and outflow at Phuket, Thailand, Marine Geology 248, 179–192.

DENGLER, L.A., and MAGOON, O., (2005), The 1964 Tsunami in Crescent City, California: The 40-year retrospective, Proceedings from Solutions to Coastal Disasters, American Society of Civil Engineers, 639–648.

DENGLER, L., USLU, B., BARBEROPOULOU, A., BORRERO, J., and SYNOLAKIS, C., (2008), The vulnerability of Crescent City, California to tsunamis generated by earthquakes in the Kuril Islands region of the northwestern Pacific, Seismological Research Letters 79(5), 608–619.

DENGLER, L., USLU, B., BARBEROPOULOU, A., YIM, S.C., and KELLY, A., (2009), Tsunami damage in Crescent City, California from the November 15, 2006 Kuril event, Pure Appl. Geophys. 166(1–2), 37–53.

DENGLER, L., ADMIRE, A., CRAWFORD, G., USLU, B., MONTOYA, J., and WILSON, R., (2011), Observed and modeled tsunami current velocities on California's north coast, [abstract] International Union of Geodesy and Geophysics Meeting 2011, 28 June–7 July.

DENGLER, L., and USLU, B., (2011), *Effects of harbor modifications on Crescent City, California's tsunami vulnerability*, Pure Appl. Geophys. *168*(6–7), 1175–1185.

DONOHO, D.L., and JOHNSTONE, I.M., (1994), *Ideal spatial adaptation by wavelet shrinkage*, Biometrika, *81*, 425–455.

EMERY, W.J., and THOMSON, R.E., (2004), Data Analysis Methods in Physical Oceanography (2nd edition): Amsterdam, The Netherlands, Elsevier B.V., 638 p.

FRITZ, H.M., PETROFF, C.M., CATALAN, P.A., CIENFUEGOS, R., WINCKLER, P., KALLIGERIS, N., WEISS, R., BARRIENTOS, S.E., MENESES, G., VALDERAS-BERMEJO, C., EBELING, C., PAPADOPOULOS, A., CONTRERAS, M., ALMAR, R., DOMINGUEZ, J.C., and SYNOLAKIS, C.E., (2011), *Field survey of the 27 February 2010 Chile tsunami*, Pure Appl. Geophys. *168*, doi:10.1007/s00024-011-0283-5.

FRITZ, H. M., PHILLIPS, D.A., OKAYASU, A., SHIMOZONO, T., LIU, H., MOHAMMED, F., SKANAVIS, V., SYNOLAKIS, C.E., and TAKAHASHI, T., (2012), *The 2011 Japan tsunami current velocity measurements from survivor videos at Kesennuma Bay using LiDAR*, Geophys. Res. Lett. *39*, doi:10.1029/2011GL050686.

GICA, E., SPILLANE, M., TITOV, V.V., CHAMBERLIN, C.D., and NEWMAN, J.C., (2008), Development of the forecast propagation database for NOAA's Short-term Inundation Forecast for Tsunamis (SIFT), NOAA Tech. Memo., OAR PMEL-139, NTIS: PB2008-109391, 89 pp.

JAFFE, B., BUCKLEY, M., RICHMOND, B., STROTZ, L., ETIENNE, S., CLARK, K., WATT, S., GELFENBAUM, G., and GOFF, J., (2011), *Flow speed estimated by inverse modeling of sandy sediment deposited by the 29 September 2009 tsunami near Satitoa, east Upolu, Samoa*, Earth-Science Reviews *107*(1–2), 23–37.

LACEY, J.R., RUBIN, D.M., and BUSCOMBE, D., (2012), *Currents, drag, and sediment transport induced by a tsunami*, J. Geophys. Res. *117*, C09028, doi:10.1029/2012JC007954.

LAMBERSON, R.H., GRIMES, S., and SCARR, D., (1998), A tsunami simulation and shoreline inundation model for Humboldt Bay, pilot study, Technical report, Report to PG&E Geosciences Department.

LYNETT, P.J., BORRERO, J.C., WEISS, R., SON, S., GREER, D., and RENTERIA, W., (2012), *Observations and modeling of tsunami-induced currents in ports and harbors*, Earth and Planetary Science Letters 327–328, 68–74.

NATIONAL GEOPHYSICAL DATA CENTER (NGDC), (2013), NOAA/WDC Historical Tsunami Database, Boulder, Colorado, http://www.ngdc.noaa.gov/hazard/tsu.shtml.

NORTEK AS, (2008), Aquadopp current profiler user guide: Nortek AS, Rud, Norway, 83 p.

PAWLOWICZ, R., BEARDSLEY, B., and LENTZ, S., (2002), *Classical tidal harmonic analysis including error estimates in MATLAB using T_TIDE*, Computers and Geosciences 28, 929–937, Available from: http://champs.cecs.ucf.edu/Library/Journal_Articles/pdfs/matlab_t_tide.pdf.

SHIMAMOTO, T., TSUTSUMI, A., KAWAMOTO, E., MIYAWAKI, M., and HIROSHI, S., (1995), *Field survey report on tsunami disasters caused by the 1993 Southwest Hokkaido earthquake*, Pure Appl. Geophys. *144*(3–4), 665–691.

SYNOLAKIS, C.E., BERNARD, E.N., TITOV, V.V., KÂNOĞLU, U., and GONZALEZ, F.I., (2008), *Validation and verification of tsunami numerical models*: Pure Applied Geophysics, v. *165*, no. 11–12, p. 2197–2228.

TANG, L., TITOV, V.V., and CHAMBERLIN, C.D., (2009), *Development, testing, and applications of site-specific tsunami inundation models for real-time forecasting*, J. Geophys. Res, *114*(C12), C12025.

TANG, L., TITOV, V.V., BERNARD, E., WEI, Y., CHAMBERLIN, C., NEWMAN, J.C., MOFJELD, H., ARCAS, D., EBLE, M., MOORE, C., USLU, B., PELLS, C., SPILLANE, M.C., WRIGHT, L.M., and GICA, E., (2012), *Direct energy estimation of the 2011 Japan tsunami using deep-ocean pressure measurements*, J. Geophys. Res. *117*, C08008, doi:10.1029/2011JC007635.

TIDES and CURRENTS, (2013), NOAA NOA/CO-OPS, Silver Spring, Maryland, http://tidesandcurrents.noaa.gov/index.shtml.

TITOV, V., and GONZALEZ, F., (1997), Implementation and testing of the method of splitting tsunami (MOST) model, NOAA Technical Memorandum ERL PMEL-112.

TITOV, V., and SYNOLAKIS, C.E., (1998), *Numerical modeling of tidal wave runup*, J. Waterw., Port, Coastal, Ocean Eng. *124*, 157-171.

TSUTSUMI, A., SHIMAMOTO, T., KAWAMOTO, E., and LOGAN, J., (2000), *Nearshore flow velocity of Southwest Hokkaido earthquake tsunami*, J. Waterw., Port, Coastal, Ocean Eng. *126*(3), 136–143.

USLU, B., BORRERO, J.C., DENGLER, L.A., and SYNOLAKIS, C.E., (2007), *Tsunami inundation at Crescent City, California generated by earthquakes along the Cascadia Subduction Zone*, Geophys. Res. Lett. *34*, L20601, doi:10.1029/2007GL030188.

WEI, Y., BERNARD, E., TANG, L., WEISS, R., TITOV, V., MOORE, C., SPILLANE, M., HOPKINS, M., and KÂNOĞLU, U., (2008), *Real-time experimental forecast of the Peruvian tsunami of August 2007 for U.S. coastlines*, Geophys. Res. Lett. *35*, L04609, doi:10.1029/2007GL032250.

WHITMORE, P.M., (1993), *Expected tsunami amplitudes and currents along the North American coast for Cascadia Subduction Zone earthquakes*, Natural Hazards 8, 59–73.

WILSON, R.I., DENGLER, L.A., LEGG, M.R., LONG, K., and MILLER, K.M., (2010), *The 2010 Chilean Tsunami on the California Coastline*, Seismol. Res. Lett. *81*(3), 545–546.

WILSON, R.I., DENGLER, L.A., GOLTZ, J.D., LEGG, M.R., MILLER, K.M., RITCHIE, A., and WHITMORE, P.M., (2011), *Emergency Response and Field Observation Activities of Geoscientists in California (USA) during the September 29, 2009, Samoa Tsunami*, Earth-Sci Rev doi:10.1016/j.earscirev.2011.01.010.

WILSON, R. I., ADMIRE, A.R., BORRERO, J.C., DENGLER, L.A., LEGG, M.R., LYNETT, P., McCRINK, T.P., MILLER, K.M., RITCHIE, A., STERLING, K., and WHITMORE, P.M., (2012a), *Observations and impacts from the 2010 Chilean and 2011 Japanese tsunamis in California (USA)*, Pure Appl. Geophys. doi:10.1007/s00024-012-0527-z.

WILSON, R.I., DAVENPORT, C., and JAFFE, B., (2012b), *Sediment scour and deposition within Harbors in California (USA), Caused by the March 11, 2011 Tohoku-oki Tsunami*, Journal of Sedimentary Geology, *282*, 228–240.

YOUNG, R., (2013), Personal Communication (101 Citizen's Dock Road, Crescent City, CA 95531).

(Received March 29, 2013, revised January 24, 2014, accepted February 10, 2014, Published online February 25, 2014)

Reprinted from the journal

Pure Appl. Geophys. 171 (2014), 3405–3419
© 2013 Springer Basel
DOI 10.1007/s00024-013-0731-5

Excitation of Basin-Wide Modes of the Pacific Ocean Following the March 2011 Tohoku Tsunami

Mohammad Heidarzadeh[1,2] and Kenji Satake[2]

Abstract—This study is an attempt towards understanding the sources of long oscillations observed within the Pacific Ocean following the 11 March 2011 Tohoku earthquake. We present evidence that extremely long modes of the Pacific Ocean in the range of 2–48 h were excited by this giant tsunami. A numerical approach was employed to calculate the basin-wide modes of the Pacific Ocean, resulting in 49 modes in the range of 2–48 h. We studied 15 tide-gauge records around the Pacific Ocean in order to extract basin-wide modes of the Pacific Ocean excited by this transoceanic tsunami. Spectral analysis of these tide-gauge records showed that some of the calculated basin-wide modes were indeed excited by the Tohoku tsunami. The observed modes ranged from 2 to 49.8 h. We attributed the long oscillations of the Pacific Ocean during the 2011 Tohoku tsunami to the excitation of these basin-wide modes, which can be grouped into global modes (15–48 h) and regional modes (2–15 h). We classified the signals on the tide gauges into three groups: (1) basin-wide modes (>1.5 h), (2) the tsunami source periods (20–90 min), and (3) local bathymetric effects (<20 min). The average contributions to the total tsunami energy were 6.4 % for the basin-wide mode, 64.1 % for the tsunami source, and 29.5 % for the local bathymetry, although the ratios varied from station to station. Simulations suggest that the amount of contribution of basin effects to the total tsunami energy depends on the location of the tsunami source.

Key words: 11 March 2011 Tohoku earthquake, Pacific Ocean, free oscillation, spectral analysis, basin-wide mode, numerical modeling.

1. Introduction

On 11 March 2011, tsunami waves generated by an *Mw* 9.0 earthquake off the coast of northeast Japan destructively attacked the Pacific coast of Japan and affected almost all other coastlines bordering the Pacific Ocean (Fig. 1). The devastating waves, which ran up as high as about 40 m (Mori *et al.* 2012), caused extensive destruction and a death toll of nearly 20,000 in the near-field along the Pacific coast of Japan (Satake *et al.* 2013). With two casualties in the far-field, the March 2011 Tohoku tsunami was the first tsunami to produce far-field fatalities in the Pacific Ocean since the 1964 Alaskan tsunami.

Analysis of sea level records of this giant tsunami showed that it was associated with some unusual phenomena. According to Saito *et al.* (2013), sea level oscillations induced by this tsunami lasted for about 4–5 days in the Pacific Basin. Borrero *et al.* (2013) reported that the maximum wave height of this tsunami arrived 30–40 h after the first arrival of tsunami waves in some tide-gauge stations in the Pacific Basin. By using a numerical modeling approach to analyze the 2011 Tohoku tsunami, Heidarzadeh and Satake (2013a) reported several tsunami wave reflections from different coastlines within the Pacific Basin and explained their roles in long-lasting tsunami oscillation.

Given such long-lasting oscillations caused by the March 2011 Tohoku tsunami, the purpose of this research was to study the origins of these long oscillations. In general, several factors may contribute to the extended oscillations of a basin-wide tsunami, among which are: reflections of the tsunami waves from different coasts in the region (e.g., Satake *et al.* 1988a), tsunami wave scattering by submarine seamounts (e.g., Mofjeld *et al.* 2001), wave trapping in the shelf region (e.g., Rabinovich *et al.* 2011; Yanuma *et al.* 1998; Van Dorn 1984), excitation of the basin-wide modes of the Pacific Basin due to the tsunami (e.g., Satake *et al.* 1988b; Rabinovich 2009; Heidarzadeh *et al.* 2012), and harbor resonance at the location of tide gauges inside harbors (e.g., Zelt *et al.* 1990).

[1] Cluster of Excellence "The Future Ocean", Christian-Albrechts University of Kiel, Otto-Hahn-Platz 1, 24118 Kiel, Germany. E-mail: mheidarzadeh@geomar.de
[2] Earthquake Research Institute (ERI), The University of Tokyo, Tokyo, Japan.

Free oscillation is an important source of long-lasting oscillations and sloshing inside enclosed or semi-enclosed basins. This is typical of harbors and lakes. However, when the source area is large enough, like the one for the March 2011 Tohoku tsunami, it may excite the basin-wide modes of large basins such as the Pacific Basin. An example of this phenomenon is the free oscillations of the Sea of Japan due to the tsunami waves generated by the May 1983 earthquake (Mw 7.9). According to SATAKE et al. (1988a), this tsunami excited long periods of 208, 104 and 62 min, which were calculated to be the basin-wide modes of the Sea of Japan. CARBAJAL et al. (2002) reported resonant oscillations having a dominant period of 36 min inside the Manzanillo Lagoon during the October 1995 Manzanillo tsunami. Resonance oscillations were reported in the Marmara Sea due to the destructive Izmit earthquake and tsunami on August 1999 in this nearly enclosed basin (YALCINER et al. 2007). The September 2009 Samoa tsunami caused resonance oscillations that were responsible for localized damage along the Samoan Islands (ROEBER et al. 2010).

The Tohoku tsunami of March 2011 presents a unique opportunity to study the basin-wide modes of the Pacific Ocean that can only be excited by such extremely large earthquakes. To examine such a possibility, we first applied a numerical algorithm to estimate the basin-wide modes of the Pacific Basin. We then studied the spectral characteristics of 15 tide-gauge records of the March 2011 Tohoku tsunami (Fig. 1) to examine whether some of the calculated modes of the Pacific Basin are present in the spectra of the observed tide gauge records. A comparative study was performed to determine the contribution of free oscillations of the Pacific Basin to the total energy of the 2011 Tohoku tsunami.

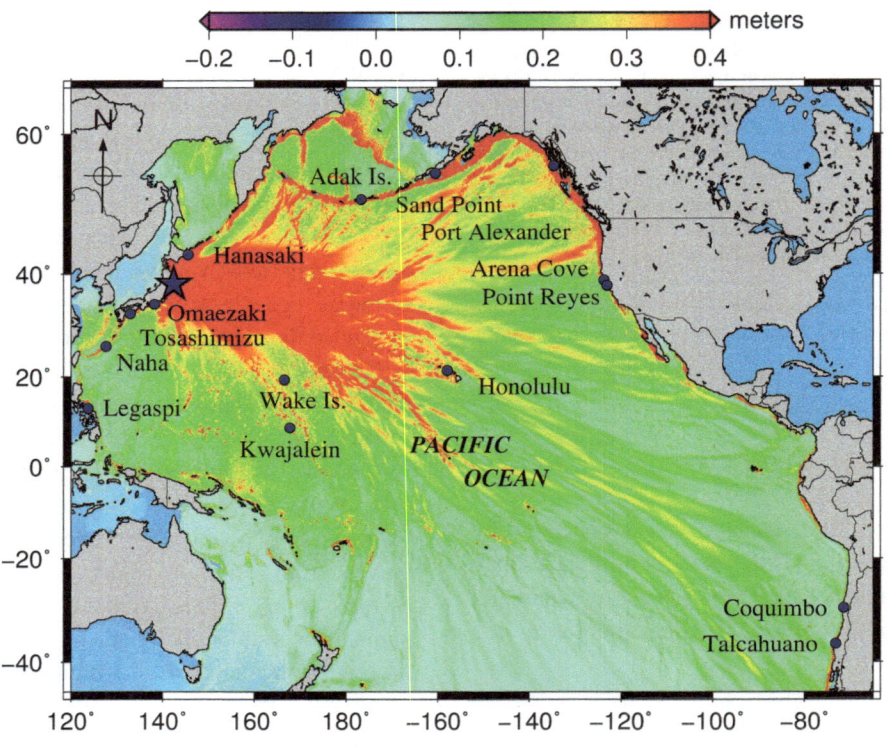

Figure 1

General location map of the Pacific Basin showing the epicenter of the 11 March 2011 Mw 9.0 large Tohoku earthquake (*asterisk*) and distribution of the maximum wave height of the resulting tsunami (modified from HEIDARZADEH and SATAKE 2013a). The *solid circles* show the locations of the tide-gauge stations used in this study

2. Calculation of Free Oscillations

The calculation of free oscillations of basins, either enclosed or semi-enclosed, has been studied by numerous ocean scientists using analytical, experimental and numerical methods (e.g., SATAKE et al. 1987; LEE 1971; ZELT et al. 1990; RAICHLEN et al. 1991; YAO 1999; YALCINER et al. 2007). The method used here for calculation of the free oscillation of the Pacific Basin is based on a numerical algorithm proposed by YALCINER et al. (2007), consisting of the following steps:

1. Disturbing the basin with an arbitrary initial wave
2. Simulating the water oscillation using a numerical model
3. Recording time histories of water level oscillations at different locations
4. Calculating spectra of the wave time histories recorded at different locations
5. Free oscillation modes are those periods that are common in most of the spectra

As shown in the following, first we examined the performance of the above algorithm on a rectangular test basin, before applying it to the Pacific Ocean.

2.1. Free Oscillations of a Rectangular Test Basin

The method described above for the calculation of free oscillations of basins was previously tested by YALCINER et al. (2007) for both enclosed and semi-enclosed basins, and was successfully applied to the Marmara Sea in Turkey. Despite this, to show the efficiency of the method, we applied this algorithm to a squared-shaped enclosed basin with a constant water depth of 100 m and a length of 5,000 m. For this test, the basin was disturbed with an initial rectangular crest wave whose length, initial height, and velocity were 300 m, 1 m and 0 m/s, respectively. The conditions of this test case were similar to those of tectonic tsunamis. Snapshots of wave propagation at different times along with spectral analysis for the time histories of waves at selected locations are shown in Fig. 2. A nonlinear, shallow-water numerical code known as TUNAMI was applied for modeling the propagation of the waves (YALCINER et al. 2004). For this test basin, the grid

spacing, simulation time-step and total simulation time were: 20 m, 0.2 s, and 5 h, respectively.

The periods of free oscillations of this test basin can be calculated using the following analytical equation (RABINOVICH 2009):

$$T_{mn} = \frac{2}{\sqrt{m^2 + n^2}} \frac{L}{\sqrt{gd}}, \qquad (1)$$

in which L is the basin's length, g is the gravitational acceleration, d is water depth, m, $n = 0, 1, 2...$ and T_{mn} is the mnth mode. Using this analytical equation, some first modes of free oscillations for this test basin are: 319.3, 159.6, 142.8, 112.9, 106.4, 88.6, 79.8, 77.4, 71.4, 63.9, 59.3, and 53.2 s. Based on Fig. 2, the employed numerical approach was successful in reproducing all of the theoretical values. Figure 2 also shows that the results (i.e., spectral peaks) were not sensitive to the location of the numerical gauges, though the level of spectral power varied from one location to another.

2.2. Free Oscillations of the Pacific Basin

Based on the algorithm presented above, the Pacific Basin was agitated by two initial disturbances consisting of a half-spherical source with a maximum initial height of 8 m at the center, and a uniform rectangular crest wave with an initial height of 8 m. These initial waves were placed at two different locations: one at the center (source 1, Fig. 3) and the other at the northwestern corner of the Pacific Ocean in the vicinity of the actual source of the 2011 Tohoku tsunami (source 2, Fig. 3). The second source seems slightly similar to the actual source of the 2011 Tohoku tsunami. A nonlinear, shallow water numerical code, known as TUNAMI, was applied for modeling the propagation of the generated waves across the Pacific Ocean (YALCINER et al. 2004). An 8-min bathymetric grid extrapolated from the 1-min GEBCO digital atlas (IOC et al. 2003) was used, resulting in a grid of around 15 km × 15 km. The time-step for the numerical modeling was 10 s and simulations were conducted for a total time of 10 days for each case.

Figure 3 presents snapshots of wave simulations at different times. Time histories of sea level oscillations were recorded at 30 numerical wave

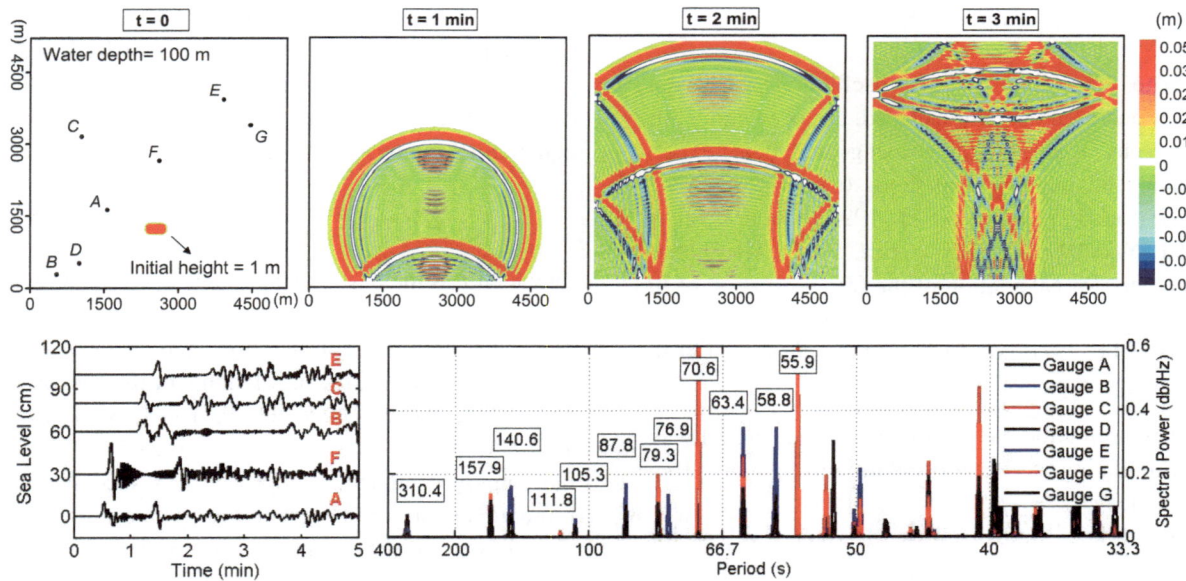

Figure 2

Top snapshots of wave propagation at different times in a square-shaped, enclosed basin with a constant water depth of 100 m, excited by an initial rectangular crest wave with an initial height of 1 m. *Bottom* time histories of wave oscillations and spectral analysis of wave time histories at selected numerical gauges. The locations of the numerical gauges A–G are shown in the *top panel*

gauges. Numerical gauges 1–15 were located at the same locations of the actual tide gauges shown in Fig. 1. The locations of the numerical wave gauges were chosen with the purpose of recording sea level oscillations that were later used for spectral analysis. We show below that this method is not sensitive to the locations of the gauges within the basin.

Results of spectral analysis for these computational sea level records are presented in Fig. 4. The spectra are shown for the periods larger than 1.5 h since basin-wide periods are normally long, of the order of several hours. Figure 4 suggests that the basin-wide peak periods were almost similar for both sources, though the amount of spectral power was different for the two sources. For example, Fig. 4 shows that the spectral powers for periods more than 7 h were larger for source 1 compared to those for source 2 (note that the vertical scales are different in Fig. 4a, b). It seems natural that the amount of energy in a particular period is a function of the location of the initial source. For source 1, because the initial disturbance was located in the middle of the basin, it seems to have been more efficient in exciting basin-wide modes. However, it needs to be noted that only the peak periods are important in this method and the

amount of spectral power in a particular peak period is not important. In fact, in this method, we are interested in discovering all of the available oscillation modes for a basin; it does not matter if those modes are strong or weak.

Different parts of the Pacific Basin may have their own natural periods because of the irregular geometrical shape of the Pacific Basin and the presence of many islands, seamounts, and other aerial or submarine barriers within the Pacific Ocean (Fig. 1). For example, the Hawaiian Islands reflect back part of the tsunami waves generated in offshore Japan, and hence an oscillation may be generated between Japan and Hawaii (USA). Such reflections were evidenced by basin-wide simulations of the March 2011 Tohoku tsunami (NOAA 2013). According to HEIDARZADEH and SATAKE (2013a), the waves generated by the March 2011 Tohoku tsunami arrived in Hawaii 7–8 h after the earthquake, indicating a period of around 14–16 h for the oscillations between Japan and Hawaii. With similar reasoning, we may expect a period of around 18–20 h for the oscillations between California (USA) and Japan because the March 2011 tsunami arrived in California 9–10 h after the earthquake. Although this estimation is rather simplistic, it gives some clues

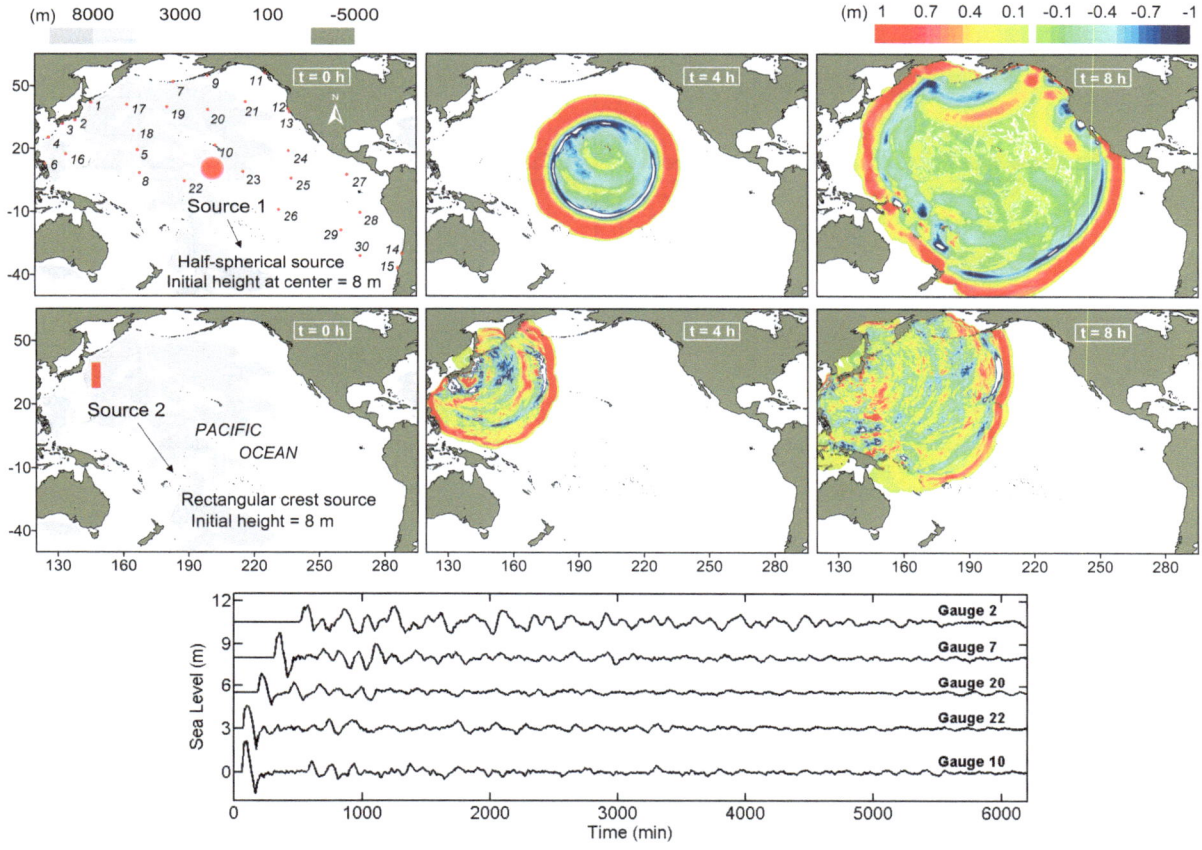

Figure 3
Excitation of the Pacific Basin using a semi-spherical initial source (*top*) and a linear crest initial source (*middle*), and snapshots of the wave evolution at different times. The numbers on the *top–left panel* show the locations of numerical wave gauges. *Bottom* time histories of wave oscillations at selected numerical wave gauges due to source 1

about the order of the modes expected from such a large basin. It is clear that we may expect several modes for the entire Pacific Ocean.

Based on the spectra presented in Fig. 4 for two different sources at 30 different locations, the most common spectral peaks can be considered as the basin-wide modes of the Pacific Ocean. In total, we found 49 basin-wide modes for the entire Pacific Basin, ranging between 2 and 48 h. These modes are: 48.0, 40.0, 34.0, 30.0, 26.7, 24.0, 21.8, 20.0, 15.0, 16.0, 14.1, 13.3, 12.6, 10.9, 10.4, 9.6, 9.2, 8.9, 8.0, 7.7, 7.5, 6.9, 6.7, 6.3, 6.0, 5.9, 5.7, 5.5, 5.3, 5.2, 4.9, 4.4, 4.1, 3.9, 3.8, 3.5, 3.4, 3.3, 3.2, 3.1, 3.0, 2.9, 2.8, 2.7, 2.6, 2.5, 2.4, 2.3, 2.2, 2.1, and 2.0 h. As shown in Fig. 4, the modes above 10 h are relatively weaker than those below 10 h, which seems natural because such long cycles occur only few times during the

whole life of wave oscillations. The longest calculated oscillation mode of the Pacific Ocean is 48 h. Through numerical modeling of the March 2011 Tohoku tsunami, SAITO et al. (2013) presented evidence that this extremely long mode was excited during the aforesaid giant tsunami. Figure 4 shows that many spectral peaks are almost the same at different locations, indicating that the results are not influenced significantly by the locations of the numerical gauges. Hence, these locations can be selected arbitrarily.

It seems difficult to precisely determine the origins of each of the basin-wide modes presented above. However, observations from past tsunamis may give us some insights into their origins. In this context, we may classify the basin-wide modes presented above into two groups: regional (2–15 h)

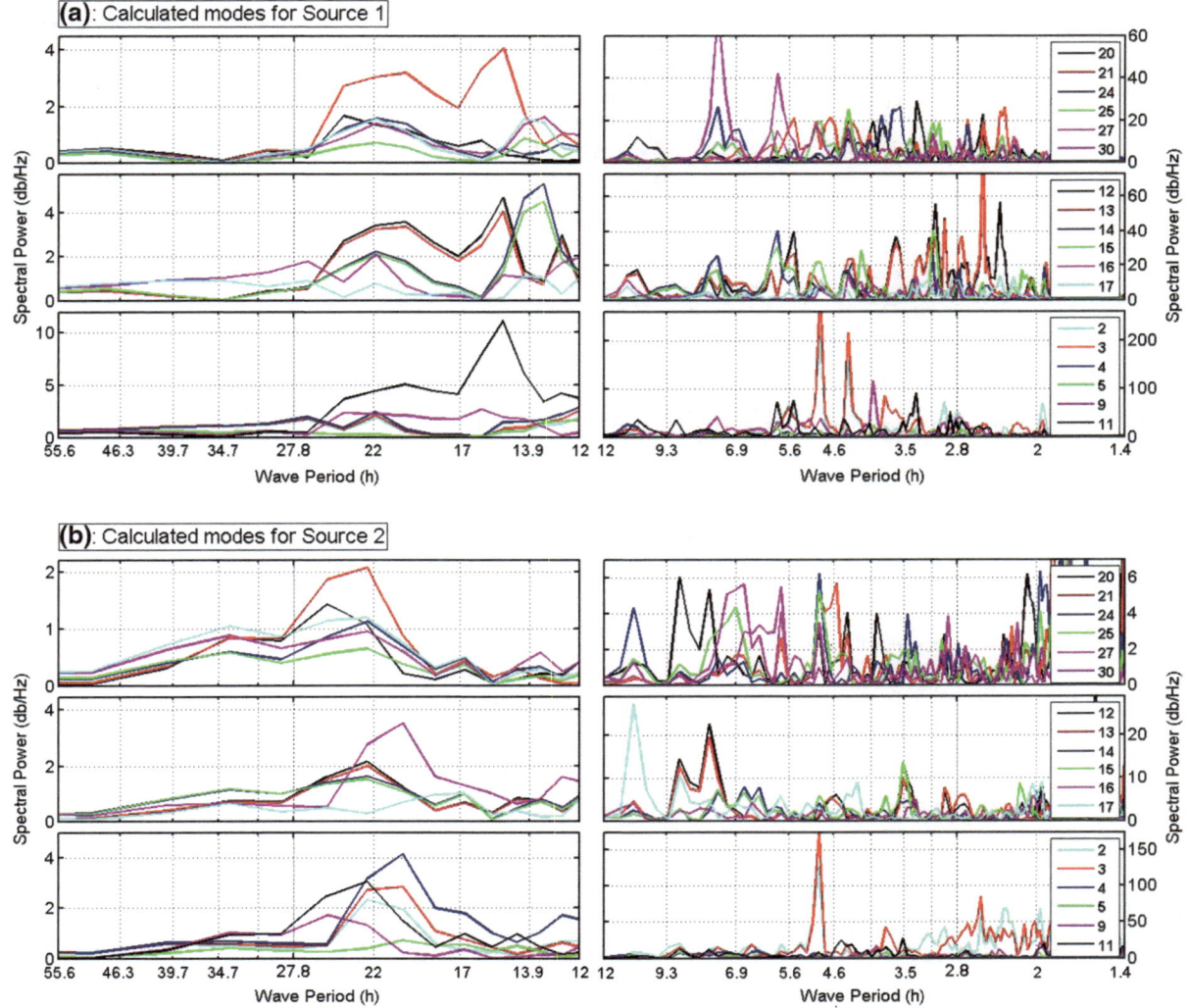

Figure 4
Results of spectral analysis for the wave records of the numerical wave gauges due to a semi-spherical source at the center (**a**) and a linear crest source at the northwest corner of the Pacific Ocean (**b**). Note that the *vertical scales* are different between (**a**) and (**b**)

and global modes (15–48 h). Regional modes are those that belong to different parts of the Pacific Basin, such as the region between Japan and Hawaii, as discussed earlier. Regional modes may have oscillation periods of around 7–15 h. Some other modes in this category with periods in the range of 2–7 h seem not to be the results of reflections from the coasts. These modes are likely the results of wave scattering and reflection from submarine seamounts and other bathymetric features (e.g., wave trapping in shelves). Intensive interactions of tsunami waves with ocean bathymetry are evidenced by basin-wide

simulations of the 2011 Tohoku and 2004 Sumatra tsunamis, suggesting that part of the waves are reflected back and scattered by submarine features before reaching the coastal areas. This is because of the nature of tsunamis, which is a series of long waves having significant interactions with the ocean bottom. Global modes are those belonging to the whole Pacific Basin and are the results of shore-to-shore oscillations. For example, the periods of 21.8 and 24.0 h are among the global modes that likely belong to the oscillations between the West Coast of the US and Japan. As another example, the modes in the range of 40–48 h are likely to represent the oscillations between the Japanese and South American coasts because the tsunami waves of the March 2011 Tohoku tsunami arrived in Peru and Chile 21–24 h after generation (HEIDARZADEH and SATAKE 2013a).

3. Analysis of the Observed Sea Level Data during the 2011 Tohoku Tsunami

3.1. Methodology

The methodology used here was based on the spectral analysis of the observed sea level records of the March 2011 Tohoku tsunami in order to extract the existing signals and to examine whether or not these signals are the basin-wide modes of the Pacific Basin. To remove the effects of astronomical tides from the sea level spectra, the sea level records were carefully de-tided before spectral analysis. In summary, our methodology consisted of the following steps, which will be briefly explained later:

1. Preparation of the sea level records of the March 2011 Tohoku tsunami,
2. Removing the tidal signal,
3. Spectral analysis of the de-tided signals,
4. Calculating the spectral ratio (tsunami spectra/ background spectra), and
5. Identifying basin-wide modes as those modes showing peaks in both spectral plots and spectral-ratio plots

The tide-gauge data used in this study were part of the data previously reported by HEIDARZADEH and SATAKE (2013a). The sea level data were provided through the National Oceanographic and Atmospheric Administration (NOAA, USA), and the UNESCO Intergovernmental Oceanographic Commission (IOC). The locations of these stations are shown in Fig. 1. The data are all of digital type and were sampled at time intervals of 1 min.

In their analysis of the sea level oscillations caused by the March 2011 Tohoku tsunami, HEIDARZADEH and SATAKE (2013a) simply filtered all of the signals having periods larger than 4–5 h from the sea level records. This practice is reasonable because their study was aimed at the detection of the tsunami source periods, which are normally <2 h. However,

the purpose of our study was to detect the basin-wide modes of the Pacific Basin that were possibly excited by the 2011 Tohoku tsunami. These modes are of the order of a few hours to 10 h and greater for a large basin like the Pacific Ocean. Therefore, all of the basin-wide modes could be removed from the sea level records by simple filtering. A band-pass filtering of the data would have been suitable, however, we chose a potentially more appropriate method of precisely calculating the tide using a sophisticated software, and then removing the tidal signal from the original records.

For predicting tidal signal, the tidal analysis package TASK (Tidal Analysis Software Kit, developed at the Proudman Oceanographic Laboratory, UK) was used (BELL et al. 2000). TASK, which has been widely used around the world for tidal analysis, is a collection of Fortran procedures providing full harmonic analysis of the observed tide-gauge data (HEIDARZADEH and SATAKE 2013b). In our tidal analysis, we used 55 major harmonic constituents in order to provide an accurate tidal prediction. The length of the tide-gauge data was 15 days, which means around 21,600 data points, taking into account a sampling interval of 1 min. Fast Fourier Transform (FFT) was used for spectral analysis in this study, for which the Matlab function FFT was applied (MATHWORKS 2013).

3.2. De-tiding of the Observed Signals

Figure 5 presents the de-tided signals of the March 2011 Tohoku tsunami for 15 tide-gauge records from across the Pacific Ocean. For all of the signals, the tide signal was precisely computed using the TASK software and then subtracted from the original signal. Based on Fig. 5, it can be seen that the tidal prediction was precise enough and was able to appropriately remove the tidal effects from the sea level records.

3.3. Averaged Root-Mean-Square (ARMS) of the Data

Different numbers have been reported for the duration of the March 2011 Tohoku tsunami within the Pacific Ocean, ranging from 3 to 6 days (e.g., SAITO et al. 2013; HEIDARZADEH and SATAKE 2013a;

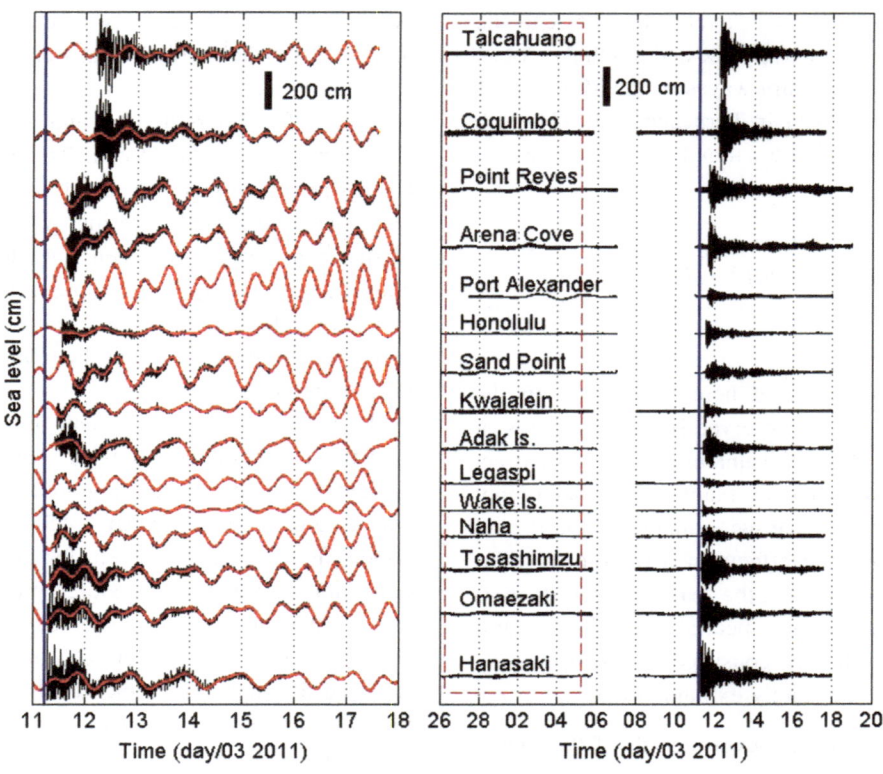

Figure 5

Left observed tide-gauge records of the March 2011 Tohoku tsunami along with the computed tidal signal (the *thick-red lines*). The locations of the tide-gauge stations are shown in Fig. 1. *Right* the de-tided tsunami records after subtracting the computed tidal signal from the original observed tide-gauge records. The *blue vertical line* represents the time of the earthquake occurrence. The *dashed-rectangle* shows the background part of the sea level data used for the calculation of the spectral ratio (color figure online)

BORRERO *et al.* 2013). These estimations were made using a visual look at the filtered tide-gauge records. However, to have a relatively more accurate estimation of the tsunami duration, we calculated the averaged root-mean-square (ARMS) for the 15 de-tided tsunami signals used in this study (Fig. 6a). We defined tsunami duration as the time interval between tsunami arrival (point A, Fig. 6a) and the time that sea level oscillations reached the level of oscillations before the tsunami arrival (point B, Fig. 6a). The results are shown in Fig. 6b, suggesting that the average duration of the tsunami waves was 4.9 days for the 15 tide-gauge records examined.

3.4. Observed Basin-wide Modes of the Pacific during the March 2011 Tsunami

Spectral analysis was performed on the 15 de-tided records of the March 2011 Tohoku tsunami

(Fig. 7a). Based on HEIDARZADEH and SATAKE (2013a), the source periods of this tsunami were in the range of 20–90 min. Figure 7a demonstrates that the signals originating from the tsunami source were the most powerful signals in all of the spectra. An exception is the spectrum for the Port Alexander station, in which the signal with the period of 38.5 h was more powerful than the tsunami source signals. However, it will be shown later that the signal at 38.5 h in Port Alexander was not associated with the tsunami and was possibly a background signal. It can be seen in Fig. 7a that a wide range of signals are available in the sea level spectra; from short-period waves of some minutes to waves with periods of some 10 h.

To examine whether the spectral peaks shown in Fig. 7a were tsunami-related signals or not, we calculated the spectral ratios (tsunami spectra/background spectra) for all of the stations (Fig. 7b). The concept behind spectral ratio (RABINOVICH 1997) is

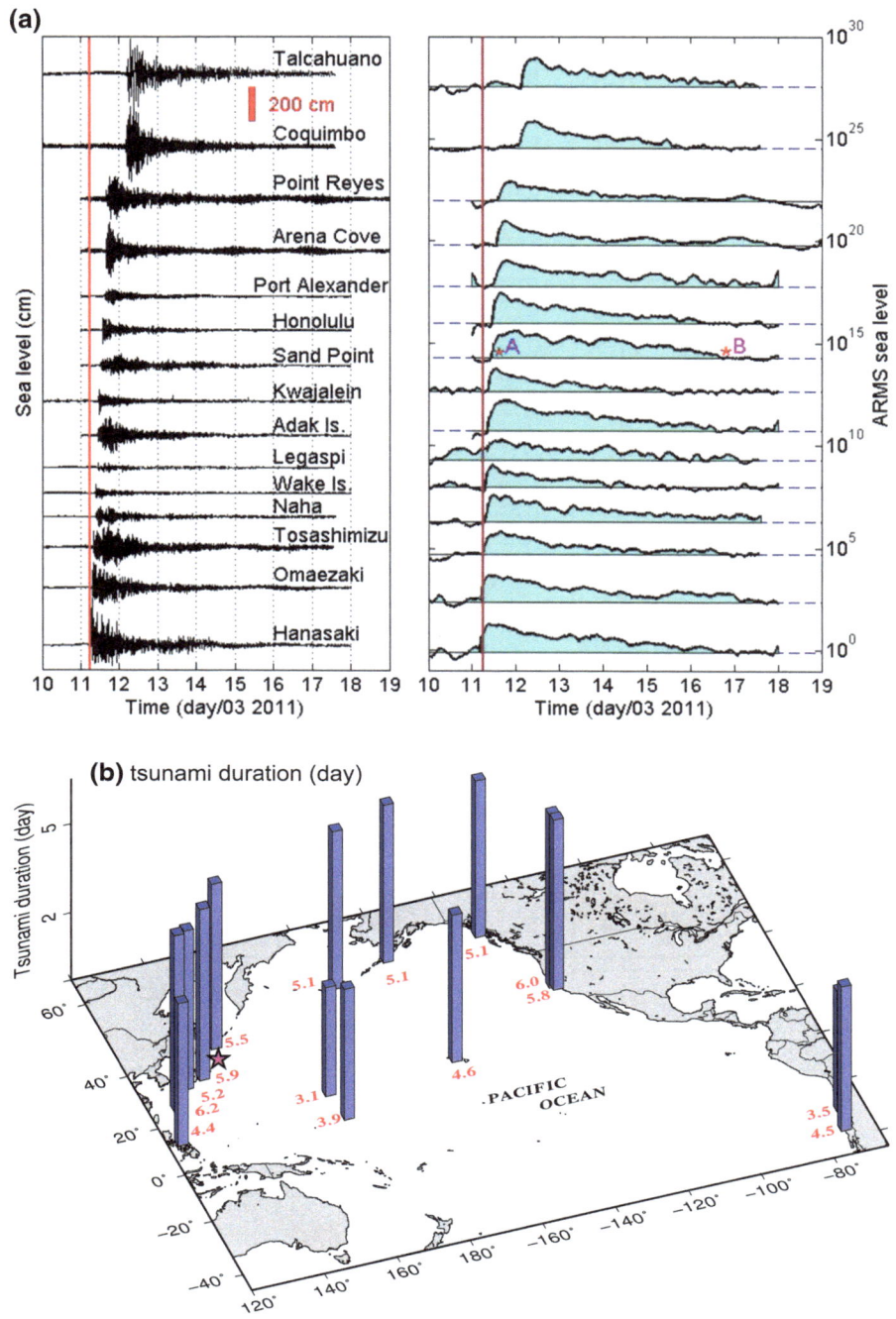

Figure 6

a, *Left panel* de-tided sea level oscillations of the March 2011 Tohoku tsunami, recorded at several tide-gauge stations in the Pacific. **a**, *Rght panel* the respective Averaged-Root-Mean-Square (ARMS) diagrams. The *red line* represents the earthquake time. The *blue-dashed lines* help to calculate tsunami duration. **b** Geographical distribution of the tsunami duration values across the Pacific Ocean (color figure online)

straightforward: if a signal was present in the basin before the tsunami occurrence, the spectral ratio does not show any peak at that frequency because that

signal is counteracted by a similar signal existing in the tsunami waveform. Spectral ratio is free from any local, regional, or global bathymetric effects, and thus

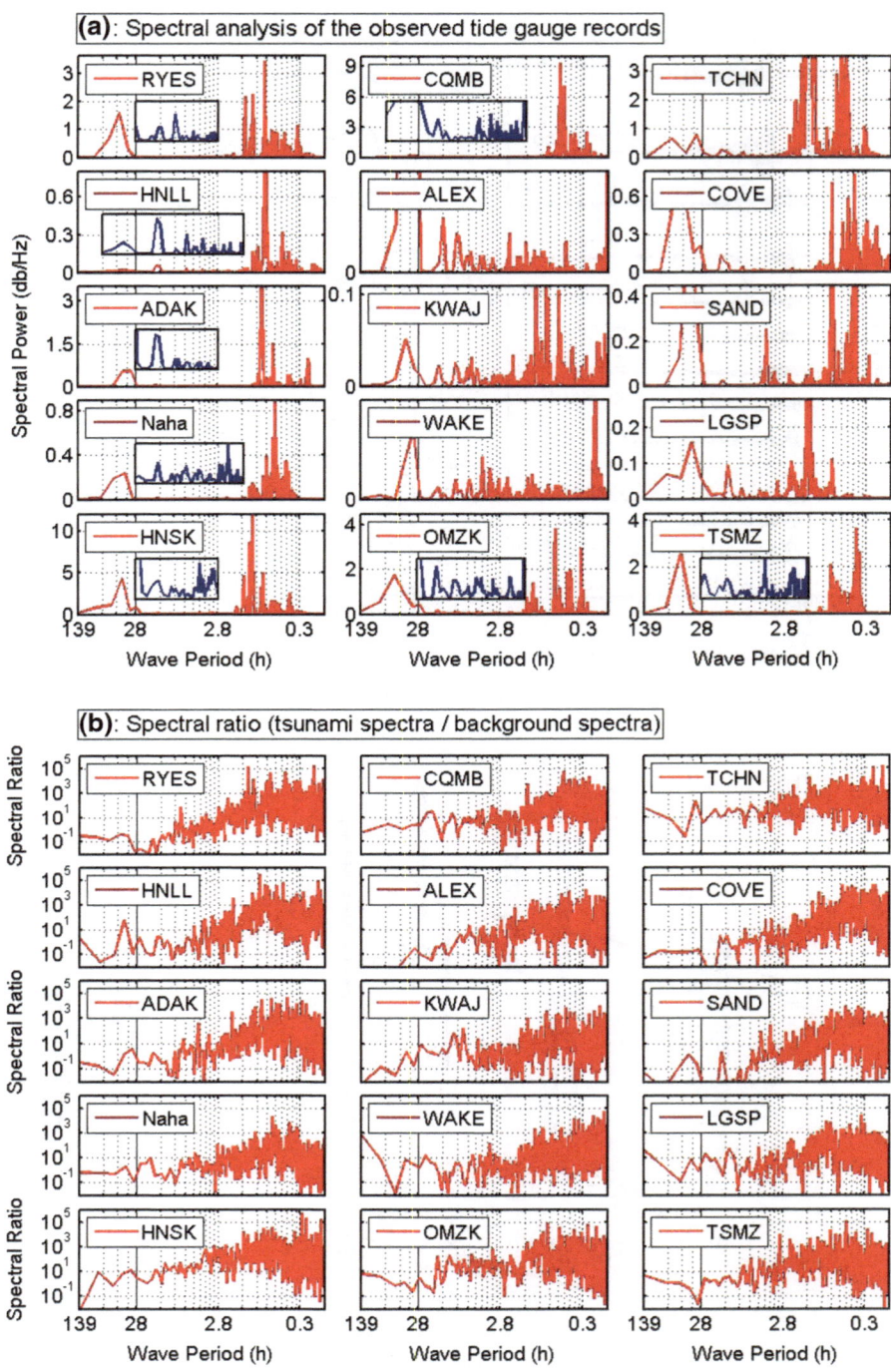

Figure 7

a Results of spectral analysis for actual observed tide-gauge records of the March 2011 Tohoku tsunami. The *insets* show part of the spectra with a better vertical resolution. **b** Spectral ratio (tsunami spectra/background spectra) for the tide-gauge records of the March 2011 Tohoku tsunami. *HNSK* Hanasaki, *OMZK* Omaezaki, *TSMZ* Tosashimizu, *WAKE* Wake Island, *LGSP* Legaspi, *ADAK* Adak Island, *KWAJ* Kwajalein, *SAND* Sand Point, *HMLL* Honolulu, *ALEX* Port Alexander, *COVE* Arena Cove, *RYES* Point Reyes, *CQMB* Coquimbo, *TCHN* Talcahuano

Table 1

Spectral peak periods of the tide-gauge records of the March 2011 Tohoku tsunami. The locations of the tide gauges are shown in Fig. 1. The periods are in hours

Tide-gauge stations

	HNSK	OMZK	TSMZ	Naha	WAKE	LGSP	ADAK	KWAJ	SAND	HNLL	ALEX	COVE	RYES	CQMB	TCHN
Peak periods	2.1	2.1	2.1	2.1	2.1	2.1	2.1	2.1	2.1	2.1	2.1	2.1	2.1	2.1	2.1
(h)	2.3	2.3	2.2	2.5	2.3	2.2	2.3	2.4	2.2	2.2	2.2	2.2	2.3	2.2	2.2
	2.5	2.5	2.3	2.6	2.4	2.3	2.4	2.5	2.4	2.4	2.3	2.5	2.5	2.4	2.3
	2.8	2.7	2.6	2.7	2.5	2.5	2.5	2.7	3.4	2.8	2.4	2.8	2.7	2.5	2.4
	3.1	2.9	2.8	2.9	2.6	2.8	2.7	3.0	3.6	3.1	2.7	2.9	3.0	2.8	2.5
	3.3	3.0	3.0	3.2	2.8	2.9	2.8	3.2	4.6	3.3	3.0	3.0	3.2	3.0	2.7
	3.4	3.2	3.3	3.6	2.9	3.4	3.0	3.3	5.0	3.7	3.1	3.5	3.4	3.2	2.8
	3.6	3.4	3.8	3.8	3.0	4.2	3.2	3.7	6.5	3.9	3.4	3.8	3.7	3.4	2.9
	3.8	3.7	4.3	4.2	3.2	4.6	3.4	4.3	7.1	4.3	3.7	4.5	5.0	3.7	3.2
	4.3	3.8	4.7	4.5	3.5	4.9	3.7	4.6	9.7	5.2	4.2	4.8	5.5	4.7	3.8
	4.6	4.2	5.0	4.8	3.8	5.5	4.6	4.9	15.5	5.5	4.5	5.2	6.8	5.2	4.3
	5.2	4.7	5.3	5.3	4.3	7.4	4.9	5.4	38.9	6.7	5.0	6.5	7.7	6.9	5.2
	5.8	5.4	7.5	6.8	4.7	9.2	6.6	6.3		9.1	5.7	7.4	9.3	8.7	6.8
	6.4	5.8	8.3	8.8	5.0	13.4	7.5	7.1		15.4	6.4	8.8	14.7	10.0	8.6
	7.3	6.2	9.3	10.0	5.5	16.3	8.3	7.8		38.7	7.3	10.4	18.5	13.0	10.0
	9.0	6.7	13.6	14.9	6.6	36.8	9.3	9.8			9.6	16.1	44.0	16.3	16.2
	10.1	7.7	16.6	24.9	8.4		15.8	15.7			14.0			32.6	32.4
	14.7	10.1	24.9	37.4	9.9		31.5	39.3							
	27.0	13.4	49.8		13.2		39.4								
	40.3	16.1			15.9										
		27			31.7										

HNSK Hanasaki, *OMZK* Omaezaki, *TSMZ* Tosashimizu, *WAKE* Wake Island, *LGSP* Legaspi, *ADAK* Adak Island, *KWAJ* Kwajalein, *SAND* Sand Point, *HMLL* Honolulu, *ALEX* Port Alexander, *COVE* Arena Cove, *RYES* Point Reyes, *CQMB* Coquimbo, *TCHN* Talcahuano

is an ideal criterion for identifying tsunami-related signals. The concept of spectral ratio was successfully applied by V<small>ICH</small> and M<small>ONSERRAT</small> (2009) to the sea level data of the May 2003 tsunami in the western Mediterranean Sea.

We refined the spectra presented in Fig. 7a using the results of the spectral ratios (Fig. 7b). Part of the sea level data used as the background signal is shown with a dashed-rectangle in Fig. 5, and the results of the spectral-ratio calculations are shown in Fig. 7b. For this refinement, we ignore the spectral peaks in Fig. 7a, for which no peaks were available in the spectral-ratio plots (Fig. 7b). In general, most of the observed peaks in the spectra (Fig. 7a) also showed a peak in the spectral-ratio plots (Fig. 7b). However, there were some exceptions. As an example, the large peak at the period of 38.5 h in Port Alexander did not repeat at the spectral-ratio plot for this station, indicating that it was not likely to be associated with the tsunami. As other examples, the peak periods of 64.8 and 58.9 h, observed at the spectra of Talcahuano

and Arena Cove, respectively, do not show any peak at the spectral-ratio plots of these stations and thus were considered as non-tsunami signals.

Table 1 presents all of the signals with periods longer than 2 h available in the tsunami spectra, shown in Fig. 7a after refinement using spectral-ratio plots. We assume that these signals are those that were excited by the 2011 Tohoku tsunami in the Pacific Ocean for two reasons: first, the tidal signals have been carefully removed from the sea level oscillations. Second, other possible non-tsunami signals have been refined from the sea level data using the concept of spectral ratio. Some of these signals were directly excited by the tsunami source (those with periods in the range of 20–90 min), some of them were the results of regional and global oscillations, and others were generated by scattering and reflections from bathymetric features, e.g., wave trappings in shelves and harbor resonance. It can be seen that some of the calculated modes (Sect. 2.2) were observed during the March 2011 Tohoku

(a) Spectral peaks for all 15 tide gauge records

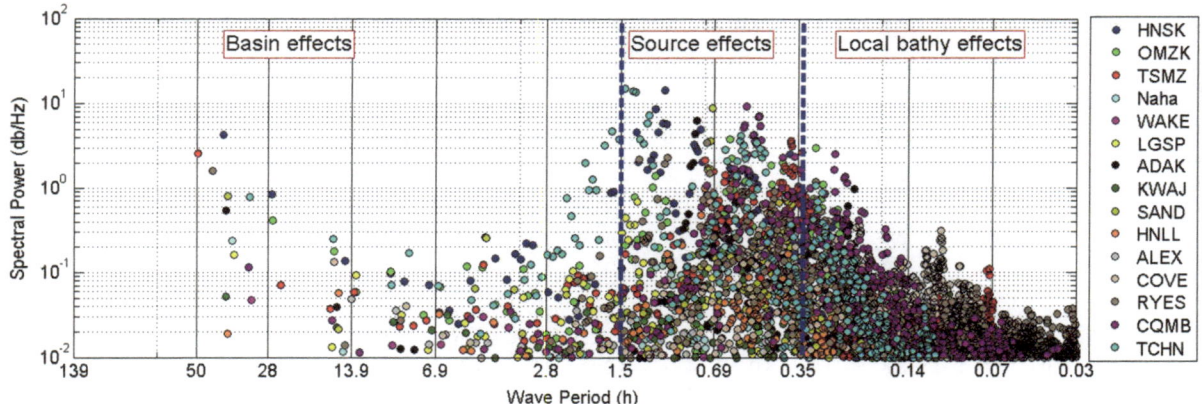

(b) Contribution of different effects on total tsunami energy

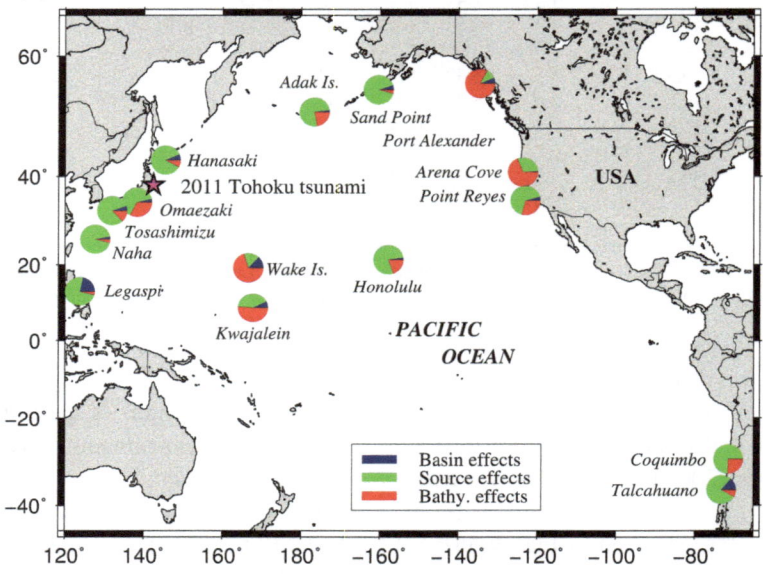

Figure 8

a Spectral peaks for all 15 tide-gauge records of the March 2011 Tohoku tsunami. The *thick-dashed lines* show the border between tsunami energy from local bathymetric effects, tsunami source, and basin effects. **b** Geographical distribution of contribution of different effects on total tsunami energy using *pie graphs*. *HNSK* Hanasaki, *OMZK* Omaezaki, *TSMZ* Tosashimizu, *WAKE* Wake Island, *LGSP* Legaspi, *ADAK* Adak Island, *KWAJ* Kwajalein, *SAND* Sand Point, *HMLL* Honolulu, *ALEX* Port Alexander, *COVE* Arena Cove, *RYES* Point Reyes, *CQMB* Coquimbo, *TCHN* Talcahuano

tsunami (Table 1). Based on Table 1, long modes ranging from 2 to 48 h were excited by the Tohoku tsunami.

4. Discussions

Figure 8a presents the spectral peaks for all 15 tide-gauge records of the March 2011 Tohoku tsunami. In general, the pattern of tsunami energy distribution over frequency domain is bell-shaped, with its maximum energy occurring at the period dictated by the tectonic source. According to HEI-DARZADEH and SATAKE (2013a), the main tsunami periods of the Tohoku tsunami were 37 and 67 min. Figure 8a shows that the maximum tsunami energy occurred around the period band of 30–70 min and then decreased in both sides.

Although it is difficult to classify different signals observed on a tide gauge into certain groups, we attempted to estimate the contribution of different effects to the total tsunami energy. We classified the spectral peaks shown in Fig. 8a as belonging to one of the three groups: (1) signals generated by basin effects with periods longer than 1.5 h, which include global modes (15–48 h) and regional modes (2–15 h), (2) signals generated by the main tsunami source with periods in the range of 20–90 min, and (3) signals generated by local bathymetric effects with periods shorter than 20 min. As an example, the spectral analysis of sea level records in Wake Island station (Fig. 7a) showed that the peak tsunami energy occurred at the period of around 10 min in this station, indicating that local bathymetric effects such as harbor resonance were responsible for that. In other words, some local modes were excited by the incoming tsunami waves and the amount of energy generated by these local modes through persistent oscillations was higher than that generated by direct tsunami waves.

To calculate the contribution of each of the above effects to the total tsunami energy, we summed the spectral powers for all signals lying in each category. Table 2 presents a summary of the contribution of the three effects to the total energy of the Tohoku tsunami for all 15 tide-gauge records. The geographical distribution of the data is shown in Fig. 8b using pie graphs. According to Table 2 and Fig. 8b, most of the tsunami energy came from its tectonic source, and the contributions by other effects to the total energy were relatively small. On the average, the contributions of basin effects, tsunami source, and local bathymetry to the total tsunami energy were 6.4, 64.1 and 29.5 %, respectively. The contribution of basin effects to the total tsunami energy ranged from 0.4 % in Coquimbo to 21.8 % in Legaspi. In most of the examined tide-gauge records, the contribution of basin effects was smaller than those of tsunami source and local bathymetry, which is expected due to the limited number of cycles passed by such long modes during the whole life of the Tohoku tsunami. For all records, the contribution of basin effects to the total tsunami energy was smaller than that of tsunami source. In three stations of Legaspi, Talcahuano and Sand Point, the contribution of basin effects to the total tsunami energy was larger than that of local effects.

Table 2

Contribution of local bathymetry, tsunami source, and basin effects to the total tsunami energy of the March 2011 Tohoku tsunami

Station name	Amount of energy (db/Hz)			Contribution to tsunami energy (%)		
	Basin effects[a] (db/Hz)	Tsunami[b] source (db/Hz)	Local bathy[c] (db/Hz)	Basin effects (%)	Tsunami source (%)	Local bathy (%)
Hanasaki	9.2	117.2	10.7	6.7	85.5	7.8
Omaezaki	2.1	36.8	18.8	3.7	63.7	32.6
Tosashimizu	3.8	59.5	9.2	5.2	82.1	12.7
Naha	0.4	13.4	0.6	2.8	92.9	4.2
Wake Is.	0.3	0.4	1.5	13.1	16.6	70.3
Legaspi	1.1	3.6	0.2	21.8	74.3	3.9
Adak Is.	0.7	25.4	7.5	2.1	75.6	22.3
Kwajalein	0.4	2.3	2.8	7.4	41.3	51.3
Sand Point	1.4	24.6	1.1	5.2	90.8	4.1
Honolulu	0.3	9.9	2.5	2.3	78.1	19.6
Port Alexander	0.3	0.5	4.0	6.5	10.7	82.8
Arena Cove	0.4	13.1	30.1	0.9	30.1	69.0
Point Reyes	2.3	37.2	16.2	4.2	66.7	29.0
Coquimbo	0.5	98.1	33.1	0.4	74.5	25.2
Talcahuano	21.3	120.1	12.8	13.8	77.9	8.3
Average	3.0	37.5	10.1	6.4	64.1	29.5

[a] Signals with periods longer than 1.5 h

[b] Signals with periods between 1.5 h and 20 min

[c] Signals with periods shorter than 20 min

Table 2 gives useful information about the complex behavior of tsunami waves in large basins. Out of the 15 tide-gauge records studied here, the contributions of local bathymetry to the total tsunami energy were the largest in the four stations of Wake Island, Kwajalein, Port Alexander, and Arena Cove (Table 2; Fig. 8b). For Wake Island, Table 2 shows that 70.3 % of the total tsunami energy was generated by local bathymetric effects, while direct tsunami waves and basin effects contributed 16.6 and 13.1 % to the total tsunami energy, respectively. As briefly discussed above, significant energy in the period band of 0–20 min can be attributed to the resonance of the harbor in which the tide gauge is located or other bathymetric features like wave focusing and refracting in coastal areas. However, harbor resonance and other coastal bathymetric effects due to incident tsunami waves are not new and were evidenced during some past tsunamis. Long-lasting oscillations in Port Salalah, Oman, were reported following the 2004 Indian Ocean tsunami (OKAL et al. 2006). Our study shows that the energy generated by local bathymetric effects was larger than that generated by direct tsunami waves in 4/15 studied stations.

The results of wave simulations for two different sources, one at the center of the Pacific Ocean and the other at a corner (Fig. 3), show that the strength of basin-wide modes possibly depends on the location of the source. The spectral powers calculated for 30 locations within the Pacific Ocean for two different sources show that the spectral powers are noticeably larger for the source located at the center of the Pacific Ocean (Fig. 4a) compared to the other source (Fig. 4b). In other words, the contribution of basin effects to the total tsunami energy, which was 6.4 % for the March 2011 Tohoku tsunami, may be different for large Pacific tsunamis occurring in different tsunamigenic locations within the Pacific Ocean.

5. Conclusions

To understand the origins of long Pacific Ocean oscillations during the March 2011 Tohoku tsunami, we calculated basin-wide modes of the Pacific Basin using a numerical approach and then studied 15 tide-gauge records of this tsunami to examine which modes were excited by this giant tsunami. Our main findings were:

1. Using a numerical approach, 49 basin-wide modes were calculated for the Pacific Ocean, ranging from 2 to 48 h.
2. An analysis of 15 tide-gauge records of the March 2011 Tohoku tsunami showed that some of the basin-wide modes of the Pacific Basin were excited by this giant tsunami.
3. To measure the contribution of basin-wide modes to the total tsunami energy, we classified all signals available in a tide-gauge record into three groups: (1) signals generated by basin effects with periods longer than 1.5 h, (2) signals generated by the main tsunami source with periods in the range of 20–90 min, and (3) signals generated by local bathymetric effects with periods shorter than 20 min. On average, the contributions of basin effects, tsunami source, and local bathymetry to the total tsunami energy were 6.4, 64.1, and 29.5 %, respectively.
4. The results of wave simulations for two different sources, one at the center of the Pacific Ocean and the other at a corner, show that the strength of basin-wide modes and their contributions to total tsunami energy possibly depends on the location of the tsunami source.

Acknowledgments

The sea level data used in this study were provided through the USA National Oceanographic and Atmospheric Administration (NOAA), and the UNESCO Intergovernmental Oceanographic Commission (IOC). Some figures were drafted using the GMT software (WESSEL and SMITH 1991). This article benefitted from detailed, constructive review comments from two anonymous reviewers, for which we are sincerely grateful. The first author was partially supported by the Alexander von Humboldt Foundation in Germany.

REFERENCES

BELL, C., VASSIE, J.M., and WOODWORTH, P.L. (2000). POL/PSMSL Tidal Analysis Software Kit 2000 (TASK-2000), Permanent Service for Mean Sea Level. CCMS Proudman Oceanographic Laboratory, UK, 22p.

BORRERO, J.C., and GREER, S.D. (2013). *Comparison of the 2010 Chile and 2010 Japan tsunamis, in the Far-field*, Pure App. Geophys., *170*, 6-8, 1249-1272.

CARBAJAL, N., and GALICIA-PEREZ, M.A. (2002). *Earthquake induced Helmoltz resonance in Manzanillo lagoon, Mexico*. Revista Mexicana de Fisca, *40*, 3, 192–196.

HEIDARZADEH, M., and SATAKE, K. (2012). *Free mode excitation of the Pacific Basin during the 2011 large Tohoku tsunami*. Proceedings of the Japan Geoscience Union Meeting, Paper No. HDS06-P02, 20-25 May 2012, Makuhari, Chiba, Japan.

HEIDARZADEH, M., and SATAKE, K. (2013a). *Waveform and Spectral Analyses of the 2011 Japan Tsunami Records on Tide Gauge and DART Stations Across the Pacific Ocean*. Pure App. Geophys., *170*, 6-8, 1275-1293.

HEIDARZADEH, M., and SATAKE, K. (2013b). *The 21 May 2003 Tsunami in the Western Mediterranean Sea: Statistical and Wavelet Analyses*. Pure App. Geophys., *170* (9-10), 1449-1462.

IOC, IHO, and BODC (2003). *Centenary edition of the GEBCO digital atlas*, published on CD-ROM on behalf of the Intergovernmental Oceanographic Commission and the International Hydrographic Organization as part of the general bathymetric chart of the oceans. British oceanographic data centre, Liverpool.

LEE, J., (1971). *Wave induced oscillations in harbors of arbitrary geometry*. J. Fluid Mechanics, *45*, 375–394.

MATHWORKS (2013). *MATLAB user manual*, The Math Works Inc., MA, USA, 282 p.

MOFJELD, H.O., TITOV, V.V., GONZALEZ, F.I., and NEWMAN, J.C. (2001). *Tsunami scattering provinces in the Pacific Ocean*. Geophys. Res. Lett., *28*, 335–337.

MORI, N., TAKAHASHI, T, and the 2011 Tohoku Earthquake Tsunami Joint Survey Group (2012). *Nationwide Post Event Survey and Analysis of the 2011 Tohoku Earthquake Tsunami*, Coastal Eng. J., *54*, 1, 1-27.

NOAA (2013). *Japan (East Coast of Honshu) Tsunami, March 11, 2011*. Pacific Marine Environmental Laboratory, National Oceanic and Atmospheric Administration. Available at: http://nctr.pmel.noaa.gov/honshu20110311/.

OKAL, E. A., FRITZ, H. M., RAAD, P. E., SYNOLAKIS, C., AL-SHIJBI, Y., and AL-SAIFI, M. (2006). *Oman field survey after the December 2004 Indian Ocean tsunami*. Earthquake Spectra, *22*, S3, 203-218.

RABINOVICH, A.B. (1997). *Spectral analysis of tsunami waves: Separation of source and topography effects*, J. Geophys. Res., *102*, 12, 663–676.

RABINOVICH, A. B., CANDELLA, R. N. and THOMSON, R. E. (2011). *Energy Decay of the 2004 Sumatra Tsunami in the World Ocean*. Pure App. Geophys., *168*, 1919-1950.

RABINOVICH, A.B. (2009). *Seiches and harbor oscillations*, in Handbook of Coastal and Ocean Engineering (edited by Y.C. Kim), Chapter 9, World Scientific Publ., Singapore, 193-236.

RAICHLEN, F., and LEE, J. (1991). *Oscillation of Bays, Harbors, and Lakes*. In: Herbich, J. (Ed.), Handbook of Coastal and Ocean Engineering. Gulf Publishing Co, Houston.

ROEBER, V., YAMAZAKI, Y., and CHEUNG, K.F. (2010). *Resonance and impact of the 2009 Samoa tsunami around Tutuila, American Samoa*. Geophysical Research Letters, 37, L21604, doi:10.1029/2010GL044419.

SAITO, T., INAZU, D., TANAKA, S., and MIYOSHI, T. (2013). *Tsunami Coda across the Pacific Ocean Following the 2011 Tohoku-Oki Earthquake*. Bull. Seismol. Soc. Am., *103*, 2B, 1429-1443.

SATAKE, K., FUJII, Y., HARADA, T., NAMEGAYA, Y. (2013). *Time and space distribution of coseismic slip of the 2011 Tohoku earthquake as inferred from tsunami waveform data*. Bull. Seismol. Soc. Am., *103*, 2B, 1473-1492.

SATAKE, K., and SHIMAZAKI, K. (1987). *Computation of tsunami waveforms by a superposition of normal modes*. J. Phys. Earth, *35*, 5, 409-414.

SATAKE, K., and SHIMAZAKI, K. (1988a). *Free oscillation of the Japan Sea excited by earthquakes-I. Observation and wave-theoretical approach*. Geophys. J. Int., *93*, 3, 451-456.

SATAKE, K., and SHIMAZAKI, K. (1988b). *Free oscillation of the Japan Sea excited by earthquakes- II. Modal approach and synthetic tsunamis*. Geophys. J. Int., *93*, 3, 457-463.

VAN DORN, W. G. (1984). *Some tsunami characteristics deducible from tide records*. J. Phys. Oceanogr., *14*, 2, 353-363.

VICH, M.M., and MONSERRAT, S. (2009). *Source spectrum for the Algerian tsunami of 21 May 2003 estimated from coastal tide gauge data*, Geophys. Res. Lett., 36, L20610.

WESSEL, P. and SMITH, W. H. F. (1991). *Free software helps map and display data*, EOS Trans. AGU 72, 441.

YALCINER, A.C., PELINOVSKY, E., TALIPOVA, T., KURKIN, A., KOZEL-KOV, A., and ZAITSEV, A.(2004). *Tsunamis in the Black Sea: comparison of the historical, instrumental, and numerical data*. J. Geophys. Res., *109*, C12023.

YALCINER, A.C., and PELINOVSKY, E.F. (2007). *A short cut numerical method for determination of periods of free oscillations for basins with irregular geometry and bathymetry*, Ocean Eng., *34*, 5-6, 747–757.

YAO, L.S. (1999). *A resonant wave theory*. J. Fluid Mechanics, *395*, 237-251.

YANUMA, T., and TSUJI, Y. (1998). *Observation of edge waves trapped on the continental shelf in the vicinity of Makurazaki Harbor, Kyushu, Japan*. J. Oceanography, *54*, 1, 9-18.

ZELT, J.A., and RAICHLEN, F. (1990). *A Lagrangian model for wave-induced harbour oscillations*. J. Fluid Mechanics, *213*, 1, 203-225.

(Received May 17, 2013, accepted October 29, 2013, Published online November 28, 2013)

Pure Appl. Geophys. 171 (2014), 3421–3435
© 2014 Springer Basel
DOI 10.1007/s00024-014-0782-2

❚ Pure and Applied Geophysics

Observations and Modeling of the August 27, 2012 Earthquake and Tsunami affecting El Salvador and Nicaragua

Jose C. Borrero,[1,2] Nikos Kalligeris,[2] Patrick J. Lynett,[2] Hermann M. Fritz,[3] Andrew V. Newman,[4] and Jaime A. Convers[4]

Abstract—On 27 August 2012 (04:37 UTC, 26 August 10:37 p.m. local time) a magnitude $M_w = 7.3$ earthquake occurred off the coast of El Salvador and generated surprisingly large local tsunami. Following the event, local and international tsunami teams surveyed the tsunami effects in El Salvador and northern Nicaragua. The tsunami reached a maximum height of ∼6 m with inundation of up to 340 m inland along a 25 km section of coastline in eastern El Salvador. Less severe inundation was reported in northern Nicaragua. In the far-field, the tsunami was recorded by a DART buoy and tide gauges in several locations of the eastern Pacific Ocean but did not cause any damage. The field measurements and recordings are compared to numerical modeling results using initial conditions of tsunami generation based on finite-fault earthquake and tsunami inversions and a uniform slip model.

1. Introduction

Situated along a major subduction zone plate boundary, the Pacific coast of Central America (Fig. 1) is vulnerable to near-field, regional, and far-field tsunamis (FERNANDEZ *et al.* 2004; Álvarez-Gómez *et al.* 2013), with risk dependent strongly on coastal exposure and population. The US National Geophysical Data Center World Data Service for Geophysics (NGDC/WDS) database lists 21 tsunami events affecting El Salvador and seven events

Electronic supplementary material The online version of this article (doi:10.1007/s00024-014-0782-2) contains supplementary material, which is available to authorized users.

[1] eCoast Ltd., 47 Cliff St., Raglan 3225, New Zealand. E-mail: jose@ecoast.co.nz
[2] Tsunami Research Center, Sonny Astani Department of Civil and Environmental Engineering, University of Southern California, Los Angeles, CA 90089, USA.
[3] School of Civil and Environmental Engineering, Georgia Institute of Technology, Atlanta, GA 30322, USA.
[4] School of Earth and Atmospheric Sciences, Georgia Institute of Technology, Atlanta, GA 30332, USA.

affecting Nicaragua. While there are several older events with questionable validity, it is clear that these countries have been affected by tsunamis in the past. Notable events in El Salvador prior to 2012 include near-field tsunamis in 1859 ($M_l \sim 7.6$; WHITE *et al.* 2004) and 1902 (no observed earthquake shaking; WHITE *et al.* 2004), the latter of which reportedly caused 185 deaths in the western portion of the country; however, this event is questionable and may have been meteorological in origin. Regional sources, such as the 1906 Ecuador earthquake (KELLEHER 1972), have also caused tsunamis affecting El Salvador. Damaging tsunami waves were reported as a result of the 9 March 1957 earthquake in the Aleutian Islands, which reportedly caused deaths and damage in Acajutla (FERNANDEZ *et al.* 2004), although this death toll is not reflected in the NGDC database. Non-damaging tsunami waves were also observed from other major transpacific tsunamis such as the 1960 and 2010 tsunamis from Chile and the 2011 Tohoku, Japan, tsunami. Nicaragua on the other hand, has scant reports of effects from far-field tsunamis. Nicaragua is best known for the tsunami of 2 September 1992, which is particularly notable in that it was caused by a slow earthquake with moment magnitude, M_w 7.7 (KANAMORI and KIKUCHI 1993). That tsunami had a maximum run-up height of ∼10 m and caused 170 deaths and USD $30 million in damage (ABE *et al.* 1993; SATAKE *et al.* 1994).

2. Earthquake Details

On 26 August 2012 at 10:37 p.m. local time (27 August 2012, 0437 UTC), an earthquake with M_w 7.3 (W-phase) occurred off the coast of El Salvador and

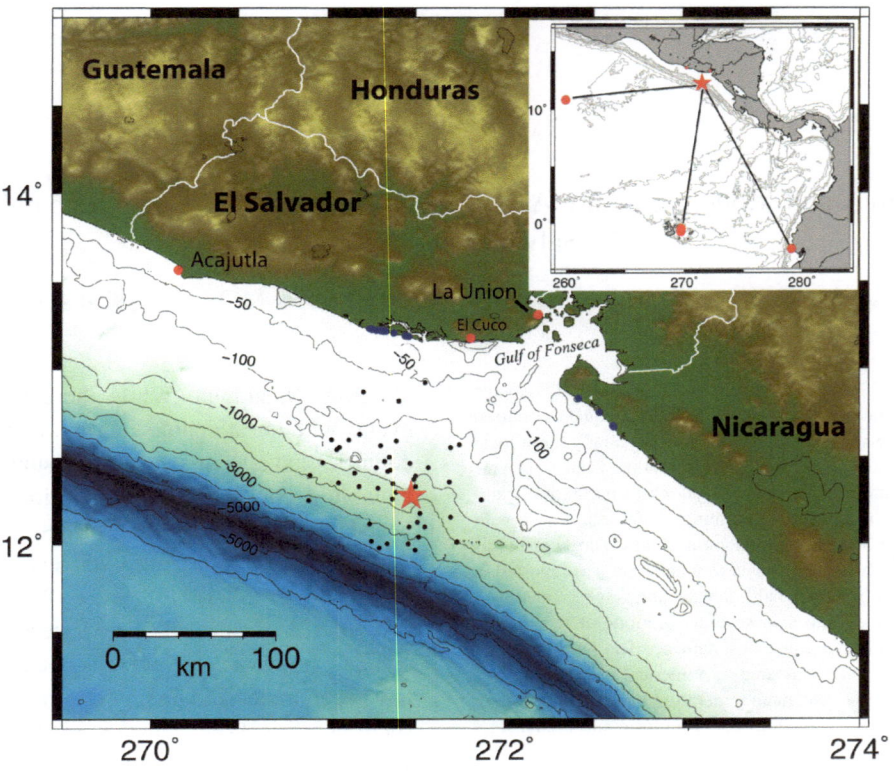

Figure 1
The offshore bathymetry of El Salvador and northern Nicaragua, and the location of the USGS-defined earthquake epicenter (*red star*). *Black dots* correspond to epicenters of aftershocks through 11 September 2012. Depth *contours* labeled in meters. *Blue dots* are locations where tsunami field data were recorded (inset map). Location of El Salvador and the earthquake epicenter relative to (from west to east) Deep-ocean Assessment and Reporting of Tsunamis (DART) tsunameter 43413 and the Galapagos Islands and La Libertad, Ecuador tide gauge stations (*red dots*)

Nicaragua. The epicenter, as reported by the US Geological Survey (USGS), was located approximately 75 km due south of the El Salvador coastline (Fig. 1). Approximately 50 aftershocks with magnitudes between 5.5 and 4.2 occurred in the vicinity of the main event between 27 August and 11 September 2012. The initial assessment of the earthquake by the Pacific Tsunami Warning Center (PTWC) determined that the earthquake was significant due to the strength of the seismic signal and the long-period nature of the initial seismic waves. Within 10 min of the main shock, additional analysis by the PTWC suggested that the earthquake could be characterized as a 'slow' earthquake. This was indicated by Θ ($=\log_{10}(E/M_0)$) values (NEWMAN and OKAL 1998) in the range of -6.5 to -6.0 as computed by the PTWC. Typical values of Θ for 'typical' thrust earthquakes are generally larger, in the range of -4.7. Additionally, Θ values

derived by the West Coast Alaska Tsunami Warning Center (WCATWC) were even lower at -7.0, further suggesting a very slow event. Analysis provided by the Real-Time Earthquake Energy and Rupture Duration project at the Georgia Institute of Technology also identified the event as slow and 'weak' 11 min after the initiation of the main shock.

3. Deficiency in Radiated Seismic Energy

Following BOATWRIGHT and CHOY (1986), and with corrections for real-time implementation by NEWMAN and OKAL (1998), both the radiated seismic energy and rupture durations of global earthquakes since 1997 with $M_w > 6.7$ have been estimated by CONVERS and NEWMAN (2011). Events since early 2009 were analyzed using a set of programs called 'RTerg'

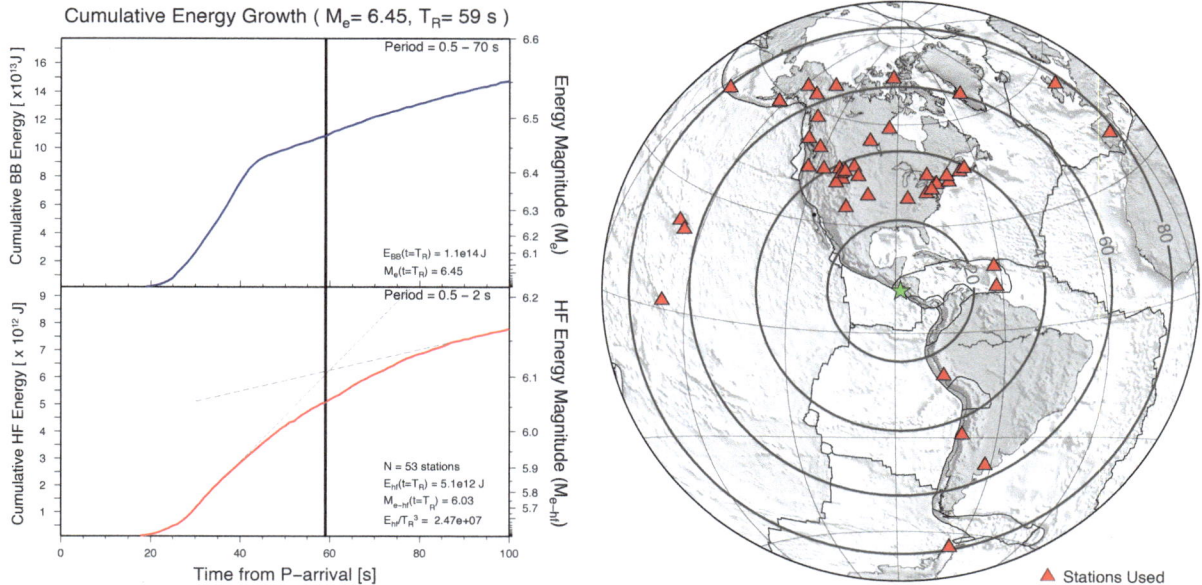

Figure 2
(*left*) Cumulative energy in broadband (0.5–70 s top trace, *blue*) and high-frequency (period 0.5–2 s, *bottom* trace, *red*) energy. The high-frequency energy is used to approximate T_R and evaluate the broadband energy E following CONVERS and NEWMAN (2011). (*right*) Teleseismic stations between 25° and 80° used to calculate the radiated seismic energy

(NEWMAN *et al.* 2011). The results of broadband energy, high-frequency energy, and rupture duration (Fig. 2) help to characterize strong shaking and tsunami potential, especially in the case of tsunami earthquakes, which are identified as being deficient at radiating seismic energy (KANAMORI 1972). Using the energy-to-moment ratio parameter as a discriminant for tsunami earthquakes (NEWMAN and OKAL 1998), such events can be identified by their low Θ value, thereby qualifying them as tsunami earthquakes (NEWMAN and OKAL 1998; CONVERS and NEWMAN 2011). While most events have a Θ value between -4.0 and -5.0, slow tsunami earthquakes have $\Theta \le -5.7$ (NEWMAN and OKAL 1998). The El Salvador earthquake, with a radiated energy of 1.1×10^{14} J (Fig. 2), released about 25 times less seismic energy than other earthquakes of the same size, and showed an energy-to-moment value of $\Theta = -6.0$ (Fig. 2), clearly lower than average and classifying it as a slow-rupturing tsunami earthquake.

The observed rupture duration (T_R) of 59 s was approximately three times longer than expected using a typical duration-cubed scaling relationship with seismic moment (HOUSTON 2001). The extended duration, unlike other earthquakes of the same

magnitude, is typical of slow-source tsunami earthquakes, along with their deficient rupture high-frequency energy. The threshold of $E_{hf}/T_R^3 < 5 \times 10^7$ J/s, is also used as a discriminator for tsunami earthquakes (NEWMAN *et al.* 2011, Fig. 3), where, similar to other tsunami earthquakes, the El Salvador event stands out with an anomalous T_R and deficiency in its radiated high-frequency energy (E_{hf}). Individuals in the area at the time of the earthquake described the ground shaking as 'light', a feature that corroborates the seismologic evidence for a slow rupture. Such was the case for the 1992 Nicaragua event and other tsunami earthquakes, where many who felt the event thought the earthquake was considerably smaller and less dangerous than it actually was (e.g. KANAMORI 1972; KANAMORI and KIKUCHI 1993; SATAKE *et al.* 1994; HILL *et al.* 2012).

4. Summary of Tsunami Effects

The earthquake generated a moderate tsunami observed by eyewitnesses and instruments located in both the near field (Acajutla and La Unión) and far field (Baltra and Santa Cruz, Galapagos Islands; La

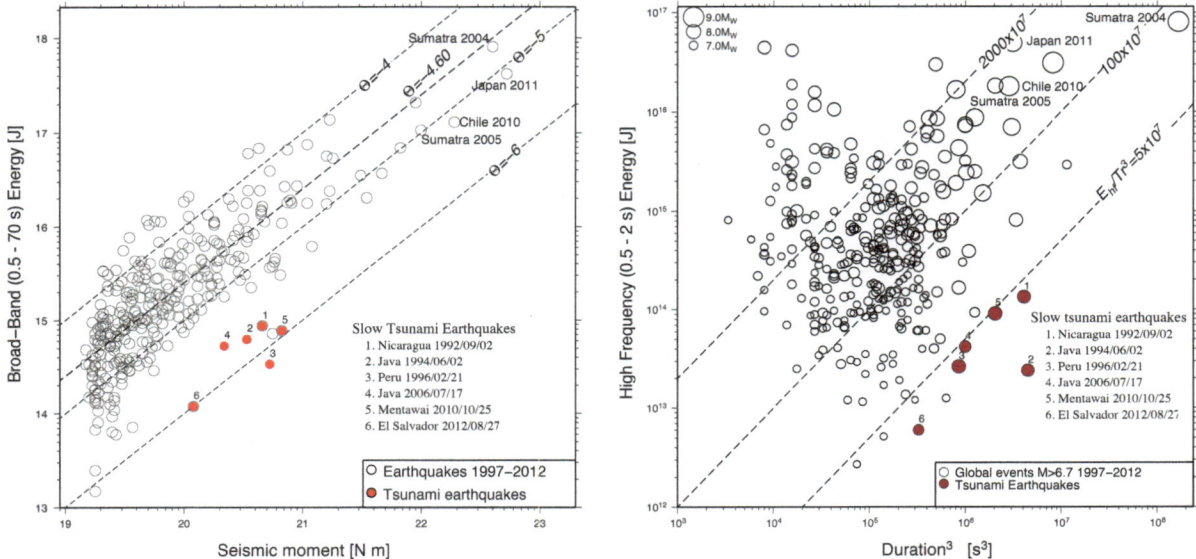

Figure 3
(*left*) Global energy-to-seismic moment comparison for earthquakes greater than M_w 6.7. The *dashed line* represents the global average and constant values of the energy-to-moment ratio ($\log_{10}E/M_0$) (NEWMAN and OKAL 1998). Energy-deficient earthquakes tend to the *bottom right*, while strong-rupturing earthquakes tend to the *top left*. Highlighted are known and identified tsunami earthquakes (events of this type before 1997 are from CONVERS and NEWMAN (2011) and are included for comparison). (*right*) Relationship between high-frequency energy and the cube of the duration for the same dataset; earthquakes with low radiated high-frequency energies and disproportional long durations are grouped in the *bottom right*

Table 1

PTWC Summary of tide gauge recordings from the El Salvador tsunami

Station	Country	Lat. (°)	Long. (°)	Arrival (hours)	Z2P (m)	P2T (m)	Period (mm:ss)
Acajutla	El Salvador	13.57	−89.84	0:52	0.11	0.21	8:00
DART 43413	n/a	10.84	−100.08	1:36	0.01	0.02	8:00
La Unión	El Salvador	13.31	−87.81	1:40	0.03	0.04	9:00
Baltra	Ecuador (Galapagos Is)	−0.44	−90.28	2:30	0.35	0.70	10:00
Santa Cruz	Ecuador (Galapagos Is)	−0.72	−90.31	2:49	0.22	0.39	13:20
La Libertad	Ecuador	−2.22	−80.91	3:36	0.21	0.37	11:30

Z2P zero-to-peak wave amplitude, *P2T* peak-to-trough wave height

Libertad, Ecuador; Easter Island; and DART station 43413, see Fig. 1). A summary of water level measurements provided by the PTWC shortly after the event is provided in Table 1. Due to the location of the earthquake and the fact that tsunamis primarily radiate wave energy perpendicular to the trench axis, the tide gauge in Acajutla was not ideally located to receive the direct tsunami signal, while El Salvador's other tide gauge at La Unión is located several kilometers from the open ocean in the shallow and sheltered Gulf of Fonseca (Fig. 1). In contrast, the

Galapagos Islands situated approximately 1,400 km away along a 190° (SSW) path from the source region (Fig. 1) are ideally located to receive a strong tsunami signal. As a result, the two stations in the Galapagos recorded a very strong, clear tsunami that arrived approximately 2.5 h after the earthquake. Following the initial wave packet, both stations also responded with a secondary (and in the case of Santa Cruz, a tertiary) wave packet with amplitudes nearly as large as the initial wave. A similar extended duration and resurgence of wave height was also observed on these stations during the 11 March 11 2011 Tohoku tsunami (LYNETT *et al*. 2013). Located farther off-axis from the main beam of tsunami energy propagation were DART 43413 (1,200 km at 264°, see Fig. 1), which recorded a single tsunami wave pulse with a peak-to-trough (P2T) height of 0.024 m, and the La Libertad, Ecuador station (1,800 km at 153°, see Fig. 1) characterized by long-period non-tsunami oscillations present before the tsunami arrival. The tsunami itself appears clearly some 3.5 h after the earthquake, with the largest signal occurring some 5 h after the tsunami arrival. Additional details of these tsunami records, including detailed spectral analysis, are discussed in HEIDAR-ZADEH and SATAKE (2014).

In the days immediately following the event, representatives from El Salvador's Ministry for the Environment and Natural Resources (MARN) conducted an initial survey of the tsunami-affected area focused on attending to immediate needs and disseminating information to local residents. This was followed by the International Tsunami Survey Team (ITST), which visited the affected areas of El Salvador on 5–7 September 2012 and collected the majority of the quantitative data. A third survey

subsequently visited sites in northern Nicaragua in response to reports of moderate tsunami inundation occurring in several coastal villages. A detailed description of these surveys is available in BORRERO (2012) and is provided as supplementary material.

During the ITST survey, the team visited 11 separate sites, focusing primarily on the San Juan del Gozo Peninsula (Fig. 4), and recording measurements of tsunami height, run-up height, flow direction, and inundation distance using established protocols (SYNOLAKIS and OKAL 2005; UNESCO 2013). Measured data are presented relative to the tide level at the time of tsunami arrival (Fig. 4; Table 2) and are divided into flow depths, tsunami heights, and run-up heights. Because the topography landward of the dune ridge sloped downward, run-up heights are generally lower than the maximum tsunami heights.

The strongest tsunami effects were experienced along the central section of the San Juan del Gozo Peninsula. At the time of the earthquake (~10:37 p.m.) it was relatively dark with a quarter moon low over the horizon and a moonset near 1 a.m. The earthquake occurred just after high tide when scores of people were present on the beach collecting

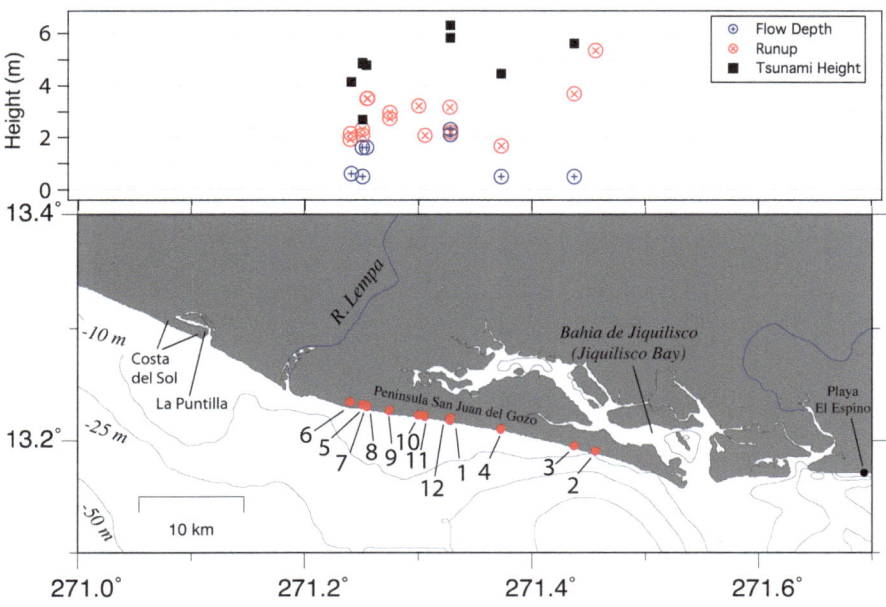

Figure 4
Summary of data collected during the ITST field survey. (*lower panel*) Map showing the survey locations, (*upper panel*) measured tsunami data broken down into run-up, flow depth, and tsunami height values

Table 2

Tsunami field data measurements from the 2012 tsunami affecting El Salvador and Nicaragua

Site	Lat. (°N)	Long (°E)	Terrain z (m)	Flow Depth h (m)	Tsunami height z + h (m)	Run-up R (m)	Distance (m)	Watermark	Description
El Salvador									
1	13.2183	−88.6716	4.2	2.1	6.3	–	82.0	Mud line inside	Eyewitness confirmed house pole
1	13.2186	−88.6716	3.5	2.3	5.8	–	121.6	Broken branch	Eyewitness confirmed wrapped in sheet metal
1	13.2206	−88.6713	2.2	–	–	2.2	340.9	Wrack line	Eyewitness confirmed
2	13.1907	−88.5433	5.3	–	–	5.3	45.8	Wrack line	Eyewitness confirmed on top of dune tree log
3	13.1950	−88.5623	5.1	0.5	5.6	–	45.9	Eyewitness	Dune overtopped ($h = 0.5$ m guessed) eyewitness confirmed
3	13.1953	−88.5622	3.7	–	–	3.7	83.9	Wrack line	Eyewitness confirmed brown grass
4	13.2101	−88.6270	4.0	0.5	4.5	–	96.4	Damaged trim line	Wooden palm leaf hut
4	13.2107	−88.6268	1.7	–	–	1.7	155.5	Wrack line	Wrack line
5	13.2315	−88.7495	3.2	1.6	4.8	–	46.0	Damaged trim line	Hut with sheet metal eyewitness confirmed Manglaron Monte Alto
5	13.2315	−88.7495	3.3	1.6	4.9	–	46.0	Damaged trim line	House siding eyewitness confirmed
5	13.2319	−88.7494	2.2	0.5	2.7	–	85.1	Raft debris	Debris in fence
5	13.2325	−88.7494	2.3	–	–	2.3	150.6	Wrack line	Brown vegetation eyewitness confirmed
5	13.2325	−88.7493	2.1	–	–	2.1	153.5	Wrack line	Brown vegetation eyewitness confirmed
6	13.2339	−88.7593	3.5	0.6	4.1	–	37.2	Damaged trim line	Hut with sheet metal eyewitness
7	13.2348	−88.7600	2.1	–	–	2.1	115.6	Wrack line	Embankment next to mangroves
7	13.2349	−88.7600	1.9	–	–	1.9	119.1	Wrack line	Wrack line next to mangroves
8	13.2305	−88.7458	3.2	1.6	4.8	–	92.1	Broken branch	Tree on top of dune
8	13.2310	−88.7447	3.5	–	–	3.5	139.9	Wrack line	Field next to fence
8	13.2310	−88.7453	3.5	–	–	3.5	141.2	Wrack line	Field next to fence
9	13.2274	−88.7252	3.0	–	–	3.0	164.7	Wrack line	
9	13.2274	−88.7253	2.7	–	–	2.7	165.7	Wrack line	
10	13.2234	−88.6997	3.2	–	–	3.2	170.4	Wrack line	Brown vegetation eyewitness confirmed
11	13.2226	−88.6941	2.1	–	–	2.1	204.2	Wrack line	Brown vegetation eyewitness confirmed
12	13.2198	−88.6720	3.2	–	–	3.2	242.9	Wrack line	Brown vegetation eyewitness confirmed
Nicaragua									
1	12.833	−87.583		0.5	–	–	106.0	n/a	n/a
2	12.672	−87.387		n/a	–	–	0.0	n/a	n/a
3	12.657	−87.376		0.5	–	–	120.0	n/a	n/a

sea turtle eggs for conservation projects. A worker at one of the hatcheries was in a ramada (a small shed with wooden posts, and walls and roof made from aluminum siding) located on the beach at the crest of the dunes approximately 70 m from the water when the earthquake occurred. This witness came out of the ramada when he heard people crying out and was subsequently caught in the wave and dragged some 90 m inland where he was ultimately suspended in a tree branch at a height of 2.1 m above the ground. This witness reported that there were three tsunami waves and that the flow depth at the ramada was just below the roof (approx. 2.5 m). During the tsunami, the walls of the ramada were torn off of the posts that are deeply embedded in the sand; the posts themselves were not pulled out of the ground, but some were pushed over by the force of the water. Other witnesses related similar stories, and in total there were more than 40 injured people, with three injuries requiring medical attention.

The effects of the tsunami along the San Juan del Gozo peninsula were relatively uniform along 25 km of mostly undeveloped coastline (Fig. 5). It is important to point out that significant tsunami effects appear to be constrained to this area. During the initial survey by MARN, the more developed tourist area of Costa del Sol some 10 km to the west did not report any inundation or wave activity. This was even true at La Puntilla, a dense cluster of several waterfront restaurants at the eastern end of the Costa del Sol sand spit. These restaurants are regularly affected by high tides (Fig. 5), however, the tsunami did not cause any damage or noticeable effects here. Individuals in other parts of El Salvador were interviewed by telephone and did not report any significant tsunami activity. At Playa El Espino (Fig. 4), a local resident and beachfront restaurant owner reported that there were no observable tsunami effects and that the local police had moved into the peninsula of San Juan del Gozo to help assist people affected in that area. Farther east at Playa

Figure 5

Images from the post-event field survey. *Clockwise* from *top left* tsunami flow depths indicated by broken branches and by an eyewitness. The vulnerable structures at La Puntilla and Costa del Sol were not affected by the tsunami. Inundation extent was clearly visible from the air and in debris *lines* on the ground. The aerial perspective also revealed evidence of sand deposition by tsunami overwash

El Cuco (Fig. 1), beachfront hotel workers did not report any tsunami activity. Local boat captains said that the boats left parked on the beach overnight were not noticeably moved or disturbed in any way and that activities of the next day resumed normally. In Nicaragua, a survey team recorded tsunami effects at three seaside villages (Fig. 1) where the tsunami caused mild inundation and flow depths on the order of 0.5 m (Norwin Acosta, *pers. comm.* report included in supplementary material).

5. Tsunami Modeling

We modeled the tsunami propagation and inundation using the MOST suite of integrated numerical codes capable of simulating tsunami generation, transoceanic propagation, and its subsequent inundation in the coastal area (TITOV 1997). The model uses a finite-difference numerical scheme to solve the $2 + 1$ nonlinear shallow-water (NSW) equations in characteristic form, accounting for nonlinearity, but not for frequency dispersion. NSW calculations are extended using a moving boundary for run-up and inundation on the dry bed, and bottom friction is included. The bathymetry used for the modeling grids was derived from the relatively coarse GEBCO 30-s global bathymetry and topography data. The offshore bathymetry of the tsunami-affected area was manually adjusted to remove a large-scale bathymetric depression situated directly offshore of the survey site that does not appear on local navigational charts. Four levels of nested grids of sequentially finer resolution (300, 100, 50, and 10 m) were used for the near-field model. A fifth grid at approximately 900-m resolution was created to assess the far-field propagation of the tsunami. Models of co-seismic slip along the earthquake fault were converted to sea floor deformation through the method of OKADA (1985) and implemented as an initial condition by translating the instantaneous static deformation to the sea surface.

6. Tsunami Source Models

We developed a suite of eight tsunami source models to initialize the hydrodynamic computation

Table 3

General description and characteristics of the different source models

Source	Details
S1	USGS finite fault model, no scaling
S2	USGS finite fault model, 1.92 scale factor based on a slow rupture in mechanically softer material (see NEWMAN et al. 2011 and "Appendix" section)
S3	USGS finite fault model, 2.55 scale factor to match amplitude of leading wave at DART 43413
S4	Moment magnitude-constrained custom inversion based on a least-squares regression between modeled tsunami waveforms produced by unit (1 m) slip on shallow subfaults in the source region and recorded data at DART 43413
S5	Same as 4, but magnitude is unconstrained
S6	Rectangular fault model. 60×30 km, 2.2 m slip, dip 15°, rake 81°, strike 287°, determined from macro-scale seismic parameters (HEIDARZADEH and SATAKE 2014)
S7	YE et al. (2013) finite fault model, no scaling ($V_r = 2.0$ km/s)
S8	YE et al. (2014) finite fault model, with modified rupture velocity ($V_r = 1.5$ km/s)

(Table 3). Five sources were based on teleseismic inversions, two sources were based on a hydrodynamic inversion of the DART water level signal and one source was based on earthquake magnitude and assumed rupture extents. For the first three models, we used the finite-fault model of JI et al. (2002), as reported by the US Geological Survey shortly after the event (USGS 2012). Source 1 used the unscaled slip distribution across a source area of 210×128 km with a maximum slip amount of approximately 1 m. Strike and dip angles of the fault were fixed to 296° and 16°, respectively, and the rake angle is variable. For Source 2, we scaled the teleseismically-determined slip distribution by a uniform factor of 1.92 using the approach of NEWMAN et al. (2011) in their modeling of the 2010 Mentawai Islands, Indonesia earthquake and tsunami (see "Appendix" section). The third source applied a scaling factor of 2.55 to the finite fault slip distribution and was determined by scaling modeled wave heights from the unscaled finite fault source to match the measured leading wave amplitude at DART 43413.

Sources 4 and 5 are based on an inversion of the tsunami water level time series recorded at DART 43413 by modeling 'unit' tsunamis generated from 1 m of coseismic slip on each of 50 subfaults

(20 × 16 km). Strike and dip angles, location, depth, and rigidity values (depth-dependent, ranging from 3.12×10^{10} to 6.75×10^{10} N/m^2) of the sub-faults were adopted from the USGS finite-fault source, and the rake angle was fixed to 90° (for maximum vertical displacement). The slip distribution is determined by summing scaled individual wave-forms and using a linear optimization algorithm (MATHWORKS 2012) that results in the smallest least-square difference to the measured tsunami time series. Dispersive effects were taken into account only in the form of arrival time corrections applied to each source waveform

using the fully dispersive linear wave speed (e.g., DEAN and DALRYMPLE 1991) to determine the "proper" arrival time. Source 4 was constrained by the moment magnitude ($M_w \sim 7.35$, as reported by CMT) while source 5 was not constrained and was equivalent to a slightly larger M_w of 7.43. As indicated in Fig. 6, the location of the high-slip area compares well to that of the USGS seismic inversion. A sixth source model was based on a 60 × 30-km rectangular fault plane with an average slip of 2.2 m. The fault dimensions are based on the aftershock area and the slip amount computed from the seismic

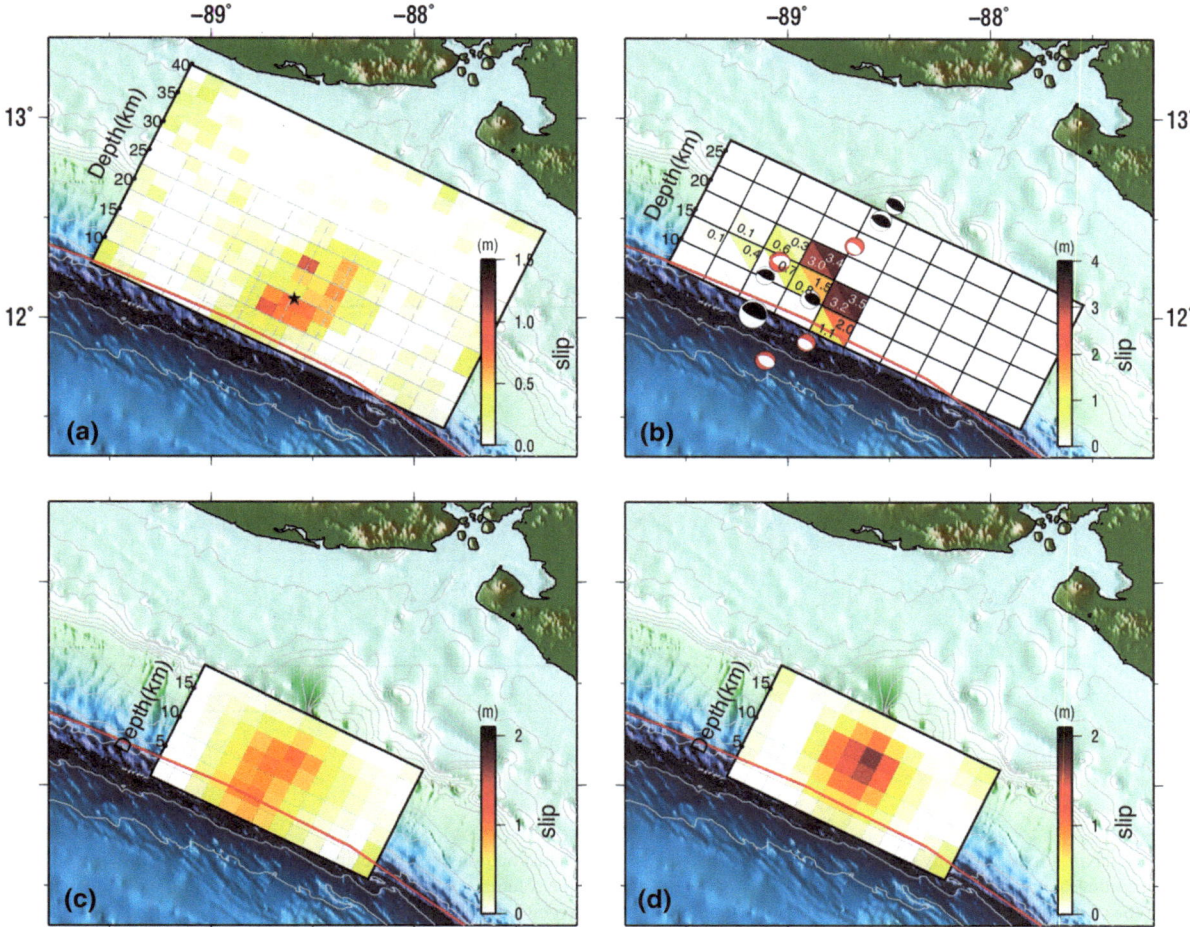

Figure 6
Slip distribution for **a** the USGS finite-fault solution (S1), **b** the solution derived from the moment magnitude constrained (S4, *lower left triangles*) and unconstrained (S5, *upper right triangles*) tsunami source inversions, **c** the YE *et al.* (2013) (S7) and **d** the modified YE *et al.* (2013) (S8) finite-fault solutions. The distribution and fault mechanism of the main and significant aftershocks from the CMT catalog are also shown. Focal-mechanism diagrams are scaled with earthquake magnitude with the largest and smallest corresponding to the main shock (M_w 7.3) and an M_w 4.8 earthquake. Normal faulting events are drawn in *red. Black star* in **a** indicates the location of the USGS-determined main shock epicenter

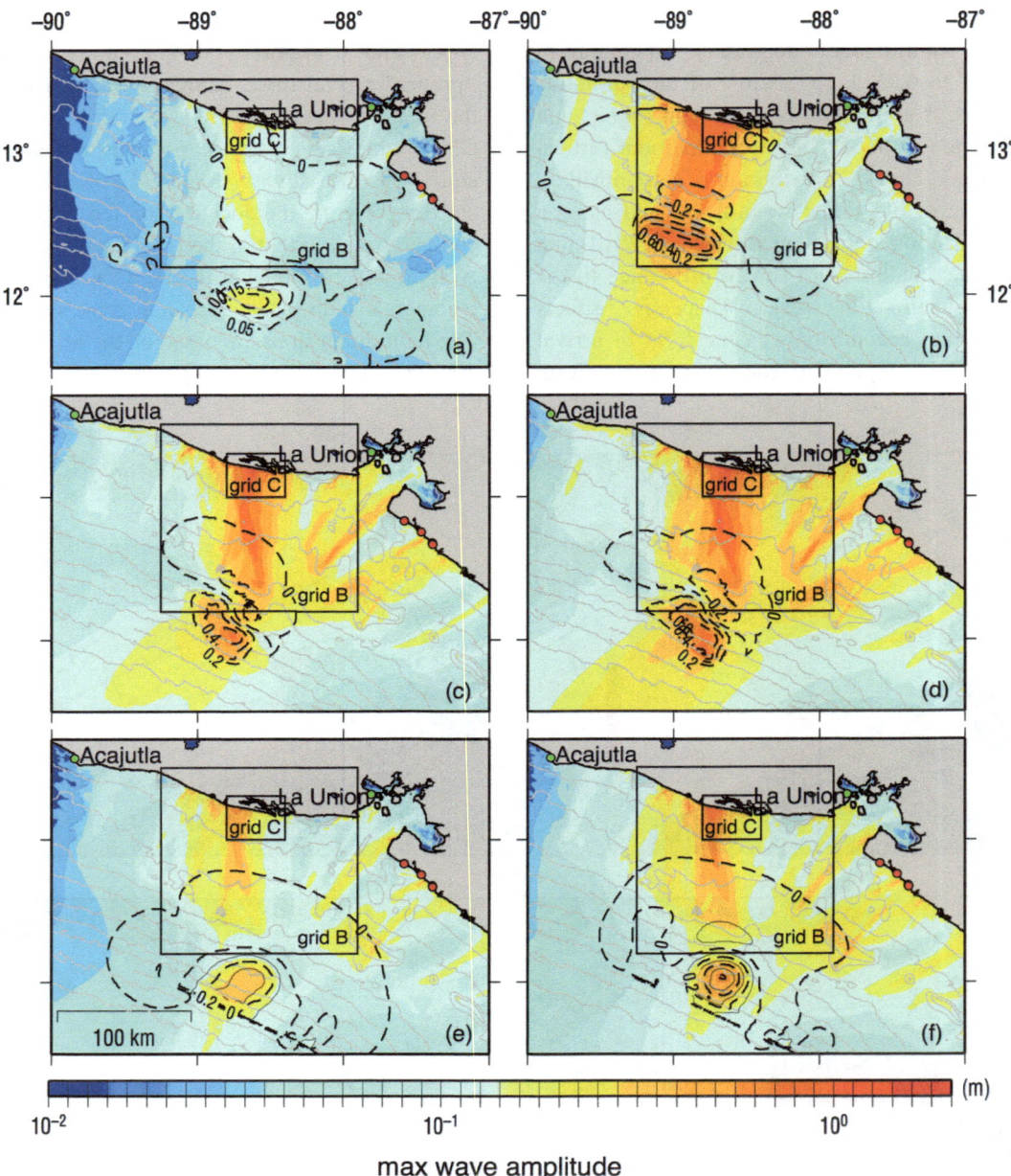

Figure 7
Modeled maximum tsunami wave heights for **a** the unscaled USGS finite-fault solution (S1), **b** rectangular fault plane with uniform slip (S6), **c** the magnitude constrained (S4), **d** the unconstrained (S5) tsunami inversion solutions, **e** the YE *et al.* (2013) (S7), and **f** the modified YE *et al.* (2013) (S8) finite-fault solutions. *Dashed lines* show contours of the static deformation. Grid A area is shown in Fig. 9. *Plots* for sources S2 and S3 (the linearly-scaled finite-fault distribution) are included as supplemental figures

moment, and was used by HEIDARZADEH and SATAKE (2014) to model the tsunami response at near- and far-field tide gauges.

Finally, we consider two models based on the teleseismic-derived finite-fault source of YE *et al.*

(2013). The source area is 130×70 km, discretized in 10×10 km subfaults. Strike and dip angles of the fault were fixed to 296° and 16°, respectively, the rake angle is variable, and the hypocenter was set at 12 km (at the center of the

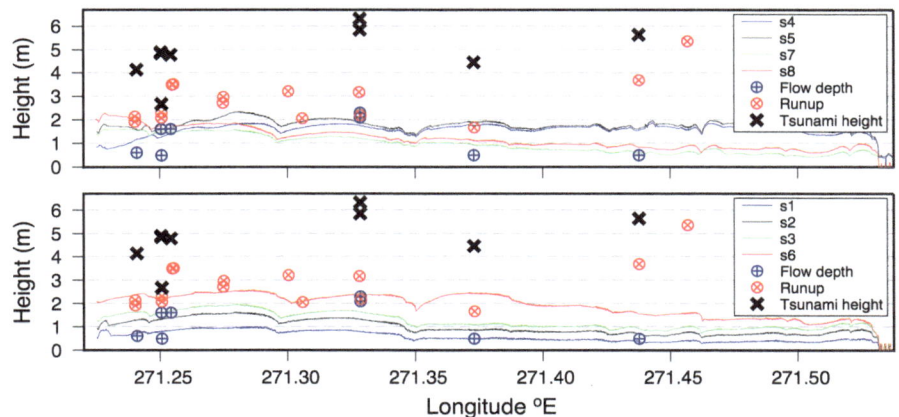

Figure 8
Tsunami heights modeled on land from the eight different sources compared to field measurements

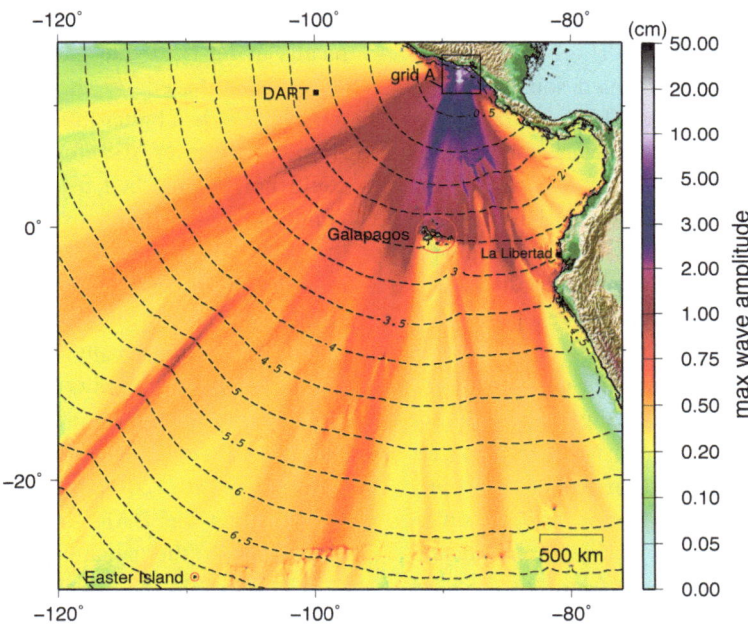

Figure 9
Modeled far-field maximum wave heights using the unscaled USGS finite-fault solution as a tsunami source. *Dashed contours* indicate tsunami travel time in hours

fault rupture area). The fault rupture area is significantly smaller than the USGS finite-fault source, allowing for more slip for a given seismic moment. Source 7 corresponds to the mechanism presented in YE *et al.* (2013), which used a rupture velocity of 2 km/s in the seismic inversion. Source 8 has the rupture velocity set to 1.5 km/s (T. Lay, *pers comm.*) resulting in a source with a nearly

equivalent seismic moment, but with more slip concentrated in a smaller area.

7. Model Results and Discussion

In the near field, the model results show strong focusing of wave energy toward the western end of

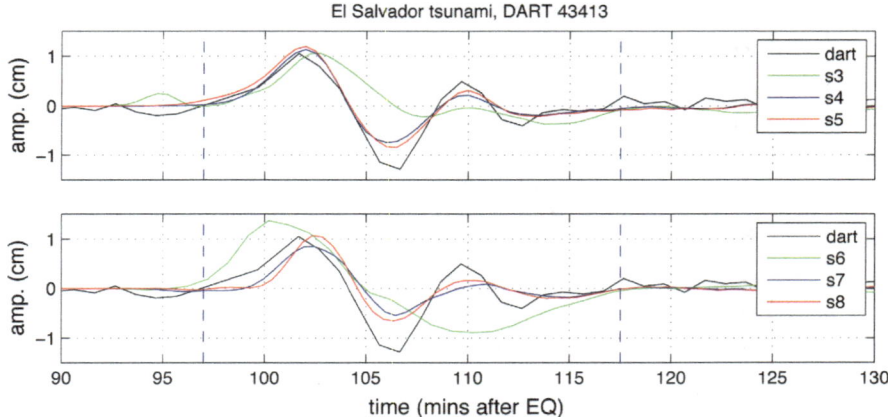

Figure 10
Recorded water level time series at DART 43414 compared to model results for six different tsunami sources. The *vertical dashed lines* denote the portion of the data record used to invert for tsunami sources *4* and *5*

the San Juan del Gozo Peninsula (Fig. 7). This feature is present in all of the modeled sources and is caused by bathymetric focusing. The beam of energy coincides with the section of the coast that experienced the strongest tsunami effects. For all of the cases, the computed tsunami heights on land (Fig. 8) are deficient relative to measured tsunami heights of 2–5 m. This deficiency may partly be a result of fine-scale bathymetric features that are not resolved in the model bathymetry and topography. The rectangular fault model (Source 6) produces the overall highest tsunami height distribution, although it is smaller than the two DART inverted and the two YE *et al.* (2013) sources towards the eastern end of the survey area. The scaled USGS finite-fault models result in the greatest underprediction of the measured tsunami heights. Even when scaled by a factor of 2.55 (to match the peak leading wave amplitude at DART 43413) the high-slip areas are relatively small compared to the other sources.

The simulated maximum far-field wave height distribution from the initial condition based on the unscaled USGS finite-fault solution (Fig. 9) shows a concentrated beam of wave energy directed towards the Galapagos Islands, with strong secondary beams of energy heading towards Ecuador and northern Peru. This is a feature that is present in all of the far-field simulations regardless of the source model, however, we did not explore the details of the model results at far-field tide gauge stations. As expected, the sources based on the inversion of the DART record provide a better fit to the measured time series (Fig. 8). However, the amplitude of the wave trough could not be matched using shallow dipping thrust faults as initial conditions. These sources also show the most pronounced focusing of wave energy towards northern Nicaragua, where the tsunami caused small-scale inundation. Inspection of Fig. 10 clearly shows that the predicted wave forms from the USGS teleseismically-derived source and the rectangular fault model have a noticeably longer period than the measured data. This is not the case with the two YE *et al.* (2013) sources, which produce a waveform at the DART station with a comparable wavelength to the tsunami recording.

The fact that the direct application of the USGS Finite Fault model as an initial condition for the tsunami hydrodynamics yields deficient results should not come as a surprise. Indeed, in the case of the 2010 Mentawai Islands earthquake and tsunami, hydrodynamic simulations initialized with a direct application of the finite-fault slip amounts also severely underpredicted the observed wave heights (HILL *et al.* 2012). In order to match the observed wave effects, it was necessary to scale the slip amounts by an average value of 5.6 (NEWMAN *et al.* 2011). Applying the same scaling technique here, however, still results in deficient model results.

Scaling the finite-fault solution with a greater factor (2.55) produces the correct positive amplitude at the DART station, but does not significantly improve the near-field results.

The rectangular fault model produced the closest match to the measured tsunami heights in El Salvador, but it does not appear to project sufficient wave energy toward Nicaragua to explain the tsunami effects there. Additionally, the period of the modeled signal at DART 43413 from this source has a much longer period than the measured tsunami, although it matches the leading wave amplitude. The YE et al. (2013) teleseismically-derived sources compare well to the DART recording, in terms of matching the first crest amplitude and wavelength, but the inundation modeling results show that they are deficient in terms of tsunami heights in the near field. Source 8, which produces a more concentrated slip due to the lower modeled rupture velocity, matches the tsunami observations better than source 7.

A more vexing problem is the inability of the models to match the tsunami wave trough seen on the DART record. A possible explanation for the pronounced trough is that the tsunami was partially generated by the release of gravitational potential energy, as suggested by MCKENZIE and JACKSON (2012) for tsunami earthquakes with normal faulting aftershocks, such as was the case in this event (see Fig. 6b—CMT catalogue, normal faulting aftershock magnitudes ranged from M_w 4.9 to M_w 5.3). A second possible explanation for the missing trough is the lack of wave generation by horizontal motion of the sea floor. However, calculations indicate that the amplitude associated with this mechanism is one order of magnitude less than the vertical motion component (TANIOKA and SATAKE 1996). Of course, an ad-hoc landslide source could always be used to explain the DART discrepancy. It should be noted that the period of the leading tsunami wave is on the order of 6–7 min, and thus requires a highly spatially-resolved source model, which may not be possible or justified in this and similar earthquakes. The short period also indicates that frequency dispersion should modify the waveform, and this effect is neglected in the shallow-water MOST model. Lastly, the data from DART 43413 used to derive candidate source mechanisms is not ideal, since the tsunameter was located outside the main and secondary beams of tsunami energy propagation.

8. Summary and Conclusions

The earthquake of 27 August 2012 offshore of El Salvador and Nicaragua generated a tsunami with peak run-up of 5.4 m and a maximum tsunami height of 6.3 m, as measured during a post-event field survey of the affected area. The tsunami run-up was generally in the 3-m range with consistent 150 m of inundation along a 25-km stretch of largely uninhabited coastline along the San Juan del Gozo peninsula. Seismological analysis of the earthquake revealed 'slow' rupture characteristics commonly associated with enhanced tsunami effects, and this event was indeed a tsunami earthquake. A numerical modeling analysis of the tsunami using different tsunami source models highlights the difficulties encountered in accurately modeling relatively small events, particularly in the absence of accurate nearshore bathymetry or detailed onshore topography. The modeling effort was further complicated by the unusual nature of the earthquake source mechanism.

Each of the tested models produced wave focusing in the direction of the area that experienced the strongest tsunami effects, however, the computed run-up heights were lower than the measured tsunami heights. While the simplest rectangular, uniform slip source provided the highest modeled tsunami heights on shore, the distribution of modeled tsunami heights was better represented using a source model derived by inverting the tsunami wave form recorded on a nearby DART tsunameter. The preliminary USGS source model derived from teleseismic observations did not adequately capture either the near-field run-up and tsunami heights or the size and character of the signal measured at the DART station. The YE et al. (2013) teleseismically-derived sources produced tsunami signals comparable to the DART recoding and somewhat larger nearfield tsunami heights than the scaled USGS source, but were still deficient compared to the field measurements.

The fact that this was the second tsunami in this region in 20 years caused by a slow earthquake highlights the hazard posed by such events there. Hazard mitigation efforts in Central America should be sure to include information highlighting the tsunami threat from earthquakes that do not 'feel' strong by including factors such as the duration or perceived character of the ground motion.

Acknowledgments

We would like to acknowledge Francisco Gavidia-Medina, Jeniffer Larreynaga-Murcia, Rodolfo Torres-Cornejo, Manuel Diaz-Flores, Fabio Alvarado and rest of the staff at MARN for logistical support, assistance during the field survey and excellent hospitality. Norwin Acosta of INETER conducted the survey of the sites in Nicaragua. Other field survey participants included Nicolas Arcos, Diego Arcas, and Julie Leonard. Laura Kong and the International Tsunami Information Center provided logistical and organizational support to the field survey. The Fuerza Aérea Salvadoreña provided the opportunity for an aerial survey by helicopter of the tsunami-affected area.

Appendix

For Source 2, we use a scaling approach similar to that used in NEWMAN et al. (2011) for the 2010 earthquake and tsunami in the Mentawai Islands offshore Sumatra, Indonesia. Following this, an appropriate scaling parameter for the displacement along the fault plane can be determined from the following relationship:

$$D/D_0 = (V_{S-ref}/V_S)^2, \qquad (1)$$

where D/D_0 is the ratio of the scaled to original slip, V_{S-ref} is a reference shear wave velocity for the region where the earthquake occurred, and V_S is the actual shear wave velocity deduced from teleseismic observations of the event and is calculated from the earthquake rupture velocity (V_R) by assuming that $V_S = 1.25 V_R$. In general, since V_R can be highly variable, individual V_S values are calculated for each subfault in the finite fault solution, However, inspection of the finite fault model from this event reveals a consistent rupture velocity (V_R) of approximately 2.25 km/s, yielding a shear wave velocity (V_S) 2.81 km/s. From the Crust 2.0 model (BASSIN et al. 2000) used in the finite fault analysis, the V_{S-ref}

along the sections of greatest slip is assumed to be 3.9 km/s. Using these values in Eq. 1 leads to an average scale factor of 1.92. We note that this factor is much less than what was deduced from the 2010 Mentawai event where rupture velocities in the main slip zones were often around 1.5 km/s. In using this approach, we emphasize that this is only a first-order approximation, as pointed out in NEWMAN et al. (2011). This is primarily due to the fact that seismic excitation at teleseismic distances includes some upper-plate signal that cannot be fully deconvolved leading to an overamplification of the result. Secondly, because the solution is entirely based on the square of the determination of rupture velocity, any inaccuracies in the solution of V_R can have a very large impact on the final scale factor.

REFERENCES

ABE, K., TSUJI, Y., IMAMURA, F., KATAO, H., IIO, Y., SATAKE, K., BOURGEOIS, J., NOGUERA, E., and ESTRADA, F. (1993), Field Survey of the Nicaragua Earthquake and Tsunami of September 2, 1992, Bulletin of the Earthquake Research Institute of the University of Tokyo (in Japanese), 68, 23–70.

ÁLVAREZ-GÓMEZ, J. A., ANIEL-QUIROGA, Í., GUTIÉRREZ-GUTIÉRREZ, O. Q., LARREYNAGA, J., GONZÁLEZ, M., CASTRO, M., GAVIDIA, F., AGUIRRE-AYERBE, I., GONZÁLEZ-RIANCHO, P., and CARREÑO, E. (2013), Tsunami hazard assessment in El Salvador, Central America, from seismic sources through flooding numerical models, Natural Hazards and Earth System Science, 13, 2927–2939. doi:10.5194/nhess-13-2927-2939.

BASSIN, C., LASKE, G., and MASTERS, G. (2000), The Current Limits of Resolution for Surface Wave Tomography in North America, EOS Trans AGU, 81, F897.

BOATWRIGHT, J., and CHOY, G. L. (1986), Telesesismic estimates of the energy radiated by shallow earthquakes, Journal of Geophysical Research-Solid Earth and Planets, 91(B2), 2095–2112.

BORRERO, J. C. (2012), Field Survey Report of Tsunami Effects Caused by the August, 2012 Offshore El Salvador Earthquake, UNESCO/IOC report.

CONVERS, J. A., and NEWMAN, A. V. (2011), Global Evaluation of Large Earthquake Energy from 1997 Through mid-2010, J. Geophys. Res., 116, B08304, doi:10.1029/2010JB007928.

DEAN, R. G., and DALRYMPLE, R. A., Water Wave Mechanics for Engineers and Scientists (World Scientific, Teaneck, NJ, 1991).

FERNANDEZ, M., ORTIZ-FIGUEROA, M., and MORA, R. (2004), Tsunami Hazards in El Salvador, in Rose, W., Bommer, J., Lopez, D., Carr, M., and Major, J. eds. Natural Hazards in El Salvador, Boulder Colorado. Geological Society of America, Special Paper 375. pp. 435–444.

HEIDARZADEH, M., and SATAKE, K. (2014), The El Salvador and Philippines tsunami of August 2012: insights from sea level data analysis and numerical modeling. Pure and Applied Geophysics. doi:10.1007/s00024-014-0790-2

HILL, E, BORRERO, J., HUANG, Z., QIU, Q., BANERJEE, P., NATA-WIDJAJA, D., ELOSEGUI, P., FRITZ, H., PRANANTYO, I., LI, L., MACPHERSON, K., SKANAVIS, V., SYNOLAKIS, C., and SIEH, K. (2012), *The 2010 Mw 7.8 Mentawai earthquake: Very shallow source of a rare tsunami earthquake determined from tsunami field survey and near-field GPS.* Journal of Geophysical Research, Vol *117*, B06402, doi:10.1029/2012JB009159.

HOUSTON, H. (2001), *Influence of depth, focal mechanism, and tectonic setting on the shape and duration of earthquake source time functions*, J. Geophys. Res.-Solid Earth, *106*(B6), 11137–11150.

JI, C., WALD, D. J. and HELMBERGER, D. V. (2002), *Source description of the 1999 Hector Mine, California earthquake; Part I: Wavelet domain inversion theory and resolution analysis*, Bull. Seism. Soc. Am., Vol *92*, No. 4. pp. 1192–1207.

KANAMORI, H. (1972), *Mechanism of Tsunami Earthquakes, Physics of the Earth and Planetary Interiors*, *6*, 346–359.

KANAMORI, H. and KIKUCHI, M. (1993), *The 1992 Nicaragua earthquake: a slow tsunami earthquake associated with subducted sediments*. Nature, *361*, 714–716.

KELLEHER, J. A. (1972), *Rupture zones of large South-American earthquakes and some predictions*, J. Geophys. Res., *77*(11), 2087–2103, doi:10.1029/JB077i011p02087.

LYNETT, P., WEISS, R., RENTERIA, W., DE LA TORRE MORALES, G., SON, S., ARCOS, M. and MACINNES, B. (2013), *Coastal Impacts of the March 11th Tohoku, Japan Tsunami in the Galapagos Islands*. Pure Appl. Geophys. *170*(6–8):1189–1206, doi:10.1007/s00024-012-0568-3.

MATLAB and STATISTICS TOOLBOX RELEASE 2012, The MathWorks, Inc., Natick, Massachusetts, United States.

MCKENZIE, D., and JACKSON J. (2012), *Tsunami earthquake generation by the release of gravitational potential energy*, Earth and Planetary Science Letters, 345–348, 1–8.

NEWMAN, A. V., HAYES G., WEI Y., and CONVERS J. (2011), *The 25 October 2010 Mentawai tsunami earthquake, from real-time discriminants, finite-fault rupture, and tsunami excitation*, Geophys. Res. Lett., *38*, L05302, doi:10.1029/2010GL046498.

NEWMAN, A. V., and OKAL, E. A. (1998), *Teleseismic estimates of radiated seismic energy: The E/M0 discriminant for tsunami earthquakes*, J. Geophys. Res., *103*, 26885–26898.

OKADA, Y. (1985), *Surface deformation due to shear and tensile faults in a half-space*, Bulletin of the Seismological Society of America, *75*, 1135–1154.

SATAKE, K., BOURGEOIS, J., ABE, K., ABE, K., TSUJI, Y., IMMURA, F., IIO, Y., KATAO, H., NOGUERA, E., and ESTRADA, F. (1994), *Tsunami Field Survey of the 1992 Nicaragua Earthquake*, Eos: Trans. Am. Geophys. Un. 74,13, pp 145, 156–157.

SYNOLAKIS, C. E., and OKAL, E. A. (2005), 1992–2002: Perspective on a decade of post-tsunami surveys, in: Tsunamis: Case studies and recent develop- ments, ed. K. Satake, Adv. Nat. Technol. Hazards, 23:1–30.

TANIOKA, Y., and SATAKE, K. (1996), *Tsunami generation by horizontal displacement of ocean bottom*, Geophysical Research Letters, Volume *23*, Issue 8, pp. 861–864.

TITOV, V., and GONZÁLEZ, F. I. (1997), Implementation and testing of the Method of Splitting Tsunami (MOST) model, NOAA Tech. Memo. ERL PMEL-112 (PB98-122773), 11 pp., Pac. Mar. Environ. Lab., NOAA, Seattle, Wash.

UNESCO (2013), International Tsunami Survey Team (ITST), Post-Tsunami Survey Field Guide. 2nd Edition. Dominey-Howes, D. and Dengler, L. eds., IOC Manuals and Guides No. 37, Paris: (English).

US GEOLOGICAL SURVEY (2012), http://earthquake.usgs.gov/earthquakes/eqinthenews/2012/usc000c7yw/finite_fault.php.

WHITE, R. A., LÍGORIA, J. P., and CIFUENTES, I. L. (2004), "Seismic history along the Middle America subduction zone along El Salvador, Guatemala and Chiapas, Mexico: 1526-2000". In Rose, William Ingersol (et al.) (Eds). Natural Hazards in El Salvador. Geological Society of America, Special Paper 375. pp. 379–96. ISBN:0-8137-2375-2.

YE, L., LAY T., and KANAMORI, H. (2013), *Large earthquake rupture variations on the Middle America megathrust*, Earth and Planetary Science Letters, *381*, pp. 147–155.

(Received October 21, 2013, revised January 18, 2014, accepted January 21, 2014, Published online February 23, 2014)

Reprinted from the journal

Pure Appl. Geophys. 171 (2014), 3437–3455
© 2014 Springer Basel
DOI 10.1007/s00024-014-0790-2

❙ Pure and Applied Geophysics

The El Salvador and Philippines Tsunamis of August 2012: Insights from Sea Level Data Analysis and Numerical Modeling

MOHAMMAD HEIDARZADEH[1,2] and KENJI SATAKE[1]

Abstract—We studied two tsunamis from 2012, one generated by the El Salvador earthquake of 27 August (*Mw* 7.3) and the other generated by the Philippines earthquake of 31 August (*Mw* 7.6), using sea level data analysis and numerical modeling. For the El Salvador tsunami, the largest wave height was observed in Baltra, Galapagos Islands (71.1 cm) located about 1,400 km away from the source. The tsunami governing periods were around 9 and 19 min. Numerical modeling indicated that most of the tsunami energy was directed towards the Galapagos Islands, explaining the relatively large wave height there. For the Philippines tsunami, the maximum wave height of 30.5 cm was observed at Kushimoto in Japan located about 2,700 km away from the source. The tsunami governing periods were around 8, 12 and 29 min. Numerical modeling showed that a significant part of the far-field tsunami energy was directed towards the southern coast of Japan. Fourier and wavelet analyses as well as numerical modeling suggested that the dominant period of the first wave at stations normal to the fault strike is related to the fault width, while the period of the first wave at stations in the direction of fault strike is representative of the fault length.

Key words: Tsunami, earthquake, DART, tide gauge, spectral analysis, Fourier analysis, wavelet analysis, numerical modeling, El Salvador earthquake, Philippines earthquake.

1. Introduction

We studied two small tsunamis occurring in the Pacific basin in August 2012, generated by submarine earthquakes offshore El Salvador and Philippines (Fig. 1). According to the United States Geological Survey (USGS 2012a), the El Salvador earthquake occurred on 27 August 2012 at 04:37:20 GMT. The epicenter of this *Mw*-7.3 earthquake was at 12.278°N and 88.528°W at the depth of around 20 km, producing almost no damage or casualty in the region (REUTERS 2012a). However, it generated a small tsunami in the Pacific Ocean whose wave amplitude was reported around 10 cm along the coastlines (CNN 2012a). A Field survey by BORRERO *et al.* (2014) showed that the tsunami generated a maximum runup of 6 m in the near-field, injuring several people. The first information bulletin about this tsunami was issued around 8 min after the earthquake occurrence, by the Pacific Tsunami Warning Center (PTWC) (PTWC 2012a). Following this early information, a tsunami warning was issued at 04:58 GMT (around 20 min after the earthquake) for the region (PTWC 2012b), and was cancelled around 110 min after the earthquake (PTWC 2012c).

The Philippines tsunami was slightly larger than the El Salvador one, having been produced by a slightly larger earthquake (*Mw* 7.6). According to the USGS (2012b), the origin time of the Philippines earthquake was at 12:47:34 GMT on 31 August 2012. The epicenter was at 10.838°N and 126.704°E at a depth of around 35 km (USGS 2012b). This offshore earthquake caused little damage and one death in the Philippines (REUTERS 2012b). Following this large submarine earthquake, a tsunami warning was issued at 12:55 GMT (around 8 min after the earthquake) for the region (PTWC 2012d). No damage was reported from the tsunami. Finally, the tsunami warning was cancelled at 14:54 GMT, about 2 h after the earthquake (PTWC 2012e).

In the following, we perform statistical, Fourier and wavelet analyses on the sea level records of these tsunamis in order to characterize the tsunami waves. In addition, numerical modeling of tsunami is conducted to give us insights into the propagation pattern of tsunami waves in the Pacific Basin.

[1] Earthquake Research Institute (ERI), The University of Tokyo, 1-1-1 Yayoi, Bunkyo-ku, Tokyo 113-0032, Japan. E-mail: mheidar@eri.u-tokyo.ac.jp
[2] Cluster of Excellence "The Future Ocean", Christian-Albrechts University of Kiel, Kiel, Germany.

2. Sea level data

We report and analyze 25 sea level records of the aforesaid tsunamis consisting of both tide gauges and Deep-ocean Assessment and Reporting of Tsunamis (DART) records (Fig. 1; Tables 1, 2).

The sea level data were provided by the USA National Oceanographic and Atmospheric Administration (NOAA 2012) and the UNESCO Intergovernmental Oceanographic Commission (IOC 2012). The sampling interval for all of the sea level data is 1 min.

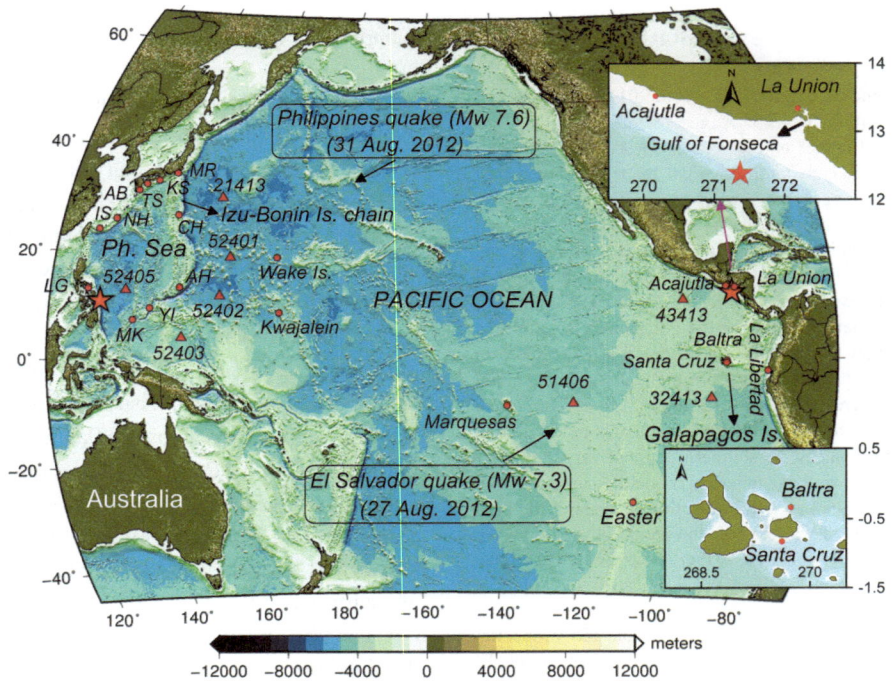

Figure 1

General location map of the Pacific Basin showing the locations of the El Salvador tsunami of 27 August 2012 (*right asterisk*) and the Philippines tsunami of 31 August 2012 (*left asterisk*). *Solid circles* and *triangles* represent the locations of tide gauge and DART stations, respectively. Abbreviations are: *LG* Legaspi, *MK* Malakal, *YI* Yap Island, *AH* Apra Harbor, *IS* Ishigakijima, *NH* Naha, *TS* Tosashimizu, *AB* Aburatsu, *KS* Kushimoto, *CH* Chichijima, *MR* Mera, and *Ph. Sea* Philippine Sea

Table 1

Sea level stations used to study the El Salvador tsunami of 27 August 2012

No.	Sea level station	Country (state)	Longitude	Latitude	Distance to the source (km)
Tide gauge stations					
1	La Union	El Salvador	87.82°W	13.31°N	140
2	Acajutla	El Salvador	89.84°W	13.57°N	206
3	Baltra	Ecuador	90.28°W	00.43°S	1,433
4	Santa Cruz	Ecuador	90.31°W	00.75°S	1,468
5	La Libertad	Ecuador	80.91°W	02.22°S	1,829
6	Easter	Chile	109.45°W	27.15°S	4,985
DART stations					
7	43413	SW of Acapulco, Mexico	99.85°W	11.07°N	1,271
8	32413	NW of Lima, Peru	93.50°W	07.40°S	2,267

Table 2

Sea level stations used to study the Philippines tsunami of 31 August 2012

No.	Sea level station	Country (state)	Longitude	Latitude	Distance to the source (km)
Tide gauge stations					
1	Legaspi	Philippines (Albay)	123.76°E	13.15°N	417
2	Malakal	Palau Islands	134.45°E	07.33°N	949
3	Yap Island	Micronesia	138.12°E	09.51°N	1,283
4	Ishigakijima	Japan (Okinawa)	124.10°E	24.20°N	1,520
5	Naha	Japan (Okinawa)	127.67°E	26.22°N	1,721
6	Aburatsu	Japan	131.41°E	31.58°N	2,376
7	Chichijima	Japan	142.19°E	27.09°N	2,507
8	Tosashimizu	Japan (Kochi)	132.97°E	32.78°N	2,548
9	Kushimoto	Japan	135.77°E	33.48°N	2,724
10	Mera	Japan	139.83°E	34.92°N	3,063
11	Wake Island	USA (Wake Island)	166.62°E	19.28°N	4,557
12	Kwajalein	Marshall Islands	167.73°E	08.73°N	4,588
DART stations					
13	52405	West of Pacific Ocean, Guam	132.33°E	12.88°N	669
14	52403	West of Pacific Ocean, Truk	145.59°E	04.05°N	2,241
15	52402	NW of Kwajalein	154.04°E	11.87°N	3,055
16	52401	NE of Saipan	155.77°E	19.26°N	3,380
17	21413	SE of Tokyo, Japan	152.12°E	30.52°N	3,590

3. Methods of Waveform Analyses

3.1. Tsunami Waveform Detection

The following steps were employed to detect the tsunami signal from the sea level records: (1) quality control, and (2) removing low-frequency signals like tides by high-pass filtering. The Butterworth IIR digital filter (MATHWORKS 2012) was employed to remove low frequency signals for which a cut-off frequency of 0.0003 Hz (about 1 h) was chosen. To examine whether the Butterworth digital filter generates a spurious leading depression wave or not in our analyses, we performed tidal harmonic analysis for selected waveforms (Sect. 4.1).

3.2. Statistical Analysis

We calculated the following physical parameters of the tsunami (HEIDARZADEH and SATAKE 2013a): (1) the arrival time of the first tsunami wave, (2) the polarity of the first wave, (3) the arrival time of the maximum wave, (4) the maximum trough-to-crest wave height, (5) the number of waves before the arrival of the maximum wave, and (6) duration of tsunami.

3.3. Spectral Analysis

The spectral content of a tsunami record is mostly dictated by two factors: (1) the effect of local and regional bathymetric features like continental shelves and coastal harbors/bays, and (2) the effect of tsunami source and dimensions of the seafloor deformation. In this context, a tide gauge record of a tsunami usually includes both of the aforesaid effects, because tide gauges are deployed in coastal areas and hence bathymetric features play an important role in their records. In contrast, DART records mostly reflect tsunami source characteristics, as they are deep-water instruments and hence are free from shallow bathymetric effects. To separate tsunami source and bathymetric effects in a tide gauge record, RABINOVICH (1997) proposed to compare spectral peaks computed from tsunami water level data recorded at different locations and then to pick the common spectral peaks contained within different water level spectra from a particular tsunami. We applied this method to separate source and bathymetric effects in our tide gauge data. In this study, we applied two methods to investigate the spectral content of the tsunami signals: Fourier and wavelet analyses.

3.3.1 Fourier Analysis

For Fourier analysis, we used Welch's averaged-modified-periodogram method of spectral estimation by considering Hamming window and overlaps (WELCH 1967). Calculation of the signal spectrum was done using the Welch algorithm from the signal processing toolbox in Matlab program (MATHWORKS 2012).

3.3.2 Wavelet Analysis

Wavelet analysis, also known as the frequency-time (f–t) analysis, has been used in tsunami research for studying the temporal changes of tsunami energy (RABINOVICH and THOMSON 2007; HEIDARZADEH and SATAKE 2013a; BORRERO and GREER 2013). Since a tsunami is a non-stationary signal, its spectral content varies in strength and peak frequency over time. Wavelet analysis presents the distribution of tsunami energy in different frequency bands (f) over time (t). In other words, wavelet analysis shows in which period band tsunami energy is concentrated at a particular time, and hence can be considered as a complementary analysis for Fourier analysis. Because a tsunami is a non-stationary signal, the use of wavelet analysis allows for the analysis of the frequency content of a tsunami wave train as it changes over time. More details about wavelet analysis is given in TORRENCE and COMPO (1998) and HEIDARZADEH and SATAKE (2013a, b).

4. Results of the Analyses

4.1. Tsunami Waveforms

Figure 2 presents the original and filtered signals for the El Salvador tsunami of 27 August 2012 (Fig. 2a) and the Philippines tsunami of 31 August 2012 (Fig. 2b, c). It can be seen that the tsunami signal is clear in most of the analyzed tide gauge records. The El Salvador tsunami was recorded on five coastal tide gauges and two DART stations (DARTs 43413 and 32413). The Philippine tsunami was recorded on ten coastal tide gauges and on two DART stations of 52405 and 52403 with 1-min temporal resolution. While the tsunami was possibly recorded on DART stations 52402, 52401 and 21413, the tsunami arrival occurred after the station had been switched from 1-min sampling to 15 min sampling. Under-sampled traces of seismic waves are also present in the DART records of both events and are evident in the figures.

Figure 2d compares the results of Butterworth digital filter (middle panel) with those of tidal harmonic analysis (right panel) for four stations recording the Philippines tsunami. We applied the tidal analysis package Tidal Analysis Software Kit (TASK) for tidal analysis (BELL et al. 2000). Figure 2d shows that both of the analyses yield almost the same results, indicating that the applied digital filter is not producing spurious leading depression waves.

4.2. Statistical Properties of the Tsunamis

Based on the filtered tsunami signals (Fig. 2), the main physical properties are summarized in Tables 3 and 4.

4.2.1 El Salvador Tsunami

Among the analyzed sea level stations for the El Salvador tsunami, the Acajutla tide gauge was the first one to receive the tsunami. The first arrival in La Union can be distinguished by taking into consideration the fact that the period of tsunami is larger than that of noise signals. The time interval between the first tsunami peak (asterisk in Fig. 2a) and the next peak is around 16 min for the La Union record. It will be shown later that the 16-min signal is one of the governing signals of the El Salvador tsunami. The largest trough-to-crest wave height is 71.1 cm recorded at the Baltra tide gauge station at a distance of around 1,400 km from the tsunami source (Table 3). At other far-field stations like Santa Cruz and La Libertad, wave heights of 34.4 and 35.6 cm were recorded, respectively, though the Santa Cruz station is located in the far-field on the lee side of the Galapagos island system relative to the direction of incidence of the tsunami (Fig. 1). The far-field wave heights are about five times larger than the near-field wave heights (e.g., La Union station). Comparison of average values for the DART and tide gauge

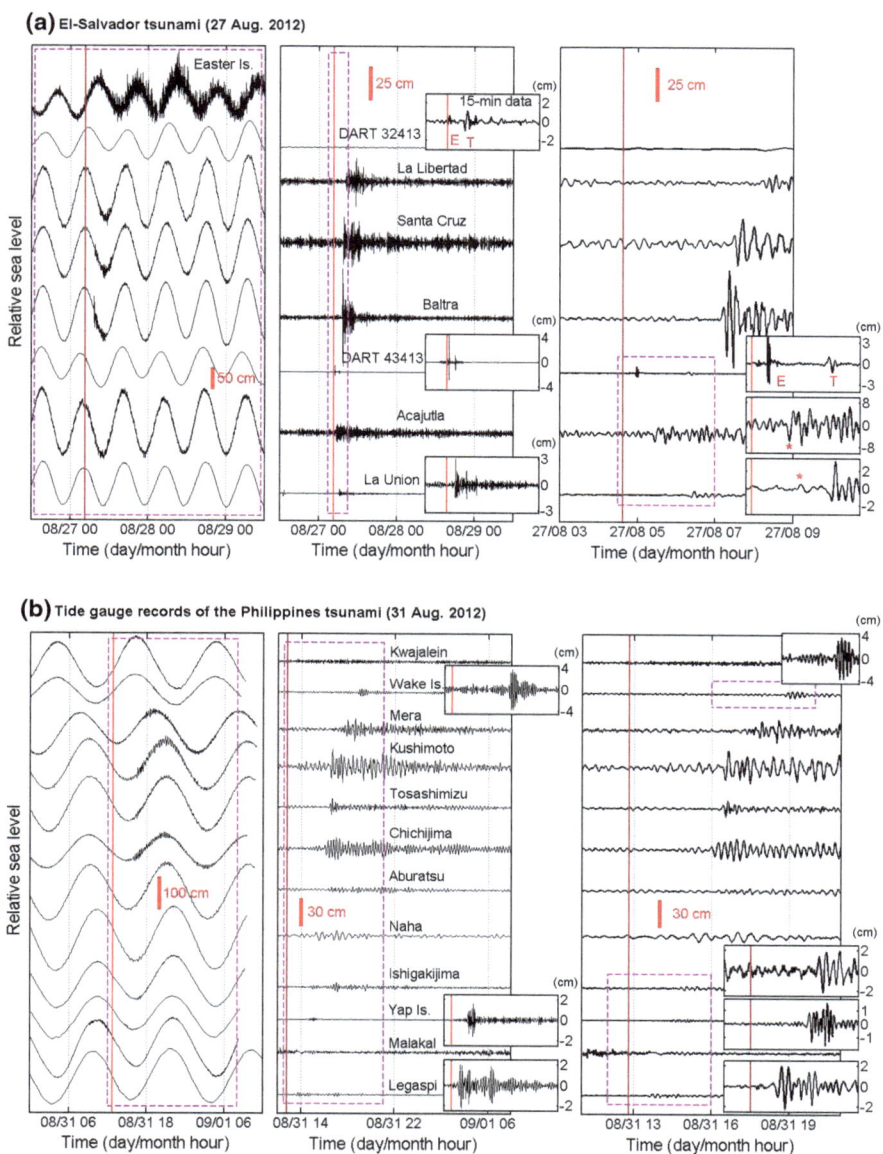

Figure 2

Sea level records of the El Salvador tsunami of 27 August 2012 (**a**) and the Philippines tsunami of 31 August 2012 (**b, c**). The *left* and *middle panels* show the original and filtered signals, respectively. In the *right panel*, only a small part of the filtered signal is enlarged. The letters *E* and *T* represent Earthquake and Tsunami, respectively. The *dashed rectangle* shows part of the data enlarged in the neighboring panel. *Small insets* show the respective waveforms with a better resolution. *Asterisks* show the arrival times of the first wave. **d** Comparison of the results of waveform analysis performed by the Butterworth digital filter (*middle panel*) and the tidal harmonic analysis (*right panel*). Tide predictions are shown by *blue curves* in the *left panel*. The *red-vertical line* represent the time of the earthquake occurrence

waveforms (Table 3) indicate three main differences between tsunami records on DARTs and tide gauges: (1) Duration of tsunami oscillations on tide gauges is about four times longer than that on DARTs, (2) the maximum wave height on tide gauges is 14 times larger than that on DARTs, and (3) the first wave is the largest one in DART records, whereas the second, third or later wave is the largest in tide gauge records.

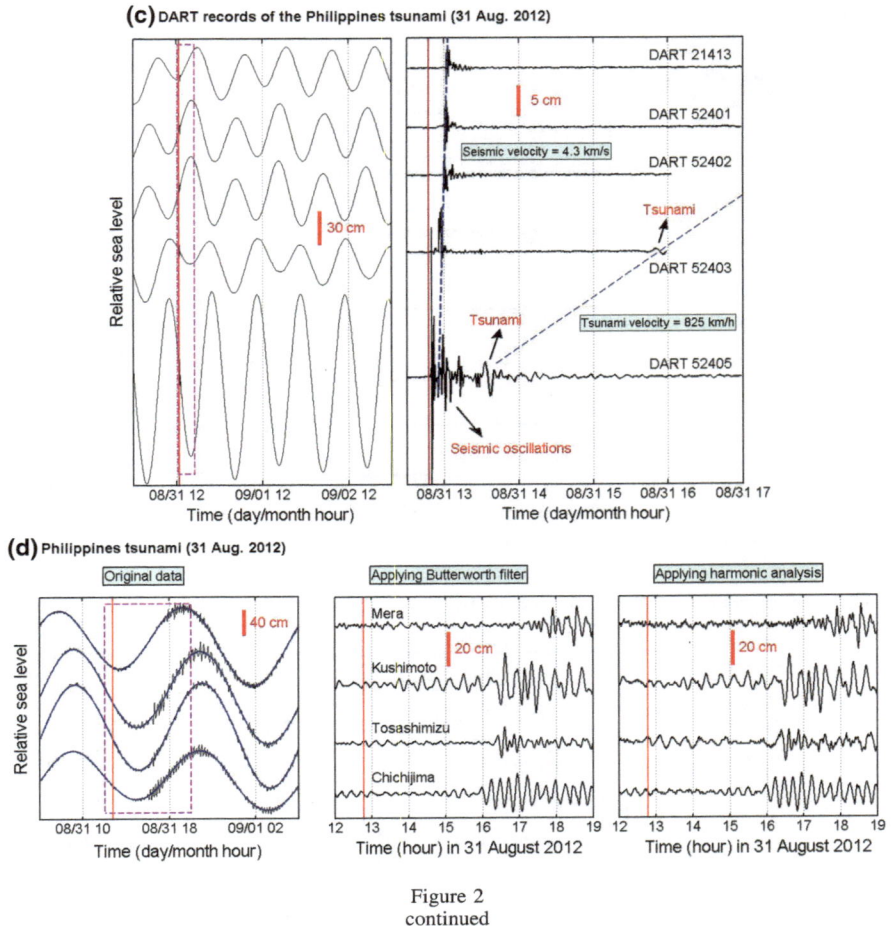

Figure 2
continued

4.2.2 Philippines Tsunami

The largest wave height was 30.5 cm, recorded at Kushimoto (Japan) for the Philippines tsunami. Similar to the El Salvador tsunami, among the analyzed sea level stations for the Philippines tsunami (Table 4), the near-field wave heights are significantly smaller than the far-field ones. According to Table 4, deep water tsunami waves recorded at DART 52405 have smaller amplitudes and last for a shorter time compared to those recorded by tide gauges.

4.3. Spectral Peaks

Fourier analysis of the tsunami waveforms identified a few peaks, which are shown by asterisks in Fig. 3 and are summarized in Tables 5, 6. We performed Fourier analysis for both the entire tsunami waveform (Fig. 3/left panels) and for the first 2 h (Fig. 3/right panels), and then the peak periods were picked from the latter spectra. It is expected that the 2-h-long spectra are less influenced by coastal bathymetric features. In general, a clear difference can be seen between the tsunami spectra of DARTs and those of tide gauges. Tsunami spectrum is usually simple and consists of only one peak for DARTs, whereas it is relatively complicated and includes several peaks for tide gauges.

4.3.1 El Salvador Tsunami

The analysis suggests spectral peaks at approximately 19 and 9 min for the El Salvador tsunami (Fig. 3a;

Table 3

Statistical properties of the El Salvador tsunami of 27 August 2012

No.	Sea level station	First wave			Maximum wave				Duration (h)
		Arrival time (GMT) (dd/mm-HH:MM)	Travel time[a] (h:min)	Sign[b]	Arrival time (GMT) dd/mm-HH:MM	Travel time[a] (h:min)	Observed max. wave height (cm)[c]	No. of the max. wave	
1	La Union	27/08-05:39	01:02	(+)	27/08-06:11	01:34	4.5	3	10
2	Acajutla	27/08-05:23	00:46	(−)	27/08-09:29	04:52	12.5	23	14
3	Baltra	27/08-07:08	02:31	(+)	27/08-07:21	02:44	71.1	2	10
4	Santa Cruz	27/08-07:25	02:48	(+)	27/08-07:34	02:57	34.4	2	7
5	La Libertad	27/08-08:11	03:44	(+)	27/08-11:23	06:46	35.6	17	10
6	Easter	NA[d]	NA	NA	NA	NA	NA	NA	NA
Average							31.6	9	10.2
7	DART 43413	27/08-06:15	01:38	(+)	27/08-06:19	01:42	2.3	1	2
8	DART 32413	NA	NA	NA	27/08-08:15	03:38	2.2	1	2.8
Average							2.25	1	2.4

[a] Compared to the earthquake origin time (27/08/2012 04:37′:20″ GMT)

[b] (−) and (+) represent leading depression and elevation waves, respectively

[c] Maximum trough-to-crest wave height

[d] Not Applicable

Table 4

Statistical properties of the Philippines tsunami of 31 August 2012

No.	Sea level station	First wave			Maximum wave				Duration (h)
		Arrival time (GMT) (dd/mm-HH:MM)	Travel time[a] (h:min)	Sign[b]	Arrival time (GMT) (dd/mm-HH:MM)	Travel time[a] (h:min)	Observed max. wave height (cm)[c]	No. of the max. wave	
1	Legaspi	31/08-13:23	00:36	(+)	31/08-13:43	00:56	4.2	2	9
2	Malakal	NA[d]	NA	NA	NA	NA	NA	NA	NA
3	Yap Island	31/08-14:28	01:41	(+)	31/08-15:04	02:17	3.2	6	2
4	Ishigakijima	31/08-14:45	01:58	(+)	31/08-16:59	04:12	9.8	11	10
5	Naha	31/08-15:05	02:18	(+)	31/08-17:00	04:13	10.5	5	5
6	Aburatsu	NA	NA	NA	31/08- 20:53	08:06	7.3	NA	6
7	Chichijima	31/08-15:52	03:07	(+)	31/08-16:57	04:10	24	5	13
8	Tosashimizu	31/08-16:19	03:32	(+)	31/08-16:37	03:50	17.9	2	2
9	Kushimoto	31/08-16:25	03:38	(+)	31/08-16:38	03:51	30.5	2	10
10	Mera	31/08-17:36	04:49	(+)	31/08-18:34	05:47	23.9	5	6
11	Wake Island	31/08-18:36	05:49	(+)	31/08-19:05	06:18	7.5	3	3
12	Kwajalein	NA	NA	NA	NA	NA	NA	NA	NA
Average							13.9	5	6.6
13	DART 52405	31/08-13:31	00:44	(+)	31/08-13:33	00:46	5.8	1	1
14	DART 52403	31/08-15:50	03:03	(+)	31/08-15:50	03:03	1.01	1	–

[a] Compared to the earthquake origin time (31/08/2012 12:47':34" GMT)

[b] (−) and (+) represent leading depression and elevation waves, respectively

[c] Maximum trough-to-crest wave height

[d] Not applicable

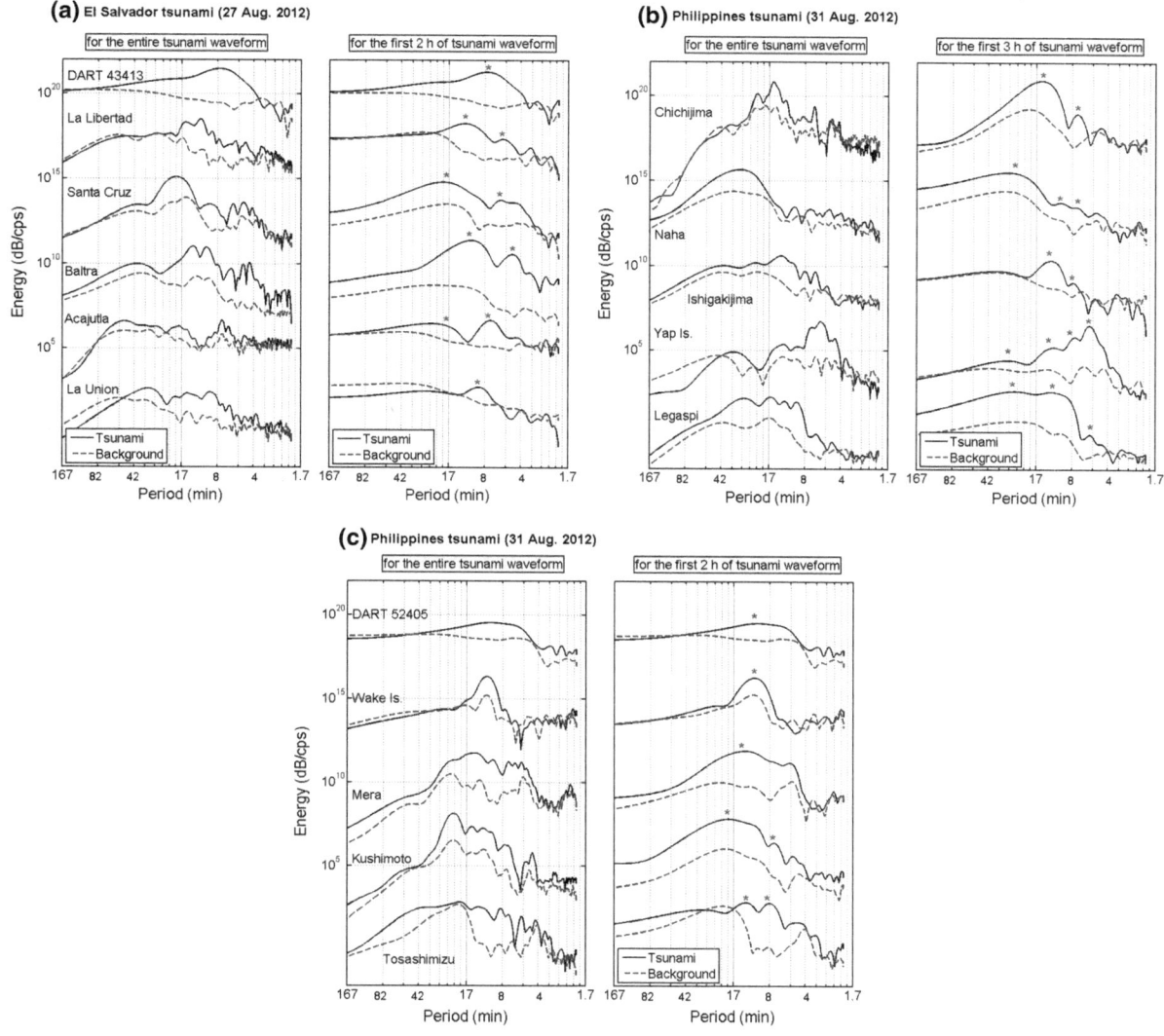

Figure 3
Spectra for the sea level records of the El Salvador tsunami of 27 August 2012 (**a**) and the Philippines tsunami of 31 August 2012 (**b, c**). Fourier analysis is performed for both the entire tsunami waveforms (*left panels*) and for the first 2 h of the tsunami waveforms (*right panels*). *Asterisks* show some of the peak periods in each spectrum. *Solid* and *dashed lines* show the spectra of tsunami and background signals, respectively

Table 5); these are possibly the main tsunami source periods. The most governing signal is possibly the one with the period of around 9 min, because this signal is the one that is most common in Table 5.

4.3.2 Philippines Tsunami

For the Philippines tsunami, the governing periods are 29, 12, and 8 min (Fig. 3b, c; Table 6), among which the 12-min period seems to be the strongest

one because it is the most common signal in different stations (Table 6). The period of the tsunami recorded at the DART 52403 is also 12 min (Fig. 2c). Some unusual observations can be seen in the tsunami spectra shown in Fig. 3b, c. For example, during the Philippines tsunami, the most powerful signal is the 5.9-min signal in Yap Island, whereas either 29- or 12-min signals are the governing ones in all of the other stations. The 5.9-min period is possibly one of the resonance modes

271

Table 5

Governing periods of the El Salvador tsunami of 27 August 2012 based on the spectra of the first 2 h of the tsunami waveforms (Fig. 3a/right panel)

No.	Station	Peak periods (min)
1	La Union	9.4
2	Acajutla	19.3, 7.5
3	DART 43413	7.9
4	Baltra	10.8, 4.9
5	Santa Cruz	18.5, 6.2
6	La Libertad	12.2, 5.9

Table 6

Governing periods of the Philippines tsunami of 31 August 2012 based on the spectra of the first 2 h of the tsunami waveforms (Fig. 3b, c/right panel)

No.	Station	Peak periods (min)
1	Legaspi	25.6, 12.4, 5.9
2	Yap Island	28.9, 11.8, 7.9, 5.9
3	Ishigakijima	12.5, 8.2
4	Naha	28.9, 10.4, 7.8
5	Chichijima	15.0, 7.5
6	Tosashimizu	13.1, 8.3
7	Kushimoto	18.3, 7.7
8	Mera	13.2
9	Wake Island	11.1
10	DART 52405	10.4

of the harbor/bay, in which the Yap Island tide gauge is located.

4.4. Temporal Changes of Dominant Periods

Wavelet analyses for the sea level records of the El Salvador and Philippines tsunamis are shown in Fig. 4, as is the global wavelet spectrum, which is the time-averaged spectral energy over all times. It should be noted that Fourier analysis (Fig. 3) gives the power of signals with different periods over the entire tsunami record, whereas wavelet analysis gives the time evolution of wave energy. Hence, the Fourier analysis given by global wavelet spectrum (Fig. 4) is slightly different from that of normal Fourier analysis performed by Welch algorithm (Fig. 3). This is why there are differences between the amount of peak energy given by Fourier (Fig. 3) and wavelet analyses (Fig. 4) although the peak periods are almost the same in both.

4.4.1 El Salvador Tsunami

For the El Salvador tsunami recorded at Santa Cruz, most of the tsunami energy is distributed in the narrow period band of 11–24 min over the entire tsunami oscillations. The Acajutla's $f–t$ plot shows that the level of wave energy before the earthquake occurrence is as high as that after the earthquake for the period band of around 45 min, indicating that this period band cannot be attributed to the tsunami. There is almost no energy at the period bands of around 7–10 and 18–20 min before the earthquake occurrence, whereas the wave energy at these period bands is higher after the earthquake. This indicates that both of the aforesaid periods may belong to the tsunami. On the other hand, we may conclude that the peak period of 45 min at the La Libertad station (Fig. 4a) belongs to non-tsunami sources (e.g., local and regional bathymetric effects).

4.4.2 Philippines Tsunami

Based on the $f–t$ plots for the Philippines tsunami (Fig. 4b, c), it can be seen that most of the wave energy is distributed at three different period bands of around 10–13, 25–30, and 45 min. However, the period band of 45 min most probably belongs to non-tsunami sources, because the level of energy in this band is high before the arrival of tsunami in many of the examined stations. This idea is supported by the tsunami and background spectra shown in Fig. 3b, c because the levels of energy in both spectra are almost the same around the period of 45 min. Distinct wave trains with high energy contents are clear at Naha. According to the wavelet plot of the Ishigakijima record (Fig. 4b), by neglecting the 45-min signal, the governing period of tsunami is around 10–12 min for the first few hours after the tsunami arrival; then the governing period switches to the period of around 20–28 min. On the contrary, the 28-min signal is the first signal arriving at the Legaspi station (Fig. 4b). At Wake Island (Fig. 4c), most of the tsunami energy occurs at the period of around 10–12 min and almost no energy is present at the period band of 20–28 min. For Mera (Fig. 4c),

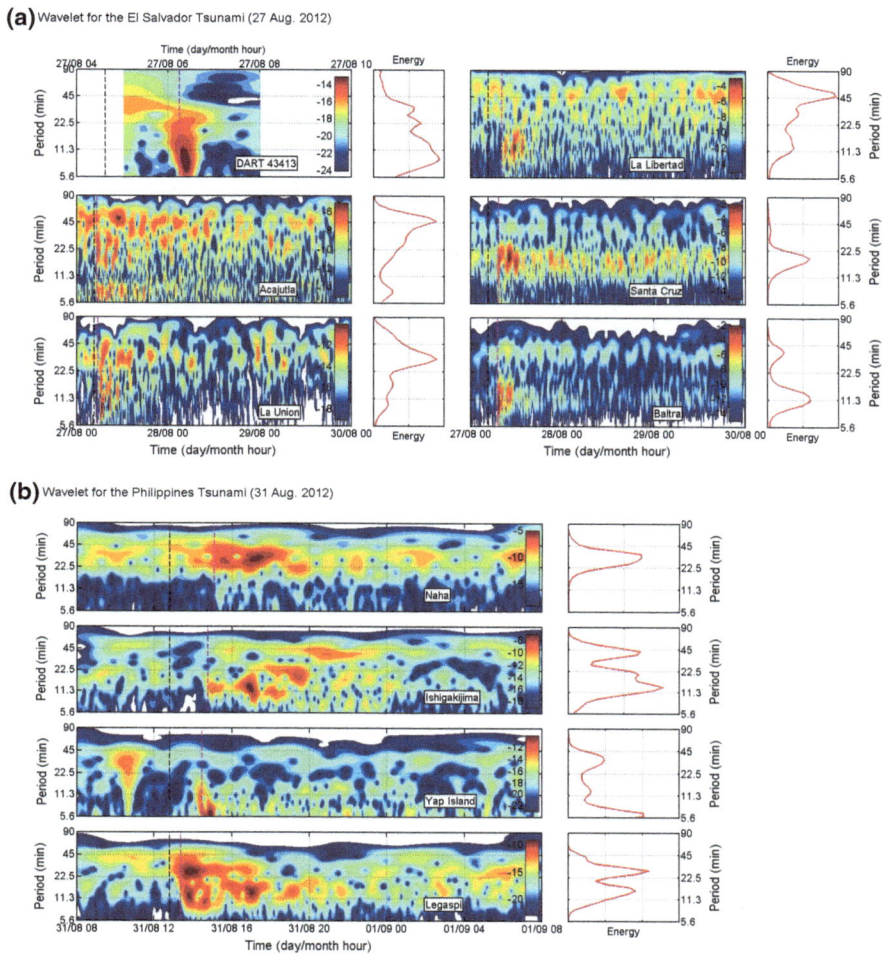

Figure 4

Wavelet analysis for the sea level records of the El Salvador tsunami of 27 August 2012 (**a**) and the Philippines tsunami of 31 August 2012 (**b**, **c**). *Dashed-vertical dark* and *purple lines* represents the earthquake occurrence and tsunami arrival times, respectively. The *small plots*, at the *right side* of each wavelet plot, show the global wavelet spectrum

tsunami energy is distributed over a wide period band of 5–25 min. The governing period is around 10–15 min for Chichijima, while it is around 20–25 min for Kushimoto (Fig. 4c).

4.5. Summary of the Sea Level Data Analysis

The analysis of the El Salvador and Philippines tsunamis of August 2012 using sea level data showed some unusual observations, such as: (1) for the El Salvador tsunami, relatively large wave height of 34.4 cm was observed on the lee side of the Galapagos island system relative to the direction of incidence of the tsunami; (2) although the Philippines earthquake (*Mw* 7.6) was larger than the El Salvador one (*Mw* 7.3), the maximum wave height generated by the El Salvador tsunami was almost two times bigger than that generated by the Philippines tsunami; (3) the largest wave heights produced by the Philippines tsunami were observed in the far-field especially along the Japan coast; and (4) the largest wave heights of the El Salvador tsunami occurred in the far-field. We perform numerical modeling of both tsunamis in order to shed some light on the above unusual observations.

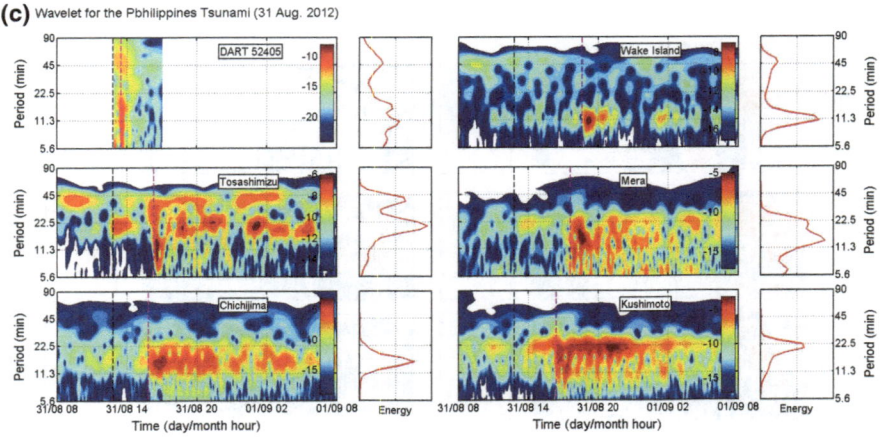

Figure 4
continued

5. Numerical Modeling of Tsunami Waves and Validation

5.1. Method and Validation

The analytical formulas by OKADA (1985) were employed to calculate the seafloor deformation due to the parent earthquake using the seismic parameters of the earthquake. The earthquake fault parameters are taken from the GLOBAL CMT (2012) and the USGS centroid moment solution (USGS 2012a, b) summarized in Table 7. Figure 5a shows the result of the seafloor deformation modeling. The 1-min bathymetry grid provided through the GEBCO digital atlas was used here for numerical modeling of tsunami (IOC et al. 2003). The numerical model used here is the same as that described in YALCINER et al. (2004), solving non-linear shallow water equations using a leap frog scheme on a staggered grid system. A time step of 3.0 s is applied and tsunami inundation on dry land is not included.

To validate the results of tsunami modeling, we compare time histories of the simulated waves with actual ones observed on DARTs and tide gauges. Figure 5b presents the results of this comparison, in which five actual sea level records are compared with the simulated ones for each tsunami. Although the grid spacing is 1 min, good agreement can be seen between the simulated and observed waveforms. The agreement is highly satisfactory for DART records. In Legaspi, the discrepancy between the observed and

simulated wave height is relatively large; however, the arrival times are almost the same. It seems that the tsunami simulations performed here are accurate enough for this study, which is aimed at studying the propagation pattern of the two tsunamis.

5.2. The El Salvador Tsunami

Snapshots of the El Salvador tsunami in Fig. 6a show that tsunami waves experience two major reflections from Isla-Del-Coco and the Galapagos Islands, which occur about 1.5 and 2.5 h after the earthquake, respectively. It is clear from Fig. 6a that each reflection acts as a new source for tsunami waves, introducing new wave trains into the tsunami wave field. Late wave trains in tsunami waveforms (Fig. 2) may be attributed to these reflected waves that arrive some hours after the earthquake generation. It is long known that a tsunami is a series of long waves; not a single wave. This is clear from the snapshot of the El Salvador tsunami at the time of 3 h (Fig. 6a) where a chain of successive wave crests (C) and troughs (T) can be seen.

Figure 6c shows the distribution of the maximum wave height of the El Salvador tsunami. Most of the tsunami energy is directed towards the Galapagos Islands. This is due to the directivity of tsunami waves in the far-field. Proposed by BEN-MENAHEM and ROSENMAN (1972), directivity of tsunami indicates that most of the tsunami energy travels

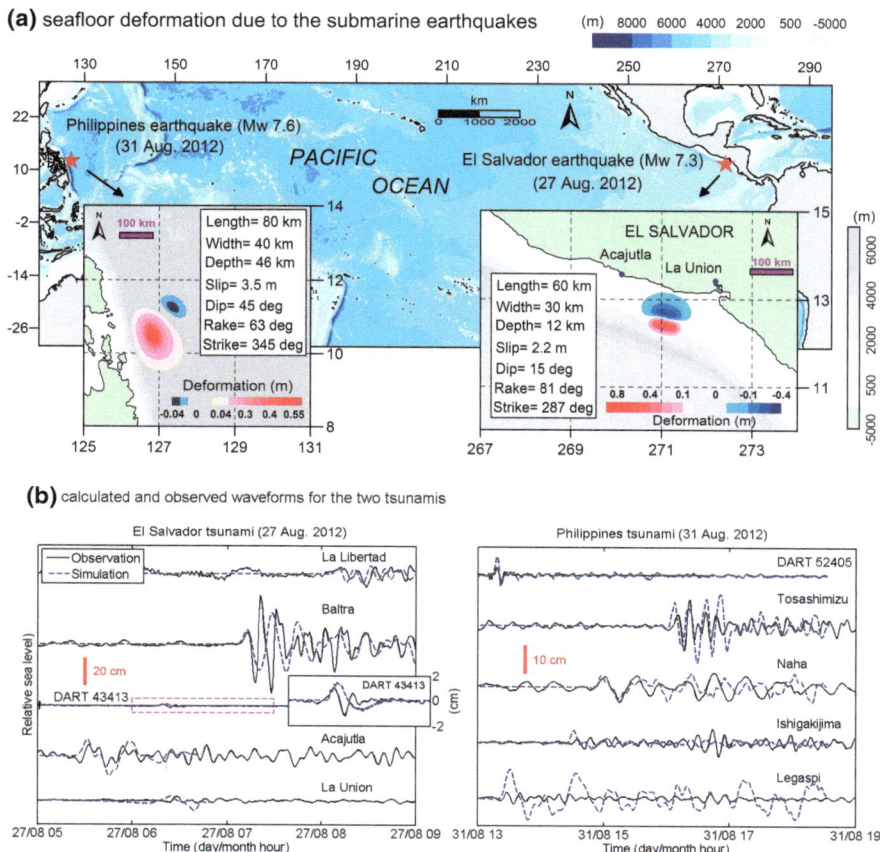

(a) seafloor deformation due to the submarine earthquakes

(b) calculated and observed waveforms for the two tsunamis

Figure 5
Results of numerical modeling of tsunami. **a** Seafloor deformation due to the El Salvador and Philippines tsunamis. Numbers on axes are in degrees. *Asterisks* show the epicenters of the earthquake. **b** Simulated and observed waveforms for the two tsunamis

perpendicular to the source strike in the far-field. Considering that the strike of the fault responsible for the El Salvador tsunami is SE–NW (Fig. 5a), the largest tsunami waves travel in the NE–SW direction (Fig. 6c). Hence, observation of the maximum wave height of the El Salvador tsunami at the Baltra station can be explained by the directivity effect. Snapshots of the El Salvador tsunami simulation around the Santa Cruz station (Fig. 6d) reveal that two different waves arrived almost at the same time from opposite sides, and are superpositioned into a single wave. One wave train comes from around the Isabela Island to the west and another wave train from the east side of the Santa Cruz Island. This phenomenon may explain the relatively large wave height of around 35 cm in Santa Cruz.

5.3. The Philippines Tsunami

Snapshots of the Philippines tsunami in Fig. 6b reveal that the tsunami wave field is more complicated than that of the El Salvador tsunami; this can be attributed to the presence of the Izu-Bonin Island chain and other Pacific Islands around the tsunami source area. Tsunami waves are reflected from this island chain and can hardly exit the Philippine Sea region. This is also supported by Fig. 6c, where the distribution of the maximum wave height of tsunami is presented. Figure 6c reveals that most of the tsunami energy is confined within the Philippine Sea region and only a small part of the tsunami is able to exit from the Philippine Sea.

The other fact in Fig. 6c is that a significant part of the Philippines tsunami energy is directed towards the

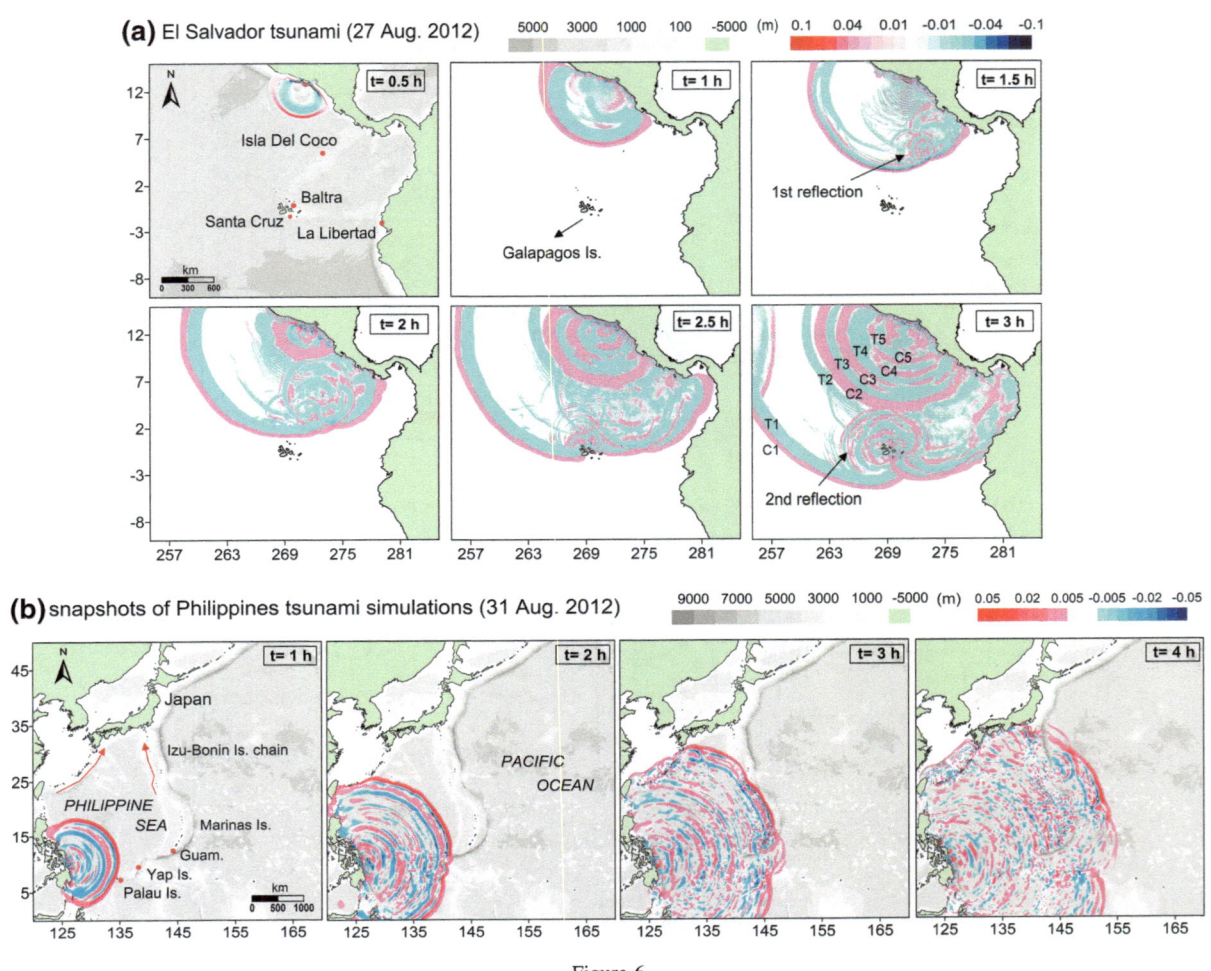

Figure 6

Results of numerical modeling of tsunami. **a** Snapshots of tsunami simulations for the El Salvador tsunami. *C* and *T* represent crest and trough waves, respectively. **b** Snapshots of tsunami simulations for the Philippines tsunami. **c** Distribution of the maximum wave amplitudes of tsunami for the two tsunamis. **d** Snapshots of the El Salvador tsunami around the Santa Cruz station. Numbers on axes are in degrees

southern coast of Japan (arrow B in Fig. 6c). In fact, the energy of the Philippines tsunami is partitioned into two parts: (1) the first part of tsunami energy travels perpendicular to the source strike (arrow A in Fig. 6c), which can be explained by the directivity effect; and (2) the second part travels towards the southern coast of Japan (arrow B in Fig. 6c), which can be explained by the effect of bathymetry on far-field propagation of tsunamis proposed by SATAKE (1988). Arrows in Fig. 6b show how the Izu-Bonin Island chain funnels the tsunami waves towards southern Japan. Observation of relatively large wave heights in Japanese coastlines due to the Philippines tsunami may be explained by this effect.

5.4. Comparing the Two Tsunamis

According to Tables 3 and 4, among the examined sea level records, the average wave height produced by the El Salvador tsunami is about two times larger than that produced by the Philippines tsunami, although the former earthquake (*Mw* 7.3) was smaller than the latter earthquake (*Mw* 7.6). The El Salvador earthquake was more efficient for tsunami generation than the Philippines one for three reasons. First, the former earthquake occurred at a shallower depth. Second, the former earthquake has a larger dip-slip component, although the dip angle is smaller (Table 7). Due to these reasons, the maximum

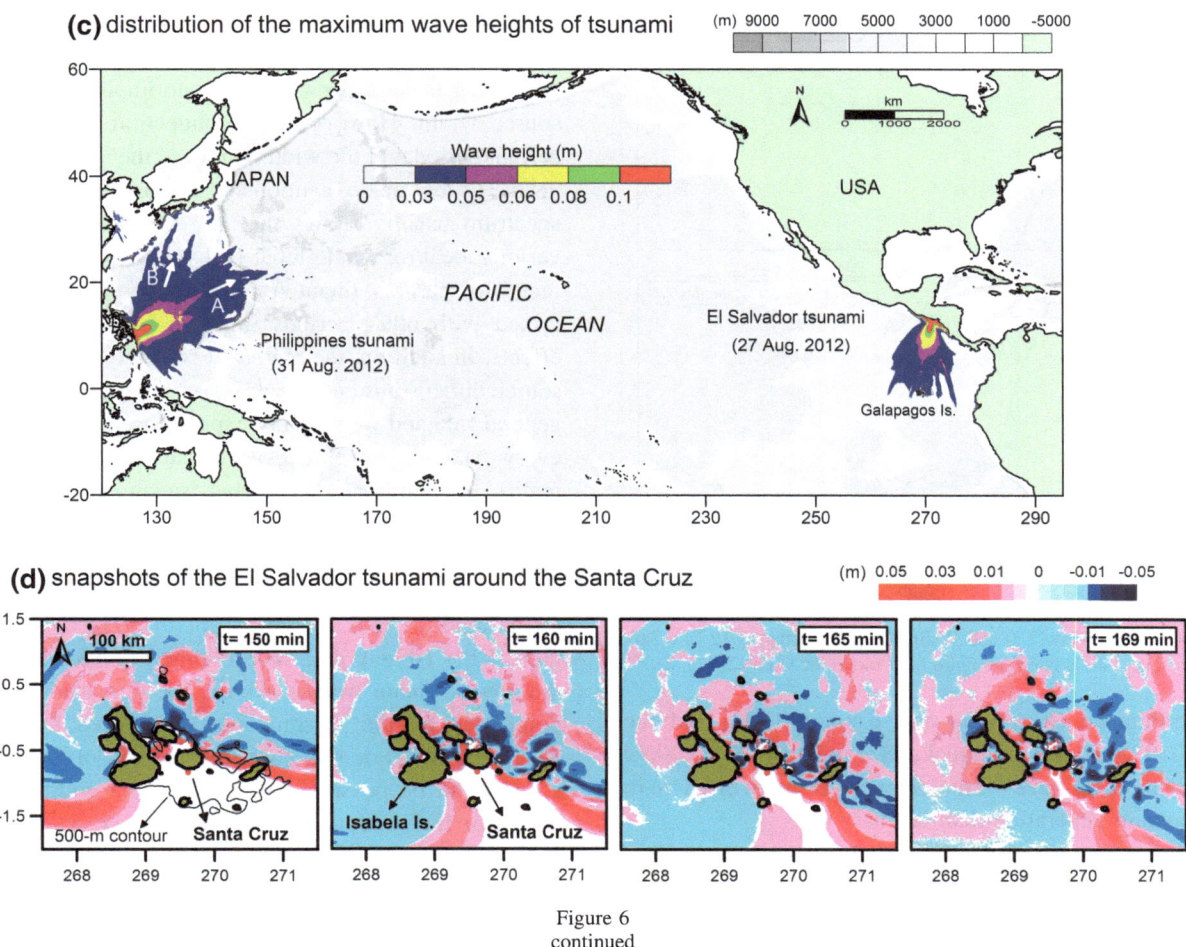

(c) distribution of the maximum wave heights of tsunami

(d) snapshots of the El Salvador tsunami around the Santa Cruz

Figure 6
continued

seafloor uplift due to the El Salvador earthquake is larger than that of the Philippines earthquake (Fig. 5a). It is evident that the larger the seafloor uplift is, the stronger the tsunami. Third, based on the source-time functions of both earthquakes shown in Fig. 7, the El Salvador earthquake was a slow earthquake. The source duration of the El Salvador earthquake was around 70 s; this is comparable to some other slow-tsunamigenic earthquakes (known as "tsunami earthquakes"), like the 1992 Nicaragua

Table 7

Tectonic parameters used for modeling seafloor deformation at the tsunami source according to GLOBAL CMT (2012) and the USGS (2012a, b)

Event name	M_0^a (dyn. cm)	Mw^b	Fault start point		Fault end point		Length (km)	Width (km)	Slip (m)	Depth (km)	Dip (deg)	Rake (deg)	Strike (deg)
			Lon (°)	Lat (°)	Lon (°)	Lat (°)							
El Salvador (27 Aug 2012)	1.18×10^{27}	7.3	88.70 (°W)	12.3 (°N)	89.22 (°W)	12.46 (°N)	60	30	2.2	12	15	81	287
Philippines (31 Aug 2012)	3.34×10^{27}	7.6	126.80 (°E)	10.20 (°N)	126.63 (°E)	10.90 (°N)	80	40	3.5	46	45	63	345

[a] Seismic moment

[b] Moment magnitude

277

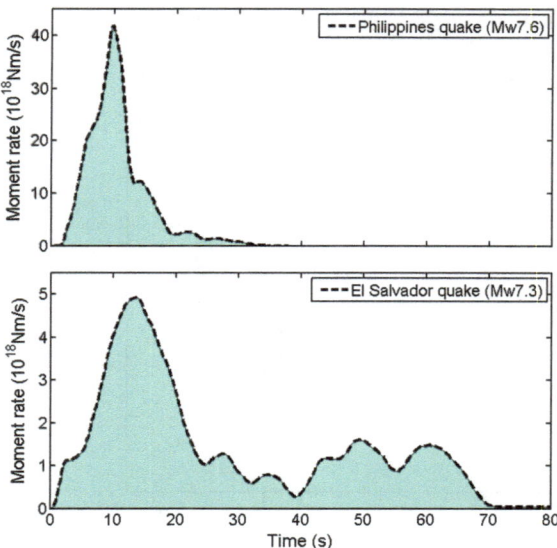

Figure 7
Source-time functions for the two earthquakes of El Salvador (27 August 2012-Mw 7.3) and Philippines (31 August 2012-Mw 7.6). The data is from the finite fault model of USGS (2012a, b)

earthquake (KANAMORI and KIKUCHI 1993) and the 2010 Mentawai earthquake (SATAKE et al. 2013). Slow earthquakes have been reported to be more efficient in tsunami generation than ordinary ones.

6. Discussions

6.1. Insights into the Tsunami Source Dimension Using Wavelet Analysis

It is already known that the main governing periods of tsunamis are normally dictated by the dimensions of the earthquake fault (i.e., length/width), as schematically shown for the Philippines tsunami in Fig. 8 (HEIDARZADEH and SATAKE 2013a). The sea level stations located normal to the fault strike (e.g., Wake Is. in Fig. 8) are mainly influenced by the tsunami period determined by the source width. For other stations located at different sides of the tsunami source (i.e., the lateral stations like Legaspi in Fig. 8), the tsunami signal controlled by the source length is usually the first wave to arrive at these stations. As most of the tsunami energy is usually concentrated towards normal to the source strike (i.e., directivity effect), the period of tsunami is mainly controlled by the width of the source rupture.

Therefore, even for the stations receiving first waves imposed by the source length, we may expect that the governing tsunami signal is the one imposed by the source width. However, any connection between a certain period and the width/length of the source fault needs to be made cautiously, because a tsunami spectrum usually shows multiple peaks due to the various local/regional/global bathymetric features. In fact, the periods dictated by the source fault are mixed with other periods imposed by bathymetric effects. In addition, the aforesaid connection between source dimensions and tsunami governing periods is a general rule and we may not expect it to hold true for every tsunami waveform, as tsunami source is, in reality, a heterogeneous one and tsunami waveforms are affected by several factors.

The Fourier analysis for the Philippines tsunami showed that the governing periods of this tsunami are around 12 and 29 min (Table 6). The wavelet plot of the Ishigakijima record (Fig. 4b) shows that the first tsunami wave arriving in this station is dominated by the 11-min signal, followed by the 25-min signal. For this case, the periods of 11 and 25 min can be possibly attributed to the width and length of the seafloor rupture, respectively (Fig. 8). In Chichijima (Fig. 4c), located almost normal to the fault strike, the 15-min signal is the first one to arrive at this station and is dominating for almost the entire tsunami lifetime. This 15-min signal seems to be dictated by the width of the tsunami source. We note that we do not necessarily expect exactly the same signal resulting from source width at different stations. Taking into account the complex wave field generated by a tsunami due to irregular bathymetry, as well as tsunami source heterogeneity, the two signals of 11 and 15 min can be considered close enough to each other as both originated from the source width. In Naha, located at an angle of around 45 degrees relative to the source strike, the first and the dominating signal is the one at the period of 28 min, which seems to be dictated by the length of the tsunami source. For the two middle-class stations of Naha and Ishigakijima, i.e. neither located pure normal relative to the source strike nor pure lateral, it can be seen that in one of them the first signal is at 11 min and in the other it is at 28 min. In Wake Island, located truly

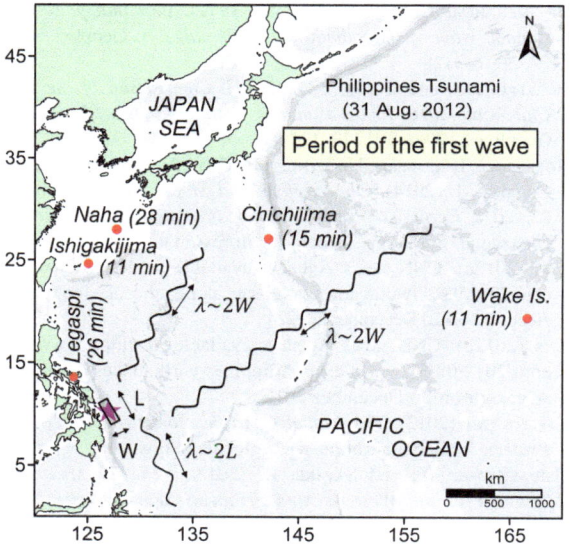

Figure 8
Sketch showing an approximation of the source location of the Philippines tsunami of 31 August 2012 (*rectangle*), and the first tsunami signals traveling towards the tide gauges located normal (e.g., Chichijima and Wake Island) and parallel (e.g., Legaspi) to the strike of the tsunami source. *W* and *L* represent width and length of the tsunami source, respectively. λ is the tsunami wavelength. *Solid circles* show the locations of the tide gauges

normal to the fault strike, the only tsunami signal is the 11-min signal, meaning that the spectra of the Wake Island station is mostly influenced by the width of the tsunami source rupture. For the Legaspi station, a pure lateral one, a first 26-min tsunami signal was recorded (Fig. 4b).

6.2. Tsunami Hazards for Southern Japan from Philippines Tsunamis

Analysis of sea level records of the 31 August 2012 Philippines tsunami revealed that the largest wave heights of this tsunami were recorded on tide gauges along the southern coast of Japan. Using numerical modeling of tsunami, it was shown above that a part of the waves generated by the Philippines tsunami was funneled towards southern Japan due to the bathymetric features in the Philippine Sea. According to LANDER *et al.* (2003), a similar observation was reported following the 3 May 1998 tsunami in the Philippine Sea (*Mw* 7.5), which was recorded in some Japanese

costal sites such as Naha and Ishigakijima. Based on these observations, a large subduction earthquake offshore the Philippines is likely to generate damage along the southern coast of Japan, and hence, possible tsunami hazards for southern Japan from submarine earthquakes offshore the Philippines need to be considered.

7. Conclusions

The El Salvador tsunami of 27 August 2012 and the Philippines tsunami of 31 August 2012 were studied using 25 sea level records of these tsunamis and numerical modeling of tsunami waves. The main findings are as follows:

1. Among the analyzed sea level records for the El Salvador tsunami, the largest wave height was observed in Baltra (71.1 cm) at a distance of about 1,400 km from the tsunami source. Near-field stations of La Union and Acajutla recorded wave heights of 4.5 and 12.5 cm, respectively. Numerical modeling showed that most of the tsunami energy is directed towards the Galapagos Islands (including the Baltra sea level station), and possibly this is the reason for observing the maximum wave height of this tsunami in Baltra. Fourier and wavelet analyses revealed that the main tsunami peak periods are around 9 and 19 min for this tsunami, and the 45-min peak observed at some stations does not represent the tsunami source. The 9-min signal seems to be the main tsunami signal, as it was observed in many sea level spectra.

2. For the Philippines tsunami, the maximum wave height was observed at the Kushimoto station (30.5 cm), a Japanese tide gauge station located 2,700 km away from the source. Legaspi tide gauge station located in the near-field recorded a wave height of 4.2 cm. Relatively large wave heights were observed at tide gauges located along the coast of Japan. Numerical modeling showed that tsunami waves were funneled towards the southern coast of Japan. Numerical modeling also revealed that most of the Philippines tsunami energy was confined within the Philippine Sea

region. The main tsunami peak periods were estimated at around 8, 12 and 29 min. The strongest signal was at the period of around 12 min, as it was more common on the examined sea level spectra.

3. A connection was made between tsunami governing periods and source dimensions using wavelet analysis. Results may suggest that the dominant period of the first wave at stations normal to the fault strike usually reflects the fault width, while the period of the first wave at stations in the direction of fault strike most probably reflects the fault length.

Acknowledgments

The sea level data used in this study were provided through the USA National Oceanographic and Atmospheric Administration (NOAA), and UNESCO Intergovernmental Oceanographic Commission (IOC). We express our sincere gratitude to our colleagues from the sea level data centers at both NOAA and IOC for their invaluable efforts regarding the preparation, processing and timely supply of sea level data which has greatly contributed to tsunami science in the last decade. Figure 1 was drafted using the GMT software (WESSEL and SMITH 1991). The wavelet package by TORRENCE and COMPO (1998) was used in this study. This article benefitted from detailed and constructive reviews from two anonymous reviewers, for which we are sincerely grateful. We would like to thank Dr. Hermann Fritz (Georgia Institute of Technology, USA), the guest editor of this issue, for his assistance during the revision process of this article. This study was supported by the Alexander von Humboldt Foundation in Germany and the Japan Society for Promotion of Science (JSPS) in Japan.

REFERENCES

BELL, C., VASSIE, J.M., and WOODWORTH, P.L. (2000). *POL/PSMSL Tidal Analysis Software Kit 2000 (TASK-2000), Permanent Service for Mean Sea Level*. CCMS Proudman Oceanographic Laboratory, UK, 22p.

BEN-MENAHEM, A., and ROSENMAN, M. (1972). *Amplitude patterns of tsunami waves from submarine earthquakes*, J. Geophys. Res., 77, 3097–3128.

BORRERO, J.C. and GREER, S.D. (2013). *Comparison of the 2010 Chile and 2010 Japan tsunamis in the Far-field*. Pure App. Geophys., *170*(6–8):1249–1274.

BORRERO, J. KALLIGERIS, N., LYNETT, P., FRITZ, H., NEWMANN, A. and CONVERS, J. (2014). *Observations and Modelling of the August 27, 2012 Earthquake and Tsunami affecting El Salvador and Nicaragua*. Pure App. Geophys., (in review).

CNN (2012a). CNN news Agency, available at: http://www.cnn.com/2012/08/27/world/americas/el-salvador-earthquake/index.html. Accessed on 20 December 2012.

CNN (2012b). CNN news Agency, available at: http://www.cnn.com/2012/08/31/world/asia/philippines-earthquake/index.html. Accessed on 20 December 2012.

GLOBAL CMT (2012). *The Global Centroid-Moment-Tensor (CMT) Project*, available at: http://www.globalcmt.org/.

HEIDARZADEH, M., and SATAKE, K. (2013a). *The 21 May 2003 Tsunami in the Western Mediterranean Sea: Statistical and Wavelet Analyses*. Pure App. Geophys., *170*(9–10):1449–1462.

HEIDARZADEH, M., and SATAKE, K. (2013b). *Waveform and Spectral Analyses of the 2011 Japan Tsunami Records on Tide Gauge and DART Stations Across the Pacific Ocean*. Pure App. Geophys., *170*(6–8):1275–1293.

IOC (Intergovernmental Oceanographic Commission), 2012, Sea Level Station Monitoring Facility, available at: http://www.ioc-sealevelmonitoring.org/map.php. Accessed on 20 December 2012.

IOC, IHO, BODC (2003). Centenary edition of the GEBCO digital atlas, published on CD-ROM on behalf of the Intergovernmental Oceanographic Commission and the International Hydrographic Organization as part of the general bathymetric chart of the oceans. British oceanographic data centre, Liverpool.

KANAMORI, H., and KIKUCHI, M. (1993). *The 1992 Nicaragua earthquake: a slow tsunami earthquake associated with sub-ducted sediments*. Nature *361*(6414):714–716.

LANDER, J.F., WHITESIDE, L.S., LOCKRIDGE, P.A. (2003). *Two decades of global tsunamis 1982–2002*, Sci. Tsunami Hazards, *21*(1), 3–88.

MATHWORKS (2012). MATLAB user manual, The Math Works Inc., MA, USA, 282 p.

NOAA (National Oceanographic and Atmospheric Administration), 2012, Center for Operational Oceanographic Products and Services (CO-OPS), National Ocean Service (NOS), at: http://tidesandcurrents.noaa.gov/tsunami/. Accessed on 22 December 2012.

OKADA Y. (1985). *Surface deformation due to shear and tensile faults in a half-space*, Bull. Seismol. Soc. Am., *75*(4), 1135–1154.

PTWC (2012a), Pacific Tsunami Warning System- NOAA's National Weather Service, available at: http://ptwc.weather.gov/ptwc/text.php?id=pacific.TIBPAC.2012.08.27.0445. Accessed on 18 December 2012.

PTWC (2012b), Pacific Tsunami Warning System- NOAA's National Weather Service, available at: http://ptwc.weather.gov/text.php?id=pacific.TSUPAC.2012.08.27.0458. Accessed on 18 December 2012.

PTWC (2012c), Pacific Tsunami Warning System- NOAA's National Weather Service, available at: http://ptwc.weather.gov/text.php?id=pacific.TSUPAC.2012.08.27.0627. Accessed on 18 December 2012.

PTWC (2012d), Pacific Tsunami Warning System- NOAA's National Weather Service, available at: http://ptwc.weather.gov/text.php?id=pacific.TSUPAC.2012.08.31.1255. Accessed on 18 December 2012.

PTWC (2012e), Pacific Tsunami Warning System- NOAA's National Weather Service, available at: http://ptwc.weather.gov/text.php?id=pacific.TSUPAC.2012.08.31.1454. Accessed on 18 December 2012.

RABINOVICH, A.B. (1997). *Spectral analysis of tsunami waves: Separation of source and topography effects*, J. Geophys. Res., *102* (12), 663–676.

RABINOVICH, A.B., and THOMSON, R.E. (2007). *The 26 December 2004 Sumatra Tsunami: Analysis of Tide Gauge Data from the World Ocean Part 1. Indian Ocean and South Africa*, Pure App. Geophys., *164*, 261–308.

REUTERS (2012a). Reuters News Agency, available at: http://www.reuters.com/article/2012/08/27/us-elsalvador-quake-idUSBRE87Q06E20120827 . Accessed on 18 December 2012.

REUTERS (2012b). Reuters News Agency, available at: http://www.reuters.com/article/2012/08/31/us-quake-philippines-idUSBRE87U0L720120831 . Accessed on 18 December 2012.

SATAKE, K. (1988). *Effects of bathymetry on tsunami propagation: application of ray tracing to tsunamis,* Pure App. Geophys., *126*(1):27–36.

SATAKE, K., NISHIMURA, Y., PUTRA, P. S., GUSMAN, A. R., SUNENDAR, H., FUJII, Y., TANIOKA, Y., LATIEF, H. and YULIANTO, E. (2013). *Tsunami source of the 2010 Mentawai, Indonesia earthquake inferred from tsunami field survey and waveform modeling,* Pure App. Geophys., *170*(9–10):1567–1582.

TORRENCE, C., and COMPO, G. (1998), *A Practical Guide to Wavelet Analysis*, Bull. Am. Met. Soc., *79*: 61–78.

USGS (2012a). US Geological Survey—Global earthquake search, available at: http://earthquake.usgs.gov/earthquakes/eqinthenews/2012/usc000c7yw/. Accessed on 18 December 2012.

USGS (2012b). US Geological Survey - Global earthquake search, available at: http://earthquake.usgs.gov/earthquakes/eqinthenews/2012/usc000cc5m/. Accessed on 18 December 2012.

WELCH, P. (1967). *The use of fast Fourier transform for the estimation of power spectra: A method based on time averaging over short, modified periodograms,* IEEE Trans. Audio Electroacoust *AE-15*:70–73.

WESSEL, P. and SMITH, W. H. F. (1991). *Free software helps map and display data*, EOS Trans. AGU 72:441.

YALCINER, A.C., PELINOVSKY, E., TALIPOVA, T., KURKIN, A., KOZELKOV, A., ZAITSEV, A. (2004). *Tsunamis in the Black Sea: comparison of the historical, instrumental, and numerical data.* J. Geophys. Res., *109*, C12023.

(Received February 25, 2013, revised January 10, 2014, accepted February 3, 2014, Published online February 19, 2014)

Pure Appl. Geophys. 171 (2014), 3457–3465
© 2014 Her Majesty the Queen in Right of Canada
DOI 10.1007/s00024-014-0775-1

An Overview of the 28 October 2012 M_w 7.7 Earthquake in Haida Gwaii, Canada: A Tsunamigenic Thrust Event Along a Predominantly Strike-Slip Margin

JOHN F. CASSIDY,[1] GARRY C. ROGERS,[1] and ROY D. HYNDMAN[1]

Abstract—The boundary between the Pacific and North America plates along Canada's west coast is one of the most seismically active regions of Canada, and is where Canada's two largest instrumentally recorded earthquakes have occurred. Although this is a predominantly strike-slip transform fault boundary, there is a component of oblique convergence between the Pacific and North America plates off Haida Gwaii. The 2012 M_w 7.7 Haida Gwaii earthquake was a thrust event that generated a tsunami with significant run up of over 7 m in several inlets on the west coast of Moresby Island (several over 6 m, with a maximum of 13 m). Damage from this earthquake and tsunami was minor due to the lack of population and vulnerable structures on this coast.

Key words: Thrust earthquake, Haida Gwaii, tsunamigenic, oblique convergence, strike-slip margin.

1. Introduction

The 03:04 28 October, 2012 (20:04 27 October local time) M_w 7.7 earthquake that occurred off the west coast of Haida Gwaii, Canada, (formerly the Queen Charlotte Islands) was the first major thrust event recorded along this predominantly strike-slip margin. The Pacific-North America boundary along the British Columbia and Southeast Alaska coast is mainly a transform fault plate boundary, with ocean crust seaward and continental crust landward. It is dominated by the right lateral Queen Charlotte Fault (QCF), an underwater seafloor feature that extends for more than 800 km from the triple junction region north of Vancouver Island to Cross Sound in the Alaska Panhandle, where it transitions into the continental Fairweather Fault (Fig. 1). The QCF is the northern equivalent of the San Andreas Fault (with

the Cascadia subduction zone between these two major transform faults). This margin accommodates between 50 and 60 mm/year relative motion between the Pacific and North America plates (e.g., DEMETS *et al.* 2010; STOCK and MOLNAR 1988; DEMETS and DIXON 1999). Canada's two largest instrumentally recorded earthquakes have occurred here—the 1949 M 8.1 strike-slip event and the 2012 M 7.7 thrust event discussed herein. The northernmost portions of the QCF ruptured in a M 7.6 earthquake near Sitka in 1972 and a M 7.5 earthquake to the west of Ketchikan in 2013, both strike-slip with minimal tsunami generation. This article provides a summary of the tectonics and earthquake history of this region, and an overview of the results to date analysing the 2012 M 7.7 Haida Gwaii earthquake that generated the largest tsunami recorded in the world in 2012.

2. Tectonics

This region is dominated by strike-slip tectonics with an overprint of convergent tectonics. The near-vertical Queen Charlotte Fault accommodates about 52 mm/year of right-lateral motion in the direction N338 between the Pacific and North American plates (e.g., DEMETS *et al.* 2010; STOCK and MOLNAR 1988; DEMETS and DIXON 1999). Recent high-resolution seafloor imaging of the fault shows that the surface expression of the fault exhibits the characteristics of a nearly pure strike-slip fault trace (BARRIE *et al.* 2013). North of Graham Island, Haida Gwaii, the trace of the Queen Charlotte fault is almost aligned with the Pacific-North America relative plate motion (Fig. 1). Along the southern portion of this fault off Moresby Island, however, there is a change in margin orientation to a difference of 15°–20° compared to the direction of relative plate motion, thus requiring a

---
[1] Natural Resources Canada, Geological Survey of Canada, Sidney, BC, Canada. E-mail: jcassidy@nrcan.gc.ca

component of convergence. This difference decreases gradually to the north, off Graham Island. As has been observed in other parts of the world, this type of oblique convergence is usually accommodated by a combination of thrusting nearly orthogonal to the margin and strike-slip faulting further landward, i.e., a forearc sliver (e.g., FITCH 1972; JARRARD 1986; McCAFFREY 1992). The high, steep topography along the west coast of Haida Gwaii is probably the result of the initiation of oblique convergence and underthrusting initiating at ~6 Ma (HYNDMAN and HAMILTON 1993). In the Haida Gwaii region, the component of convergence is interpreted to be accommodated mainly by oblique underthrusting of the seafloor beneath Moresby and Graham Island (e.g., YORATH and HYNDMAN 1983; HYNDMAN and HAMILTON 1993). Especially important evidence comes from receiver function studies (SMITH et al. 2003; BUSTIN et al. 2007; CASSIDY et al. 2014) that show an eastward dipping 10-km-thick low-velocity zone interpreted as the subducted oceanic crust.

A series of structural and other geophysical studies have been used to develop three main models to explain how the oblique convergence is accommodated across the Haida Gwaii margin through a combination of thrust and strike-slip faulting or through crustal shortening (e.g., DEHLER and CLOWES 1988; PRIMS and GOVERS 1997; ROHR and CURRIE 1997; ROHR et al. 2000; ROHR and DIETRICH 1992; SMITH et al. 2003). These models include; (a) simple oblique subduction with the strike-slip QCF in the overlying continental plate, i.e., a forearc sliver; (b) oblique subduction with the nearly vertical strike-slip QCF cutting the subducting oceanic plate; and (c) much of the convergence accommodated by crustal shortening (see Fig. 3 of SMITH et al. 2003). The 2012 thrust event has provided evidence that the first model is mostly correct. There are separate thrust and strike-slip faults accommodating the orthogonal and parallel components of relative plate motion (Fig. 1). The now preferred model is discussed below with the M_w 7.7 2012 event.

3. Seismicity

The Haida Gwaii region is one of Canada's most seismically active areas (Fig. 2a). There have been four large (M >7) earthquakes and 18 M >6 earthquakes during the past eight decades. The modern seismic recording history of this region began in 1898 when a seismograph was deployed in Victoria, British Columbia (~800 km to the south of Haida Gwaii) that could detect earthquakes of M >7 in the Haida Gwaii region. With the deployment of a seismic station at Sitka, Alaska in 1904 (~500 km to the north of Haida Gwaii), and improvements in the global earthquake monitoring network, earthquakes of M > 6 could be detected and roughly located along this plate boundary region. Since the 1950s, it has been possible to locate most M > 5 earthquakes in this region. Between 1982 and 1996, a temporary network of 12 analog seismographs was operated on Haida Gwaii and the adjacent mainland that allowed location of earthquakes as small as M 0.3 in some locations. Since the mid-1990s, a network of six digital stations has operated on Haida Gwaii, and since the 2012 earthquake, other temporary digital stations have been installed.

Several detailed studies of microseismicity have been conducted that located and examined earthquakes as small as M −0.5 in the Haida Gwaii region during limited time periods. Studies of composite earthquake mechanisms have provided important information on the regional stress and strain directions along the Haida Gwaii margin indicated by the earthquakes, as discussed below.

3.1. The Largest Earthquakes

The Haida Gwaii region is home to Canada's two largest instrumentally recorded earthquakes—the 1949 M 8.1 strike slip event that initiated off the northwest tip of Graham Island (e.g., BOSTWICK 1984; ROGERS 1983), and the 2012 M 7.7 thrust earthquake that occurred off the west coast of Moresby Island (Fig. 1). There are very important differences between these two large earthquakes; the 1949 event was a nearly pure strike-slip mechanism that involved slip along the QCF [and generated a small tsunami (SOLOVIEV and GO 1984)], whereas the 2012 earthquake was a thrust event, with no evidence uncovered so far that it involved any movement along the strike-slip QCF, and it

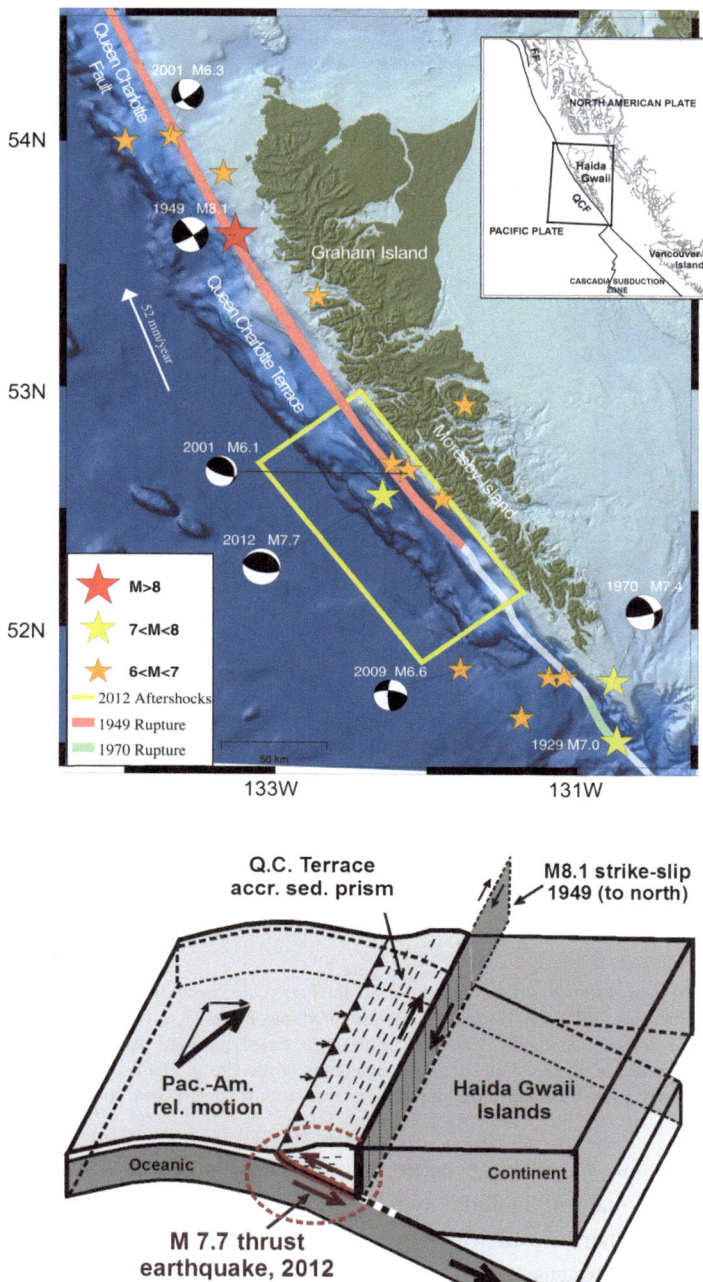

Figure 1

Tectonic setting in the Haida Gwaii region. *White arrow* indicates the direction of relative motion between the Pacific and North America plates. Locations and focal mechanisms (if available) of the largest historical earthquakes (*stars*) are shown, heavy *pink* and *green lines* indicate rupture zones of the two previous largest historical earthquakes (1949 and 1970). The section of the QCF between the southern end of the 1949 rupture (*pink line*) and the northern end of the 1970 rupture (*green line*) is the "seismic gap" identified by ROGERS (1986). The approximate extent of the aftershock zone of the 2012 thrust event is shown as a *yellow box*. Inset shows the regional tectonics, *QCF* is the Queen Charlotte Fault and *FF* is the Fairweather Fault. Cartoon shows the location of the thrust fault beneath the Queen Charlotte Terrace along which the 2012 earthquake occurred, the relative plate motions in this area, and the location of the near-vertical Queen Charlotte fault

(a)   **(b)**

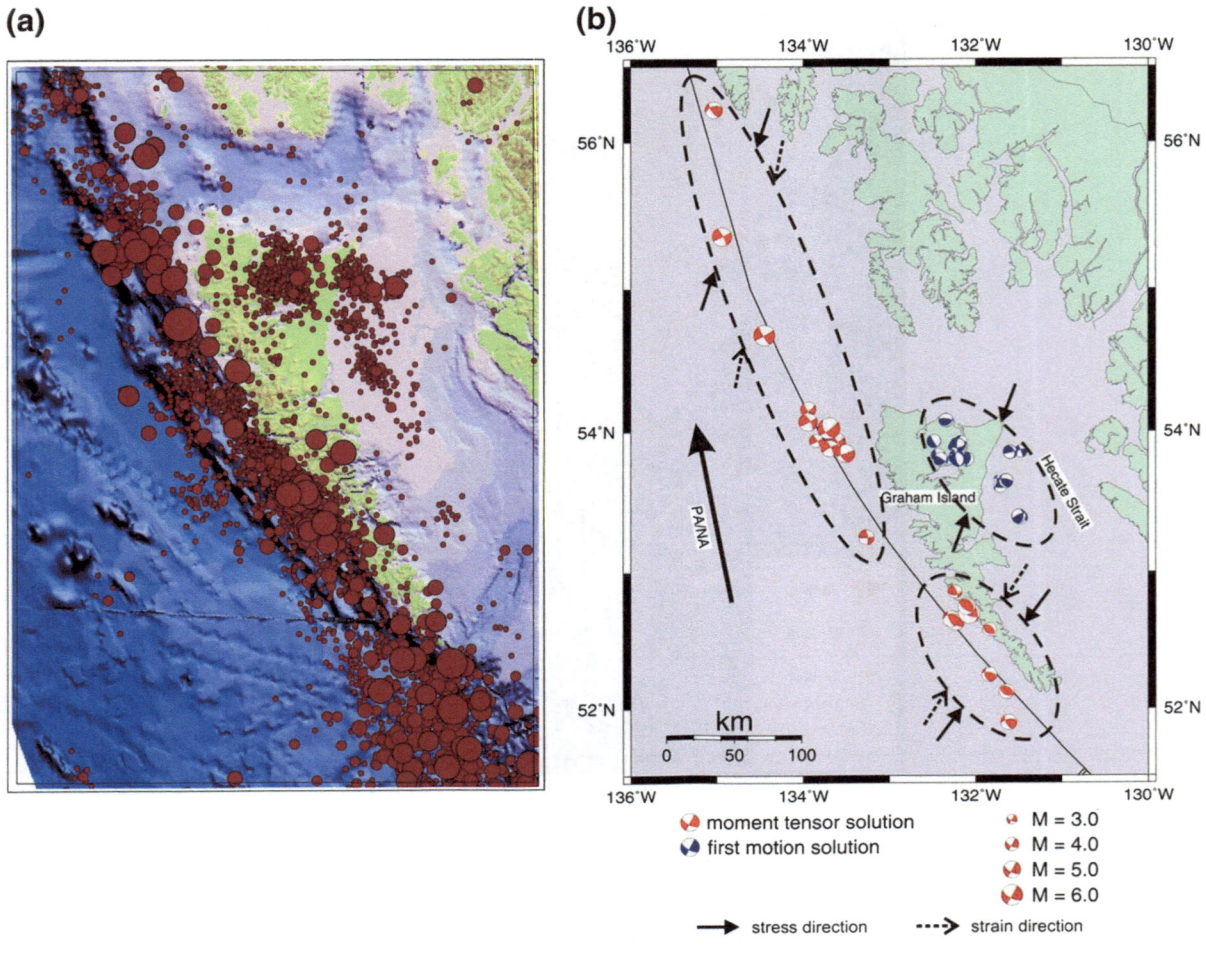

Figure 2

a Seismicity of the Haida Gwaii region (1900–2012). Earthquakes (red dots) are scaled according to magnitude and range from M < 1 to M 8.
b Previous earthquake focal mechanisms (1982–2004), stress and strain directions in the vicinity of Haida Gwaii modified from RISTAU *et al.*
(2007). The southernmost ellipse shows the region of mainly high-angle thrust mechanisms with a nearly margin-normal maximum
compressive stress direction. The northern ellipse shows the area of mainly strike-slip mechanisms (consistent with the strike of the QCF
trace) and a more northerly (margin-oblique) maximum compressive stress direction

generated a significant tsunami. Other large earthquakes (Fig. 1) along this margin include two
M ~7 earthquakes off the southern tip of Moresby
Island in 1970 and in 1929 (ROGERS 1986). The
section of the QCF between those two earthquakes
and the southern end of the 1949 rupture was
identified as a potential seismic gap by ROGERS
(1986). Since 2001, there have been four M >6
earthquakes off Haida Gwaii. The 2001 M_w 6.1
thrust event generated a small tsunami recorded on
tide gauges on the west coast of Vancouver Island to
the south of Haida Gwaii (RABINOVICH *et al.* 2008).

3.2. Earthquake Focal Mechanisms and Stress and Strain Directions

RISTAU *et al.* (2007) determined the regional
moment tensor solutions for 15 M >4 earthquakes
along the Queen Charlotte Fault (Fig. 2b). Such
studies provide average crustal strain directions as
accommodated by the earthquakes, and the inferred
average stress direction. Off the west coast of
Graham Island and to the north, the mechanisms are
mainly strike-slip (consistent with the strike of the
QCF trace), often with a small thrust component.
However, off the west coast of Moresby Island in the

south, the solutions are mainly high-angle thrust mechanisms. Ristau et al. (2007) show a change in maximum compressive stress direction from nearly margin normal off Moresby Island to more northerly (margin oblique) off Graham Island. Although the thrust solutions off Moresby Island indicate convergence, they are mainly on high-angle faults and so not on the main thrust that generated the large 2012 earthquake. The focal mechanisms of the largest earthquakes and ongoing GPS survey results from Haida Gwaii (Mazzotti et al. 2003) also agree with the proposed tectonic models of convergence and underthrusting along the southern portion of the Haida Gwaii margin (e.g., Hyndman and Hamilton 1993; Ristau et al. 2007).

3.3. Microseismicity Studies

The first microseismicity study of the Haida Gwaii region was undertaken by Hyndman and Ellis (1981) who deployed a temporary array of on-land seismic stations on Moresby Island and three ocean bottom seismographs just to the west. They were able to accurately locate 11 earthquakes (M 0.4–2.1) during a 9-day period. Ten of those were located (within uncertainties) beneath the surface trace of the strike-slip Queen Charlotte Fault mostly at depths between 15 and 20 km. The rate of seismicity was consistent with the long-term average for larger earthquakes in the region.

During the summer of 1983 a total of 22 seismographs (19 on land and three ocean bottom seismographs) were deployed for a larger-scale microseismicity study of Haida Gwaii (Bérubé et al. 1989). During this 9-week survey, 317 earthquakes were recorded, of which 109 were well located. Key results from the Bérubé et al. (1989) study include:

1. Most of the microseismicity along the west coast occurs in the vicinity of the Queen Charlotte Fault, beneath the inner bathymetric slope (and about 15 km east of the main surface trace of the fault).
2. Seismicity is within the top 20 km, and composite focal mechanisms reveal a NE-oriented maximum horizontal compressive stress.
3. Microseismicity along the QCF is higher within the aftershock zone of the 1949 earthquake compared to adjacent areas. In contrast, only two earthquakes occurred in the "seismic gap" off southern Moresby Island.

Another detailed microseismicity study with a temporary array of seismographs between 1982 and 1996 (Bird 1997) examined more than 2,600 earthquakes and revealed a variation in intensity of seismicity in the vicinity of the Queen Charlotte Fault. A paucity of activity was identified along the QCF just to the south of the 1949 M 8.1 epicentre, and also in the "seismic gap" off southern Moresby Island. First motion focal mechanisms in the region show a mixture of strike-slip and high angle thrust faulting, and also reveal a north-northeast compressional stress regime (Bird 1997). Further focal mechanism solutions were subsequently obtained as noted above (Ristau et al. 2007).

3.4. Previous Tsunami Generating Earthquakes in the Haida Gwaii Region

Prior to 2012, only two earthquakes in the Haida Gwaii region are known to have generated a tsunami. The 1949 M 8.1 strike-slip earthquake generated a small tsunami recorded on tides gauges in Sitka, Alaska (7.5 cm) and Hawaii (10 cm) (Leonard and Hyndman 2010; Soloviev and Go 1975) and the 2001 M_w 6.1 thrust earthquake that produced a small tsunami (maximum 23 cm) recorded on Vancouver Island (Rabonovich et al. 2008). This lack of tsunami observations is, in part, due to the fact that the plate boundary in this region is predominantly strike-slip, and tsunami-generating earthquakes are relatively rare. Other factors are the short observing time, the lack of settlements along the west coast of Haida Gwaii, and the fact that there are no tide gauges on the outer coast of Haida Gwaii (with the exception of Henslung Cove, on the very northern tip of Moresby Island).

The nearest tide gauges have been on northern Vancouver Island more than 400 km to the south in a direction of expected low amplitude for thrusting orthogonal to the Haida Gwaii margin. There is one tide gauge station on the landward side of Haida Gwaii, again where little amplitude is expected for thrust earthquakes off the west coast.

4. The Oct 2012 M 7.7 Thrust Earthquake

On 28 October 2012 at 0304 UTC (8:04 p.m. October 27 local time), Canada's second largest instrumentally recorded earthquake occurred off the west coast of Moresby Island (JAMES et al. 2013; SZELIGA 2013; LAY et al. 2013). Strong shaking was experienced on Haida Gwaii, but fortunately no significant damage resulted, as the region adjacent to the fault rupture is an uninhabited National Park and the large tsunami was limited to the west coast of the islands where there are no settlements or significant coastal structures. The closest community, the village of Queen Charlotte, is about 50 km from the estimated rupture surface (JAMES et al. 2013). This earthquake was felt as far away as 1,500 km in Alberta, Yukon, Washington State, and Montana.

For tsunami generation and modeling, the earthquake focal mechanism and slip distribution are key factors. The mechanism of the mainshock was thrust faulting (H. Kao, personal communication, 2013; http://earthquake.usgs.gov, http://globalcmt.org, LAY et al. 2013) along a shallow eastward dipping plane (Fig. 1). Estimates of dip in the various focal mechanism solutions vary from $17°$ to $25°$. Finite fault analysis suggests a maximum slip of over 5 m (http://earthquake.usgs.gov, LAY et al. 2013). LAY et al. (2013) estimate a maximum slip of 7.7 m and an average slip of 3.3 m. The overall rupture zone is about 150 km long as defined by the aftershocks (Fig. 3), and about 30 km wide—which is the approximate width of the Queen Charlotte Terrace. It is noteworthy that there is no indication that this slip represents any movement along the strike-slip QCF, but rather represents movement on a shallow thrust fault beneath the Queen Charlotte Terrace (see Fig. 1), to the west of the QCF. The preferred (Geological Survey of Canada) epicenter (Fig. 1) that utilizes both local (Haida Gwaii) data and regional waveforms, is located offshore (~ 25 km SSW of the initial USGS epicenter). The exact landward limit of rupture is still under investigation, but is approximately to the location of the strike-slip fault near the coast at a depth of 15–20 km. Thermal models of the thrust zone also suggest this is about the landward limit of rupture (SMITH et al. 2003). The recent slip model (Wang, personal communication; NYKOLAISHEN et al. 2013) incorporates the latest GPS coseismic deformation data, improved hypocenter and fault geometry information, as well as the expected landward limit based on thermal modeling. It shows in the offshore region a vertical uplift of as much as 3 m, and a maximum horizontal slip of up to 6–8 m.

The October 2012 earthquake generated a large tsunami [with significant local run-ups exceeding 7 m in some of the small inlets along the west coast of Haida Gwaii (possible maximum of 13 m)—see papers by LEONARD and BEDNARDSKI (2014), and FINE et al. (2014) in this volume]. The nearest tide-gauges on northern Haida Gwaii and Vancouver Island were not exposed to the main focus of the tsunami energy and recorded amplitudes up to 0.5 m. In Hawaii (more than 4,000 km to the southwest, but in the focus of the tsunami energy), the maximum amplitude recorded on a tide gauge was 0.8 m.

This earthquake has had a very rich aftershock sequence, with tens of thousands of recorded aftershocks. Figure 3 shows the best-located aftershocks ($M > 2$) as of 1 May 2013. There are two concentrations—one over 50 km offshore, west of the Queen Charlotte Terrace, and the other concentrated just offshore and east of the surface trace of the Queen Charlotte fault. Most of the largest aftershocks are located in the farthest offshore concentration in the Pacific plate and show normal faulting (FARAHBOD et al. 2013; LAY et al. 2013) (Fig. 3). These events are west of the thrust rupture surface of the mainshock. It is common to observe outer rise normal faulting aftershocks seaward of large megathrust earthquake in subduction zones. The concentration of aftershocks closest to land have a larger depth range and appear to be in both in the North American plate and Pacific, and have a variety of focal mechanisms. Thus far, no aftershocks with low angle thrusting mechanisms similar to the mainshock have been identified (Honn Kao, personal communication 2013, FARAHBOD et al. 2013; LAY et al. 2013). GPS data from Haida Gwaii reveal up to 1.2 m of co-seismic southwestward surface displacement on the west coast of Moresby Island (JAMES et al. 2013; NYKOLAISHEN et al. 2013). Vertical coseismic subsidence of up to 30 cm was measured at near-coastal GPS sites on Moresby Island. GPS observations reveal

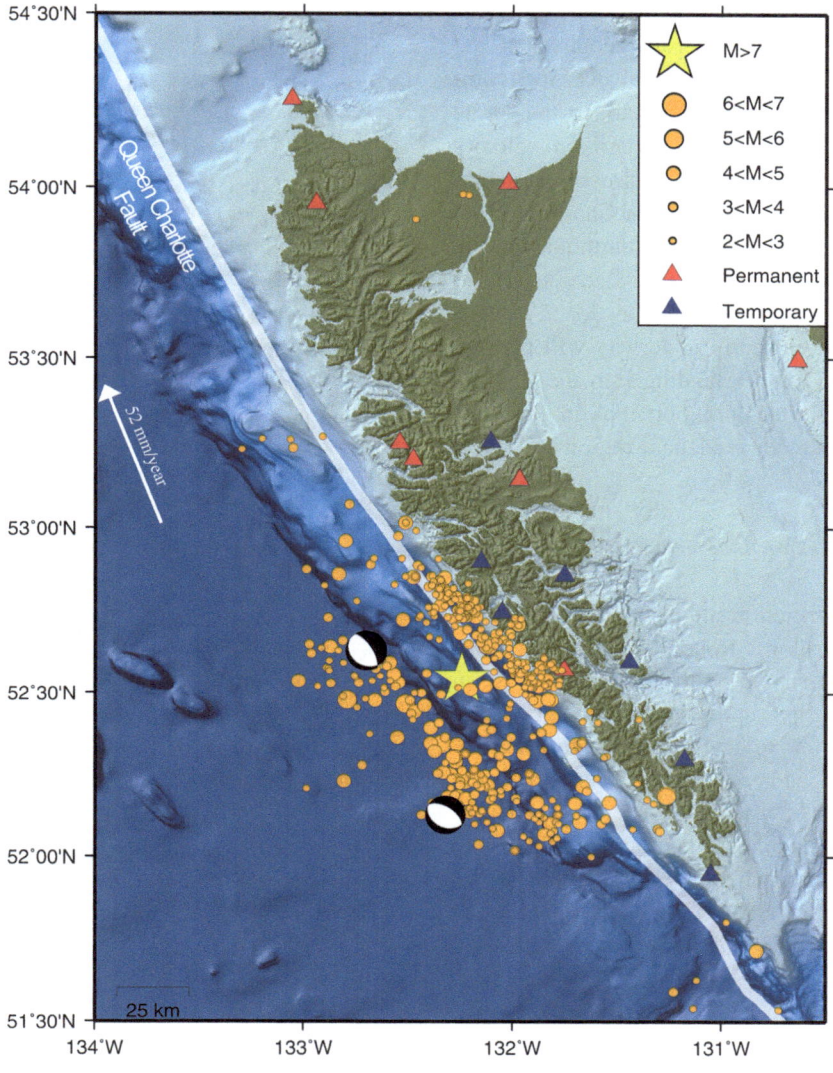

Figure 3

Haida Gwaii aftershocks (M >2) with the highest-quality solutions from October 2012 to 1 May 2013. The focal mechanisms (normal faulting) of the two largest aftershocks are shown. *Yellow star* denotes the epicentral location of the mainshock (note that the focal mechanism of the mainshock is shown on Fig. 1). *Grey line* is the trace of the near-vertical strike-slip Queen Charlotte Fault. Locations of seismic stations (both permanent stations of the Canadian National Network and temporary stations set up to monitor aftershocks) are indicated by *triangles*

postseismic motions that are still ongoing as of December 2013 (14 months after the earthquake).

A large thrust earthquake like this one at this location is a relatively rare event. If we use a margin normal convergence estimate for the southern Haida Gwaii region of 6–10 mm/year (MAZZOTTI *et al.* 2003) and the maximum slip (about 8 m) and average slip (about 3.5) resulting from finite fault slip modeling (http://earthquake.usgs.gov, LAY *et al.* 2013), repeat times ranging from 350 to 1,300 years result.

5. Summary

The M 7.7 2012 Haida Gwaii earthquake is the largest recorded thrust earthquake along the predominantly strike-slip Pacific-America plate boundary. It generated a substantial tsunami with run-ups of over 7 m (with a possible maximum of 13 m) in several inlets on the west coast of Moresby Island [see paper in this volume by LEONARD and BEDNARSKI (2014)]. An event of this nature was expected based on our understanding of the tectonics and previous small thrust earthquakes. Numerous geophysical and seismic studies over the past several decades, combined with analysis of recent small to moderate earthquakes and analysis of GPS derived velocity vectors, have clearly shown the compressional nature of the southern portion of the Queen Charlotte Fault and the potential for large thrust earthquakes and their accompanying tsunamis (e.g. LEONARD et al. 2012). The detailed studies of the M 7.7 2012 earthquake that are currently underway will provide valuable insight and understanding into the subduction earthquake and tsunami potential in both the Haida Gwaii and Vancouver Island regions.

Acknowledgments

This is ESS Contribution number 20130320. We thank Camille Brillon, Robert Kung, and Lisa Nykolaishen for assisting with figures, and Kelin Wang for providing his rupture model. We gratefully acknowledge helpful reviews and suggestions by Alison Bird, two anonymous reviewers, and the Associate Editor. The field response to this earthquake would not have been possible without the prompt and ready assistance of many people within the Geological Survey of Canada (Canadian Hazard Information Service and Public Safety Geoscience). We are grateful for the assistance of the Council of the Haida Nation, Parks Canada, Fisheries and Oceans Canada, the BC Ministry of Forests, and the captain and crew of the CCG vessel John P. Tully.

REFERENCES

BARRIE, J.V., CONWAY, K.W., and HARRIS, P.T. (2013), *The Queen Charlotte Fault, British Columbia: seafloor anatomy of a transform fault and its influence on sediment processes*, Geo-Mar Lett., doi:10.1007/s00367-013-0333-3.

BÉRUBÉ, J., ROGERS, G.C., ELLIS, R.M., and HASSELGREN, E.O. (1989), *A microseismicity study of the Queen Charlotte Islands region*, Can. J. Earth Sci., 26, 2556–2566.

BIRD, A.L. (1997), Earthquakes in the Queen Charlotte Islands region: 1982-1996, M.Sc. thesis, University of Victoria, Victoria, BC, 123 pp.

BOSTWICK, T.K. (1984), A re-examination of the August 22, 1949 Queen Charlotte earthquake, M.Sc. Thesis, University of B.C., Vancouver, 115p.

BUSTIN, A.M.M., HYNDMAN, R.D., KAO, H., and CASSIDY, J.F. (2007), *Evidence for underthrusting beneath the Queen Charlotte Margin, British Columbia, from teleseismic receiver function analysis*, Geophys. J. Int., 171, 1198–1211, doi:10.1111/j.1365-246X.2007.03583.x.

CASSIDY, J.F., GOSSELIN, J., and DOSSO, S.E. (2014), *Shear Wave Velocity Structure Beneath Haida Gwaii, Canada, in the Vicinity of the 2012 Mw 7.7 Earthquake (abstract)*, Seis. Res. Lett. (in press).

DEHLER, S.A., and CLOWES, R.M. (1988), *The Queen Charlotte Islands refraction project. Part I. The Queen Charlotte Fault Zone*, Can. J. Earth Sci., 25, 1857–1870.

DEMETS, C. and DIXON, T.H. (1999), *New kinematic models for Pacific-North American motion from 3 Ma to present, I: evidence for steady motion and biases in the NUVEL-1A model*, Geophys. Res. Lett., 26, 1921–1924.

DEMETS, C., GORDON, R.G., and ARGUS, D.F. (2010): *Geologically current plate motions*, Geophys. J. Int. v. 181, p. 1–80, doi:10.1111/j.1365-246X.2009.04491.x.

FARAHBOD, A.M., KAO, H., and SHAN, S.-J. (2013). A Mini-Megathrust Event in an Incipient Subduction Zone: The 2012 M_w7.8 Haida Gwaii Earthquake Sequence, Am. Geophys. U. Ann. Fall Meeting, San Francisco, CA, Dec. 9-13, 2013.

FINE, I.V., CHERNIAWSKY, J.Y., THOMSON, R.E., RABINOVICH, A.B., and KRASSOVSKI, M.V. (2014), *Observations and numerical modeling of the 2012 Haida Gwaii tsunami off the coast of British Columbia*, Pure App. Geophys. (this volume).

FITCH, T.J., (1972). *Plate convergence, transcurrent faults, and internal deformation adjacent to Southeast Asia and the western Pacific*, J. Geophys. Res., 77, 4432–4460, doi:10.1029/JB077i023p04432.

HYNDMAN, R. D., and ELLIS, R.M. (1981), *Queen Charlotte fault zone: Microearthquakes from a temporary array of land stations and ocean bottom seismographs*, Can. J. Earth Sci., 18, 776–788.

HYNDMAN, R.D. and HAMILTON, T.S. (1993), *Queen Charlotte area Cenozoic tectonics and volcanism and their association with relative plate motions along the northeastern Pacific margin*, J. Geophys. Res., 98, 14,257–14,277.

JAMES, T., ROGERS, G.C., CASSIDY, J.F., DRAGERT, H., HYNDMAN, R.D., LEONARD, L., NYKOLAISHEN, L., RIEDEL, M., SCHMIDT, M., and WANG, K. (2013), *Field studies target 2012 Haida Gwaii earthquake*, Eos Trans., 94 (22), 197–198, doi:10.1002/2013EO220002

JARRARD, R. D. (1986). *Terrane motion by strike-slip faulting of forearc slivers*. Geology, 14, 780–783.

LAY, T., L. YE, H. KANAMORI, Y. YAMAZAKI, K.-F. CHEUNG, K. KWONG, and KOPER, K.D. (2013), *The October 28, 2012 M7.8 Haida Gwaii underthrusting earthquake and tsunami: Slip partitioning along the Queen Charlotte Fault transpressional plate boundary*, Earth and Planetary Sci. Lett., doi:10.1016/j.epsl.2013.05.005.

LEONARD, L.J., and J.M. BEDNARDSKI (2014), *Field survey following the 27 October 2012 Haida Gwaii tsunami*, Pure App. Geophys. (this volume).

LEONARD, L., G.C. ROGERS, and HYNDMAN, R.D. (2010), *Annotated bibliography of references relevant to tsunami hazard in Canada*, Geol. Survey of Canada, Open File 6552, 269p.

LEONARD, L.J., G.C. ROGERS and S. MAZZOTTI (2012), *A preliminary tsunami hazard assessment of the Canadian coastline*, Geological Survey of Canada, Open File 7201, 119p, doi:10.4095/292067.

MAZZOTTI, S., HYNDMAN, R.D., FLÜCK, P., SMITH, A.J., and SCHMIDT, M. (2003). *Distribution of the Pacific-North America motion in* the Queen Charlotte Islands-S. Alaska Plate boundary zone, Geophys. Res. Lett., 30, doi:10.1029/2003GL017586.

McCAFFREY, R. (1992). Oblique plate convergence, slip vectors, and forearc deformation. J. Geophys. Res., 97, 8905–8915.

NYKOLAISHEN, L., DRAGERT, H., WANG, K., SCHMIDT, M. LU, Y., and SCHOFIELD, B. (2013). GPS-Observed Displacements for the M7.7 October 27, 2012, Haida Gwaii Earthquake, 2013 Am. Geophys. U. Ann. Fall Meeting, San Francisco, CA, Dec. 9-13, 2013.

PRIMS, J., K.P. FURLONG, K.M.M. ROHR, and GOVERS, R. (1997). Lithospheric structure along the Queen Charlotte margin in western Canada: constraints from flexural modeling Geo-Marine Lett., 17, 94–99.

RABINOVICH, A.B., THOMSON, R.E., TITOV, V.V., STEPHENSON, F.E., and ROGERS, G.C. (2008), Locally generated tsunamis recorded on the coast of British Columbia, Atmosphere-Ocean, 46, no. 3, p. 343–360, doi:10.3137/ao.460304.

RISTAU, J., ROGERS, G.C., and CASSIDY, J.F. (2007), Stress in western Canada from regional moment tensor analysis, Can. J. Earth Sci., 44, 127–148, doi:10.1139/E06-057.

ROGERS, G.C. (1983), Seismotectonics of British Columbia, PhD Thesis, University of British Columbia, 247p.

ROGERS, G.C. (1986), Seismic gaps along the Queen Charlotte Fault, Earthq. Predict. Res., 4, 1–11.

ROHR, K. M. M., and CURRIE, L. (1997), Queen Charlotte basin and Coast Mountains: Paired belts of subsidence and uplift caused by a low-angle normal fault, Geology, 25, 819–822.

ROHR, K.M.M. and DIETRICH, J.R. (1992), Strike-slip tectonics and development of the Tertiary Queen Charlotte Basin, offshore western Canada: evidence from seismic refection data, Basin Res., 4, 1–19.

ROHR, K.M.M., SCHEIDHAUER, M., and TREHU, A. (2000), Transpression between two warm mafic plates: the Queen Charlotte Fault revisited, J. Geophys. Res., 105, 8147–8172.

SMITH, A.J., HYNDMAN, R.D., CASSIDY, J.F., and WANG, K. (2003), Structure, seismicity, and thermal regime of the Queen Charlotte Transform Margin, J. Geophys. Res., 108 (B11, 2539, doi:10.1029/2002JB002247.

SOLOVIEV, S. L. and GO, N. (1975), Catalogue of tsunamis of the eastern shore of the Pacific Ocean (1513-1968), Nauka Publishing House, Moscow, 204p.

SOLOVIEV, S. L. and GO, N. (1984) Catalogue of tsunamis of the eastern shore of the Pacific Ocean (1513-1968), Canadian translation of Fisheries and Aquatic Sciences, 5078.

STOCK, J.M. and MOLNAR, P. (1988), Uncertainties and implications of the Late Cretaceous and Tertiary position of North America relative to the Farallon, Kula, and Pacific plates, Tectonics, 7, 1339–1384.

SZELIGA, W. (2013), 2012 Haida Gwaii quake: insight into Cascadia's subduction extent, Eos Trans., 94 (9), 85–86, doi:10.1002/2013EO09001.

YORATH, C.J., and HYNDMAN, R.D. (1983), Subsidence and thermal history of Queen Charlotte Basin, Can. J. Earth Sci., 20, 135–159.

(Received July 21, 2013, revised January 2, 2014, accepted January 3, 2014, Published online February 14, 2014)

Reprinted from the journal

Pure Appl. Geophys. 171 (2014), 3467–3482
© 2014 Her Majesty the Queen in Right of Canada
DOI 10.1007/s00024-014-0792-0

Field Survey Following the 28 October 2012 Haida Gwaii Tsunami

L. J. LEONARD[1,2] and J. M. BEDNARSKI[2]

Abstract—This article documents the near-field effects of the largest tsunami of 2012 (globally), which occurred following Canada's second-largest recorded earthquake, on a thrust fault offshore western Haida Gwaii on October 28 (UTC). Despite a lack of reported damaging waves on the coast of British Columbia (largest amplitudes were recorded in Hawaii), three field surveys in the following weeks and months reveal that much of the remote unpopulated, uninstrumented coastline of western Haida Gwaii was impacted by significant tsunami waves that reached up to 13 m above the state of tide. Runup exceeded 3 m at sites spanning ∼200 km of the coastline. Greatest impacts were apparent at the heads of narrow inlets and bays on western Moresby Island, where natural and manmade debris with a clear oceanward origin was found on the forest floor and caught in tree branches, inferring flow depths up to 2.5 m. Bays that see regular exposure to storm waves were generally less affected; at these sites a storm origin cannot be ruled out for the debris surveyed. Logs disturbed from their apparent former footprints on the forest floor at the head of Pocket Inlet provide evidence of complex runup, backwash and oblique flow patterns, as noted in other tsunamis. Discontinuous muddy sediments were found at a few sites; sedimentation was not proportional to runup. Lessons learned from our study of the impacts of the Haida Gwaii tsunami may prove useful to future post-tsunami and paleotsunami surveys, as well as tsunami hazard assessments.

Key words: Tsunami, runup, Haida Gwaii, Canada, Pacific Ocean, earthquake.

1. Introduction

On 28 October 2012 at 03:04 UTC (20:04 Oct 27 local time) an M_w 7.7 thrust earthquake occurred off western Moresby Island, Haida Gwaii, British Columbia (Fig. 1). This was the second-largest earthquake ever recorded in Canada; its rupture area (defined by

aftershocks) was mainly to the south of the 1949 M 8.1 strike-slip event on the near-vertical Queen Charlotte fault that initiated off Graham Island and produced only a small tsunami (e.g., CASSIDY et al. 2014).

A tsunami generated by the 2012 Haida Gwaii earthquake was recorded in the far field on tide gauges and Deep-ocean Assessment and Reporting of Tsunamis (DART) buoys all around the Pacific Ocean, as well as by NEPTUNE Canada bottom-pressure sensors offshore Vancouver Island (FINE et al.—this volume). The largest recorded peak-to-trough height of 1.52 m occurred at Kahului, Maui, HI, compared with only 0.52 m at Henslung Cove, Langara Island (Fig. 2; NGDC/WDS Global Historical Tsunami Database 2013), the closest tide gauge (in terms of unrestricted tsunami travel path) at ∼135 km from the northern end of the rupture area. This is not surprising for a tsunami generated on a northwest-striking thrust fault, where the greatest energy is directed perpendicular to the fault rather than along strike. Greatest amplitudes would be expected on the west coast of Moresby Island, which forms part of Gwaii Haanas National Park Reserve, but there were no witnesses or instruments in this unpopulated region to record tsunami impacts.

The Haida Gwaii earthquake occurred in a region where no large thrust earthquakes had occurred historically, but oblique subduction (e.g., HYNDMAN and ELLIS 1981) was one of two end-member models proposed to accommodate convergence across the Pacific-North America margin that increases from ∼8 to 15 mm/year from northern to southern Haida Gwaii; the alternate model is that convergence is accommodated by internal deformation of both plates (e.g., ROHR et al. 2000). Compression of the North America plate by ∼5 mm/year is inferred by Global Positioning System (GPS) data, leaving 6–10 mm/year residual convergence (MAZZOTTI et al. 2003).

---
[1] School of Earth and Ocean Sciences, University of Victoria, Victoria, BC, Canada. E-mail: lleonard@uvic.ca
[2] Natural Resources Canada, Geological Survey of Canada, Sidney, BC, Canada.

Prior to the 2012 earthquake, evidence for subduction included the presence of a bathymetric trench and accretionary prism (Queen Charlotte trough and terrace), receiver function analysis showing a dipping oceanic slab, and gravity and heat flow data (e.g., BUSTIN et al. 2007). The 2012 earthquake is most consistent with this model (SZELIGA 2013; CASSIDY et al. 2014).

A recent tsunami hazard assessment (LEONARD et al. 2012a, 2014) calculates an average recurrence of ∼713 years for M_w 7.7 ± 0.3 thrust earthquakes off Haida Gwaii, and uses empirical relations by ABE (1995) to estimate that such events could produce runup exceeding 3 m above the state of tide (locally up to ∼7 m) for coastlines within ∼40 km of the rupture zone. In the near field, potentially damaging runup (locally exceeding 1.5 m) could be expected up to 150 km beyond the rupture.

Although people and infrastructure were not ultimately at risk from this tsunami due to its location

and timing, field survey investigation of tsunami runup and other effects can provide critical constraints to models of earthquake rupture and tsunami generation on this little-understood fault zone. The earthquake and tsunami provide an invaluable case study for better understanding of the hazard and potential risk associated with similar events off northern Vancouver Island (Explorer plate subduction) and the much larger great earthquakes and tsunamis on the Cascadia subduction zone (Fig. 1) that represent the greatest overall tsunami hazard to populated areas throughout coastal British Columbia (e.g., LEONARD et al. 2014), Washington, Oregon and northern California.

2. Survey Logistics

Post-tsunami reconnaissance was carried out in November 2012, with follow-up surveys in February

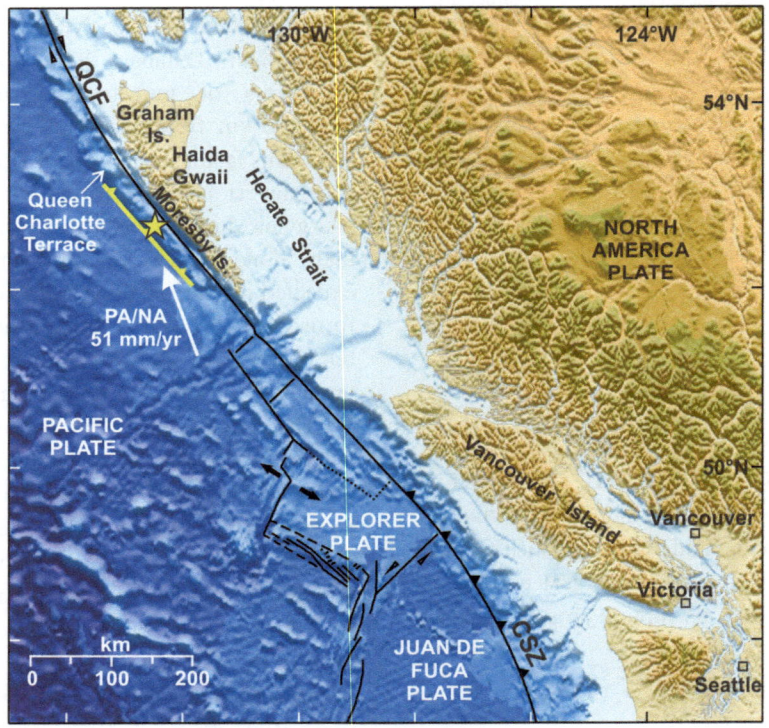

Figure 1
Regional tectonic setting of Haida Gwaii and epicentre (*yellow star*) of the 28 October 2012 M_w 7.7 thrust earthquake. *Yellow line* location of Queen Charlotte trench along approximate rupture length of the 2012 earthquake. *CSZ* Cascadia subduction zone, *QCF* Queen Charlotte transform fault. Pacific/N. America relative plate motion vector from MORVEL (DEMETS et al. 2010)

and June 2013. The survey was part of a post-earthquake field effort by Natural Resources Canada that included deployment of high-precision GPS equipment to constrain coseismic and postseismic deformation, and of land and ocean-bottom seismometers to measure aftershocks (JAMES *et al.* 2013).

The post-tsunami survey team faced a number of challenges, outlined below. (1) No eyewitness accounts were available; these are commonly used in post-tsunami surveys (e.g., BORRERO *et al.* 2009; FRITZ *et al.* 2011). (2) Access to this remote and rugged coastline was feasible only by helicopter, limiting the extent of the survey on a constrained budget. (3) A large storm occurred on 4 November, prior to the post-tsunami survey. Any tsunami evidence left below the reach of this storm would have been overprinted. The west coast of Haida Gwaii (particularly Moresby Island) is punctuated by long inlets, many of which have multiple corners or constrictions; the heads of these inlets are protected from storm waves, but other bays are more exposed, making it generally difficult to distinguish between tsunami and storm effects. (4) The west coast of Haida Gwaii has a tidal range of ∼4.5 m (lowest to highest tides; WWW Tide/Current Predictor, accessed at http://tbone.sc.edu/tide). The event occurred during relatively low tide (∼1 m below mean sea level; see Table 1); at the heads of inlets, tsunami runup had to exceed 3 m in order to leave evidence above the high tide line. On exposed parts of the outer coast, runup would have to be considerably higher to leave evidence above the reach of storm waves. (5) Reconnaissance solely from the air could not assess the extent of runup due to thick forest cover that extends close to the high tide line in most inlets, obscuring any tsunami evidence. Even where debris appeared from the air to be particularly high, landing was not possible on the outer coast, but only at the heads of inlets. Forest cover also proved problematic on the ground for the initial survey equipment (GPS).

The initial November survey focused on the area closest to the offshore rupture zone, i.e., the west coast of Moresby Island from Englefield Bay in the north to Gowgaia Bay in the south (Fig. 2). Based on a combination of preliminary tsunami modelling (e.g., Fig. 3; LEONARD *et al.* 2012b; for updated modelling see FINE *et al.*—this volume) and initial near-ground observations from the helicopter, the team landed and surveyed probable tsunami runup evidence at six locations: the heads of Sunday, Pocket and Mike Inlets, Kwoon and Puffin Coves, and Goski Bay (in northern Gowgaia Bay). Horizontal and vertical positions of debris were surveyed using equipment that included an Ashtech Z-Xtreme GPS receiver and an Ashtech antenna on a 2-m staff with attached legs to ensure a stable setup. Measurement of the vertical height of debris above the ground (e.g., seaweed in tree branches) provided flow depth data. Positional data were recorded continuously at 1 Hz, with surveying of waypoints for several minutes. The data were later processed relative to a nearby high-precision GPS campaign site (above the head of Barry Inlet; later surveys used other nearby sites as available). Unfortunately, many key runup sites were located under thick tree cover, where poor satellite coverage and probable multipath effects led to large positional uncertainties. Data corrections and uncertainties are detailed in Table 1.

On 18–19 February 2013, the team returned to four sites (Goski Bay, Sunday, Pocket and Mike Inlets) to re-survey key runup points, using a Total Station (Nikon Nivo C Series) tied into GPS base stations in open beach locations to ensure better precision.

In March 2013, the authors were notified by two sources (P. Haeussler and R. Witter, U.S. Geological Survey; B. Rahier, Fisheries and Oceans Canada) of possible tsunami evidence at a number of additional sites, mostly north of the area previously surveyed (including Husband Harbour, Davidson Inlet, Saunders Island, and Gudal Bay; Fig. 2). A third survey was carried out 19–24 June 2013, following the methodology of the February survey. Twelve sites were targeted, of which nine provided quantitative runup data. These sites span a distance of 230 km from Otard Bay, Graham Island to Gilbert Bay, Kunghit Island (Fig. 2).

3. Tsunami Survey Observations

Due to the challenges outlined in the previous section, tsunami runup could not be quantified along much of the west coast of Haida Gwaii, but generally only at the heads of long inlets. Some of these inlets

Table 1

Tsunami runup, flow depth and inundation data

Lat (°N)	Long (°W)	Ortho Ht (m above msl)	Flow depth (m above ground)	Runup (m above tide)[a]	Uncertainty (m)[b]	Inundation (m from shore)[c]	Date surveyed	Watermark (debris surveyed)
Sites sheltered from storm waves								
Seal Inlet								
53.53665	132.73343	2.45		3.59	0.25	25	2013-06-24	Plastic fishing float
53.53651	132.73318	2.48		3.62	0.25	32	2013-06-24	Plastic and wood swash line
Sunday Inlet								
52.64444	131.88916	2.66	1.43	5.00	0.24	20	2013-02-19	Seaweed in branches (Fig. 4a)
52.64453	131.88899	2.11	1.73	4.75	0.24	7	2013-02-19	Seaweed in branches
52.64440	131.88906	2.85	1.2	4.96	0.24	16	2013-02-19	Seaweed in branches
52.64429	131.88910	2.23	1.8	4.94	0.24	24	2013-02-19	Seaweed in branches
52.64424	131.88909	1.96	2	4.87	0.24	32	2013-02-19	Seaweed in branches
52.64429	131.88934	1.14	1.3	3.35	0.24	35	2013-02-19	Driftwood log on fallen tree
52.64434	131.88927	1.60	2.48	4.99	0.24	29	2013-02-19	Seaweed in branches
52.64432	131.88873	2.91	0.4	4.22	0.24	17	2013-02-19	Seaweed in branches
52.64424	131.88886	5.344		6.25	0.21	27	2012-11-19	Polystyrene
52.64433	131.88859	2.75		3.66	0.47	20	2012-11-19	Plastic and seaweed
52.64432	131.88836	3.47	1.1	5.48	0.21	33	2012-11-19	Abraded twigs on tree roots
Pocket Inlet								
52.61532	131.86168	2.03	2.15	5.09	0.28	14	2013-02-18	Seaweed in branches
52.61552	131.86180	4.91		5.82	0.28	18	2013-02-18	Plastic float and seaweed on log
52.61524	131.86140	6.24		7.15	0.28	39	2013-02-18	Seaweed at base of tree
52.61476	131.86087	3.56	1.04	5.51	0.28	28	2013-02-19	Bull kelp, other seaweed on log (Fig. 4b)
52.61486	131.86061	3.95		4.86	0.28	34	2013-02-19	Landward footprint of log (Fig. 5d)
52.61486	131.86053	4.19	1.1	6.20	0.28	33	2013-02-19	Broken-off tree mushroom (Fig. 4d)
52.61568	131.86164	1.602		2.51	0.22	38	2012-11-19	Inundated area, seaweed on ground
Mike Inlet								
52.53390	131.74715	6.71		7.62	0.30	28	2013-02-18	Polystyrene
52.53369	131.74721	4.06		4.97	0.30	26	2013-02-18	"Purex" plastic container
52.53399	131.74731	5.17		6.08	0.30	19	2013-02-18	Plastic and polystyrene
52.53455	131.74759	6.14		7.05	0.40	18	2012-11-19	Plastic debris
Puffin Cove								
52.49847	131.72337	2.489		3.40	0.21	9	2012-11-15	Debris wrack line
52.49857	131.72435	2.38		3.29	0.48	1	2012-11-15	Debris wrack line
Goski Bay								
52.44009	131.55171	3.31		4.22	0.27	8	2013-02-18	Musashi large fishing float (Fig. 6e)
52.43997	131.55174	3.59		4.50	0.27	6	2013-02-18	Fish carcass/bones (Fig. 6d)
52.44028	131.55172	3.40		4.31	0.27	25	2013-02-18	Debris concentration
52.44053	131.55168	3.39		4.30	0.27	50	2013-02-18	Max debris inundation
52.44053	131.55157	3.11		4.02	0.27	48	2013-02-18	Max debris inundation
Staki Bay								
52.27433	131.37570	2.40		3.36	0.18	25	2013-06-21	Plastic debris in forest
52.27430	131.37579	1.91		2.87	0.18	26	2013-06-21	Level of mud deposit surface (sampled)
Louscoone Inlet								
52.24340	131.29374	3.02		3.98	0.19	14	2013-06-21	Plastic debris (furthest "upstream")
52.24333	131.29363	2.71		3.67	0.19	12	2013-06-21	Plastic wrapper and bottle
52.24327	131.29353	2.91		3.87	0.19	10	2013-06-21	Plastic fragments

Table 1 *continued*

Lat (°N)	Long (°W)	Ortho Ht (m above msl)	Flow depth (m above ground)	Runup (m above tide)[a]	Uncertainty (m)[b]	Inundation (m from shore)[c]	Date surveyed	Watermark (debris surveyed)
52.24325	131.29327	2.47		3.43	0.19	15	2013-06-21	Plastic float, other plastic, polystyrene

Sites exposed to storm waves
Otard Bay

Lat (°N)	Long (°W)	Ortho Ht (m above msl)	Flow depth (m above ground)	Runup (m above tide)[a]	Uncertainty (m)[b]	Inundation (m from shore)[c]	Date surveyed	Watermark (debris surveyed)
53.77232	132.99394	6.03		7.20	0.28	35	2013-06-24	Polystyrene concentration (Fig. 6a)
53.77244	132.99390	5.11		6.28	0.28	47	2013-06-24	Polystyrene landward (lower elev)
53.77252	132.99382	5.07		6.24	0.28	57	2013-06-24	Furthest evidence of inundation
53.77206	132.99425	6.20		7.37	0.28	0	2013-06-24	Plastic debris above beach berm

Gudal Bay

Lat (°N)	Long (°W)	Ortho Ht (m above msl)	Flow depth (m above ground)	Runup (m above tide)[a]	Uncertainty (m)[b]	Inundation (m from shore)[c]	Date surveyed	Watermark (debris surveyed)
53.23196	132.56030	5.29		6.37	0.23	12	2013-06-24	Plastic floats, polystyrene etc.
53.23185	132.56383	6.33		7.41	0.23	5	2013-06-24	Plastic debris

Saunders Island

Lat (°N)	Long (°W)	Ortho Ht (m above msl)	Flow depth (m above ground)	Runup (m above tide)[a]	Uncertainty (m)[b]	Inundation (m from shore)[c]	Date surveyed	Watermark (debris surveyed)
53.03090	132.45940	6.73		7.69	0.25	30	2013-06-24	Plastic, polystyrene etc.
53.03086	132.45940	7.28		8.24	0.25	33	2013-06-24	Highest polystyrene
53.03101	132.45948	6.60		7.56	0.25	30	2013-06-24	Swash line (foam, plastic, driftwood)
53.03007	132.45909	4.74		5.70	0.25	23	2013-06-24	Debris in bay to the south

Davidson Inlet

Lat (°N)	Long (°W)	Ortho Ht (m above msl)	Flow depth (m above ground)	Runup (m above tide)[a]	Uncertainty (m)[b]	Inundation (m from shore)[c]	Date surveyed	Watermark (debris surveyed)
52.75680	132.12102	5.35	1.83	8.04	0.16	11	2013-06-19	Foam wedged in tree branches (Fig. 4c)
52.75706	132.12096	6.94	1	8.80	0.16	32	2013-06-19	Erosion in landward direction
52.75701	132.12094	5.89		6.75	0.16	26	2013-06-19	Plastic debris at approx max inundation
52.75719	132.12194	6.81		7.67	0.16	8	2013-06-19	Debris above NW end of beach
52.75656	132.12024	8.11		8.97	0.16	18	2013-06-19	Debris above south end of beach
52.75662	132.12025	8.63		9.49	0.16	18	2013-06-19	Plastic and polystyrene
52.75688	132.12025	12.12		12.98	0.16	35	2013-06-19	Highest debris landward of S. beach

Kwoon Cove

Lat (°N)	Long (°W)	Ortho Ht (m above msl)	Flow depth (m above ground)	Runup (m above tide)[a]	Uncertainty (m)[b]	Inundation (m from shore)[c]	Date surveyed	Watermark (debris surveyed)
52.62869	131.92531	5.235		6.15	0.23	4	2012-11-19	Plastic debris concentration
52.62867	131.92531	6.037		6.95	0.23	5	2012-11-19	Large polystyrene float (windblown?)
52.62852	131.92585	5.55		6.46	0.35	6	2012-11-19	Polystyrene and plastic wrack line

Ta'dasl

Lat (°N)	Long (°W)	Ortho Ht (m above msl)	Flow depth (m above ground)	Runup (m above tide)[a]	Uncertainty (m)[b]	Inundation (m from shore)[c]	Date surveyed	Watermark (debris surveyed)
52.07200	131.11415	7.13		8.09	0.23	14	2013-06-21	Polystyrene debris

Gilbert Bay

Lat (°N)	Long (°W)	Ortho Ht (m above msl)	Flow depth (m above ground)	Runup (m above tide)[a]	Uncertainty (m)[b]	Inundation (m from shore)[c]	Date surveyed	Watermark (debris surveyed)
52.03940	131.07997	4.93		5.89	0.19	26	2013-06-21	Concentration of plastic etc.
52.03944	131.07952	7.49		8.45	0.19	52	2013-06-21	Plastic bottle further landward
52.04002	131.08174	5.97		6.93	0.19	23	2013-06-21	Plastic debris above sand dunes
52.04002	131.08196	7.08		8.04	0.19	22	2013-06-21	Plastic debris above sand dunes

[a] Runup is the orthometric height (elevation above mean sea level), corrected to height above state of tide by assuming that the largest runup occurred within 15–45 min after the earthquake, and using tide predictions from the WWW Tide/Current Predictor (Accessed at http://tbone.sc.edu/tide) at nearby locations, averaged if between tide prediction locations [tide was below mean sea level by (from north to south): 1.17 ± 0.12 m at Port Louis, 1.11 ± 0.13 m at Shields Bay, 1.05 ± 0.12 m at Armentieres Channel, 0.86 ± 0.12 m at Tasu, 0.96 ± 0.13 m at the Gordon Islands]. Where the ground below a flow depth indicator was surveyed, the flow depth is added to provide a minimum runup

[b] Uncertainty is the geometric sum of the GPS vertical position uncertainty, tide correction uncertainty, stated uncertainty from CGG2010 geoid model, plus an uncertainty associated with using a control point at Queen Charlotte City (where geoid surface is 9 cm below MSL)

[c] Distance from shoreline (beach)—shoreline position estimated from orthorectified aerial photographs of the BC government Web Mapping Services (http://www.data.gov.bc.ca/dbc/geographic/connect/index.page; Accessed June 2013)

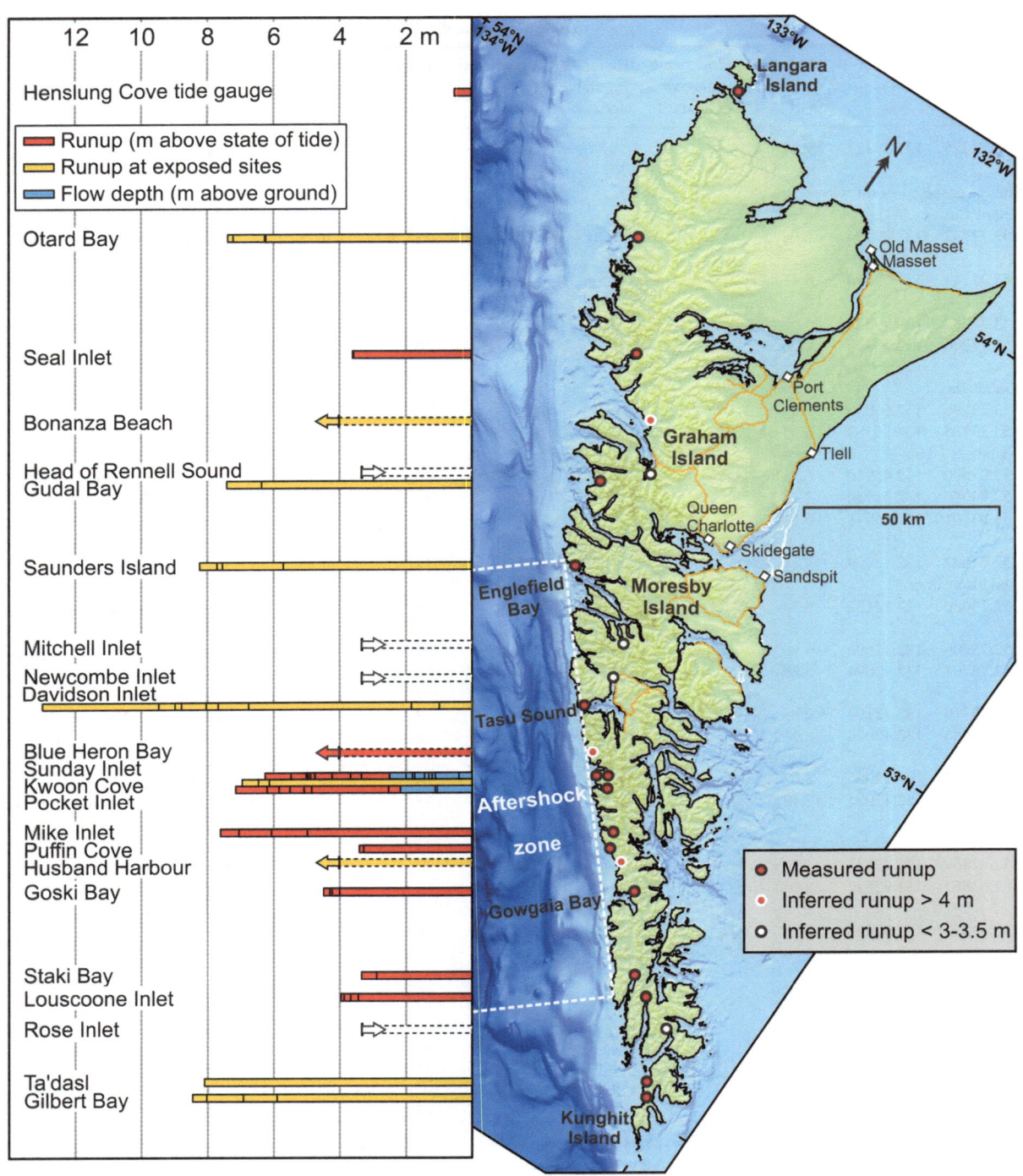

Figure 2

Haida Gwaii survey locations, runup and flow depth data (see also Table 1). *Dashed bars* indicate inferred minimum/maximum runup at unsurveyed sites; *red/orange* indicates that debris was seen to exceed the elevation of the forest edge by at least 1 m; *white* indicates that debris was not observed to exceed the forest edge elevation. "Runup" value from Henslung Cove tide gauge is the measured peak-to-trough tsunami amplitude (approximately equivalent to runup, ABE 1995)

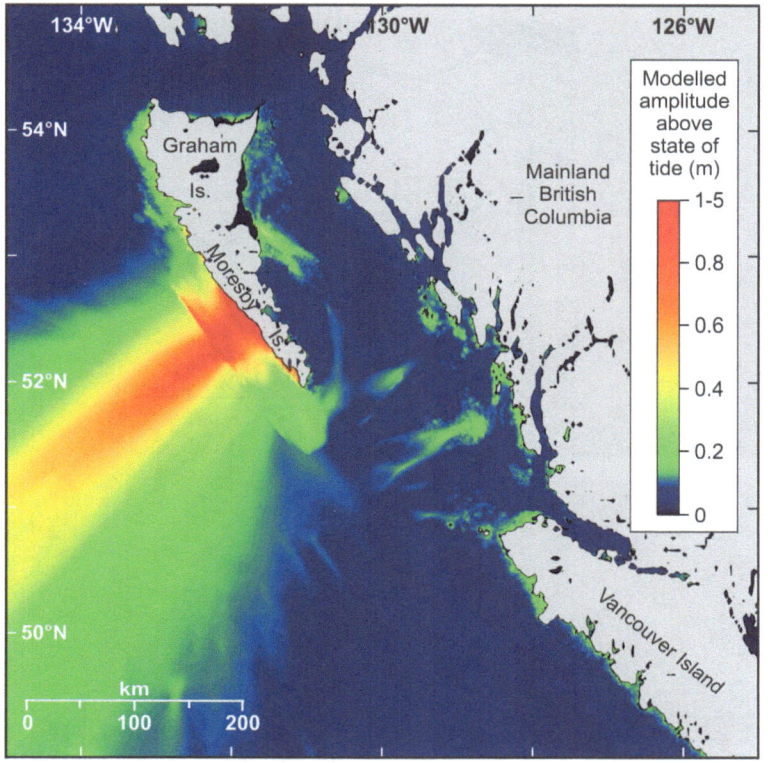

Figure 3
Preliminary (coarse-grid) tsunami model, with highest amplitudes on western Moresby Island (provided by Isaac Fine; more recent and finer-grid results shown for some regions in Fig. 9, and in FINE *et al.*—this volume)

may have natural resonant periods close to the ~30-minute period of the tsunami (FINE *et al.*—this volume), which would promote wave amplification to produce locally high tsunami runup. Debris was also surveyed landward of the forest edge at a number of bays that are regularly exposed to storm waves; in such cases, it is impossible to unequivocally determine whether the debris was deposited during the 28 October 2012 tsunami or during a major storm such as that of 4 November 2012 (see Sect. 3.2). For this reason, we make a distinction between data collected at sheltered versus exposed sites.

3.1. Observations at Sites Sheltered from Storm Waves

The heads of long inlets on the west coast of Moresby Island, particularly those with multiple corners or constrictions (Sunday, Pocket and Mike Inlets), are sheltered from storm waves, as are eastern Blue Heron Bay, Puffin Cove, Goski Bay, Staki Bay and Louscoone Inlet, as well as Seal Inlet on Graham Island (Fig. 2). Evidence of significant tsunami runup and inundation into the forest was found at all of these sites, except at Puffin Cove, where apparent tsunami evidence in the form of a wrack line of debris (driftwood, plastic, polystyrene, and seaweed) was found above the beach but seaward of the forest. Landing access to the Blue Heron Bay site was not possible, but plastic and polystyrene debris was seen to extend well above the forest edge; runup likely exceeded 4 m or more.

Runup and flow depth measurements are provided for the other eight sites in Table 1 and Fig. 2. Among these sites, a maximum apparent runup of 7.6 m above the state of tide was measured at Mike Inlet, where the grade above the high tide line is steeper than at the other sites. It is possible that this is not the maximum elevation reached by the tsunami at this site (and others); debris was often found on the landward side of obstacles such as logs, suggesting that it was

Figure 4
Flow depth indicators at sites on western Moresby Island: **a** seaweed (*Fucus spp.*) in tree branches at Sunday Inlet; **b** seaweed (mostly *Nereocystis lueteana*) draped over log at Pocket Inlet; **c** foam lodged in tree branches at Davidson Inlet; **d** mushroom scar on tree; splintered wood and broken mushroom found at base of tree at Pocket Inlet

deposited on the ebb flow. Runup at Sunday Inlet and Pocket Inlet exceeded 6 m and 7 m, respectively. At these two sites, multiple flow depth indicators were found mostly in the form of seaweed (*Fucus spp.* and *Nereocystis lueteana*) in tree branches or draped over logs (Fig. 4). Maximum flow depths of 2.5 and 2.2 m above the ground were measured at Sunday and Pocket Inlets, respectively. Maximum measured runup at other sites ranges from 3.3 m at Staki Bay to 4.5 m at Goski Bay. Pocket Inlet provided evidence that large logs on the forest floor had been disturbed by a marine influx (Fig. 5 and Sect. 3.3).

An additional four storm-sheltered sites on the west coast of Haida Gwaii examined for tsunami impact showed no evidence of debris above the high tide line. At these locations, tsunami runup must not have exceeded 3 m and may have been substantially less. Such areas include the heads of Rennell Sound, and Mitchell, Newcombe and Rose Inlets (Fig. 2).

3.2. Observations at Sites Exposed to Storm Waves

Debris above the storm beach and/or forest edge was surveyed at a number of bays that see regular

Figure 5
Movement of logs at the head of Pocket Inlet. **a** Aerial photograph image from the BC government Web Mapping Services (http://www.data. gov.bc.ca/dbc/geographic/connect/index.page; Accessed June 2013); *arrows* show tsunami flow directions inferred from the movement of logs relative to their apparent former footprints. *Arrow tails* mark the locations of sites labelled (**b–f**). **b** to **f** logs and their footprints, from which flow directions were determined (see *insets*)

exposure to storm waves. These include several bays that face southwest or south-southwest (Otard Bay, Gudal Bay, Davidson Inlet, and Gilbert Bay), two that face northwest (Kwoon Cove and Ta'dasl), and one that faces east-northeast (Eastern Saunders Island) but appears affected by wave reflection/

refraction between a number of islands in northwestern Englefield Bay (locations in Fig. 2). Southwest-facing bays may have been particularly susceptible to large waves during the storms of 14 October and 4 November 2012, which had southwesterly prevailing winds and characteristic significant peak-to-trough wave heights of 9.6 and 13 m, respectively (measured at a buoy offshore western Moresby Island; data from http://www.isdm-gdsi.gc.ca, accessed Oct 2013). A simple examination of the buoy data shows that the 4 November storm had the third largest characteristic wave height over the 23-year record (ninth-largest peak wave height). Gilbert Bay and Kwoon Cove may have experienced large waves during a storm in February 2013, which had northwesterly prevailing winds and a significant wave height of 9–10 m offshore Moresby Island; of the two sites, only Kwoon Cove was surveyed prior to this storm.

Potential tsunami evidence at these exposed sites takes a number of forms. At Kwoon Cove, debris formed a wrack line above the storm beach but below the forest edge; bull kelp (*N. lueteana*) among the highest debris (in November 2012) appeared to be at a similar state of decomposition as seen at nearby sheltered sites with more definitive tsunami evidence, and provided a similar runup measurement (6.5–7 m). The southernmost sites (Ta'dasl and Gilbert Bay) both yielded a wrack line of sparse debris only just above the forest edge that is impossible to distinguish from possible storm deposition. Otard Bay and Gudal Bay on Graham Island provided evidence for inundation of the forest floor. At Otard Bay, marine breaching of the forest at one point within the area surveyed was evidenced by a trail of polystyrene, including a large piece that appeared to have been fragmented against the landward side of a log, likely on the ebb flow, although a storm origin cannot be ruled out (Fig. 6a). Gudal Bay and Saunders Island both yielded more plentiful debris landward of the forest edge. However, highest elevations were reached at Davidson Inlet.

Davidson Inlet is a long, narrow, steep-sided inlet facing to the south-southwest that showed clear evidence, at its head, of recent erosion and deposition by a large marine influx (Fig. 7). Evidence includes uprooted trees at the forest edge, abundant debris in the forest (including plastic and polystyrene debris wedged in tree branches indicating flow depths up to 1.8 m; Figs. 4, 7), and a smaller amount of debris high above the beach at both sides. The highest surveyed debris (polystyrene) was found at ~12 m above sea level, or 13 m above the tide level at the time of the tsunami. A storm origin seems improbable, but cannot be completely ruled out at this time. It is hoped that tide gauge measurements over the winter of 2013–2014 can assess the potential for large storm waves in this inlet.

3.3. Depositional and Erosional Evidence

Although sand sheets are commonly cited as evidence for past tsunamis (e.g., CLAGUE et al. 2000), no sand layer was found to have been deposited by the tsunami at the sites surveyed. Discontinuous muddy deposits were present in some locations (Fig. 8), generally not those with highest measured runup, but they could not be used to define the extent of inundation. Instead, runup extent was defined by natural and manmade debris that had a clear oceanward origin (Fig. 6). Natural debris included seaweed, driftwood, and salmon carcasses, whereas manmade debris included a large variety of polystyrene and plastic (bottles and other containers, sheeting, fishing floats, sandals, hard hats, crates), as well as some glass bottles and light bulbs. Some debris has a clear Asian origin, and some of it may have been originally mobilised by the 2011 tsunami in northeastern Japan; a large fishing float found in the forest at the head of Goski Bay (Fig. 6e) is identical to one found on a Washington state beach that was traced to a Japanese oyster farm damaged in the 2011 tsunami (RICE 2012).

At Pocket Inlet, evidence for disturbance of large windfall logs provided indicators of runup as well as of flow directions (Fig. 5). Large logs appeared to have been floated, flipped, rotated, dragged, or completely removed from their former position, leaving gouge marks in the moss and bark-lined footprints that were often filled with seaweed. Peeled-back moss indicated a landward flow direction, while disturbed logs indicated a variety of flow directions—seaward, landward and along shore. Similar complex patterns of runup and backwash have been inferred

Figure 6
Debris at various survey sites on Graham (**a**) and Moresby (**b–f**) Islands: **a** polystyrene pieces on the landward side of a log at Otard Bay; **b** plastic sheeting suggesting shore-parallel flow at Pocket Inlet; **c** a selection of debris at Mike Inlet; **d** salmon carcass at Goski Bay; **e** "Musashi" fishing float at Goski Bay; **f** debris at Staki Bay

from the geomorphological effects of other tsunamis (DAWSON 1994). Erosive evidence was also present at Davidson Inlet, where uprooted trees were found near the forest edge, along with large root wads (Fig. 7). It appeared that this area had been stripped to reveal a former forest floor.

Figure 7
Davidson Inlet: evidence of erosion and deposition. **a** View towards the head of the inlet; **b** uprooted tree and exposed forest floor; **c** an eroded root wad on the shoreward side of the forest edge; **d** debris in the forest; **e** debris in the forest, with disturbed trees in background; **f** plastic bottle wedged between small branches

Figure 8
Mud deposits at **a** Staki Bay, Moresby Island (*landward view*) and **b** Gudal Bay, Graham Island (*seaward view*). Compass for *scale*

4. Discussion

4.1. Tsunami Runup Observations and Modelling

Field surveys following the 28 October 2012 M_w 7.7 thrust earthquake off Haida Gwaii reveal that the unpopulated west coast of the islands was impacted by the largest tsunami of 2012, globally. Debris was left over 3 m above the state of tide at sites spanning at least 170 km of the coastline, and possibly more than 230 km. Maximum measured runup was 7.6 m above the state of tide (at Mike Inlet) among sites sheltered from storm waves, and 13 m (at Davidson Inlet) among sites that are relatively open to storm waves (Table 1; Fig. 2). The maximum measured flow depth, indicated by debris caught in tree branches, was 2.5 m at Sunday Inlet.

Runup observations from storm-sheltered sites (and some exposed sites) are generally well-matched by the amplitude predictions of tsunami models that are based on a relatively simple offshore northeastward-dipping thrust fault rupture constrained by available seismological and geodetic data for the event (e.g., Fig. 9; FINE *et al.*—this volume). However, the maximum runup of 13 m at Davidson Inlet cannot be matched by the current model, which predicts a maximum amplitude of only ~4.5 m there.

There are several possible explanations for the mismatch at Davidson Inlet. The tsunami may have been amplified in this area due to a localized patch of higher-than-average slip on a shallow part of the fault, or other non-uniform slip patterns; local slip variations cannot be resolved with the available seismological or geodetic data. Alternatively, tsunami amplitudes may have been locally increased due to splay faulting or a submarine landslide triggered by earthquake shaking or seafloor rupture. Such effects have been documented for other large thrust earthquakes, including the 1964 Alaska and 2004 Sumatra great earthquakes (e.g., SULEIMANI *et al.* 2011; DEDONTNEY and RICE 2012), and cannot be discounted with data currently available. A final possibility is that the observations at Davidson Inlet (Table 1; Figs. 2, 7) can be explained in terms of wave erosion and deposition during the storm of 4 November 2012, rather than the tsunami of 28 October. It seems unlikely that storm waves of relatively short wavelength could have carried material up to 12 m above sea level and inundated the forest with debris, leaving some of it firmly wedged into tree branches (Figs. 4, 7), but the possibility cannot be discounted at this time. Temporary tide gauge measurements in Davidson Inlet during the winter of 2013–2014 may provide some insight to possible storm wave amplitudes.

4.2. Recommendations for Future Post-Tsunami Surveys

The surveys described in this paper faced a number of challenges that may be fairly unique among coastal areas at risk of impact from tsunami

305

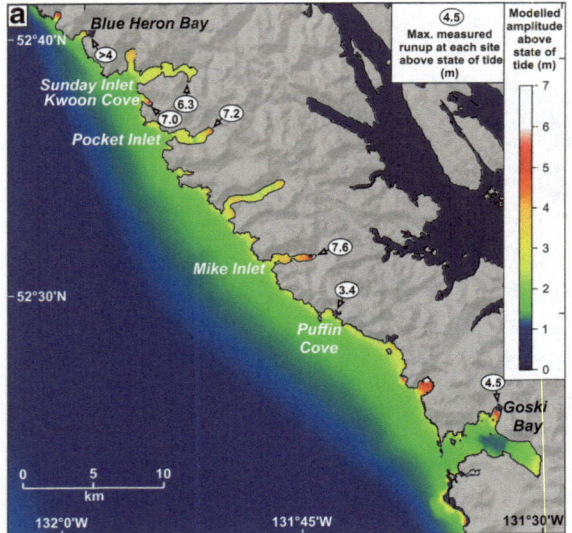

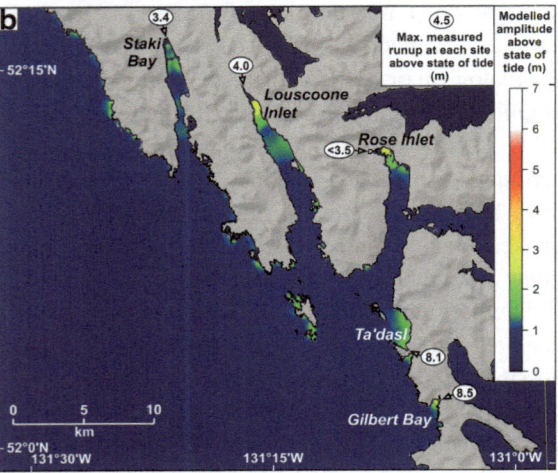

Figure 9
Comparison of model predicted tsunami amplitudes (provided by Isaac Fine; also see FINE *et al.*—this volume) with highest measured runup at sites on western Moresby and Kunghit Islands: **a** sites from Blue Heron Bay to Goski Bay, and **b** sites from Staki Bay to Gilbert Bay. Runup site locations indicated by arrowheads (also see Table 1). *Dark blue* in some areas (including Blue Heron Bay, Puffin Cove, Goski Bay, Staki Bay, Louscoone Inlet and Rose Inlet) reflects a lack of bathymetry data rather than near-zero modelled amplitude

waves. However, some lessons learned may be applicable for surveys following future tsunamis around the world. (1) For coastal regions with thick forest cover, tsunami impacts are likely to evade detection from the air and from the water. Observations must be made on the ground, to assess possible inundation of the forest. Furthermore, to measure

tsunami runup in these environments, the observer cannot rely solely on satellite-based methods, as they are compromised by tree cover. (2) Following a tsunami, a post-tsunami survey should take place as soon as possible, in order to document the evidence before it is disturbed, e.g., by humans, animals, storms or further tsunamis. Unfortunately, our first survey took place 3 weeks after the Haida Gwaii event, and following a large storm that may have disturbed tsunami evidence in exposed areas. However, the timing of our three surveys allows us to comment on the preservation potential of tsunami evidence in remote areas (i.e., where debris is unlikely to be disturbed by humans).

As expected, the greatest range of evidence (including both natural and manmade debris) was available during the first survey, but a surprising amount of evidence remained 3 months later, when sites were revisited. All manmade debris (plastic, polystyrene etc.) remained exactly in place, and most of the natural debris also remained undisturbed. Exceptions were fish carcasses and some seaweed. The fish likely rotted or were scavenged, but one partial skeleton was found in place. Much of the seaweed was still in place 4 months after the tsunami—on the ground and hanging from tree branches. However, some of it had disappeared (e.g., fallen from branches) or partially dried up— presumably this process would be more rapid in a drier season or environment than a coastal Haida Gwaii winter. At sites visited 8 months after the event, there was no sign of seaweed on the ground or in tree branches that could have provided maximum runup or flow depth data. However, driftwood and manmade debris did not appear to have been disturbed, altered or buried by natural processes. Muddy sedimentary deposits documented at this time appeared relatively fresh, with just a covering of conifer needles or sparse young vegetation (Fig. 8).

4.3. Implications for Paleotsunami Surveys

Paleotsunami data are an important contribution to tsunami and seismic hazard assessments, providing constraints on the timing and magnitude of past events (e.g., JAFFE and GELFENBAUM 2002; GONZÁLEZ *et al.* 2009). If there is a sand source available offshore, tsunami deposits typically comprise a landward-thinning, landward-fining sand layer that is preserved in near-coastal stratigraphy (e.g., CLAGUE et al. 2000). In theory, the maximum landward elevation of the sand sheet provides a measure of tsunami runup, and the absence of a sand layer in the stratigraphy could imply that maximum runup was lower than the elevation of the site. However, surveys following the massive 2004 Indian Ocean and 2011 Japan tsunamis suggest that sites where tsunami runup is less than 3 m are unlikely to preserve sedimentary evidence (SZCZUCIŃSKI 2012), and that only 50–60 % of the tsunami inundation distance is represented by sand deposits after 1–4 years (GOTO et al. 2011). Sedimentary evidence left by the Haida Gwaii tsunami also has implications for paleotsunami surveys, both on the coast of British Columbia and elsewhere.

At the sites surveyed on the west coast of Haida Gwaii, there was no sign of a sandy deposit, despite evidence that debris had been carried by a marine inundation up to 12 m above mean sea level with flow depths up to 2.5 m. The absence of sand may reflect the particular environment of western Haida Gwaii, where most inlets comprise narrow steep-sided fjords, and there is only a narrow shelf offshore that is primarily bedrock (e.g., BARRIE et al. 2013). The lack of sand may also partially reflect the low tide conditions at the time of the event, such that flow depths were relatively low at the sites surveyed and may have had insufficient energy to transport a significant amount of sand.

Discontinuous mud deposits were observed in some areas (e.g., Fig. 8), up to ~20 cm thick where measured. Similar scattered mud deposits were documented in the Kuril Islands following the 2011 Japanese tsunami, which had maximum documented runup of ~5 m in that region (RAZJIGAEVA et al. 2013). In Haida Gwaii, the presence of sedimentary deposits does not appear to correlate with runup; at sites with some of the largest runup (e.g., Mike Inlet and Pocket Inlet; >7 m), no mud deposits were observed. At Davidson Inlet, potentially the site with highest tsunami runup (13 m), mud deposits appeared to be restricted to localized pockets, e.g., in deep tree wells. Among sheltered sites, the most extensive mud deposit was found at the head of Staki Bay, where maximum runup was only 3.4 m.

The above observations imply that interpretation of paleotsunami deposits should take many factors into account, including the following. (1) The highest apparent paleotsunami deposit is unlikely to represent the maximum tsunami runup and inundation; sedimentary evidence may be absent from sites where maximum runup occurred. (2) The full tidal range should be considered; a tsunami occurring at high tide will leave deposits at higher elevations than an event of the same magnitude occurring at low tide. Despite the significant runup of the Haida Gwaii tsunami (up to 7–13 m above the state of tide), very little evidence will be preserved in the form of sedimentary deposits (excluding manmade debris).

Acknowledgments

The manuscript was improved by useful comments from two anonymous reviewers and Alexander Rabinovich. Isaac Fine and Josef Cherniawsky provided us with preliminary tsunami models that were invaluable to our field survey; subsequent discussions with them were also useful in the preparation of this paper. Peter Haeussler and Rob Witter (USGS) and Ben Rahier provided important observations leading us to expand our survey and our findings. We thank many colleagues at the Pacific Geoscience Centre who provided us with field equipment as well as logistical and other support—including Mike Schmidt, Lisa Nykolaishen, Scott Dallimore, Roger MacLeod, Michelle Côté, Joe Henton, Thomas James, John Cassidy, and Garry Rogers. We also thank Matt Barker (Department of Fisheries and Oceans) for the loan of (and training on) the Total Station. We thank Cindy Wright, Lisa Nykolaishen and Brian Schofield for assistance in the field. Finally, the post-tsunami survey would not have been possible without the permission and support of Parks Canada and the Haida Nation. This is Earth Sciences Sector contribution No. 20130462.

REFERENCES

ABE, K. (1995), Estimate of tsunami run-up heights from earthquake magnitudes, In: Tsuchiya, Y., and Shuto, N. (eds.), Tsunami: progress in prediction, disaster prevention and warning, Advances in Technological Hazards Research, Kluwer Academic Publishers, Dordrecht, the Netherlands, 4:21–35.

BARRIE, J.V., CONWAY, K.W., and HARRIS, P.T. (2013), *The Queen Charlotte fault, British Columbia: seafloor anatomy of a transform fault and its influence on sediment processes*, Geo-Mar Lett. *33*:311–318, doi:10.1007/s00367-013-0333-3.

BORRERO, J.C., WEISS, R., OKAL, E.A., HIDAYAT, R., SURANTO, ARCAS, D., and TITOV, V.V. (2009), *The tsunami of 2007 September 12, Bengkulu province, Sumatra, Indonesia: post-tsunami field survey and numerical modelling*, Geophys. J. Int. *178*:180–194, doi:10.1111/j.1365-246X.2008.04058.x.

BUSTIN, A.M.M., HYNDMAN, R.D., KAO, H., and CASSIDY, J.F. (2007), *Evidence for underthrusting beneath the Queen Charlotte margin, from teleseismic receiver function analysis*, Geophys. J. Int. *171*:1198–1211, doi:10.1111/j.1365-246X.2007.03583.x.

CASSIDY, J.F., ROGERS, G.C., and HYNDMAN, R.D. (2014), An overview of the 28 October 2012 M_w 7.7 earthquake in Haida Gwaii, Canada: a tsunamigenic thrust event along a predominantly strike-slip margin, Pure Appl. Geophys., doi:10.1007/s00024-014-0775-1.

CLAGUE, J.J., BOBROWSKY, P.T., and HUTCHINSON, I. (2000), *A review of geological records of large tsunamis at Vancouver Island, British Columbia, and implications for hazard*, Quat. Sci. Rev. *19*(9):849–863, doi:10.1016/S0277-3791(99)00101-8.

DAWSON, A.G. (1994), *Geomorphological effects of tsunami run-up and backwash*, Geomorphology *10*:83–94.

DEDONTNEY, N., and RICE, J.R. (2012), *Tsunami wave analysis and possibility of splay fault rupture during the 2004 Indian Ocean earthquake*, Pure Appl. Geophys. *169*(10):1707–1735, doi:10.1007/s00024-011-0438-4.

DEMETS, C., GORDON, R.G., and ARGUS, D.F. (2010), *Geologically current plate motions*, Geophys. J. Int. *181*:1–80, doi:10.1111/j.1365-246X.2009.04491.x.

FINE, I.V., CHERNIAWSKY, J.Y., THOMSON, R.E., RABINOVICH, A.B., and KRASSOVSKI, M.V. (this volume), Observations and numerical modeling of the 2012 Haida Gwaii tsunami off the coast of British Columbia.

FRITZ, H.M., PETROFF, C.M., CATALÁN, P.A., CIENFUEGOS, R., WINCKLER, P., KALLIGERIS, N., WEISS, R., BARRIENTOS, S.E., MENESES, G., VALDERAS-BERMEJO, C., EBELING, C., PAPADOPOULOS, A., CONTRERAS, M., ALMAR, R., DOMINGUEZ, J.C., and SYNOLAKIS, C.E. (2011), *Field survey of the 27 February 2010 Chile tsunami*, Pure Appl. Geophys. *168*(11):1989–2010, doi:10.1007/s00024-011-0283-5.

GONZÁLEZ, F.I., GEIST, E.L., JAFFE, B., KÂNOĞLU, U., MOFJELD, H., SYNOLAKIS, C.E., TITOV, V.V., ARCAS, D., BELLOMO, D., CARLTON, D., HORNING, T., JOHNSON, J., NEWMAN, J., PARSONS, T., PETERS, R., PETERSON, C., PRIEST, G., VENTURATO, A., WEBER, J., WONG, F., and YALCINER, A. (2009), *Probabilistic tsunami hazard assessment at Seaside, Oregon, for near- and far-field seismic sources*, J. Geophys. Res. *114*(C11023), doi:10.1029/2008JC005132.

GOTO, K., CHAGUÉ-GOFF, C., FUJINO, S., GOFF, J., JAFFE, B., NISHIMURA, Y., RICHMOND, B., SUGAWARA, D., SZCZUCIŃSKI, W., TAPPIN, D.R., WITTER, R.C., and YULIANTO, E. (2011), *New insights of tsunami hazard from the 2011 Tohoku-oki event*, Mar. Geol. *290*(1–4):46–50, doi:10.1016/j.margeo2011.10.004.

HYNDMAN, R.D., and ELLIS, R.M. (1981), *Queen Charlotte fault zone: microearthquakes from a temporary array of land stations and ocean bottom seismographs*, Can. J. Earth Sci. *18*(4):776–788, doi:10.1139/e81-071.

JAFFE, B.E., and GELFENBAUM, G. (2002), Using tsunami deposits to improve assessment of tsunami risk, *In:* Solutions to Coastal Disasters'02, Conference Proceedings, American Society of Civil Engineers, pp. 836–847.

JAMES, T., ROGERS, G., CASSIDY, J., DRAGERT, H., HYNDMAN, R., LEONARD, L., NYKOLAISHEN, L., RIEDEL, M., SCHMIDT, M., and WANG, K. (2013), *Field studies target 2012 Haida Gwaii earthquake*, Eos Transactions, AGU, *94*(22):197–198.

LEONARD, L.J., ROGERS, G.C., and MAZZOTTI, S. (2012a), A preliminary tsunami hazard assessment of the Canadian coastline, Geological Survey of Canada, Open File 7201, 126 p., doi:10.4095/292067.

LEONARD, L., BEDNARSKI, J., FINE, I., CHERNIAWSKY, J., and WRIGHT, C. (2012b), The Haida Gwaii tsunami of October 27, 2012, Risky Ground, Newsletter of the Centre for Natural Hazard Research, Simon Fraser University, Dec 21 2012—Winter edition, pp. 10–11.

LEONARD, L.J., ROGERS, G.C., and MAZZOTTI, S. (2014), *Tsunami hazard assessment of Canada*, Nat. Hazards *70*(1):237–274, doi:10.1007/s11069-013-0809-5.

MAZZOTTI, S., HYNDMAN, R.D., FLÜCK, P., SMITH, A.J., and SCHMIDT, M. (2003), *Distribution of the Pacific/North America motion in the Queen Charlotte Islands-S. Alaska plate boundary zone*, Geophys. Res. Lett. *30*(14):1762, doi:10.1029/2003GL017586.

National Geophysical Data Center/World Data System (NGDC/WDS) Global Historical Tsunami Database, Boulder, CO, USA. Available at: http://www.ngdc.noaa.gov/hazard/tsu_db.shtml (Last Accessed October 2013).

RAZJIGAEVA, N.G., GANZEY, L.A., GREBENNIKOVA, T.A., IVANOVA, E.D., KHARLAMOV, A.A., KAISTRENKO, V.M., and SHISHKIN, A.A. (2013) *Coastal sedimentation associated with the Tohoku tsunami of 11 March 2011 in South Kuril Islands, NW Pacific Ocean*, Pure Appl. Geophys. *170*(6–8):1081–1102.

RICE, A., 2012, Debris possibly from Japanese tsunami floating up Strait of Juan de Fuca, Peninsula Daily News, May 15 2012. Available at http://www.peninsuladailynews.com/article/20120515/news/305159993/debris-possibly-from-japanese-tsunami-floating-up-strait-of-juan-de-fuca.

ROHR, K.M.M., SCHEIDHAUER, M., and TREHU, A.M. (2000), *Transpression between two warm mafic plates: the Queen Charlotte fault revisited*, J. Geophys. Res. *105*(B4):8147–8172, doi:10.1029/1999JB900403.

SZCZUCIŃSKI, W. (2012), *The post-depositional changes of the onshore 2004 tsunami deposits on the Andaman Sea coast of Thailand*, Nat. Hazards *60*(1):115–133, doi:10.1007/s11069-011-9956-8.

SULEIMANI, E., NICOLSKY, D.J., HAEUSSLER, P.J., and HANSEN, R. (2011), *Combined effects of tectonic and landslide-generated tsunami runup at Seward, Alaska during the Mw 9.2 1964 earthquake*, Pure Appl. Geophys. *168*:1053–1074.

SZELIGA, W. (2013), *2012 Haida Gwaii quake: insight into Cascadia's subduction extent*, Eos Trans., AGU *94*(9):85–86.

(Received December 9, 2013, revised February 2, 2014, accepted February 4, 2014, Published online March 27, 2014)

Pure Appl. Geophys. 171 (2014), 3483–3491
© 2013 The Author(s)
This article is published with open access at Springerlink.com
DOI 10.1007/s00024-013-0720-8

❙ Pure and Applied Geophysics

Impact of Near-Field, Deep-Ocean Tsunami Observations on Forecasting the 7 December 2012 Japanese Tsunami

EDDIE BERNARD,[1] YONG WEI,[1,2] LIUJUAN TANG,[1,2] and VASILY TITOV[1]

Abstract—Following the devastating 11 March 2011 tsunami, two deep-ocean assessment and reporting of tsunamis (DART®) (DART® and the DART® logo are registered trademarks of the National Oceanic and Atmospheric Administration, used with permission) stations were deployed in Japanese waters by the Japanese Meteorological Agency. Two weeks after deployment, on 7 December 2012, a M_w 7.3 earthquake off Japan's Pacific coastline generated a tsunami. The tsunami was recorded at the two Japanese DARTs as early as 11 min after the earthquake origin time, which set a record as the fastest tsunami detecting time at a DART station. These data, along with those recorded at other DARTs, were used to derive a tsunami source using the National Oceanic and Atmospheric Administration tsunami forecast system. The results of our analysis show that data provided by the two near-field Japanese DARTs can not only improve the forecast speed but also the forecast accuracy at the Japanese tide gauge stations. This study provides important guidelines for early detection and forecasting of local tsunamis.

1. Introduction

Following the devastating 11 March 2011 Japanese tsunami, two papers by NOAA scientists were published: (1) far-field forecast and impact of the tsunami on the Pacific basin (TANG *et al.* 2012) and (2) the local impact on Japan (WEI *et al.* 2013). Both of these studies used the "method of splitting tsunami" model that has been previously validated and verified (SYNOLAKIS *et al.* 2008). For the far-field study, a methodology was presented for determining the energy of a tsunami using real-time, deep-ocean assessment and reporting of tsunamis (DART®) data within the NOAA tsunami forecast system (TANG *et al.* 2012). Results of this study showed that data from DART stations, near the tsunami-generation region, could help to accurately estimate the energy of a tsunami. For the local study, nested tsunami inundation models were developed that used the source information from TANG *et al.*'s (2012) far-field study as input to simulate the flooding along Japan's coastline. The modeling results for tsunami inundation in the near-field along 600 km of Japan's coastline were compared with observed tsunami time series, surveyed tsunami height and run-up, and the extent of tsunami inundation (WEI *et al.* 2013). This comparison indicated inundation-modeling accuracy was approximately 85.5 % for the affected area between latitudes 36–41°N of Japan's coastline.

Based, in part, on the forecast value of deep-ocean tsunami measurements in tsunami warnings, the Japanese Meteorological Agency (JMA), which operates the Japanese tsunami warning system, decided to install deep-ocean tsunami detection (DART) stations in Japanese waters. Two DART stations[1] were deployed at locations JP1 and JP2 (see Fig. 1), on November 23 and 24, 2012 respectively. Standard performance checks for pressure accuracy and two-way communications were conducted to verify that both stations were performing as designed. The initial performance checks indicated that both Japanese DART stations were working properly in sensing and reporting pressure changes to JMA on 7 December 2012. The 7 December 2012 tsunami was recorded at both DART stations JP2 and JP1 about 11–20 min tsunami travel time from the source.

---

[1] Pacific Marine Environmental Laboratory, NOAA Center for Tsunami Research, National Oceanic and Atmospheric Administration, 7600 Sand Point Way NE, Seattle, WA 98115, USA. E-mail: eddie.bernard@comcast.net
[2] Joint Institute for the Study of the Atmosphere and Ocean, University of Washington, Box 355672, Seattle, WA 98105, USA.

---

[1] Produced by Science Applications International Corporation (SAIC) using DART® Technology.

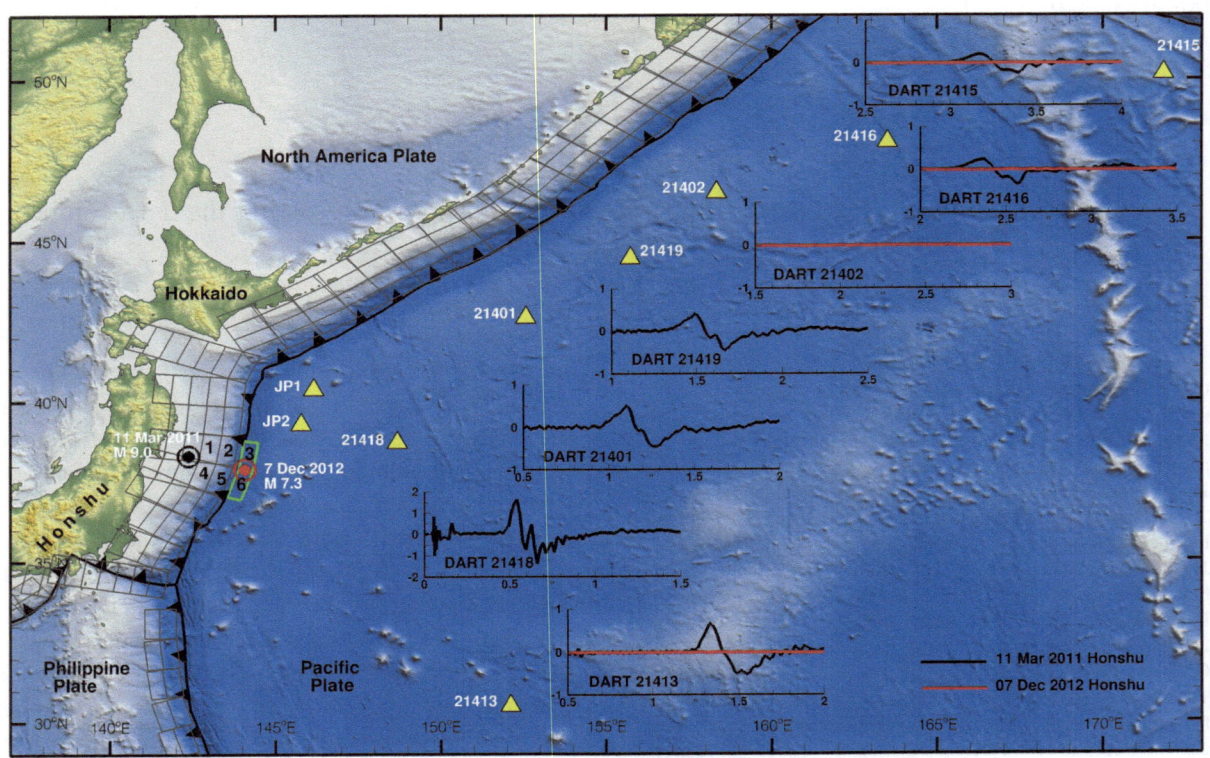

Figure 1

Locations of 6 DART stations (*yellow triangles*) that recorded the tsunami from the 7 December 2012 earthquake (*red circle*). JP1 and JP2 are Japanese-owned DART stations; 21413, 21416, and 21415 are US-owned DART stations; 21402 is a Russian-owned DART station. For comparison with the March 11, 2011 tsunami (earthquake epicenter is *black circle*), the tsunami time series plot adjacent to stations shows the tsunami for 2011 as a *black line* and for the 2012 tsunami as a *red line*. Note that DART station 21402 was deployed after the 2011 tsunami, hence no *black line*. Stations 21419, 21418 and 21401 were not operational on 7 December 2012. The *gray boxes* represent the unit tsunami sources, and the *two green boxes* are normal-faults unit sources that were developed specifically for the 7 December 2012 tsunami

Tsunami data from JP1 and JP2 were reported in real time to JMA. However, because the two Japanese DART stations were being tested for performance acceptance, no data from JP1 or JP2 were available through NOAA in real-time.

The existing DART network near Japan, consisting of DART stations owned by the United States and Russia (see Fig. 1), was not completely operational. The two DART stations closest to the 2011 tsunami source, namely DART stations 21418 (US) and 21401 (Russia), were not operational on 7 December 2012. However, US and Russian DART stations, namely 21413, 21415, 21416, and 21402, located to the southeast and northeast of the earthquake epicenter (see Fig. 1), were operational on 7 December 2102. In summary, six DART stations in the region of Japan, owned and operated by Japan, Russia, and the

US, provided valuable measurements for us to constrain the tsunami source of the 7 December 2012 event.

2. The Tsunami Source of 7 December 2012 Earthquake

According to the USGS, the 7 December 2012 M_w 7.3 earthquake east of Sendai, Japan occurred as a result of reverse faulting within the oceanic lithosphere of the Pacific plate, approximately 20 km east of the plate boundary between the Pacific and North America plates where three Pacific plates subduct beneath Japan (see Fig. 1). At the epicenter of this earthquake, the Pacific plate moves west–northwestward with respect to the North America plate at a

Table 1

Tsunami forecast sources constrained from 4-DARTs and 6-DARTs

Unit sources	Location	Strike	Dip	Rake	Depth (km)	Source 1 4-DART constrained coefficient (m)	Source 2 6-DART constrained coefficient (m)
1	142.7622E, 38.5837N	188	21	90	21.28	0.2	–
2	143.2930E, 38.5254N	188	19	90	5.0	–	0.2
3	144.4149E, 38.2976N	8	40	90	5	–	–
4	142.5320E, 37.7830N	198	21	90	21.28	−0.3	−0.1
5	143.0357E, 37.6534N	198	19	90	5.0	0.2	−0.4
6	144.1376E, 37.3656N	18	40	90	5.0	–	0.1

Each unit source is identified in Fig. 1 by *rectangles* and has a dimension of 100 km in length and 50 km in width. Positive coefficients for each source indicates a thrust fault movement, while negative coefficients indicates a normal fault movement

velocity of approximately 83 mm/year (http://earthquake.usgs.gov/earthquakes/map/). The Harvard Centroid Moment Tensor (CMT) project reported two earthquakes, the first, a thrust fault earthquake of M_w 7.2, followed 12 s later by a normal fault earthquake of equal magnitude approximately at the same location (http://www.globalcmt.org/).

A tsunami source was computed based on the real-time deep-ocean tsunami observations recorded at the four DARTs (source 1 in Table 1). Figure 2a–d (green line) show good model-data comparison at the four stations. However, at the two near-field Japanese DARTs, the modeled wave period is too long (Fig. 2e, f).

By using the Japanese retrospective DART data in addition to the data from the other four stations, a second source was derived (source 2 in Table 1). A good solution was found when the model time series were shifted 2 min behind (red line in Fig. 2). It should be noted each unit source has a spatial resolution of 100 by 50 km. If an earthquake occurs in between of two adjacent unit sources, the model could introduce a travel time error of 2–4 min, depending on the orientation and water depth. Figure 2 shows that the 6-DART source (red line) gives an improved fit to the observations, particularly the period and amplitude of the first wave at the two Japanese DART stations.

Table 1 summarizes the source parameters and coefficients for the 4- and 6-DART inverted sources. A positive source coefficient implies an initial tsunami triggered by a thrust fault rupture, while a negative coefficient indicates the cause of a normal-fault rupture. Clearly, the source coefficients obtained through both inversions indicate the complexity of the 7 December 2012 event that involved both normal- and thrust-fault ruptures, as indicated by the CMT solutions.

Development in processing technology and the availability of robust seismic measurements have identified multiple source mechanisms in recent tsunamis. For example, the 29 September 2009 Samoa tsunami was caused by a M8.1 normal faulting in the outer trench followed by two M7.8 under thrusting sub events (LAY *et al.* 2010), or thrust-fault triggered outer-rise earthquake (BEAVAN *et al.* 2010). The 12 January 2010 Haiti tsunami may have been generated by complex rupture from both strike-slip and thrust faults (HAYES *et al.* 2011; CALAIS *et al.* 2011). In addition, the 11 March 2011 Japan tsunami may have even involved contribution from a seabed failure that was responsible for the high run-up along Sanriku's coasts (GRILLI *et al.* 2012). At present, these earthquake complexities are hard to identify until rigorous, post-event, seismic analysis is performed. However, these complex earthquake processes that produce tsunamis are reflected in the tsunami wave measurements, and can be estimated through inversion of the recorded tsunami time series in real-time (TITOV 2009; WEI *et al.* 2008). The advantage of DART-inversion allows the models to capture the characteristics of the tsunami, including its energy content, in real time necessary for effective warnings (TANG *et al.* 2012).

Figure 3 illustrates the tsunami maximum offshore amplitude for the 6-DART tsunami source that

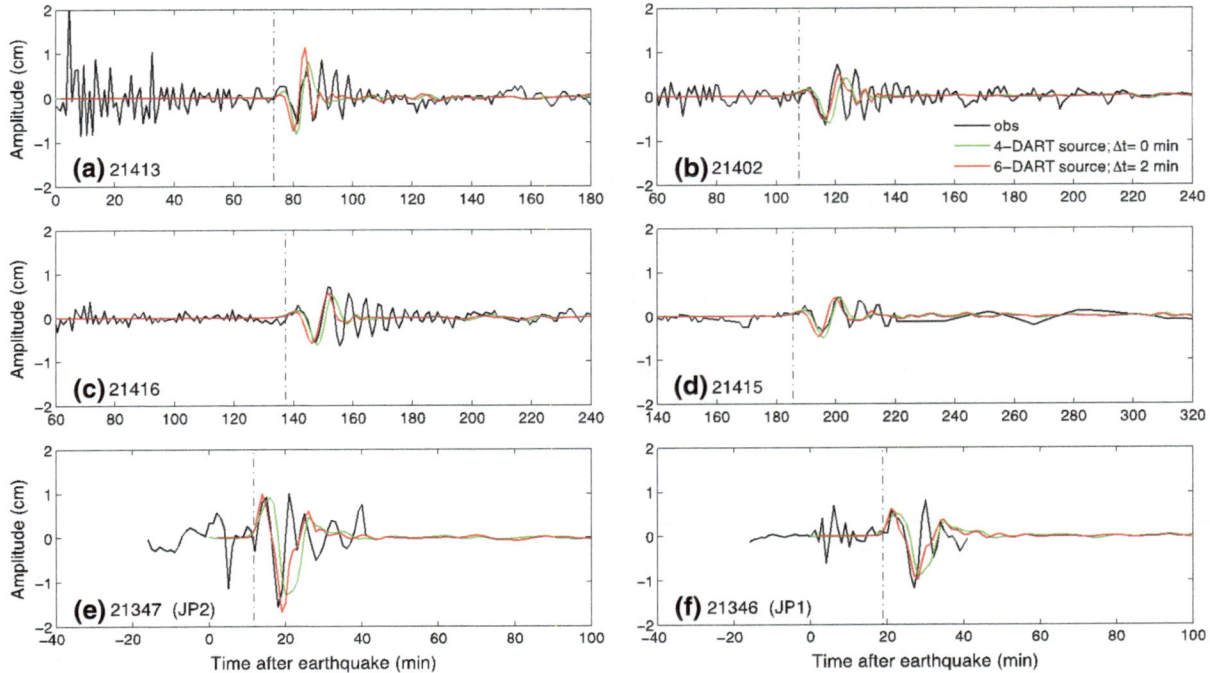

Figure 2
December 7, 2012 tsunami source as derived using two Japanese stations [21347 (JP2) and 21346 (JP1)], three US stations (21413, 21415, 21416) and one Russian station (21402). *Black line* represents observations. The *green line* represents the source using only the US and Russian stations (4-DART solution), while the *red line* represents the source including the additional two Japanese stations (6-DART solution). The *red line* indicates an improved match with observations at all six stations. The model time series are shifted Δt min. *Vertical dashed lines* represent tsunami arrival time at each DART station, which is estimated from the modeled first wave from the 4-DART source, since the signal to noise ratio of the observation is relatively low

produced the tsunami amplitude time series, represented by the red lines in Fig. 2. The maximum amplitude map is a good proxy for tsunami energy (TANG *et al.* 2012). Note in Fig. 3 that the tsunami energy distribution was primarily perpendicular to the trench. To the north of the source, the trench along Japan, the Kuril Islands, and the Alaska archipelago served as a wave guide for the tsunami. To the south of the source, DART station 21413 was in a main lobe of energy providing key data on this tsunami. The energy content from each tsunami source was computed using the methodology described in TANG *et al.* (2012). Using four US and Russian DART data, the energy content was estimated to be 2.86×10^{11} J. Using all six DART stations, the energy content was estimated to be 3.63×10^{11} J or 27 % more energy.

3. Near-Field Inundation Model

Because the tsunami was small (around 1-cm amplitude near the source), the tsunami caused no flooding or damage. The tsunami was measured along Japan's coastline at tide stations providing an opportunity to compare coastal observations with model results. To further evaluate the source of the 7 December 2012 tsunami, both sources (4- and 6-DART inversions) were used as input into tsunami inundation models developed by WEI *et al.* (2013) for the 2011 Japanese tsunami (see Fig. 4 for coastline covered by the model). Using the WEI *et al.* (2013) inundation models, results were compared with the Ofunato, Kushiro, and Hanasaki tide gauge observations (see Fig. 4b). The red lines in Fig. 4b represent the model results from 6-DART inversion that shows

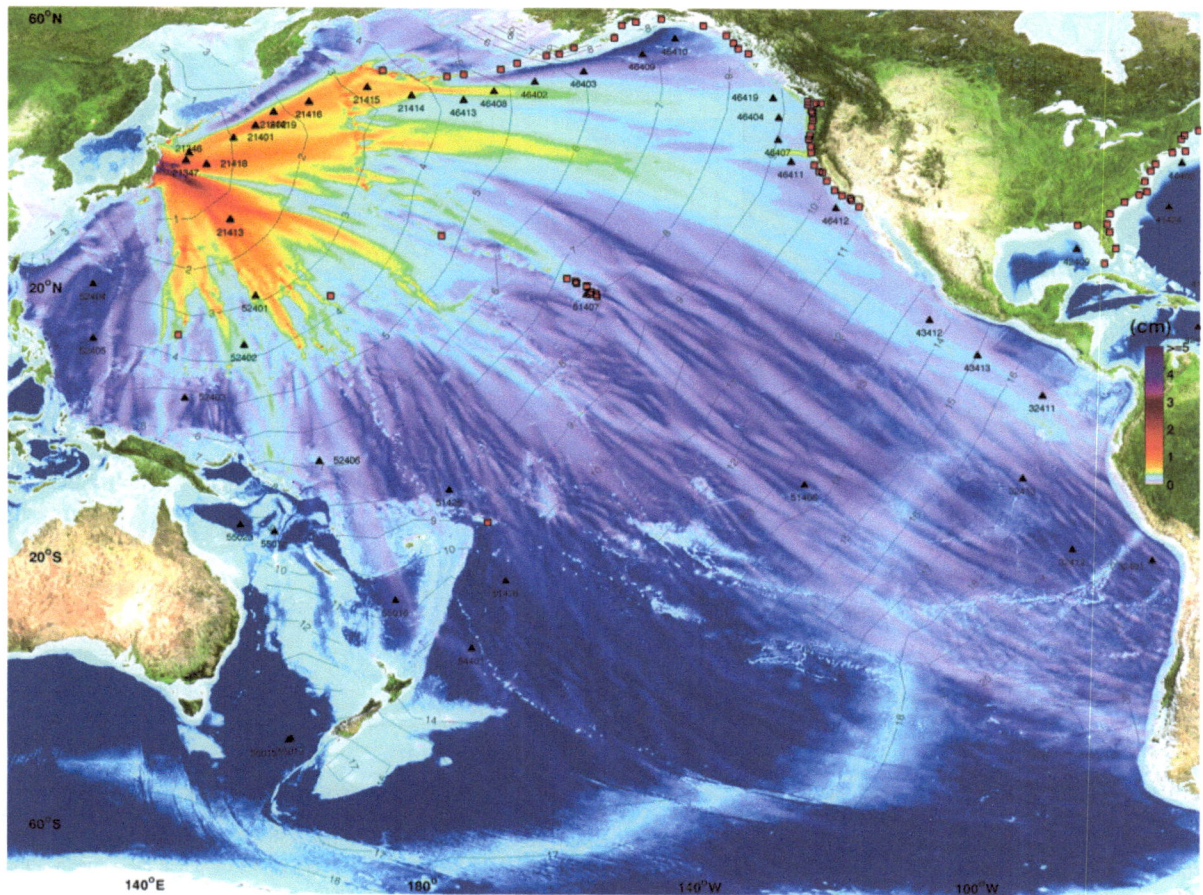

Figure 3
Model maximum offshore amplitude plot from 7 December 2012 Japanese tsunami using six DART stations. *Black triangles* represent locations of DART stations in the global array. *Red boxes* are locations of high-resolution inundation models. *Contours* indicate the travel time in hours

agreement with the tide gauge observations. Comparing to the 6-DART inversion, the 4-DART inversion (green lines in Fig. 4b) gives similar results at the Hanasaki and Kushiro, but differs in arrival time and wave period at the Ofunato. Without introducing an outer-rise unit source, the 4-DART inversion led to an 8-min early arrival time at Ofunato, while the 6-DART inversion provided improved estimate of tsunami arrival time. The model computation compares well with the measurements at Ofunato, which were mainly due to the northern rupture of the earthquake (see Fig. 4a). The spectrum analysis in Fig. 4c indicates that the model reproduced two dominant wave periods (approximately 4

and 10 min) recorded at Ofunato station, with slight overestimation for the 10 min wave period. The 4 and 10-min waves appear within the first 2 h after the earthquake, and are more related to the tsunami source itself. However, the 40- and 50-min wave period may be a result of local wave system near Japan's east coast. The spectral comparison between model and measurements in Fig. 4c (Hanasaki) also shows similar phenomenon. At Hanasaki, the model indicates the dominant wave period induced by the tsunami wave is about 16–20 min, agreeing well with measurements. At Kushiro, the measurements are dominated by waves of 30- to 70-min period, which can be clearly visualized in the time series

313

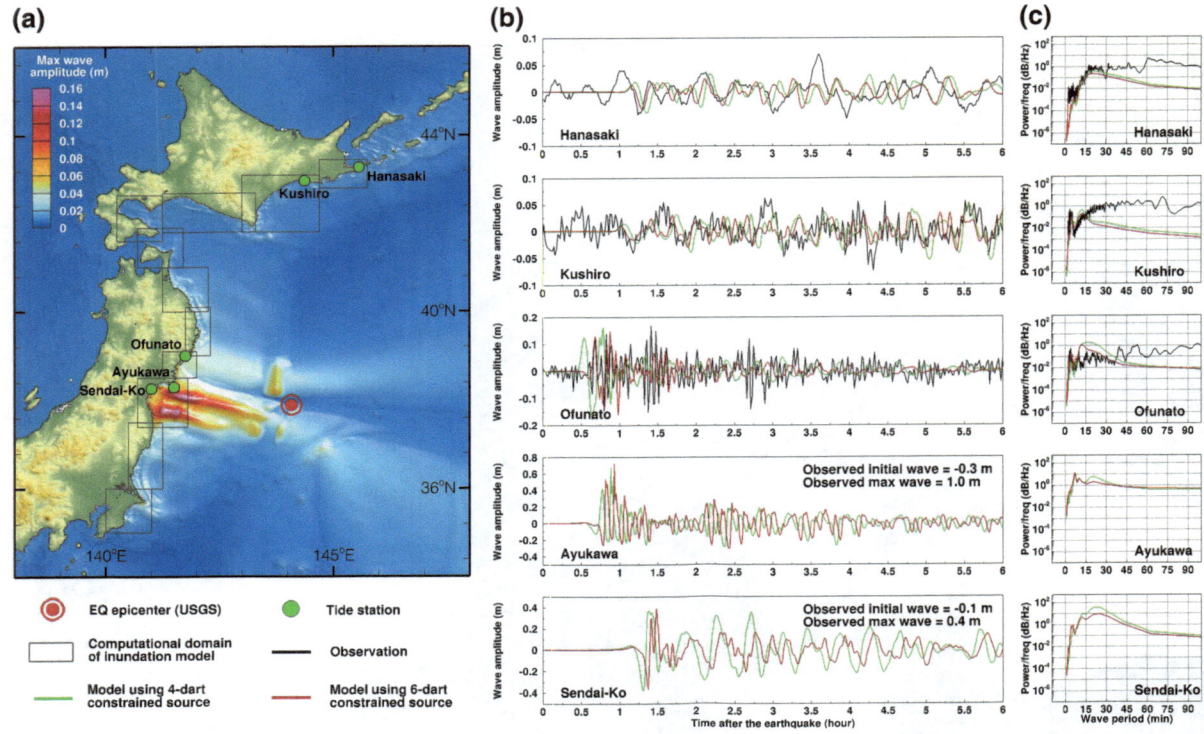

Figure 4

a Model setup. The *grey rectangles* represent the coverage of WEI *et al.* (2013) models; the *green dots* indicate the tide gauge locations; and *dot within a circle* is the USGS epicenter of the December 7, 2013 earthquake. **b** Comparison of results using the 4-dart and 6-DART sources at five tide gauge stations (*green dots* in **a**), and with observation at three tide gauge stations. In the tsunami time series, the *black line* represents observations, while the *red line* (6-DART solution) and *green line* (4-DART solution) represents model output at the same location. **c** Spectrum analysis of observed and computed time series at five tide gauge stations

comparison (Fig. 4b) before tsunami arrived. Computational results at Kushiro show the tsunami has two dominant wave periods, 5 and 13 min.

During the 7 December 2012 event, JMA's tsunami information bulletin No. 8 described the recorded wave amplitudes (initial and maximum) at Ayukawa and Sendai-Ko tide stations (the time series of observation at Ayukawa and Sendai-Ko are not available for us to be used in this study). The bulletin reported the initial waves were both depressions, −0.3 and −0.1 m at Ayukawa and Sendai-Ko, respectively. Our model computation at these two locations (Fig. 4b) fits well with the observations, although the amplitude of the depression is slightly larger than recorded at Sendai-Ko. The model results indicate that the maximum tsunami amplitude near Ayukawa is >0.7 m, and 0.4 m at Sendai-Ko, which both agree with the reported values, 1 m at Ayukawa

and 0.4 m at Sendai-Ko. These results further validate the tsunami energy projection shown in Fig. 4a that the highest tsunami waves are focusing on the coastline of Sendai Bay between Soma and Ayukawa. More interestingly, the focusing and bifurcation of the tsunami energy towards Sendai Bay (Fig. 4a) can be attributed to an interaction of two factors: the tsunami energy focusing from a strip source (KÂNO-ĞLU *et al.* 2013), and the energy focusing and bifurcation caused by bathymetric features over the continental shelf. These results highlight the value of obtaining the correct source estimate from offshore tsunami measurements.

It is worth noting that all of our model results for tide gage comparison were extracted from offshore points close to the tide gage locations. Due to the lack of accurate tide gage coordinates, the model might underestimate the wave amplitudes recorded at the tide gages. The lack of accurate bathymetric and topographic data can also affect the modeling results, since many coastal structures near tide gages, such as breakwaters and seawalls, are not reflected in the 50-m resolution model (WEI *et al.* 2013).

Hence, we are able to conclude that the two Japanese DART stations performed as designed during a tsunami and improved the accuracy of the tsunami source. The more accurate tsunami source led to more accurate model simulations of tsunami dynamics along Japan's coastlines.

4. Discussion

4.1. Our Analysis Leads Us to Two Important Findings

4.1.1 Improvements in Tsunami Forecast Accuracy in the Near and Far-Fields

The two recently deployed Japanese DART stations provided data that improved tsunami forecasting in the near and far-fields. Near-field improvements can be seen in Fig. 4 where the tsunami amplitude time series (represented by red lines) shows a better agreement to the observations at three tide gauges. Specifically at Ofunato, the shorter wave period of the model amplitude time series and the 15 cm maximum amplitude and 15 cm maximum drawdown serve as an accurate forecast of maximum and minimum amplitude range. The model results, however, did not simulate well the second tsunami packet that arrived about 1 h after the first energy packet. Improvements can also be seen in Fig. 2 in model simulations at the six DART stations that recorded the tsunami. For the two Japanese DART stations, one can see a better agreement between model results and observations at DART JP2 and JP1. Specifically, the 6-DART derived source, represented by the amplitude time series (red line), has a a shorter wave period that matches the observations of the first tsunami wave at these two stations. This shorter-period wave also improves the match with observations at the far-field DARTs, namely 21413, 21415, 21416, and 21402. At 21416 and 21402, northwest of the tsunami source location, the amplitude time series

(red line) agrees well with the observed amplitudes of the initial wave. However, at the southernmost DART station, 21413, the amplitude agreement is not as good. Overall, the 6-DART source provides a better simulation of the observed tsunami time series at six different DART stations and at least one coastal tide gauge. The 6-DART source contained 27 % more energy than the 4-DART source, and using this source provided better model agreement with observations at five tide gauges.

4.1.2 Reduced Detection Time Leads to a Faster Assessment of the Tsunami's Destructive Power

The greatest benefit of the recently deployed Japanese DART stations was the close proximity to the tsunami source. The locations of JP2 and JP1 enabled the detection of the 7 December 2012 Japanese tsunami 11–20 min after generation. The fact that the tsunami was only 1.0 cm in amplitude at DART JP2 meant that no destructive tsunami had been generated. In Fig. 1, which illustrates the comparison between the devastating 2011 and the non-destructive 2012 tsunamis, the tsunami amplitude at DART station 21413 in 2011 was measured to be 78 cm, or nearly 80 times the amplitude the 2012 tsunami. Five minutes later, the 0.5 cm amplitude tsunami was detected at DART station JP1, confirming that no destructive tsunami was approaching Japan's coastline.

At 08:51:46 UTC (33 min after the earthquake origin time) JMA had a full tsunami waveform from JP2 as shown in the red line of the lower left panel of Fig. 2. At 08:52:46 UTC (34 min after the earthquake origin time), JMA had a full tsunami waveform from JP1 as shown in the red line of the lower right panel of Fig. 2. Visual analysis of these waveforms would have confirmed that the 7 December 2012 tsunami was not destructive and posed no flooding hazard to Japan's coastline. Since the tsunami arrived at the Ofunato tide gauge 40 min after the earthquake origin time, an accurate forecast could have been provided between 7 and 23 min before the tsunami's arrival at Ofunato. Such fast, accurate assessments of tsunami danger are the foundation blocks in building confidence in tsunami warning accuracy.

The 7 December 2012 tsunami has provided a good test of the tsunami mode performance for the newly deployed Japanese DART stations JP1 and JP2. The bottom pressure sensor worked within stated accuracy while the tsunami reporting mode worked properly in acquiring and sending near real-time data to JMA within the designed time frames. Data from DART stations JP1 and JP2 have been analyzed within the NOAA forecast system and found to be working properly for tsunami forecast and warning applications. The Japanese data from DART stations JP1 and JP2 improved the accuracy of the tsunami source derived from the NOAA forecast system. This more accurate source was used to improve forecast accuracy at coastal tide gauge in Ofunato, Japan. The inclusion of DART stations JP1 and JP2 into the JMA tsunami warning system will improve tsunami forecast speed and accuracy.

4.2. Future Opportunities

The addition of two deep-water tsunami detectors by Japan to the global network of DART stations (see Fig. 3, black triangles) is an excellent example of international cooperation among tsunami threatened nations. By sharing data from these two Japanese DART stations, the Pacific coastal nations benefit from faster detection of tsunamis and more accurate forecasts. Further, the tsunami research community benefits from more data available immediately following tsunami generation. We encourage other nations to follow the leads of Australia, Chile, India, Japan, Russia, Thailand, and the United States in deploying DART stations off their coastlines and sharing their data with all nations. Such international cooperation leads to faster and more accurate tsunami warnings, which, in turn, save lives from the destructive power of tsunamis.

Acknowledgments

We thank Rob Lawson of SAIC for providing 15 s data from DART stations 21436 (JP1) and 21437 (JP2); Michael Spillane and Jean C. Newman for assistance with the propagation database. We are grateful for UNESCO and IOC sea level station monitoring facility (http://www.ioc-sealevelmonit oring.org/) that provide access to the tide gage data along Japan's coastline. We also acknowledge the hard work of JMA, Toho Mercantile, and Oyo Corporation for initiating and completing the purchase of DART stations 21436 and 21437. Special acknowledgement goes to the SAIC deployment team who successfully deployed two DART buoys in the hostile, dangerous, November seas off Japan's coastline. This publication was (partially) funded by the Joint Institute for the Study of the Atmosphere and Ocean (JISAO) under NOAA Cooperative Agreement No. NA10OAR4320148, Contribution No. 2116; PMEL Contribution 4009.

References

BEAVAN, J., X. WANG, C. HOLDEN, K. WILSON, K. POWER, G. PRASETYA, M. BEVIS, and R. KAUTOKE, (2010), *Near-simultaneous great earthquakes at Tongan megathrust and outer rise in September 2009*, Nature, *466*, 959–963.

CALAIS, E., A. FREED, G. MATTIOLI, F. AMELUNG, S. JÓNSSON, P. JASMA, S.-H. HONG, T. DIXON, C. PRÉPETIT, R. MOMPLAISIR (2011), *Transpressional rupture of an unmapped fault during the 2010 Haiti earthquake*, Nature Geoscience, *3*(11), 794–799.

GRILLI, S.T., J. HARRIS, TAPPIN, D.R., MASTERLARK, J.T. KIRBY, F. SHI and G. MA (2012), *Modeling of the Tohoku-oki 2011 tsunami coastal hazard: effects of a mixed co-seismic and seabed failure source*, EOS Trans. AGU, *93*(52), Fall Meet. Suppl., Abstract NH42A-06.

HAYES, G.P., R.W. BRIGGS, A. SLADEN, E.J. FIELDING, C. PRENTICE, K. HUDNUT, P. MANN, F.W. TAYLOR, A.J. CRONE, R. GOLD, T. ITO, and M. SIMMONS (2011), *Complex rupture during the 12 January 2010 Haiti earthquake*, Nature Geoscience, *3*(11):800–805.

KÂNOĞLU U., V.V. TITOV, B. AYDIN, C. MOORE, T.S. STEFANAKIS, H. ZHOU, M. SPILLANE, C.E. SYNOLAKIS (2013), *Focusing of long waves with finite crest over constant depth*, Proc R Soc A 20130015. http://dx.doi.org/10.1098/rspa.2013.0015.

LAY, T., C.J. AMMON, H. KANAMORI, L. RIVERA, K.D. KOPER and A.R. HUTKO (2010), *The 2009 Samoa-Tonga great earthquake triggered doublet*, Nature, *466*, 964–968, doi:10.1038/nature09214.

SYNOLAKIS, C.E., BERNARD, E.N., TITOV, V.V., KANOGLU, U., and GONZALEZ, F. (2008), *Validation and verification of tsunami numerical models*. Pure and Applied Geophysics *165*(11–12), 2197–2228. doi:10.1007/s00024-004-0427-y.

TANG, L., V.V. TITOV, E. BERNARD, Y. WEI, C. CHAMBERLIN, J.C. NEWMAN, H. MOFJELD, D. ARCAS, M. EBLE, C. MOORE, B. USLU, C. PELLS, M.C. SPILLANE, L.M. WRIGHT, and E. GICA (2012), *Direct* energy estimation of the 2011 Japan tsunami using deep-ocean pressure measurements, J. Geophys. Res., *117*, C08008, doi:10.1029/2011JC007635.

TITOV, V.V. (2009), Tsunami forecasting, Chapter 12 in *The Sea, Volume 15: Tsunamis*, Harvard University Press, Cambridge, MA and London, England, 371–400.

WEI, Y., E.N. BERNARD, L. TANG, R. WEISS, V.V. TITOV, C. MOORE, M. SPILLANE, M. HOPKINS and U. KÂNOĞLU (2008), *Real-time experimental forecast of the Peruvian tsunami of August 2007 for U.S. coastlines*, Geophys. Res. Lett. *35*, L04609, doi:10.1029/2007GL032250.

WEI, Y., C. CHAMBERLIN, V.V. TITOV, L. TANG, and E.N. BERNARD (2013), *Modeling of 2011 Japan Tsunami—lessons for near-field forecast*, Pure Appl. Geophys., *170*, 1309–1331, doi:10.1007/s00024-012-0519-z.

(Received April 8, 2013, revised August 3, 2013, accepted September 24, 2013, Published online October 23, 2013)

Pure Appl. Geophys. 171 (2014), 3493–3500
© 2013 Springer Basel (outside the USA)
DOI 10.1007/s00024-013-0674-x

| Pure and Applied Geophysics

Relationship Between Maximum Tsunami Amplitude and Duration of Signal

Yoo Yin Kim[1] and Paul M. Whitmore[1]

Abstract—All available tsunami observations at tide gauges situated along the North American coast were examined to determine if there is any clear relationship between maximum amplitude and signal duration. In total, 89 historical tsunami recordings generated by 13 major earthquakes between 1952 and 2011 were investigated. Tidal variations were filtered out of the signal and the duration between the arrival time and the time at which the signals drops and stays below 0.3 m amplitude was computed. The processed tsunami time series were evaluated and a linear least-squares fit with a 95 % confidence interval was examined to compare tsunami durations with maximum tsunami amplitude in the study region. The confidence interval is roughly 20 h over the range of maximum tsunami amplitudes in which we are interested. This relatively large confidence interval likely results from variations in local resonance effects, late-arriving reflections, and other effects.

Key words: Tsunami observations, maximum amplitude, duration of signal, confidence interval.

1. Introduction

Historical tsunami records show that tsunami time series display amplitude decays that have a variable duration. The duration has been shown to be site and event specific (Rabinovich *et al.* 2011; Shevchenko *et al.* 2011), may continue for several days (Berkman and Symons 1964; Spaeth and Berkman 1967), and often decays exponentially with time (Hayashi *et al.* 2012; Mofjeld *et al.* 2000; Munk 1963; Van Dorn 1984). Uncertainty of duration can cause problems in emergency response during tsunami warnings (Kelly *et al.* 2006; Wilson *et al.* 2012), including local populations returning to the coast before the hazard has ended.

The purpose of this study is to determine the predictability of tsunami duration based on peak amplitudes, which are of primary interest in developing tsunami hazard assessments. That is, can the duration of dangerous tsunami activity at a site be predicted based on the expected peak tsunami amplitude at that site? To address this question, all tsunamis recorded in the West Coast/Alaska Tsunami Warning Center (WCATWC) Pacific Ocean area-of-responsibility (California, Oregon, Washington, British Columbia, and Alaska) are examined. Due to the paucity of observed tsunamis over dangerous levels at individual sites, there are not enough historical recordings to attempt this study on an individual site basis and the observations from multiple sites have been grouped together.

Tsunami Warning Centers (TWCs) have provided tsunami amplitude forecasts for many years (Tatehata 1997; Titov 1997; Wang *et al.* 2012; Whitmore 2003), though they have not included timing of peak amplitudes or durations of dangerous tsunami activity. If a reasonable estimate of the duration of tsunami danger can be provided during the early stages of tsunami warnings, the problem of local populations returning to the coast before the danger is over may be partially alleviated. Even a rough estimate, which is refined later in the event after the wave has been recorded, may provide life-saving information.

2. Data Source and Processing

2.1. Data Source

Tsunami observations used in this study are taken from tide gauge recordings of tsunamis along the California, Oregon, Washington, British Columbia,

---
[1] National Oceanic and Atmospheric Administration (NOAA), National Weather Service (NWS), West Coast/Alaska Tsunami Warning Center, Palmer, Alaska, USA. E-mail: yooyin.kim@noaa.gov

Table 1

Tsunami duration of danger (h)/maximum zero-to-peak amplitude (cm)

Station	2011 Honshu	2010 Chile	2009 Samoa	2006 Kuril	2006 Tonga	1992 N.Cal.	1986 Alaska	1975 Hawaii	1965 Alaska	1964 Alaska	1960 Chile	1957 Alaska	1952 Kamch
Magnitude	9.0	8.8	8.0	8.3	7.9	7.2	8.0	7.1	8.7	9.2	9.5	9.1	9.0
Attu, AK	37.9/117									20.2/38			
Adak, AK		11.3/37					6.1/103		5.1/43			6.7/65	4.6/31
Dutch Harbor, AK	12.2/38									11.0/37			
King Cove, AK	38.7/52	3.5/64											
Kodiak, AK	11.6/34												
Nikolski, AK	49.9/80												
Port Alexander, AK	6.1/37												
Sand Point, AK	44.8/62	9.4/36											
Saint Paul, AK	12.4/62	9.2/34											
Yakutat, AK	18.5/32									17.3/113	10.5/75		
Atka, AK		8.4/43											
Sitka, AK										15.2/138	11.0/39		
Langara Point, BC	8.5/55												
Port Alberni, BC	27.7/59												
Winter Harbor, BC	15.8/90												
Tasu Sound, BC										3.5/67			
Bella Bella, BC										6.0/67			
Ocean Falls, BC										17.5/130			
Alert Bay, BC										2.9/85			
Tofino, BC										14.7/73	9.7/46		
Victoria, BC										5.0/48			
La Push, WA	14.8/69												
Toke Point, WA	2.8/33												
Neah Bay, WA	13.2/42									7.9/64			
Port Angeles, WA	14.2/55												
West Port, WA	9.4/48												
Charleston, OR	16.4/75												
Garibaldi, OR	4.3/31												
Port Orford, OR	52.7/194	21.7/36		6.3/60									
South Beach, OR	10.0/46												
Alameda, CA	3.5/50									5.5/64			
Arena Cove, CA	42.4/153	47.4/63		7.2/60									
Crescent City, CA	83.0/290		3.7/36	8.1/96	2.6/31	5.3/47							
La Jolla, CA	4.7/33	1.6/41								0.2/35			
Los Angeles, CA	16.7/49	11.7/42								3.6/47			
Monterey, CA	47.6/67												7.1/33

Table 1 *continued*

Station	2011 Honshu	2010 Chile	2009 Samoa	2006 Kuril	2006 Tonga	1992 N.Cal.	1986 Alaska	1975 Hawaii	1965 Alaska	1964 Alaska	1960 Chile	1957 Alaska	1952 Kamch
North Spit, CA	37.6/92												
Point Reyes, CA	52.2/131		4.2/33	6.1/37									
Port San Luis, CA	59.5/197		1.1/31	5.3/52				5.1/33					
Richmond, CA	1.5/32												
San Diego, CA	32.6/62	4.5/34								14.5/49			
San Francisco, CA	13.3/61									8.9/85	11.2/40		2.8/49
Santa Babara, CA	53.2/109	20.8/83		6.6/43						19.0/86			
Santa Monica, CA	40.5/84	11.6/59								20.1/84			
Rincon I, CA										5.3/36			
Alamitos, CA													
Port Hueneme, CA													6.1/69

N.Cal. Northern California, *Kamch* Kamchatka

and Alaska coasts with a peak amplitude over 0.3 m (zero-to-peak). This peak amplitude is used as it is the threshold for Tsunami Advisory level alerts used at the WCATWC. The vast majority of gages used in this study are operated by the NOAA National Ocean Service (NOS). Two types of recordings are used: data recorded in digital format from 1999 to the present and data from older events recorded on paper charts which have been digitized at a 2 to 4 min sample rate. The raw data sets are available on the web sites http://oldwcatwc.arh.noaa.gov/about/tsunamimain.php and http://oldwcatwc.arh.noaa.gov/web_tsus/pastaor_tsunamis.htm.

Thirteen different earthquakes which generated tsunamis and their corresponding 89 different tsunami recordings are examined. Table 1 summarizes the observed duration of danger and maximum amplitude (zero-to-peak) at each site. The duration is the time difference between the estimated tsunami arrival time and the time at which the signal drops and stays below 0.3 m amplitude. The ten largest recordings of the 89 tsunamis are plotted as examples in Fig. 1. These plots show that coastal tsunami signal often follows an exponential decay with time, though it also exhibits other variations which may be due to local resonance effects, late-arriving reflections, and other effects.

2.2. Data Processing

The raw data were pre-processed to remove various problems: duplicate data deletion, filling gaps, re-ordering, and spike removal by applying the variance limit of difference between two continuous points. The length of the older, digitized sea level records is in some cases small, only containing a few tide periods (not enough to estimate duration of tsunami danger), and in some cases the quality of the signal is poor. About 30 measurements of short and poor records, which come from the digitized older events, were excluded from this analysis. A fourth-order Butterworth high pass filter was applied to the pre-processed data to remove long period noise such as the tides. Ten examples of the processed results are displayed in Fig. 1. For the examination of duration of tsunami danger, record length is a critical factor because the tsunami amplitude decays quite slowly,

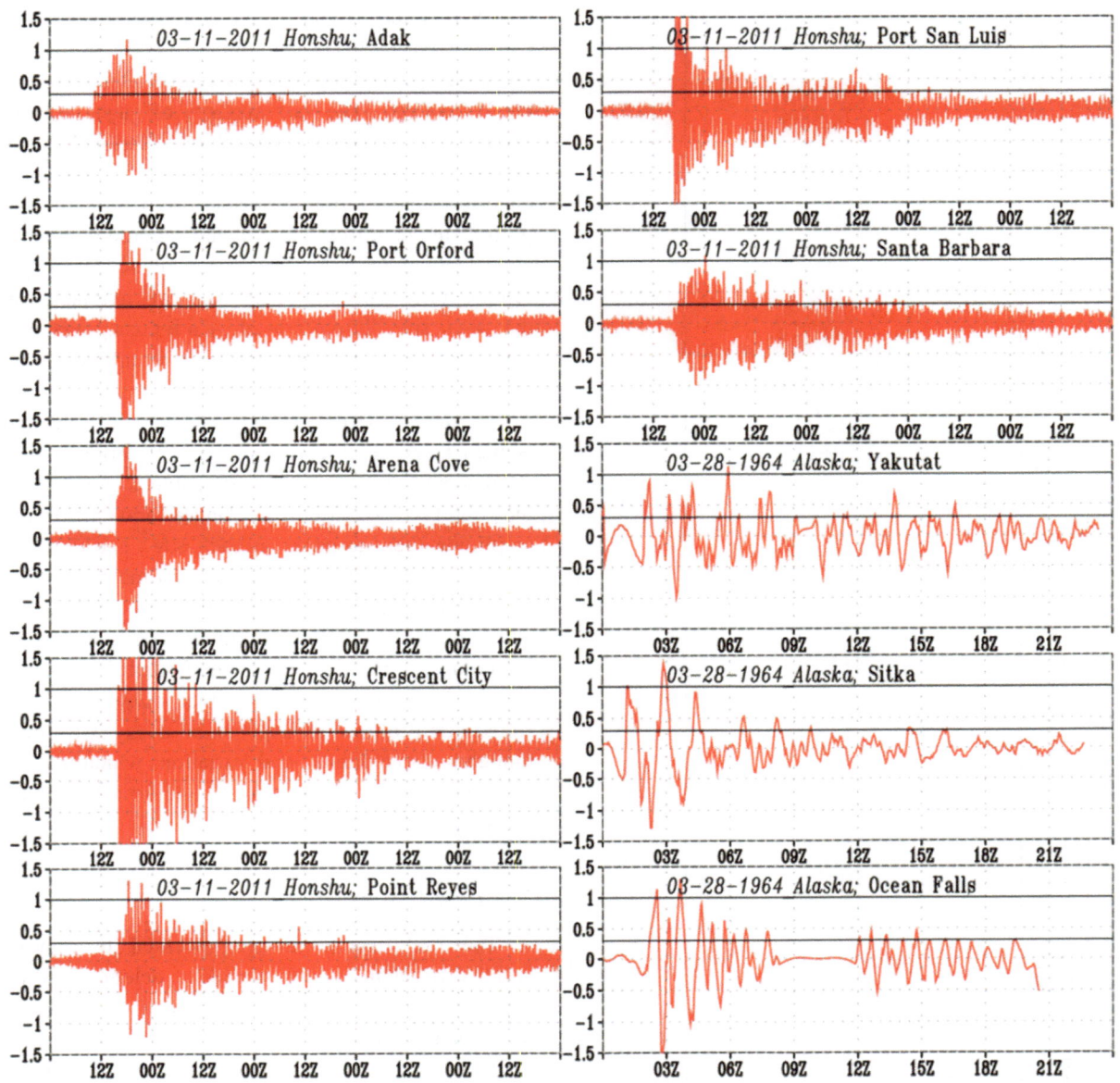

Figure 1
Examples of filtered tide gauge records—the ten largest observations in the data set

suggesting that several days of tsunami records are needed to correctly estimate the duration of tsunami danger (RABINOVICH *et al.* 2011). Eighty-nine tsunami observations were found to be of sufficient quality to use to determine amplitudes and duration; most of these were obtained from the largest Pacific Basin earthquakes such as 2011 Honshu, 2010 Chile, and 1964 Alaska (Table 1).

As an example, Fig. 2 shows how duration of danger was computed from the filtered sea level data. The duration was calculated by subtracting the estimated time of arrival (ETA) from the time of the last zero-to-peak amplitude greater than 0.3 m. The ETA is computed by the WCATWC operational software using an estimate based on Huygen's principal and is used instead of the observed time

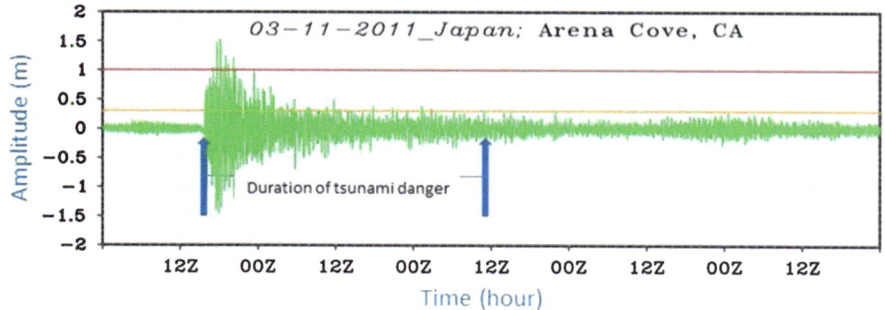

Figure 2

Filtered tsunami record from Arena Cove, California (as an example) which was generated by the March 11, 2011 Japan earthquake. The *left* and *right arrows* indicate the estimated time of arrival (ETA) and the last time of amplitude greater than tsunami advisory level (0.3 m), respectively. The time between the two *arrows* represents the duration of tsunami danger

since in many cases the initial arrival is difficult to determine from the records.

3. Methods and Results

3.1. Confidence Interval

A confidence interval represents a closed interval where a certain percentage of the population is likely to lie. A $100(1-\alpha)$ confidence interval (DRAPER and SMITH 1998) was estimated for an individual value of the observed maximum amplitude, x_p, from

$$\hat{y} \pm t_{\alpha/2,n-2} S_\varepsilon \sqrt{1 + \frac{1}{n} + \frac{(x_p - \bar{x})^2}{\sum_{i=1}^{n}(x_i - \bar{x})^2}}, \quad (1)$$

where $\bar{x}$ is the mean of all the observed values of x_p, x_i is the new observation independent of observations used to obtain the linear fit, $\hat{y}$ is the linear fitting line obtained by least squared method, and $t_{\alpha/2,n-2}$ is the t-distribution critical value with the significance level, α, degree of freedom, $n-2$, and n is the total number of the observations.

The standard deviation (S_ε) of variation of observations around the linear fit line is estimated by the sample variance, which is calculated using the following relationship:

$$S_\varepsilon = \sqrt{\frac{\frac{1}{n}\sum_{i=1}^{n}(\hat{y} - y_i)^2}{n - k - 1}}, \quad (2)$$

where y_i represents all observations, n is number of observations, and k is number of independent

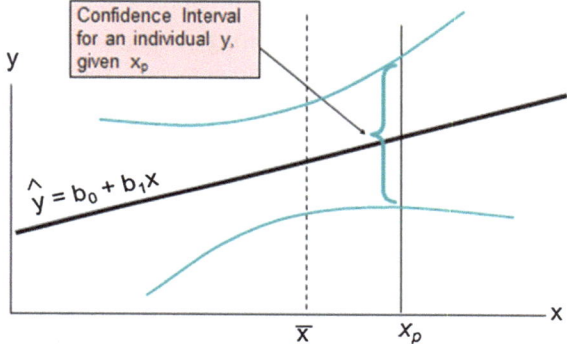

Figure 3

Confidence interval determination for different values of x. b0 and b1 represent y-intercept and slope, respectively

variables in the model ($k = 1$). The quantity in the numerator in (2) is called the mean square error. A schematic diagram of the confidence interval is shown in Fig. 3. The width of the confidence interval depends on the point value of x_i, which will be a minimum at $x_i = \bar{x}$, and will widen as $|x_i - \bar{x}|$ increases.

3.2. Relationship Between Maximum Amplitude and Duration of Danger

To show the relationship between maximum zero-to-peak tsunami amplitude and duration above the Tsunami Advisory level, a scatter diagram and corresponding linear fit by least-squares method are obtained from 89 measurements. Since tsunami heights are expected to decay exponentially with time after the maximum value (HAYASHI *et al.* 2012;

MOFJELD *et al.* 2000; MUNK 1963; VAN DORN 1984), the logarithm of the danger duration should be related to the maximum amplitude. The linear correlation coefficient between the two is 0.56 (Fig. 4) and a significance test at the 5 % level indicates that maximum tsunami amplitude is reasonably related to duration of danger. The confidence interval is also computed to describe the amount of uncertainty associated with the linear relationship. The 95 % confidence interval ($\alpha = 0.05$), along with the processed results, is also shown in Fig. 4. This figure shows that the confidence interval is roughly 20 h over the range of peak tsunami amplitudes in which we are interested. This relatively large confidence interval may be the result of a complex interaction between waves developed during propagation, oblique propagation across the continental shelf, late arriving reflected waves, and the excitation of edge

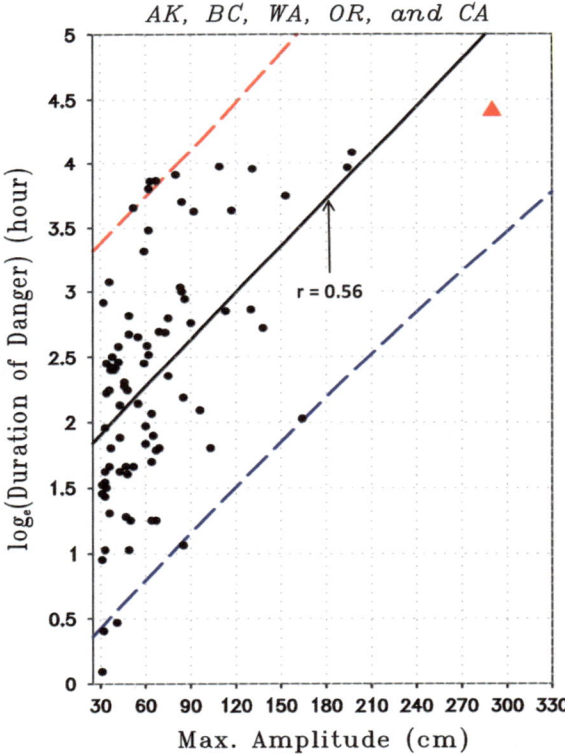

Figure 4

Linear best fit computed from the data in Table 1 using the logarithm of the duration. The 95 % confidence interval is also shown with *dashed line*; the correlation coefficient (*r*) is indicated. The *closed triangle* represents the 2011 Honshu signal at Crescent City, CA

waves caused by coastal irregularities and the subsequent scattering and resonance (GEIST 2009). The tsunami is also very persistent in time and the maximum amplitude can occur much later than the first arrival. Complex interactions are often observed at tide gauge records in Crescent City which provides more data points than any other site to this analysis. To see if the observations at Crescent City bias the results, the linear correlation between amplitude and duration with and without Crescent City included in the data set is tested. The resulting correlation coefficient is approximately equal (0.56 with Crescent City included and 0.53 with Crescent City excluded). The 2011 Honshu signal at Crescent City shows the largest maximum amplitude along with the longest duration of danger, and is located well within the confidence interval (Fig. 4).

To check if the correspondence between amplitude and duration of danger improves when analyzing just one event, 35 tsunami recordings generated by the 2011 Honshu earthquake were investigated. This comparison provides a higher linear correlation coefficient (0.67—not shown). This could imply that late arriving reflections for tsunamis generated far from the North American coast decrease the correlation. The correlation coefficient for tsunamis generated in the southern hemisphere (2010 Chile, 2009 Samoa, 2006 Tonga, and 1960 Chile earthquakes) is found to be 0.45.

4. Discussion

Eighty-nine tide gauge records of tsunamis generated by 13 earthquakes were analyzed to determine if there is a relationship between maximum tsunami amplitude and the duration of danger. The relationship reveals that there is a correspondence between the two.

Duration is also compared to other potential controlling variables such as distance from the source. The relationship between the tide station's distance to the source and signal duration is shown in Fig. 5. This relationship displays effectively no correlation between the two (correlation coefficient < 0.09). Compared to the distance from the source, the amplitude is more noticeably related to the duration.

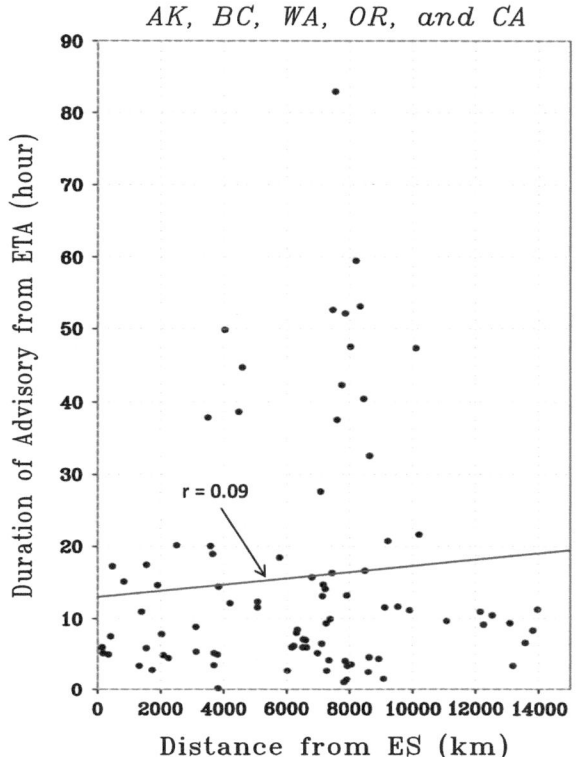

AK, BC, WA, OR, and CA

r = 0.09

Figure 5
Scatter plot illustrating the correlation (*r*) between the distance [between earthquake source (ES) and tide station] and the duration of danger

The relatively large confidence interval determined in this study likely results from several factors, such as variable local resonance effects, reflected signal from other coasts, and other effects at certain stations (GEIST 2009; ROYER and REID 1971). Tsunamis are highly sensitive to bathymetry, and the wave energy often is reflected or refracted by nearshore features and can be trapped as edge waves that travel up and down the coast parallel to the beach (KELLY *et al.* 2006). While the confidence interval determined from the data is relatively large, the maximum tsunami amplitude may still be a valuable predictor as a first order estimate of tsunami duration. That is, when amplitudes are forecasted for a site, a rough estimate of the danger time can be provided. The duration can then be updated throughout the event based on more advanced techniques which examine the recorded tsunami at specific sites (e.g., MOFJELD *et al.* 2000; NYLAND and HUANG 2013). As tsunami modeling becomes more advanced and the bathymetry grids more precise, more accurate duration estimates may be possible using model estimates.

Acknowledgments

The authors wish to thank Isaac Fine, David Salzberg, Paul Huang, David Nyland, Bohyun Bahng, and two anonymous reviewers for helpful comments on this study, and to all WCATWC staff who have helped collect and store the digital marigrams used in this study.

REFERENCES

BERKMAN, S.C., SYMONS, J.M. (1964), *The tsunami of May 22, 1960, as recorded at tide stations, Coast and Geodetic Survey*, U.S. Dept. of Commerce, Washington D.C., 79 pp.

DRAPER, N., SMITH, H. (1998), *Applied Regression Analysis*, John Wiley & Sons, Inc., New York.

GEIST, E.L. (2009), *Phenomenology of tsunamis: Statistical properties from generation to runup*, Advances in Geophysics, *51*, 107–169, doi:10.1016/S0065-2687(09)05108-5.

HAYASHI, Y., KOSHIMURA, S., IMAMURA, F. (2012), *Comparison of decay features of the 2006 and 2007 Kuril Island earthquake tsunamis*, Geophys. J. Int., *190*, 347–357.

KELLY, A., DENGLER, L.A., USLU, B., BARBEROPOULOU, A., YIM, S.C., BERGEN, K.J. (2006), *Recent tsunami highlights need for awareness of tsunami duration*, Eos Trans. AGU, *87*(50), 566–567. doi:10.1029/2006EO500004.

MOFJELD, H.O., GONZÁLEZ, F.I., BERNARD, E.N., NEWMAN, J.C. (2000), *Forecasting the Heights of Later Waves in Pacific-Wide Tsunami*, Natural Hazards, *22*(1), 71–89.

MUNK, W.H. (1963), *Some comments regarding diffusion and absorption of tsunamis*, In: D.C. COX (ed.). Proc. tsunami meetings associated 10th Pac. Sci. Cong., Honolulu, Hawaii Union Geod. Geophys. Monogr. 24, pp. 53–72.

NYLAND, D. and HUANG, P. (2013), *Forecasting wave amplitudes after the arrival a tsunami*, Pure and Applied Geophysics (submitted).

RABINOVICH, A.B., CANDELLA, R.N., THOMSON, R.E. (2011), *Energy decay of the 2004 Sumatra tsunami in the world ocean*, Pure and Appl. Geophys., *168*, 1919–1950.

ROYER, T.C., REID, R.O. (1971), *The detection of secondary tsunamis*, Tellus, *23*(2), 136–141.

SHEVCHENKO, G., SHISHKIN, A., BOGDANOV, G., LOSKUTOV, A. (2011), *Tsunami measurements in bays of Shikotan Island*, Pure and Appl. Geophys., *168*, 2011–2021.

SPAETH, M.G., BERKMAN, S.C. (1967), *The tsunami of March 28, 1964, as recorded at tide stations, Coast and Geodetic Survey*, U.S. Dept. of Commerce, Rockville, Md, 86 pp.

TATEHATA, H. (1997), *The new tsunami warning system of the Japan Meteorological Agency, in Perspectives on Tsunami Hazard Reduction*, In: G. Hebenstreit, (ed.) Kluwer, Dordrecht, The Netherlands, 175–188.

TITOV, V.V., GONZÁLEZ F.I. (1997), *Implementation and testing of the Method of Splitting Tsunami (MOST) model*, NOAA Technical Memorandum ERL PMEL-112, 11 pp.

VAN DORN, W.G. (1984), *Some tsunami characteristics deducible from tide records*, J. Phys. Oceanogr., *17*, 1507–1516.

WANG, D., BECKER, N.C., WALSH, D., FRYER, G.J., WEINSTEIN, S.A., MCCREERY, C.S., SARDINA, V., HSU, V., HIRSHORN, B.F., HAYES, G.P., DUPUTEL, Z., RIVERA, L., KANAMORI, H., KOYANAGI, K., SHIRO, B. (2012), *Real-time forecasting of the April 11, 2012 Sumatra tsunami*, Geophys. Res. Letters, *39*, doi:10.1029/2012GL053081.

WHITMORE, P.M. (2003), *Tsunami amplitude prediction during events: a test based on previous tsunamis*, Science of Tsunami Hazards, *21*, 135–143.

WILSON, R.I., ADMIRE, A.R., BORRERO, J.C., DENGLER, L.A., LEGG, M.R., LYNETT, P., MCCRINK, T.P., MILLER, K.M., RITCHIE, A., WHITMORE, P.M. (2012), *Observations and impacts from the 2010 Chilean and 2011 Japanese tsunamis in California (USA)*, Pure and Appl. Geophys., http://dx.doi.org/10.1007/s00024-012-0527-z.

(Received March 1, 2013, accepted April 15, 2013, Published online May 10, 2013)

Pure Appl. Geophys. 171 (2014), 3501–3513
© 2013 Springer Basel (outside the USA)
DOI 10.1007/s00024-013-0703-9

Pure and Applied Geophysics

Forecasting Wave Amplitudes after the Arrival of a Tsunami

DAVID NYLAND[1] and PAUL HUANG[1]

Abstract—The destructive Pacific Ocean tsunami generated off the east coast of Honshu, Japan, on 11 March 2011 prompted the West Coast and Alaska Tsunami Warning Center (WCATWC) to issue a tsunami warning and advisory for the coastal regions of Alaska, British Columbia, Washington, Oregon, and California. Estimating the length of time the warning or advisory would remain in effect proved difficult. To address this problem, the WCATWC developed a technique to estimate the amplitude decay of a tsunami recorded at tide stations within the Warning Center's Area of Responsibly (AOR). At many sites along the West Coast of North America, the tsunami wave amplitudes will decay exponentially following the arrival of the maximum wave (MOFJELD *et al.*, Nat Hazards 22:71–89, 2000). To estimate the time it will take before wave amplitudes drop to safe levels, the real-time tide gauge data are filtered to remove the effects of tidal variations. The analytic envelope is computed and a 2 h sequence of amplitude values following the tsunami peak is used to obtain a least squares fit to an exponential function. This yields a decay curve which is then combined with an average West Coast decay function to provide an initial tsunami amplitude-duration forecast. This information may then be provided to emergency managers to assist with response planning.

Key words: Tsunami coda, tsunami forecast, tsunami warning, tsunami advisory, tsunami decay.

1. Introduction

As tsunami waves travel away from their source, they are scattered and reflected off continents, islands, and bathymetric features such as seamounts (HORILLO *et al.*, 2008). When the waves approach the shore, energy is reflected back to the open ocean and diffracted by abrupt variations in the coastline. Tsunami waves may create resonance effects in bays and harbors and interactions with coastal features can generate a series of guided waves that are trapped along coastlines. These waves propagate parallel to the coastline and they may also be scattered by shoreline irregularities (GEIST, 2010). Combined, these accumulated secondary oscillations form a slowly decaying wave train, referred to as the tsunami coda, which follows the maximum peak arrival of a tsunami (HAYASHI *et al.*, 2009).

Several investigations have shown tsunamis form wave trains that may continue for several days and decay exponentially with time. Utilizing spectra from the 22 May 1960 Chilean tsunami, Munk (as cited in VAN DORN, 1984) noticed that tsunami energy in the ocean appears to decay much like sound intensity does in a closed room. He developed an acoustic equivalent decay model for the ocean and examined several decay mechanisms. VAN DORN (1984) applied a similar decay model to several tsunamis propagating in the northeast quadrant of the Pacific Ocean and his results were consistent with Munk's acoustic decay hypothesis. Using data gathered from five different tsunamis (1946, 1952, 1957, 1960, and 1964), VAN DORN came to the conclusion that after an initial diffusion period of 40 h, all major tsunamis generated in the Pacific Ocean obey a common decay law of the form:

$$E(t) = E_0\, e^{-t/\tau} \quad t > 40 \text{ h}$$

where E_0 is regarded as a tsunami energy index and τ is the e-folding decay time equal to 22 ± 0.07 h. He noted that the North Pacific is remarkably isotropic after 40 h and suggested that it would be desirable to examine energy decay for the same five tsunamis in Japan, Chile or New Zealand. MOFJELD *et al.* (2001) has pointed out that the Northeast Pacific Ocean is a region in which open ocean scattering is negligible. The lack of strong bathymetric scattering features may contribute to the uniformity of the proposed decay law. VAN DORN (1987) extended his

[1] NOAA/NWS West Coast and Alaska Tsunami Warning Center, 910 Felton St, Palmer, AK 99645, USA. E-mail: paul.huang@noaa.gov

previous work to include the tsunami response in other areas of the Pacific, Atlantic, and Indian Oceans. Plotting the decay of 5-h consecutive variances for stations in Antarctica, Japan, and Australia, he found that his measurements were consistent with his previous result that the entire Pacific becomes isotropic after 40 h and decays synchronously everywhere thereafter. He also found that, subsequent to diffusion, the isotropic energy decay was also uniformly exponential in the Atlantic and Indian Oceans except that the decay and diffusion times were different in each ocean. The computed e-folding decay time for the Indian Ocean is 14.6 h and for the Atlantic Ocean is 13.3 h.

In a study of the 2004 Sumatra tsunami, RABINO-VICH et al. (2011) examined the energy decay of tsunami waves in the Indian, Pacific, and Atlantic oceans. Their findings indicate that the decay times in a particular oceanic basin are not uniform, as previously reported, but depend on the travel time and the absorption characteristics of the continental shelf near the recording site. The decay at island tide gauge sites without shelves was found to be faster than mainland sites with shelves. Periods of increased wave energy have been found superimposed on the basic decay structure that may be due to the reflection of tsunami waves at ridges or continental boundaries. In their study, they did not calculate the diffusion times because they are difficult to properly define. MERRIFIELD et al. (2005) noted that the amplitude and frequency from the 2004 Sumatra tsunami were very different at stations located along the South African Coast and presumed that the local shelf interactions were responsible for the differences.

HAYASHI et al. (2009) studied the characteristics of the 2006 and 2007 Kuril Island tsunamis by measuring the moving-average root mean square (MRMS) time function at several tidal stations located in Japan. The MRMS values were then corrected for the background RMS noise. In their study, they calculated the amplitude decay and not the energy decay as was measured in previous studies. The decay calculations were limited to 48 h following the maximum value of the recorded tsunami, and therefore began prior to the beginning of the isotropic wave distribution ($t > 40$ h) as defined by VAN DORN (1984, 1987). Nevertheless, they concluded that the

decay of the MRMS wave amplitude is approximately exponential. To obtain an estimate of the energy–density decay, RABINOVICH et al. (2011) and VAN DORN (1984, 1987) calculated the variance in a sliding window. Hayashi's MRMS technique is equivalent to the variance method if the MRMS values are squared. Of course, this assumes that the mean value of the signal is zero and the gate lengths are the same.

MOFJELD et al. (2000) developed a method to forecast the wave heights of later waves in Pacific-wide tsunamis for locations near tide gauges that are monitored in real time. The goal is to provide reasonable duration forecasts as soon as possible following the arrival of a tsunami. They pointed out that emergency managers need wave height duration forecasts to help guide rescue and recovery operations as well as for helping to decide when to issue an all clear. Their forecast method is based on a study (MOFJELD et al., 1997) which showed that at tide gauges located in the eastern North Pacific Ocean, the envelope function can be determined from observations of earlier waves at the same location. The forecasting scheme was extended to the remainder of the Pacific by assuming that amplitudes decay exponentially in time. This form was based directly on previous studies of tsunami decay (MILLER et al., 1962; VAN DORN, 1984, 1987). Since the time of tsunami energy decay was close to 1 day, MOFJELD chose to use an amplitude decay that was fixed at 48 h. In general, the decay times will vary between individual coastal locations due to resonance effects in the bays that receive energy input from the main tsunami (MURTY, 1977). In a study of the 2004 Sumatra tsunami, CANDELLA et al. (2008) noted that there were extreme differences in tsunami wave heights from closely located tide gauges along the coast of Argentina which he attributed to the importance of local resonance effects. VAN DORN (1987) concluded that the shoreline tide gauge response consists of resonant amplification of normal modes of shelf oscillation in which 95 % of the total energy is contained in the fundamental and first harmonic. RABINOVICH et al. (2011) noted that parameters of continental shelves play a major role in the rate of the tsunami energy decay. The shelf effectively absorbs energy from a tsunami where wider and shallower continental shelves absorb a greater portion of the wave energy. He also mentioned that there was a noticeable lack of uniformity in the energy–density decay at different locations that recorded the 2004 Sumatra tsunami.

In this paper, we propose a method to calculate amplitude decay functions for tsunamis arriving at specific sites within the WCATWC Pacific AOR (California, Oregon, Washington, British Columbia, and Alaska). Amplitude durations may vary significantly between different locations and different tsunamis; nevertheless, it is possible to obtain an estimate of the local amplitude decay rate of the tsunami by calculating the analytic envelope of the first 2 h subsequent to the maximum amplitude envelope time. The calculated e-folding decay time is used to forecast amplitudes for the first 6 h following the envelope peak. After that time a pre-calculated average amplitude decay function with an e-folding time of 31.8 h is appended to the original decay function to provide a local wave-height forecast for emergency managers. In the following sections, West Coast and Alaska tide gauge data from five different tsunamis are used to demonstrate the generation of a wave height forecast (Fig. 1).

2. Data Acquisition and Processing

All sea level data were taken from stations along the Alaska, British Columbia, Washington, Oregon, and California coasts. The majority of gauges located within the United States are operated by the National Ocean Service (NOS). Tide stations located along the coast of British Columbia are maintained by the Canadian Hydrographic Service. With the exception of the gauge located at Shemya, Alaska, all tide data were recorded with a 1-min sampling interval. The raw data from Shemya were sampled every 15 s. These data were subsequently converted to a 1-min sampling rate by first applying an eighth-order low-pass Chebyshev Type I filter followed by a 4 to 1 decimation. To remove all phase distortion, the filter is applied in both the forward and reverse directions.

Data spikes were removed with the Tukey 53H algorithm (GORING and NIKORA, 2002). Using the principle that the median is a robust estimator of the

mean, a smooth time sequence is generated that can be subtracted from the original signal. Data containing small gaps of a few sample points were interpolated using a third-order spline fit. To remove the tidal contribution at each station, a fourth-order high-pass Butterworth filter with a cut-off period of 4 h was applied to the data. Again, the phase distortion was removed by applying the filter in both directions. The original data sets used in this study are available on the West Coast and Alaska Tsunami Warning Center's web site at: http://oldwcatwc.arh. noaa.gov/about/tsunamimain.php (Table 1).

3. Tsunami Warnings and Advisories

Initial tsunami warnings, watches, and advisories are based solely on seismic data. A tsunami warning is issued if the earthquake's magnitude exceeds a specific threshold level and the epicenter is located within a region that could potentially generate a tsunami. Once a large magnitude earthquake has been detected near the ocean, the Tsunami Warning Center (TWC) scientists monitor coastal tide gauges and deep ocean pressure sensors (DART buoys). Pressure data from DART sensors are used in conjunction with seismic source information in an inversion process that produces an inundation forecast for various coastal locations. The inversion algorithm utilizes a pre-computed forecast database to select the appropriate linear combination of unit sources that most closely matches the observational data. If the inundation forecasts indicate that wave amplitudes will exceed 1.0 m in a specific region, a Tsunami Warning is issued for that area. If wave amplitudes are forecasted to be between 0.3 and 1.0 m, a Tsunami Advisory is issued. These maximum wave amplitude forecasts and the predicted coastal arrival times are passed on to the appropriate government agencies. Throughout a major event, the WCATWC will issue Warning or Advisory messages at 0.5–1.0 h intervals. In addition, conference calls are held with government emergency managers every 1.0–2.0 h.

During the course of the 2011 Japan tsunami warning, the WCATWC staff issued 36 messages and held 18 conference calls over a period of 32 h. Soon after the tsunami arrived along the West Coast, emergency managers requested an estimate of the time until cancellation of the warning. The WCATWC had no means of forecasting the duration of this tsunamigenic event. Unfortunately, the numerical algorithms used to provide an initial maximum wave height forecast are not accurate enough (after the first few cycles) to provide a long term forecast. Other methods must be employed to provide a longer term amplitude forecast. To address this issue, the WCATWC developed a technique to help the scientists estimate the wave height decay at various coastal locations. The method is based on the assumption that the analytic envelope of the tsunami coda decays exponentially.

4. Amplitude Decay Estimation

The tide gauge response to a tsunami arriving on the West Coast can show significant differences between recording sites. The spectrograms shown in Fig. 2 are from Point Reyes, California, and Port Orford, Oregon. The difference in arrival time between these stations was only 36 min, so the variations in response are mainly due to local differences in topography and bathymetry. The apparent decay at Point Reyes is much longer than that at Port Orford due to the presence of multiple reflected waves. In previous studies, the energy-density decays were calculated for different seas and oceans by measurements taken over the course of several days (CANDELLA et al., 2008; VAN DORN, 1984, 1987; RABINOVICH et al., 2011). In contrast, the staff at a Tsunami Warning Center must obtain an estimate of the apparent amplitude decay at each station within a few hours following the arrival of the first wave in order to provide an amplitude duration forecast to emergency managers. From the analysis of five recent tsunamis (Table 2) impacting the WCATWC's AOR, it is evident that, in many cases, it is possible to obtain a good least squares approximation of the apparent amplitude decay envelope of the tsunami coda.

Most sites along the West Coast show multiple decay phases during the 24-hour period following the peak of the tsunami arrival. These decay phases occur during the diffusion period as defined by VAN DORN (1984). Two distinct decay rates from the 15

November 2006 tsunami arriving at Arena Cove, California, are illustrated in Fig. 3. After the tsunami wave amplitudes reach a peak, they begin to decay with an e-folding time of 6.0 h and continue at that rate for approximately 5 h. From 6 to 24 h, the amplitude envelope decays with an e-folding time of 18.5 h. The initial and secondary decay rates may vary between sites and between tsunamis at the same site as illustrated in Fig. 3. The different rates of decay measured at the same location may be due to the dependence of decay time on wave period. In a recent paper, RABINOVICH et al. (2013a) showed that there is a strong relation between the frequency content of a wave and the rate of decay. They proposed that the variation in dissipation rates is primarily due to differences in the coastal geometry and the irregular bottom topography of different source regions.

Except for Arena Cove, California, the secondary decay rates vary within a small range with an average amplitude e-folding time of 31.8 h (Fig. 4). This amplitude decay rate implies that the energy density would be decaying with an e-folding time of 15.9 h. One approach to providing an amplitude forecast is to quickly obtain an estimate of the initial amplitude decay at a particular location and combine this information with the average secondary decay rate.

Trans-Pacific tsunamis arriving at the West Coast tend to slowly build in amplitude to a peak value where they begin to rapidly decay. After several hours, the waves commence to decay at a much slower rate (Fig. 3). At some locations this secondary decay rate is difficult to measure due to the subsequent arrival of numerous reflected waves. Because there are not enough examples to calculate a unique decay time for each site, data from several locations and tsunamis were used to calculate an average secondary e-folding amplitude decay time of 31.8 h beginning approximately 5.4 h after the amplitude envelope peak (Fig. 4).

To estimate the primary decay rate, the tide data containing the arrival of the tsunami is initially filtered with a fourth-order zero phase high-pass Butterworth filter with a cutoff period of 4 h to remove the low-frequency contributions of the local tides. The Hilbert transform $f_H(t)$, known as the

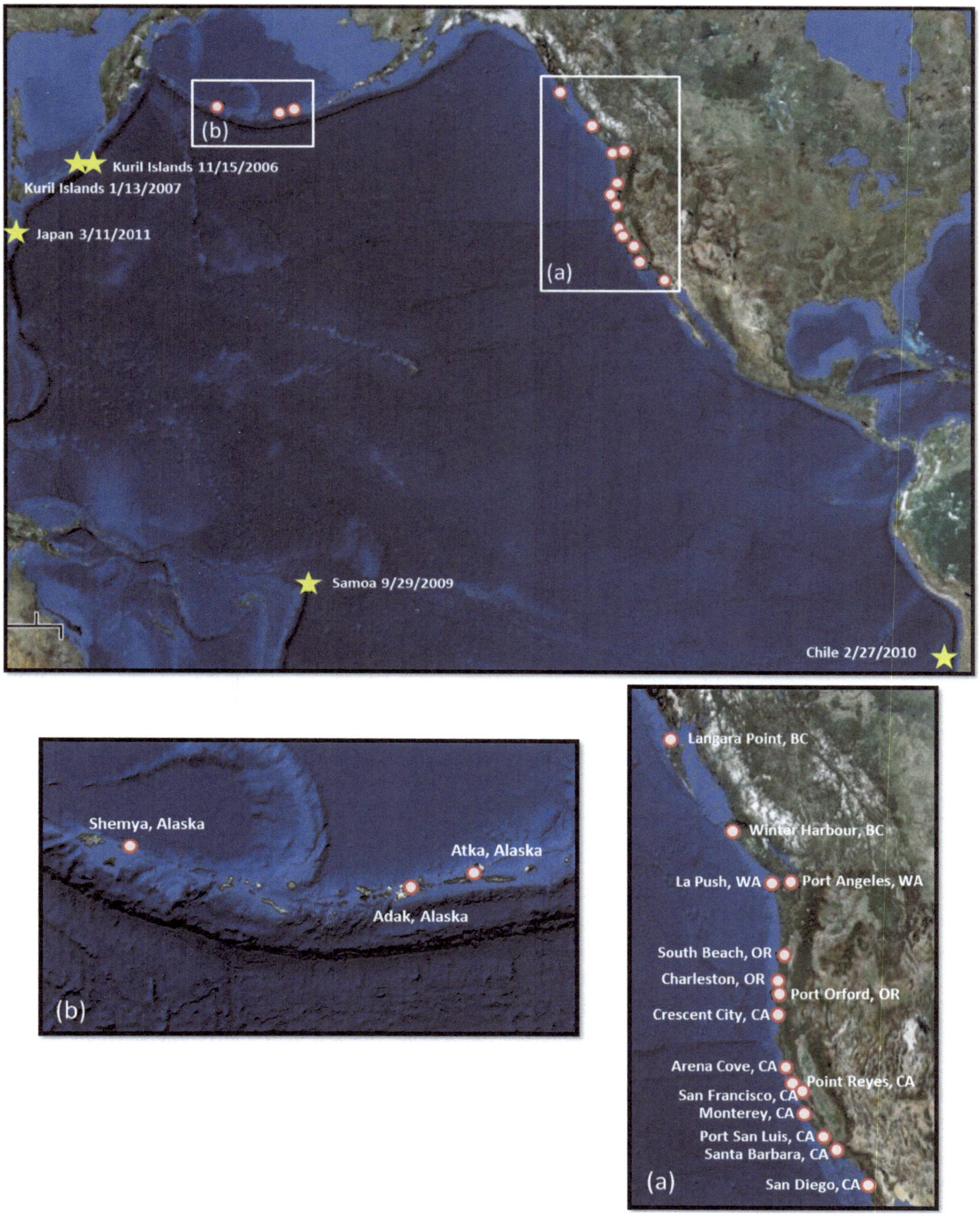

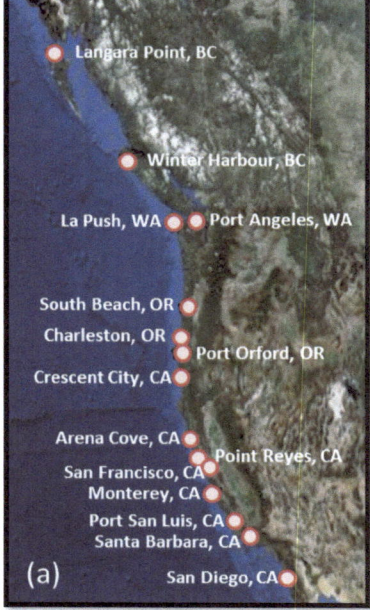

Figure 1
Map of Pacific Ocean showing the location of five recent earthquakes (*stars*) that were responsible for tsunamis impacting Alaska and the West Coast. The *circles* show the locations of tide gauges within the area of responsibility of the West Coast and Alaska Tsunami Warning Center that recorded one or more of the tsunamis

Table 1

A list of the West Coast and Alaska tide gauge sites

Station	Source	Amplitude	UTC arrival	Date	Decay	Latitude	Longitude
Adak, AK	Kuril Islands	0.3	13:44:00	11/15/2006	−441.9	51.86 N	176.63 W
Adak, AK	Chile	0.4	01:42:00	02/28/2010	−259.2	51.86 N	176.63 W
Adak, AK	Japan	1.1	10:24:00	03/11/2011	−473.0	51.86 N	176.63 W
Arena Cove, CA	Kuril Islands	0.5	19:35:00	11/15/2006	−330.6	38.91 N	123.71 W
Arena Cove, CA	Kuril Islands	0.2	12:51:00	01/12/2007	−342.0	38.91 N	123.71 W
Arena Cove, CA	Japan	1.7	15:30:00	03/11/2011	−533.2	38.91 N	123.71 W
Atka, AK	Chile	0.4	01:30:00	02/28/2010	−323.9	52.23 N	174.17 W
Charleston, OR	Japan	0.7	15:36:00	03/11/2011	−306.0	43.34 N	124.32 W
Crescent City, CA	Kuril Islands	0.9	19:47:00	11/15/2006	−263.8	41.74 N	124.18 W
Crescent City, CA	Samoa Islands	0.3	04:44:00	09/29/2009	−397.4	41.74 N	124.18 W
Crescent City, CA	Japan	2.3	15:30:00	03/11/2011	−465.1	41.74 N	124.18 W
Garibaldi, OR	Japan	0.3	15:43:00	03/11/2011	−299.0	45.56 N	123.92 W
Langara Point, BC	Japan	0.5	13:22:00	03/11/2011	−395.2	54.25 N	133.03 W
La Push, WA	Japan	0.7	15:29:00	03/11/2011	−378.6	47.91 N	124.64 W
Monterey, CA	Japan	0.7	15:52:00	03/11/2011	−312.0	36.61 N	121.89 W
Point Reyes, CA	Kuril Islands	0.4	20:03:00	11/15/2006	−912.5	38.00 N	122.98 W
Point Reyes, CA	Samoa Islands	0.4	04:56:00	09/30/2009	−659.5	38.00 N	122.98 W
Point Reyes, CA	Japan	1.3	15:57:00	03/11/2011	−603.6	38.00 N	122.98 W
Port Angeles, WA	Japan	0.6	15:51:00	03/11/2011	−444.1	48.12 N	123.44 W
Port Orford, OR	Japan	2.0	15:24:00	03/11/2011	−329.5	42.74 N	124.50 W
Port San Luis, CA	Kuril Islands	0.5	20:18:00	11/15/2006	−274.4	35.18 N	120.75 W
Port San Luis, CA	Samoa Islands	0.3	04:30:00	09/30/2009	−696.2	35.18 N	120.75 W
Port San Luis, CA	Japan	1.7	15:53:00	03/11/2011	−347.4	35.18 N	120.75 W
Saint Paul Is., AK	Japan	0.1	11:55:00	03/11/2011	−595.4	57.12 N	170.28 W
San Francisco	Japan	0.6	16:22:00	03/11/2011	−219.0	37.81 N	122.46 W
San Diego, CA	Chile	0.3	19:44:00	02/27/2010	−303.6	32.43 N	117.10 W
Santa Barbara, CA	Samoa Islands	0.3	05:39:00	09/30/2009	−713.7	34.41 N	119.69 W
Santa Barbara, CA	Chile	0.9	20:25:00	02/27/2010	−685.5	34.41 N	119.69 W
Santa Monica, CA	Chile	0.6	19:37:00	02/27/2010	387.7	34.01 N	118.50 W
Shemya, AK	Japan	1.5	09:25:00	03/11/2011	−340.7	52.73 N	174.10 W
South Beach, OR	Japan	0.4	15:47:00	03/11/2011	−614.7	44.63 N	124.04 W
Winter Harbour, BC	Japan	0.9	14:53:00	03/11/2011	−251.2	50.52 N	128.03 W

The amplitudes are measured in meters and the e-folding decay time is in minutes

quadrature function of $f(t)$, is calculated to form the imaginary portion of an analytic signal. This transform is equivalent to an unusual type of filter in which the amplitudes of the spectral components are left unchanged but the phase of all frequency components is altered by $\pi/2$ radians (see BRACEWELL, 1965). The transformed record has the same amplitude and frequency content as the original data and includes phase information that depends on the phase of the original data. An envelope time function (see CLAERBOUT, 1976) may be defined by:

$$e(t) = \sqrt{f^2(t) + f_H^2(t)}$$

where $f(t)$ is the original time function, $f_H(t)$ is the Hilbert transform of the $f(t)$, and $e(t)$ is the amplitude envelope time function.

A 120-point (2 h) zero-phase moving average filter is applied to the envelope function to remove high-frequency fluctuations. The result is then scaled to match the peak amplitude of the original signal as shown in Fig. 5.

The first 2 h of the envelope function following the maximum value (Fig. 5d) are used in a least-squares process to obtain an equation for a best-fit exponential decay. Using this equation, the envelope amplitudes are calculated for 6 h following the peak envelope function as shown by the red curve in Fig. 5e. Past this point a spline fit is used to connect the pre-computed decay curve of 31.8 h (green curve in Fig. 5f) to the original decay curve to produce the initial amplitude duration forecast (Fig. 5g).

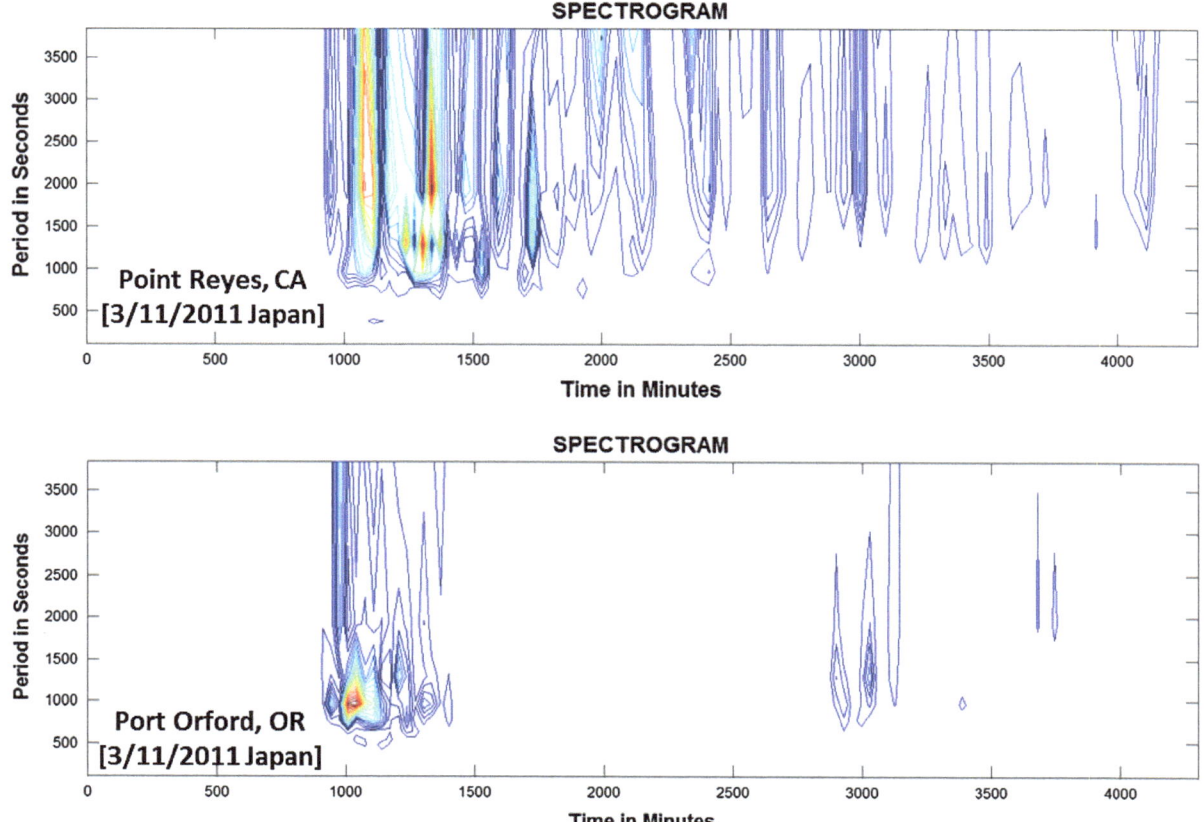

Figure 2
These spectrograms illustrate the local response at different sites on the West Coast due to the Japan tsunami of 3/11/2011. As noted by many researchers, the response is principally governed by the local topography and bathymetry

Table 2

Recent earthquakes that have generated tsunamis impacting the West Coast of the United States

Location	Date	Mag	UTC time	Latitude	Longitude	Depth (km)
Kuril Islands	11/15/2006	8.3	11:14:00	46.6 N	253.2 E	27.7
Kuril Islands	01/13/2007	8.1	04/24/00	46.3 N	154.5 E	10.0
Samoa Islands	09/29/2009	8.0	17:48:00	25.6 S	172.1 W	18.0
Chile	02/27/2010	8.8	06:34:14	25.8 S	72.7 W	35.0
Japan	03/11/2011	9.0	05:46:23	38.3 N	142.4 E	32.0

The short-term amplitude decay that is of interest to the Tsunami Warning Center will differ from the energy–density decay measured over a period of several days. In general, energy density measurements are averaged over a long window length of up to 12 h. In contrast, the Warning Center needs amplitude decay information of the direct wave arrival and coda over a much shorter time period. The Port Orford, Oregon, tide data is a good example of the amplitudes of the initial waves decaying exponentially over a short-time period (Fig. 2). However, in many locations the decay rate is not this simple. The Point Reyes, California, tide gauge data is an example of a record exhibiting multiple local reflections (Fig. 2) and for these stations, the moving-average filter length had to be increased. Using the data from the 2011 Japan tsunami arrival, the minimum filter length that produced good results for the

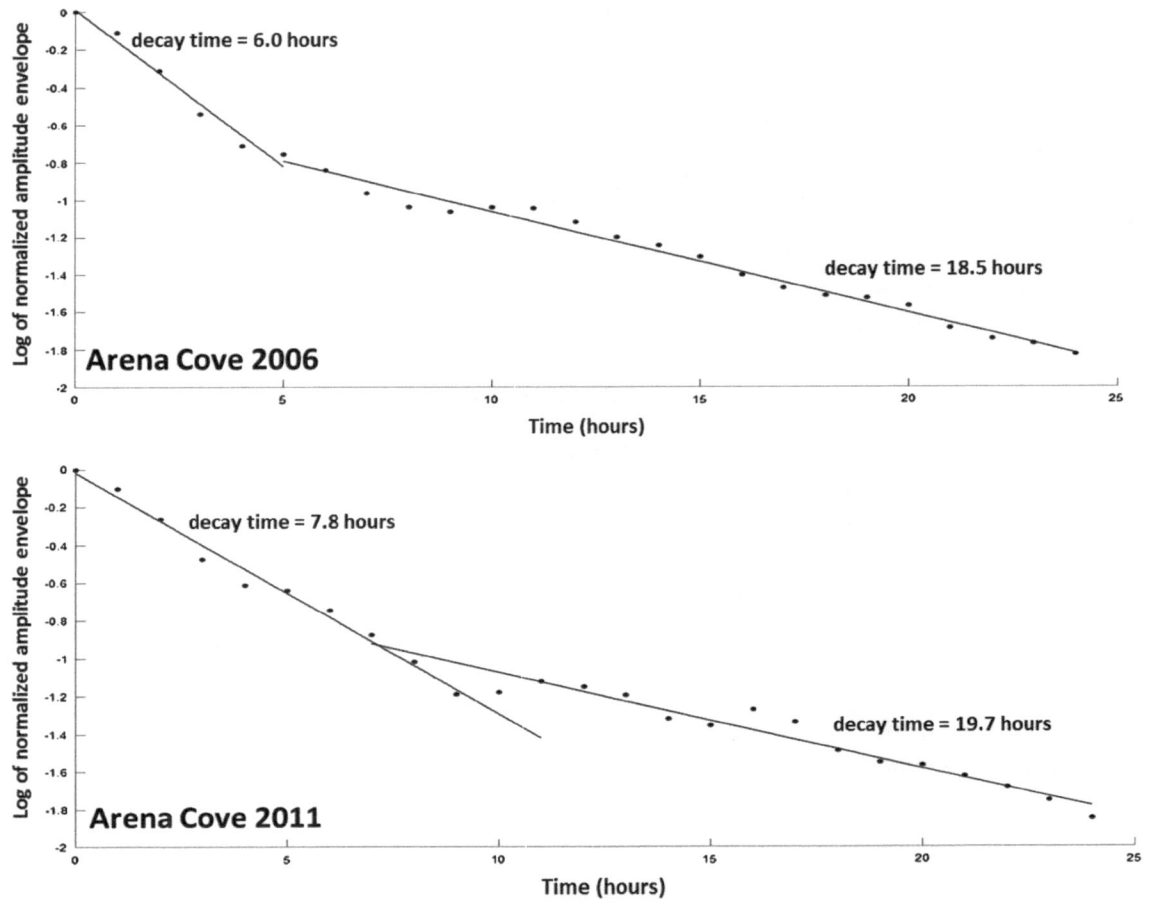

Figure 3
Illustrates both the primary and secondary decays of two different tsunamis recorded at Arena Cove, California. The amplitude envelope was
normalized to the value of 1.0 and the natural logarithm was plotted every hour following the envelope peak

Point Reyes site was found to be 180 points. This
filter was then used to process the 2006 and 2009
Point Reyes data with good results. Based on the
analysis from Point Reyes, a 180-point filter was also
used to process data from Adak, Santa Barbara, and
Santa Monica. The filter lengths for each station are
fixed and not expected to change even though different tsunamis decay at different rates due to
variations in their frequency content and propagation
paths (RABINOVICH et al., 2013a).

The rate of short-term amplitude decay, as applied
here, is principally influenced by local bathymetry
and coastal geometry; however, in some cases the
amplitudes of late arriving reflected waves may be
greater than those of the direct arrivals. This phenomenon is clearly illustrated in the Crescent City

marigraph from the 2006 Kuril Island tsunami
(Fig. 6). The peak amplitude occurring at the time
index 1,333 was not due to the direct arriving tsunami
wave, but was rather the result of the wave interacting
with the Koko Seamount and the Hess Ridge. In
addition, ocean ridges can act as waveguides that
transport energy over thousands of kilometers with
very little attenuation resulting in late-arriving, high-
amplitude waves (KOWALIK et al., 2007).

5. Discussion and Conclusions

A tide gauge that is recording the arrival of a
tsunami will also record the subsequent arrival of
many different types of secondary oscillations.

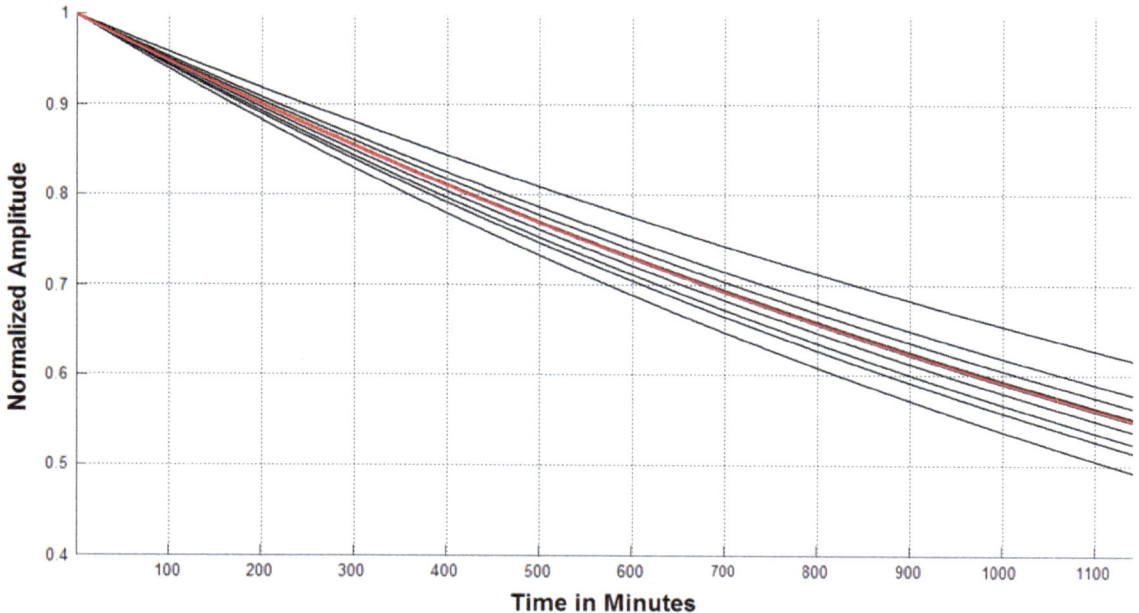

Figure 4
For sites exhibiting a good signal to noise ratio, an envelope function was computed from 6 to 25 h following the peak amplitude envelope (3/11/2011 Japan tsunami). A least squares exponential function was fit to these data and normalized to the value of 1.0. The resulting e-folding decay times ranged from 28.6 to 39.3 h with an average of 31.8 h shown in *red*. This average value was used for all other tsunamis and recording stations

In theory, it is possible to numerically model the full tsunami time series of a specific tide gauge if the time signature of an approaching tsunami is known and if high–spatial resolution regional bathymetry is accessible. Unfortunately, for most regions, high-resolution bathymetry is not generally available (GEIST, 2010). Currently, it is not possible to accurately forecast long term amplitude decay by numerical methods in real time; however, West Coast tide gauge recordings of different tsunamis show that the amplitude envelope of a tsunami may be approximated by appending two exponential decay functions.

Our technique was specifically designed for large trans-Pacific tsunamis that impact the WCATWC's AOR. Small tsunamis do not result in lengthy warnings or advisories and therefore do not require an extended forecast. Data spikes and gaps that occur in real time can have a major effect on the initial decay calculation resulting in a forecast that is not realistic. Even though automated algorithms can detect and repair most of these errors, it is necessary to carefully check the amplitude estimates. Significant errors in

the estimated amplitudes can occur due to the arrival of secondary oscillations at times greater than the length of the moving-average filter. Waves arriving several hours after the peak amplitude envelope would not be accounted for in the initial forecast calculations. An example of a major wave arriving several hours after the peak of the initial wave may be seen on the Adak, Alaska record from the 17 February 2010 Chile event (Fig. 6). The only way to handle this type of situation is to restart the estimation at the time of the late arriving wave. In another example (Fig. 6), the recording for Langara Point, British Columbia (11 March 2011) shows a major reflected wave arriving 6 h after the original peak. This wave attenuates quite rapidly and does not have a major impact on the original estimation. The proposed forecasting method is a dynamic process and new estimates will be continuously calculated and updated over the course of a major event.

Under the right conditions, it is possible for waves to become trapped between the shoreline and the shelf. These normal wave modes may be subject to resonant amplification due to repeated reflections of

335

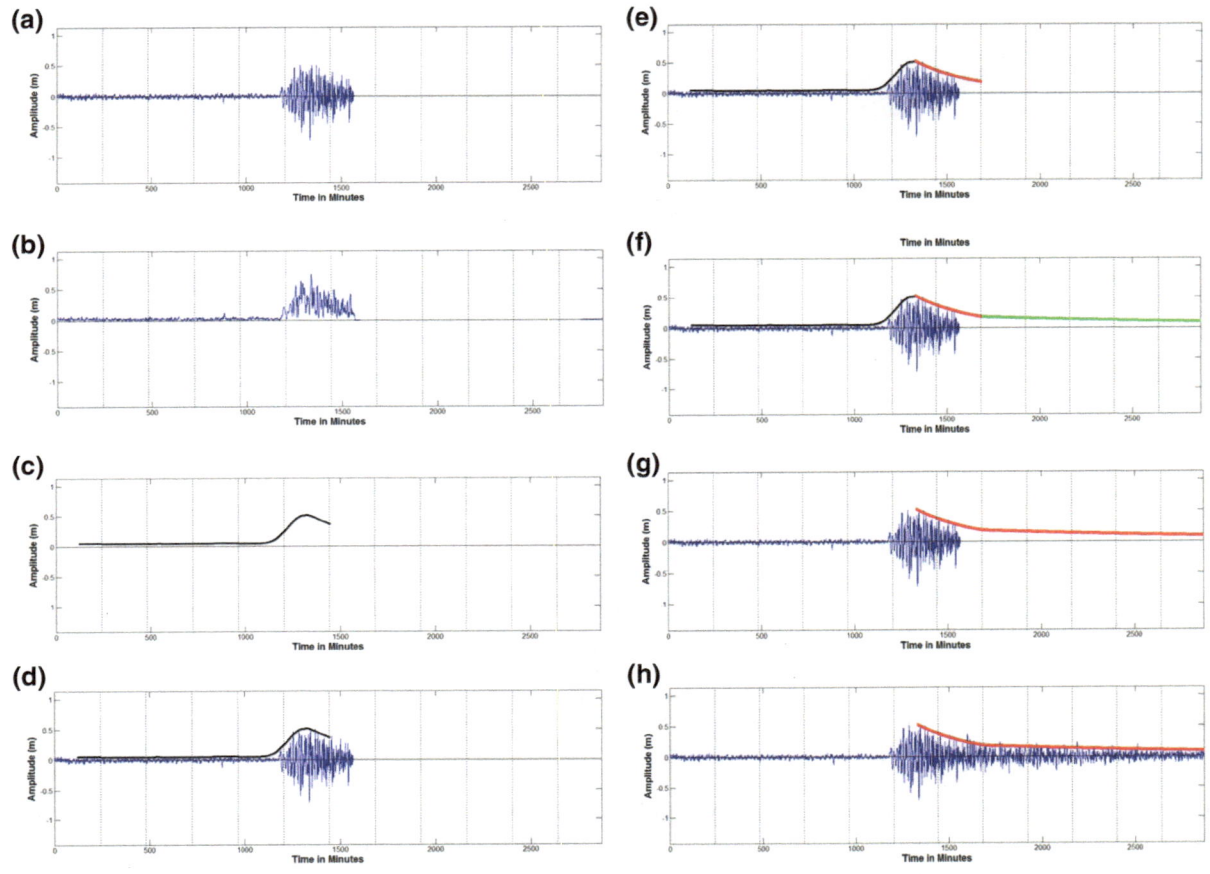

Figure 5

The 2006 Arena Cove data is used to illustrate the processing sequence used to produce an amplitude forecast. **a** The original tide data is first filtered using a zero-phase filter to remove the long-period tidal contributions. **b** The analytic envelope is calculated from the filtered data. **c** The smoothed envelope function shown in *black* is obtained by applying a 2-h moving average (MA) filter. **d** The smoothed amplitude envelope is scaled to match the peak value of the data. **e** A least squares exponential function is calculated using 2 h of data following the peak amplitude envelope. The resulting equation is used to calculate amplitudes for 6 h past the envelope peak (*red curve*). **f** An average decay function (*green curve*), computed for West Coast sites, is appended to the original decay estimate (*red*) using a spline fit. **g** The initial amplitude forecast is plotted in *red*. **h** This shows both the forecast and data

waves between the shoreline and the shelf edge (VAN DORN, 1987). The method will not work for sites that undergo strong resonance. The staff at the Warning Center would recognize this situation and inform emergency managers of the condition.

Many authors have studied tsunami decay in various oceans (CANDELLA *et al.*, 2008; FINE *et al.*, 2013; HAYASHI *et al.*, 2009; MOFJELD *et al.*, 2000; MUNK, 1963; RABINOVICH *et al.*, 2011, 2013a, b; VAN DORN, 1984, 1987). Most investigations were concerned with measuring the energy density decay over a period of several days. In contrast, the WCATWC personnel and government emergency

managers are interested in the amplitude decay over the first 36 h.

When scientists at the WCATWC decide to downgrade a particular region from a warning or an advisory, they do so by consulting with emergency managers and taking into account readings from of a large number of DART gauges and tide stations. An entire region is not kept in an advisory or warning status simply because a single tide gauge indicates that wave amplitudes are a few centimeters above a particular threshold. In a similar sense, an amplitude duration forecast is based on data from many locations within a large region and not simply from a

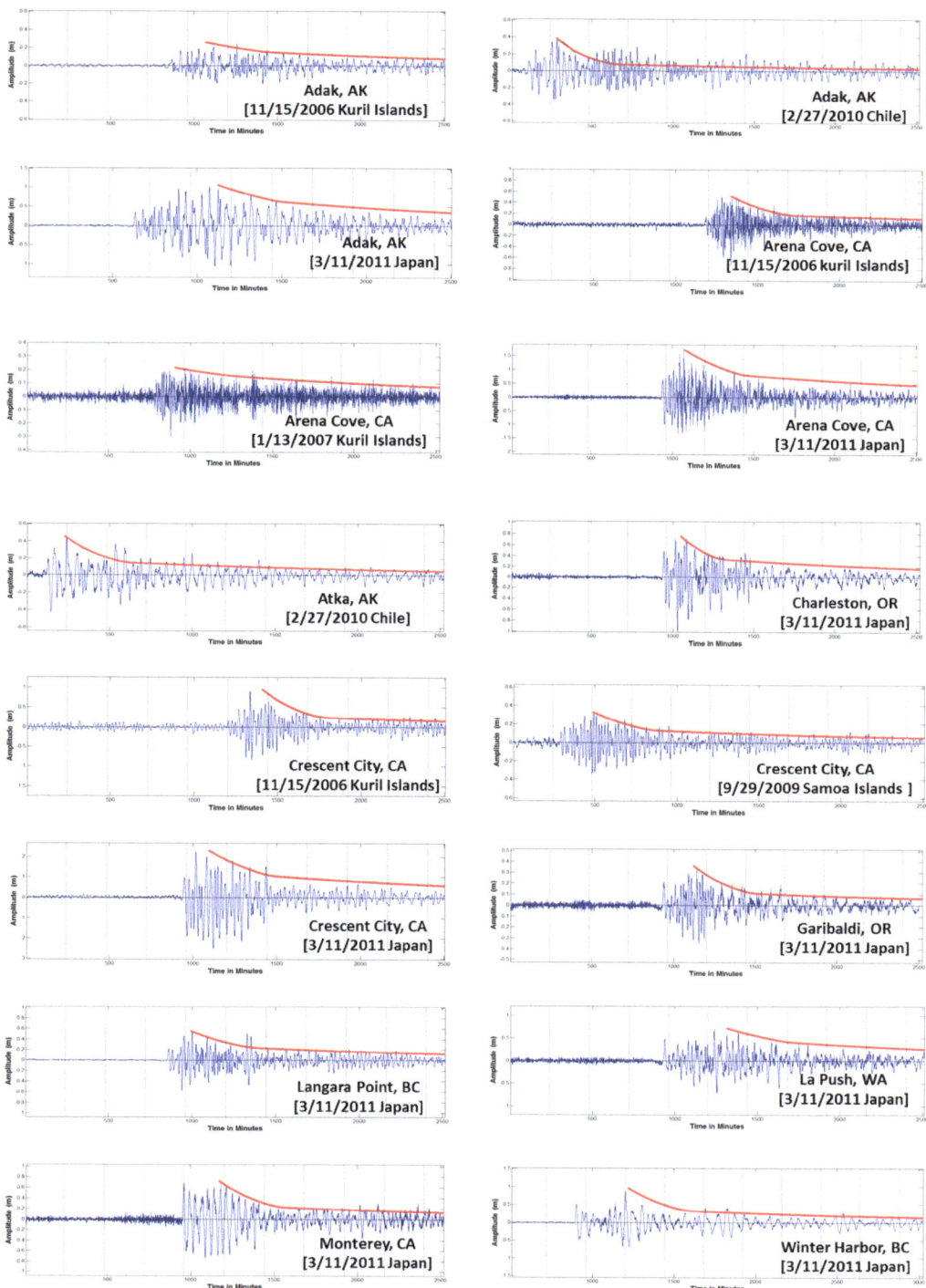

Figure 6
The amplitude forecasts, shown in *red*, are based on the method described in this paper. The actual tide data is superimposed on the amplitude forecast to illustrate how closely the forecast amplitudes follow the data

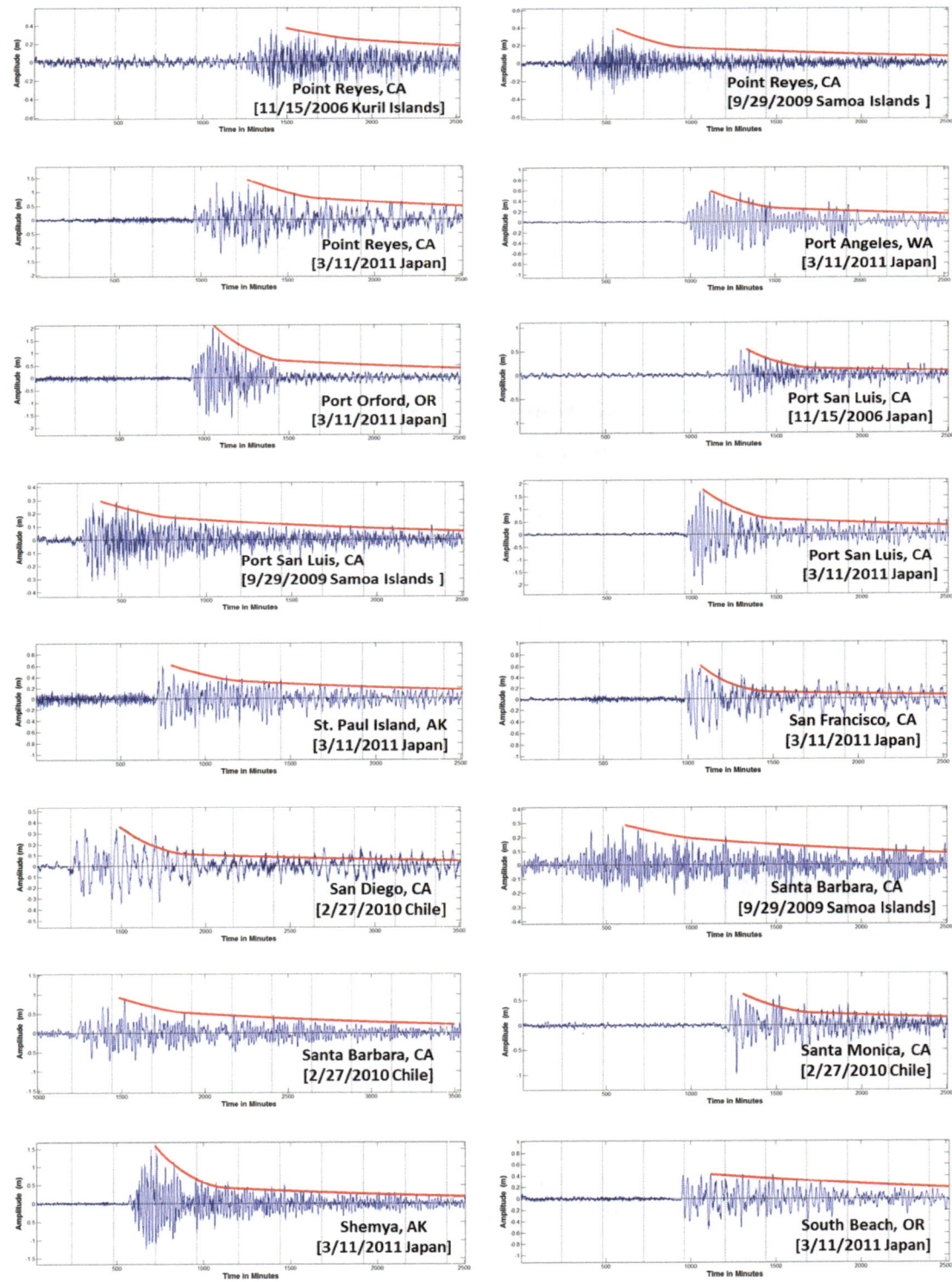

Figure 6
continued

single station. The forecast method outlined in this paper has been implemented at the WCATWC and will be employed until technological advances allow us to hydrodynamically model the decay of a tsunami at a specific site during an actual event with equal accuracy.

Acknowledgments

This work would not have been possible without the encouragement and engagement of Paul Whitmore, whose contribution is highly acknowledged. Our sincere thanks goes to the Editor Alexander Rabinovich and two anonymous reviewers who provided thorough and constructive comments that helped improve the manuscript.

REFERENCES

BRACEWELL, R. (1965), The Fourier Transform and Its Applications, McGraw-Hill, New York.

CANDELLA, R.N., RABINOVICH, A.B., and THOMSON, R.E. (2008), The 2004 Sumatra tsunami as recorded on the Atlantic coast of South America, Advances in Geophysics 14(1), 117–128.

CLAERBOUT, J.F., (1976), Fundamentals of Geophysical Data Processing: With Applications to Petroleum Prospecting, McGraw-Hill, New York.

FINE, I.V., KULIKOV, E.A., and CHERNIAWSKY, J.Y. (2013), Japan's 2011 Tsunami: Characteristics of Wave Propagation from Observations and Numerical Modeling, Pure Appl. Geophys. 170, 1295–1307.

GEIST, E., (2010), The 2010 Chilean tsunami and uncertainty in tsunami modeling, USGS Sound Waves. April, 2010.

GORING, D.G., and NIKORA V.I. (2002), Despiking acoustic Doppler velocimeter data, Journal of Hydraulic Engineering, 128 (1), 117–126.

HAYASHI, Y., KOSHIMURA, S., and IMAMURA, F. (2009), Comparison of decay features of the 2006 and 2007 Kuril Island earthquake tsunamis, Geophys. J. Int. 190, 347–357.

HORILLO, J., KNIGHT, W., and KOWALIK, Z. (2008), The Kuril Islands tsunami of November 2006 Part II: Impact at Crescent City by local enhancement. J. Geophys. Res. 113, C01021, doi:10.1029/2007JC004404.

KOWALIK, Z., KNIGHT, W., LOGAN, T., and WHITMORE, P. (2007), The tsunami of 26 December 2004: Numerical modeling and energy considerations, Pure Appl. Geophys. 164, 379–393.

MERRIFIELD, M.A., FIRING, Y.L., AARUP, T., AGRICOLE, W., BRUNDRIT, G., CHANG-SENG, D., FARRE, R., KILONSKY, B., KNIGHT, W., KONG, L., MAGORI, C., MANURUNG, P., MCCREEY, C., MITCHELL, W., PILLAY, S., SCHINDELE, F., SHILLINGTON, F., TESTUT, L., WIJERATNE, E. M. S., CALDWELL, P., JARDIN, J., NAKAHARA, S., PORTER, F.Y., TURETSKY, N. (2005), Tide gauge observations of the Indian Ocean tsunami, December 26, 2004. Geophysical Research Letters 32, L09603, doi:10.1029/2005GL022610.

MILLER, G.R., MUNK, W.H. and SNODGRASS, F.E. (1962), Long-period waves over California's borderland. Part II: Tsunamis. J. Mar. Res. 20, 31–41.

MOFJELD, H.O., GONZÁLEZ, F.I. and NEWMAN, J.C. (1997), Short-term forecasts of inundation during teletsunamis in the eastern North Pacific Ocean, In: G. HEBENSTREIT (ED.), Perspectives on Tsunami Hazard Reduction, Kluwer, Norwell, MA, pp. 145–155.

MOFJELD, H.O., GONZÁLEZ, F.I., BERNARD, E.N., and NEWMAN, J.C. (2000), Forecasting the heights of later waves in Pacific-wide tsunamis. Nat. Hazards 22, 71–89.

MOFJELD, H.O., GONZÁLEZ, F.I., and NEWMAN, J.C. (2001), Tsunami scattering provinces in the Pacific Ocean, Geophys. Res. Lett. 28, 335–338.

MUNK, W.H., (1963), Some comments regarding diffusion and absorption of tsunamis, In: Proc. Tsunami Meet., X Pacific Science Congress, IUGG Monogr. 24, Paris, 53–72.

MURTY, T.S. (1977), Seismic Sea Waves-Tsunamis. Bull. Fish. Res. Board Canada 198, Ottawa, 337 p.

RABINOVICH A.B., CANDELLA R.N., and THOMSON, R.E. (2011), Energy decay of the 2004 Sumatra tsunami in the world ocean, Pure Appl. Geophys. 168, 1919–1950.

RABINOVICH A.B., CANDELLA R.N., and THOMSON, R.E. (2013a), The open ocean energy decay of three recent trans-Pacific tsunamis, Geophys. Res. Lett, 40, doi:10.1002/grl.50625.

RABINOVICH A.B., THOMSON, R.E. and FINE I.V. (2013b), The 2011 Chilean Tsunami Off the West Coast of Canada and the Northwest Coast of the United States, Pure Appl. Geophys. 170, 1115–1126

VAN DORN, W.G. (1984), Some tsunami characteristics deducible from tide records. J. Phys. Oceanogr. 14, 353–363.

VAN DORN, W.G. (1987), Tide gage response to tsunamis, Part II: Other oceans and smaller seas, J. Phys. Oceanogr. 17, 1507–1516.

(Received March 25, 2013, revised July 4, 2013, accepted July 20, 2013, Published online August 9, 2013)

Pure Appl. Geophys. 171 (2014), 3515–3525
© 2013 Springer Basel
DOI 10.1007/s00024-013-0730-6

Displaced Water Volume, Potential Energy of Initial Elevation, and Tsunami Intensity: Analysis of Recent Tsunami Events

Mikhail A. Nosov,[1] Anna V. Bolshakova,[1] and Sergey V. Kolesov[1]

Abstract—We consider recent ocean-bottom earthquakes for which detailed slip distribution data are available. Using these data and the Okada formulae, we calculate the vector fields of co-seismic bottom deformations, which allow us to determine the displaced water volume and the potential energy of initial elevation of the tsunami source. It is shown that, in the majority of cases, the horizontal components of bottom deformation provide an additional contribution to the displaced water volume and virtually never diminish the contribution of the vertical component. The absolute value of the relative contribution of the horizontal components of bottom deformation to the displaced volume varies from 0.07 to 55 %, on average amounting to 14 %. The displaced volume and the energy of initial elevation (tsunami energy) are examined as functions of the moment magnitude, and the relevant regressions (least-squares fits) are derived. The obtained relationships exhibit good correspondence with the theoretical upper limits that had been obtained under the assumption of uniform slip distribution along a rectangular fault. Tsunami energy calculated on the basis of finite fault model data is compared with the earthquake energy determined from the energy–magnitude relationship by Kanamori. It is shown that tsunami takes from 0.001 to 0.34 % of the earthquake energy, and on average 0.04 %. Finally, we analyze the Soloviev–Imamura tsunami intensity as a function of the following three quantities: (1) the moment magnitude, (2) the decimal logarithm of the absolute value of displaced volume, and (3) the decimal logarithm of the potential energy of initial elevation. The first dependence exhibits rather poor correlation, whereas the second and third dependences demonstrate noticeably higher correlation coefficients. This gives us grounds to suggest considering the displaced volume and the energy of initial elevation as measures of the tsunamigenic potential of an earthquake.

Key words: Tsunami source, slip distribution, finite fault model, co-seismic bottom deformation, water displacement, initial elevation, tsunami energy, Soloviev–Imamura tsunami intensity.

[1] Faculty of Physics, M.V. Lomonosov Moscow State University, Leninskie Gory, Moscow 119991, Russia. E-mail: m.a.nosov@mail.ru

1. Introduction

Estimation of the degree of tsunami hazard for a coast is primarily based on statistical analysis of events that occurred in the past. Tsunamis vary in strength within wide limits: from weak waves that can be registered only with the aid of sea level gages, up to catastrophic events devastating the coast along hundreds of kilometers. Any statistical analysis must involve the quantitative characteristics of a phenomenon. The strength of a tsunami can be quantitatively determined by use of the Soloviev–Imamura tsunami intensity scale (Soloviev 1970; Levin and Nosov 2009). For decades, the Soloviev–Imamura tsunami intensity has been routinely determined for the majority of noticeable events, and these values are collected in tsunami catalogs and databases (Historical Tsunami Database for the World Ocean, NOAA/WDC Global Historical Tsunami Database at NGDC).

As for earthquakes, which represent the most widespread cause of tsunamis, a widely adopted quantitative measure of strength is the moment magnitude M_W. However, estimation of the level of tsunami hazard on the basis of earthquake moment magnitude often fails. The first reason for this is that the initial seismic analysis is not reliable enough. The second reason turns out to be much more important: tsunamis are related to earthquakes in a complex and ambiguous way. In particular, this complexity is manifested in the relationship between the tsunami intensity and the earthquake moment magnitude as a large spread in the data (e.g., Chubarov and Gusiakov 1985; Levin and Nosov 2009; Gusiakov 2011). In his recent paper, Gusiakov (2011) pointed out the following four factors that cause significant scatter of tsunami intensity for earthquakes of similar magnitude: (1) difference in the water depth in the source area, (2) difference in the earthquake source mechanism, (3) difference in the earthquake focus depth, and (4) difference in the tectonic setting of the source area.

The main mechanism of tsunami generation—nearly instant displacement of water by co-seismic deformation of the ocean bottom during strong seafloor earthquakes—is well known (e.g., YAMASHITA and SATO 1974; OKAL 1988; DOTSENKO and SOLOVIEV 1995; NOSOV and SHELKOVNIKOV 1997; NOSOV 1998; LEVIN and NOSOV 2009). The past decade has seen significant progress in reproducing the co-seismic bottom deformation of tsunami sources. Nowadays, calculation of the co-seismic bottom deformation can be carried out on the basis of the earthquake slip distribution or a finite fault model (FFM). In comparison with the model of a single rectangular fault that used to be widely employed, the FFM certainly allows more precise reproduction of the co-seismic bottom deformation in the near-field zone, i.e., at the tsunami source. Such precise determination of the tsunami source results in reasonable coincidence of in situ measured [e.g., by Deep-ocean Assessment and Reporting of Tsunamis (DART)] and simulated tsunami waveforms in the open ocean (e.g., LAVEROV et al. 2009; NOSOV et al. 2011a, 2013; MAEDA et al. 2011; LØVHOLT et al. 2012).

In this study, taking advantage of the precise determination of the co-seismic bottom deformation enabled by use of a FFM, we examine the following two quantities: (1) the water volume displaced by the co-seismic bottom deformation, and (2) the potential energy of initial elevation of the tsunami source (the tsunami energy). Being integrals of motion, this volume and energy undoubtedly play an important role in the hydrodynamic tsunami problem. Moreover, both quantities consider the factors pointed out by GUSIAKOV (2011). Thus, it is expected that the displaced volume and the energy of initial elevation will correlate with the tsunami intensity better than the moment magnitude. This is why these quantities can be used, at least in principle, as a measure of the tsunamigenic potential of an ocean-bottom earthquake. By now, FFM data are available for a few tens of recent tsunami events. This sufficiently large

number of events makes it possible to perform a sort of statistical analysis.

The tsunami energy for real events has already been studied by many researchers. Without claiming to present a full list, we cite several publications (e.g., HATORI 1970; MURTY and LOOMIS 1980; KAJIURA 1981; OKAL 2003; BOLSHAKOVA and NOSOV 2011). Although the displaced water volume also represents an important parameter, it is rarely mentioned in publications (HATORI 1970; GRILLI et al. 2007; NOSOV and KOLESOV 2009; BOLSHAKOVA and NOSOV 2011). The concept of displaced water volume has also been discussed for tsunamis associated with landslides, asteroids or explosions (e.g., FRITZ et al. 2003; LEVIN and NOSOV 2009). It is worth mentioning that the displaced volume can be determined, at least in principle, from in situ measurements of horizontal motions of water in the vicinity of the tsunami source, i.e., without any seismological analysis (NOSOV et al. 2011b, 2013; NOSOV and NURISLAMOVA 2012).

The main purpose of this study is to examine particular features of the displaced volume and the energy of initial elevation for recent tsunami events for which FFM data are available. In Sect. 2, the data on which we base our analysis are described. In Sect. 3, the methods of calculation of the displaced volume and potential energy of initial elevation are presented. In Sect. 4, first, we consider the relative contribution of the vertical and horizontal components of co-seismic bottom deformation to the displaced volume. Then, the displaced volume and the energy of initial elevation are examined as functions of the moment magnitude. Finally, we investigate how the Soloviev–Imamura tsunami intensity is related to the displaced volume and the energy of initial elevation. Sect. 5 summarizes results and primary conclusions.

2. Finite Fault Models, Bathymetry, and Tsunami Intensity: Sources of Data

In this study we consider recent tsunamigenic earthquakes for which detailed slip distribution estimates from finite fault models (FFMs) exist. In calculations of the vector field of co-seismic bottom deformation of tsunami sources, we rely on FFM data

Table 1

List of ocean-bottom earthquakes under consideration

Date YYYYMMDD	Location	Data source	M_W	I	V_{xy}, km^3	V_z, km^3	V_{xyz}, km^3	E_{TS}, J
19941004	Kuril	UCSB	8.36	2.6	0.466	5.809	6.275	1.05E+14
19950730	Antofagasta	UCSB	8.14	1.5	0.780	7.273	8.053	2.41E+13
19951203	Iturup	UCSB	7.81	0.5	0.247	1.266	1.514	4.95E+12
20001117	New Britain	UCSB	7.5	No data	0.073	0.711	0.784	1.02E+12
20010623	Peru	UCSB	8.4	2.6	1.814	11.516	13.330	9.18E+13
20041226	Sumatra	Caltech	9.15	4.5	6.970	83.751	90.721	3.14E+15
20050328	Nias	Caltech	8.5	1.5	1.807	10.443	12.250	1.42E+14
		UCSB	8.68	1.5	3.623	18.580	22.203	3.14E+14
20050615	California	UCSB	7.2	No data	−0.023	0.098	0.075	6.00E+10
20050724	Nicobar	UCSB	7.25	No data	−0.004	−0.034	−0.038	9.87E+10
20050816	Japan	UCSB	7.19	No data	0.004	0.247	0.250	9.27E+10
20060717	Java	Caltech	7.9	2	1.327	3.054	4.381	2.38E+13
20061115	Simushir	Caltech	8.3	3	2.073	5.305	7.378	4.99E+13
		USGS	8.3	3	2.884	6.636	9.520	1.13E+14
20070113	Simushir	Caltech	8.1	2	−0.874	−4.098	−4.973	1.25E+14
		USGS	8.1	2	−1.169	−5.602	−6.771	2.24E+14
20070401	Solomon	USGS	8.1	No data	0.579	4.067	4.647	2.60E+13
20070815	Peru	Caltech	8	2	0.171	4.313	4.484	2.17E+13
		USGS	8	2	−0.105	4.900	4.795	3.53E+13
20070912	Kepulauan	Caltech	7.9	No data	0.077	2.332	2.410	4.50E+12
		UCSB	7.94	No data	−0.003	2.977	2.974	2.75E+12
		USGS	7.9	No data	0.027	2.664	2.691	3.75E+12
20070912	S. Sumatra	Caltech	8.4	1	1.081	9.778	10.859	8.43E+13
		UCSB	8.4	1	1.224	6.524	7.748	4.04E+13
		USGS	8.5	1	1.182	5.757	6.939	5.34E+13
20071114	Antofagasta	UCSB	7.81	−1	0.123	1.590	1.712	9.72E+11
		USGS	7.7	−1	0.116	1.491	1.607	1.21E+12
20080220	Simeulue	Caltech	7.4	No data	0.019	0.019	0.038	1.86E+11
		USGS	7.4	No data	0.017	0.068	0.085	1.32E+11
20080929	Kermadec	USGS	7	No data	−0.001	0.106	0.106	3.29E+10
20081116	Sulawesi	Caltech	7.3	No data	0.010	0.776	0.786	8.28E+11
20090103	Papua	USGS	7.7	No data	0.076	1.072	1.147	2.27E+12
20090715	New Zealand	USGS	7.8	No data	0.062	1.308	1.370	6.88E+12
20090929	Samoa	USGS	8.1	1.5	−0.619	−3.572	−4.191	9.10E+13
20090930	S. Sumatra	Caltech	7.6	−1	0.008	0.805	0.813	2.28E+11
		Caltech	7.6	−1	0.004	0.816	0.821	2.71E+11
		USGS	7.6	−1	0.005	0.643	0.648	1.65E+11
		USGS	7.6	−1	0.005	0.666	0.671	1.65E+11
20091007	Vanuatu	Caltech	7.6	0	−0.011	0.838	0.827	1.48E+12
		USGS	7.7	0	−0.007	1.015	1.009	1.26E+12
20100112	Haiti	USGS	7	0	−0.011	0.115	0.104	4.88E+11
20100227	Chile	Caltech	8.8	3	2.031	38.955	40.986	1.85E+14
		UCSB	8.86	3	5.265	69.204	74.469	6.53E+14
		USGS	8.8	3	5.936	69.220	75.156	8.17E+14
20100406	N. Sumatra	USGS	7.8	0	0.069	1.339	1.407	2.59E+12
20100509	N. Sumatra	USGS	7.2	No data	0.000	0.344	0.344	1.14E+11
20101025	Kepulauan	UCSB	7.82	No data	0.761	0.737	1.499	5.67E+12
		USGS	7.7	No data	0.233	0.606	0.839	1.26E+12
20101221	Bonin	USGS	7.4	No data	−0.013	−0.279	−0.292	8.66E+11
		USGS	7.4	No data	−0.016	−0.304	−0.320	7.75E+11
20101225	Vanuatu	USGS	7.3	No data	−0.035	−0.238	−0.274	6.70E+11
20110309	Japan	UCSB	7.4	No data	0.066	0.231	0.297	7.92E+11
		USGS	7.3	No data	0.044	0.209	0.253	5.44E+11

Table 1 *continued*

Date YYYYMMDD	Location	Data source	M_W	I	V_{xy}, km³	V_z, km³	V_{xyz}, km³	E_{TS}, J
20110311	Tohoku	Caltech	9	4.1	33.275	63.784	97.059	3.42E+15
		UCSB	9.1	4.1	30.789	79.669	110.458	8.00E+15
		UCSB	9.1	4.1	29.678	66.200	95.878	7.54E+15
		UCSB	9.1	4.1	33.370	64.863	98.233	9.65E+15
		USGS	9	4.1	19.895	77.489	97.385	2.38E+15
20110706	Kermadec	USGS	7.6	No data	−0.110	−0.999	−1.109	3.80E+12
20110820	Vanuatu	USGS	7.1	No data	0.005	0.152	0.157	7.03E+10
20111021	Kermadec	USGS	7.4	No data	0.000	0.407	0.408	3.78E+11
20120110	N. Sumatra	USGS	7.2	No data	−0.014	−0.011	−0.025	1.40E+11
20120411	Sumatra	Caltech	8.6	No data	−1.040	4.510	3.469	7.43E+13
		UCSB	8.64	No data	−0.991	3.974	2.983	1.33E+14
		USGS	8.6	No data	−0.890	3.539	2.649	1.44E+14
20120827	Salvador	USGS	7.3	No data	0.062	0.247	0.310	2.51E+11
20120831	Philippines	USGS	7.6	No data	0.045	0.739	0.785	1.87E+12
20120905	Costa Rica	USGS	7.6	No data	0.021	0.438	0.458	2.15E+11
20121028	Canada	Caltech	7.8	No data	0.072	0.701	0.774	4.97E+12
		UCSB	7.72	No data	0.061	1.129	1.189	1.57E+13

M_W, moment magnitude; I, Soloviev–Imamura tsunami intensity; V_{xy} and V_z, water volumes displaced by the horizontal and vertical components of co-seismic bottom deformation, respectively; $V_{xyz} \equiv V_{xy} + V_z$, total displaced water volume; E_{TS}, energy of tsunami

available from the following sites: California Institute of Technology (Caltech, http://www.tectonics.caltech.edu/slip_history/), UC Santa Barbara (UCSB, http://www.geol.ucsb.edu/faculty/ji/big_earthquakes/home.html), and US geological survey (USGS, http://earthquake.usgs.gov/earthquakes/eqinthenews/).

All the events under consideration are listed in Table 1. We examine 44 ocean-bottom earthquakes that occurred during the period 1994–2012. For many seismic events several solutions have been provided. Thus, in total, we deal with 70 earthquake source models.

Calculation of the displaced volume and the potential energy of initial elevation requires not only the vector field of co-seismic bottom deformation but also the distribution of oceanic depths. Necessary bathymetric data were extracted from the 1-min digital atlas General Bathymetric Chart of the Oceans (GEBCO).

The strength of a tsunami is usually determined by use of the Soloviev–Imamura tsunami intensity scale: $I = 0.5 + \log_2 h$, where h is the average tsunami height on the coast closest to the source (SOLOVIEV 1970; LEVIN and NOSOV 2009). For this study, tsunami intensity data were extracted from the online version of the Historical Tsunami Database for the World Ocean (HTDB/WLD, http://tsun.sscc.ru/nh/tsunami.php).

3. Calculation of Displaced Volume and Potential Energy of Initial Elevation

We place the origin of the Cartesian reference frame, $0xyz$, at the unperturbed free water surface and direct the $0z$ axis vertically upward. The $0x$ axis is directed northward, and the $0y$ axis eastward. Let us assume that the bottom position, before an earthquake, is given by the function

$$z = -H(x, y), \qquad (1)$$

where H is the ocean depth. After an earthquake, due to the co-seismic deformation, the bottom adopts a new position given by the function

$$z = -H(x, y) + \eta(x, y), \qquad (2)$$

where η is the function which describes the changes of the bottom position due to the co-seismic bottom deformation.

Knowing the function η, it is possible to calculate the displaced water volume and the potential energy of initial elevation. However, when dealing with real events, one determines the vector field of the co-seismic bottom deformation, $\vec{D} \equiv (D_x, D_y, D_z)$, instead of the function η. In this study we also calculate the vector fields $\vec{D}$ from FFM data. According to the FFM, the fault plane for an earthquake is divided into a number of rectangular subfaults. The bottom deformation caused by each of these subfaults is calculated analytically by using the Okada formulae (OKADA 1985). Then, the contributions of all the elements are summed up. The Lamé constants μ and λ enter the Okada expressions in the form of the following combination: $\mu/(\lambda + \mu)$. In all cases the medium is assumed to have $\lambda = \mu$. Details of our method and examples of co-seismic bottom deformations calculated using this method can be found in our recent publications (NOSOV and KOLESOV 2009, 2011; BOLSHAKOVA and NOSOV 2011; NOSOV et al. 2013).

Now let us use the vector field, $\vec{D}$, and the distribution of ocean depth, H, to recalculate the function η. For this purpose we consider a point $P_0 = (x_0, y_0, z_0)$ situated on the unperturbed bottom surface. The coordinates of this point obey Eq. 1. As a result of the co-seismic deformation, the point P_0 moves to the new position $P_1 = (x_0 + D_x, y_0 + D_y, z_0 + D_z)$. Since the point P_1 remains on the bottom surface, its coordinates have to obey Eq. 2, i.e.,

$$z_0 + D_z = -H(x_0 + D_x, y_0 + D_y) + \eta(x_0 + D_x, y_0 + D_y). \quad (3)$$

In practice of numerical simulation of tsunamis, the distribution of ocean depth, H, the vector field, $\vec{D}$, and the sought function, η, are set on a discrete grid with a spatial increment Δ. This means that the spatial structure of these functions between the closest points of this grid remains beyond consideration. One only has to assume that, between the closest points, all the functions are smooth enough, say, linear. The spatial increment in the open ocean usually amounts to $\Delta \approx 10^3$ m (i.e., approx. 1 angular min). The amplitude of the co-seismic bottom deformation is significantly inferior to the spatial increment: $|\vec{D}| < < \Delta$. Indeed, even for such catastrophic events as the 2004 Sumatra or 2011 Tohoku earthquakes, the maximum co-seismic bottom deformation, $|\vec{D}|$, was of the order of 10 m (e.g., GRILLI et al. 2007; FUJII et al. 2011; NOSOV et al. 2011a; KOKETSU et al. 2011; SATAKE et al. 2013).

Further, in formula 3 we expand the function $H(x, y)$ in a Taylor series about the point (x_0, y_0). The relative smallness of the bottom deformation gives us

grounds to accept the following two assumptions: First, we keep only linear terms in the Taylor series; Second, we adopt the approximate formula: $\eta(x_0 + D_x, y_0 + D_y) \approx \eta(x_0, y_0)$. Ultimately, taking into account Eq. 1, we arrive at the following formula:

$$\eta = \frac{\partial H}{\partial x}D_x + \frac{\partial H}{\partial y}D_y + D_z. \quad (4)$$

Making use of physical reasoning but not a mathematical method, TANIOKA and SATAKE (1996) derived the same relation for the vertical displacement of the water surface of a tsunami source, i.e., the initial elevation. It is worth stressing that formula 4, strictly speaking, determines the vertical displacement of the bottom surface. Due to the "smoothing effect," the initial elevation and the displacement of the bottom surface may differ (e.g., KAJIURA 1963; NOSOV and KOLESOV 2009, 2011). In what follows we neglect this difference.

The volume of water displaced by the co-seismic bottom deformation can be determined as the integral of the function η taken over the entire tsunami source area. Considering formula 4, it is reasonable to introduce the total displaced volume, V_{xyz}, as the following sum:

$$V_{xyz} = V_{xy} + V_z, \quad (5)$$

$$V_{xy} = \iint\limits_S \left(\frac{\partial H}{\partial x}D_x + \frac{\partial H}{\partial y}D_y \right) dx\,dy, \quad (6)$$

$$V_z = \iint\limits_S D_z \, dx\,dy, \quad (7)$$

where V_{xy} is the volume displaced by the horizontal components of deformation of sloping bottom and V_z is the volume displaced by the vertical component of bottom deformation.

In the calculation of the potential energy of initial elevation we restrict our consideration to the traditional assumption that the co-seismic bottom deformation, η, is formed instantaneously and the initial elevation is equal to this deformation. Under this assumption the potential energy of initial elevation is determined by the formula (e.g., OKAL and SYNOLAKIS 2003)

$$E_{TS} = \frac{\rho\,g}{2} \iint\limits_{S} \eta^2 \, dx\, dy, \qquad (8)$$

where g is the acceleration due to gravity and ρ is the density of water (we assume $g = 9.8$ m/s^2 and $\rho = 1{,}000$ kg/m^3).

The integration in formulae 6–8 was performed numerically by use of the midpoint rule for double integrals. The size of subrectangles, $\Delta x \times \Delta y$, corresponds to 1×1 angular min. Each subrectangle is obviously small enough to neglect the sphericity of the Earth and to consider the Cartesian reference frame as a local reference frame. In contrast to subrectangles, the whole integration domain turns out to be so large that the sphericity of the Earth must be taken into account. Indeed, in each case the integration domain has to be chosen so that it fully covers all noticeable bottom deformations. For strong events the size of the integration domain may amount to $10° \times 10°$ or even more. While computing the integrals, the sphericity was taken into account by considering the dependence of the size of the subrectangles along the east–west direction, Δx, from the latitude.

4. Results and Discussion

The results of the calculations of the displaced volume and the potential energy of initial elevation are collected in Table 1. In the column entitled "M_W" the moment magnitude is shown. Note that, for some seismic events, the estimation of moment magnitude provided by Caltech, UCSB, and USGS can be slightly different. The next column entitled "I" contains the values of the Soloviev–Imamura tsunami intensity. The column entitled "V_{xy}, km^3" presents the volume displaced by the horizontal deformation of sloping bottom (in cubic kilometers). The column entitled "V_z, km^3" presents the volume displaced by the vertical deformation. The total displaced volume is shown in the column entitled "V_{xyz}, km^3". In the last column entitled "E_{TS}, J" the potential energy of initial elevation can be found.

Let us first examine the relative contribution of horizontal and vertical components of bottom deformation to the displaced water volume. According to

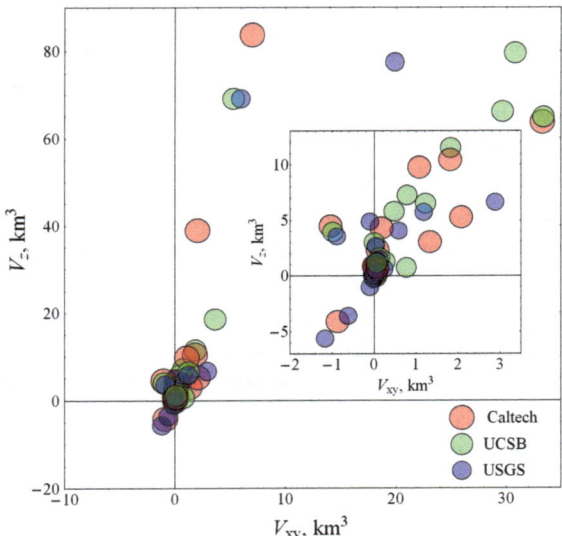

Figure 1
Volume displaced by the *vertical* component of co-seismic bottom deformation, V_z, versus volume displaced by the *horizontal* components, V_{xy}. The inset shows the near-zero domain in detail

data presented in Table 1, the absolute value of the relative contribution of the horizontal components, $|V_{xy}/V_{xyz}|$, varies from 0.07 to 55 %. The average value of the relative contribution amounts to 14 %. So, generally speaking, the contribution of the horizontal components of bottom deformation to the displaced volume is not negligible. We recall here that the role of horizontal deformations in tsunami generation has been discussed earlier in some publications (e.g., IWASAKI 1982; TANIOKA and SATAKE 1996; NOSOV and KOLESOV 2009, 2011; NOSOV et al. 2011a; SATAKE et al. 2013). Nevertheless, in numerical simulations of tsunamis, the horizontal component of bottom deformation is usually neglected.

Figure 1 shows the volume displaced by the vertical component of bottom deformation, V_z, versus the volume displaced by the horizontal components, V_{xy}. The most striking feature of this dependence is that virtually always the signs of the contributions from the horizontal and vertical components turn out to be the same. Therefore, the horizontal components provide an additional contribution to tsunami and virtually never diminish the contribution of the vertical component. This is an important reason to take into account the horizontal components of bottom deformation in tsunami simulation. By neglecting the horizontal components of bottom deformation, one will usually underestimate the tsunami. This revealed feature means that co-seismic bottom deformations are definitely correlated with bottom topography. This fact is not surprising in the light of plate tectonics theory.

Among all 70 cases under consideration, there are only 10 exceptions for which the signs of the contributions from the horizontal and vertical components turn out to be different. The exceptional character of these cases can be explained as follows: Exceptional cases of the first kind (20070912-Kepulauan-UCSB, 20080929-Kermadec-USGS, 20091007-Vanuatu-Caltech&USGS) are characterized by a negligible relative contribution of the horizontal components to the displaced volume ($\sim 1\%$ or less). In these cases, the sign may be determined erroneously due to inaccuracies in the GEBCO bathymetry or FFM data. Exceptional cases of the second kind (20070815-Peru-USGS and 20100112-Haiti-USGS) correspond to earthquakes that occurred near the shoreline, so that a significant part of the co-seismic deformation area is attributed to the coast. Exceptional cases of the third kind (20050615-California-UCSB and 20120411-Sumatra-Caltech&UCSB&USGS) correspond to strike–slip earthquakes that occurred beneath nearly flat horizontal bottom. Besides, strike–slip earthquakes themselves are not capable of generating significant tsunamis. However, one should bear in mind that strike–slip earthquakes may trigger landslides that are capable of generating huge waves, as happened, for example, in Lituya Bay on July 10, 1958 (FRITZ et al. 2009).

The absolute value of the total displaced water volume as a function of earthquake moment magnitude is shown in Fig. 2 by colored circles. The regression line (least-squares fit) associated with the data is determined by the equation

$$\log_{10}\left|V_{xyz}\right|(\mathrm{km}^3) = 1.45 M_{\mathrm{W}} - 11.2. \qquad (9)$$

The regression equation (9) is depicted in Fig. 2 as a black dashed line. Before discussing the particular features of this dependence, we point out that strong earthquakes with moment magnitude $M_{\mathrm{W}} \approx 8$ are capable of displacing a few cubic kilometers of

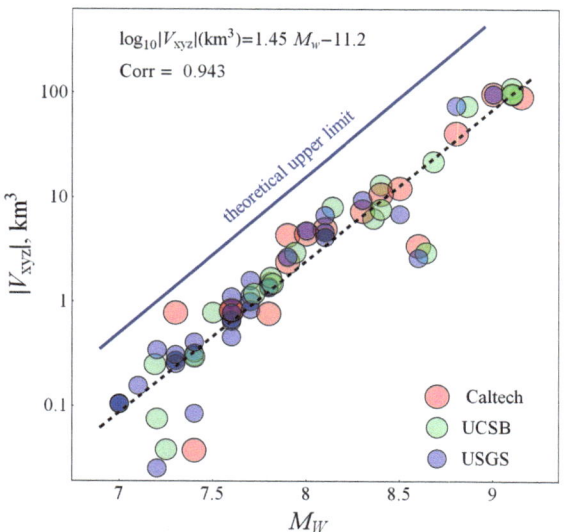

Figure 2
Absolute value of displaced water volume versus earthquake moment magnitude (*colored circles*). Linear regression is shown by *black dashed line*. Regression equation and correlation coefficient are depicted in the *left upper corner*. *Blue solid line* indicates the theoretical upper limit estimated by BOLSHAKOVA and NOSOV (2011)

water, whereas huge earthquakes of $M_{\mathrm{W}} \approx 9$ (such as the 2004 Sumatra, 2010 Chile, and 2011 Tohoku earthquakes), can displace up to 100 cubic kilometers of water.

It is seen from Fig. 2 that the displaced water volume exhibits reasonably good correlation with the moment magnitude (correlation coefficient, Corr = 0.943). However, there are two groups of points that are located noticeably below the common cluster of points. The first group corresponds to the event 20120411-Sumatra-Caltech&UCSB&USGS. This was a strike–slip earthquake which did not displace much water and, thus, did not generate significant tsunami in spite of its relatively large moment magnitude ($M_{\mathrm{W}} = 8.6$). The second group ($M_{\mathrm{W}} = 7.2$–7.4) again includes three strike–slip earthquakes (20050615-California-UCSB, 20050724-Nicobar-UCSB, 20120110-N. Sumatra-USGS), which cannot effectively displace water and generate tsunamis. Also, this second group includes two points corresponding to the event 20080220-Simeulue-Caltech&USGS. In these two cases, the small amount of displaced water volume is the consequence of the small dip angle ($7°$ or $7.7°$) and relatively large depth of the fault plane (~ 15–40 km).

347

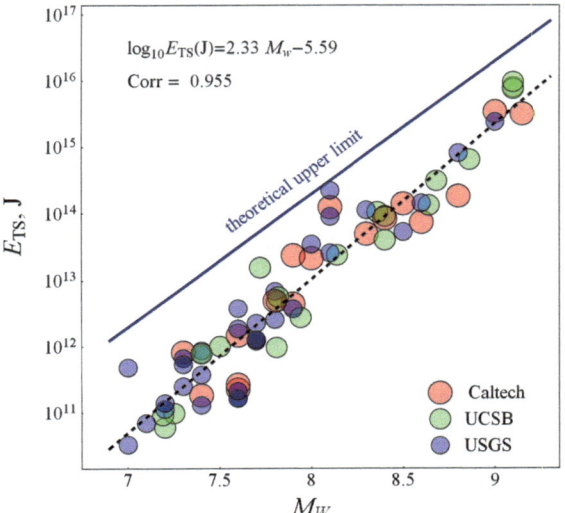

Figure 3

Potential energy of initial elevation versus earthquake moment magnitude (*colored circles*). Linear regression is shown by *black dashed line*. Regression equation and correlation coefficient are depicted in the *left upper corner*. *Blue solid line* indicates the theoretical upper limit estimated by BOLSHAKOVA and NOSOV (2011)

Under the conditions of a uniform slip distribution along a finite rectangular fault and a flat horizontal bottom, BOLSHAKOVA and NOSOV (2011) derived the following dependence for the upper limit of the displaced volume as a function of moment magnitude:

$$\log_{10}|V_{max}|(\text{km}^3) = 1.5 M_W - 10.8. \qquad (10)$$

This dependence is depicted in Fig. 2 by a blue line. It is seen that displaced volumes calculated from realistic FFM data in consideration of real bathymetry never exceed this theoretical upper limit. On average, for the real tsunamigenic earthquakes listed in Table 1, the displaced volume turns out to be 5–7 times smaller than the theoretical upper limit.

Figure 3 shows the potential energy of initial elevation (tsunami energy) as a function of earthquake moment magnitude. The regression (black dashed line) is determined by the equation

$$\log_{10} E_{TS}(\text{J}) = 2.33 M_W - 5.59. \qquad (11)$$

The tsunami energy exhibits rather good correlation with the earthquake moment magnitude (Corr = 0.955). In contrast to the displaced water volume plotted in Fig. 2, all the points shown in Fig. 3 lie more or less within a common cluster.

The theoretical upper limit of the tsunami energy derived by BOLSHAKOVA and NOSOV (2011) is determined by the formula

$$\log_{10} E_{TS}^{max}(\text{J}) = 2.0 M_W - 1.7. \qquad (12)$$

Relation (12) is depicted in Fig. 3 by a blue line. The upper limit provides a good upper estimate for the energy calculated from realistic FFM data in consideration of real bathymetry.

Comparing formula 11 with the well-known energy–magnitude relationship defined in terms of M_W by KANAMORI (1977)

$$\log_{10} E_{EQ}(\text{J}) = 1.5 M_W + 4.8, \qquad (13)$$

we readily arrive at the following estimate for the ratio between the tsunami energy and earthquake energy:

$$\log_{10} E_{TS}/E_{EQ} = 0.83 M_W - 10.39. \qquad (14)$$

According to formula 14, tsunami takes approximately from 0.0026 % ($M_W = 7$) to 0.12 % ($M_W = 9$) of the earthquake energy. The direct calculation of the fraction of tsunami energy, E_{TS}/E_{EQ}, in consideration of relationship (13) for each of the 70 cases collected in Table 1 gives the following results: tsunami takes from 0.001 to 0.34 % of the earthquake energy, and on average $E_{TS}/E_{EQ} \approx 0.04\%$. These values more or less correspond to estimations by KAJIURA (1981). The estimations of the theoretical upper limit (BOLSHAKOVA and NOSOV 2011), as expected, give approximately one order larger values for the fraction of tsunami energy.

Relationships between the Soloviev–Imamura tsunami intensity and the earthquake moment magnitude have already been examined in some publications (e.g., CHUBAROV and GUSIAKOV 1985; GUSIAKOV 2011). In particular, CHUBAROV and GUSIAKOV (1985) obtained the following theoretical dependence by means of numerical modeling:

$$I = 3.55 M_W - 27.1. \qquad (15)$$

Figure 4 shows the relationship between the Soloviev–Imamura tsunami intensity and the earthquake moment magnitude according to data listed in Table 1. Since the tsunami intensity had been provided for only 36 of the 70 cases under consideration,

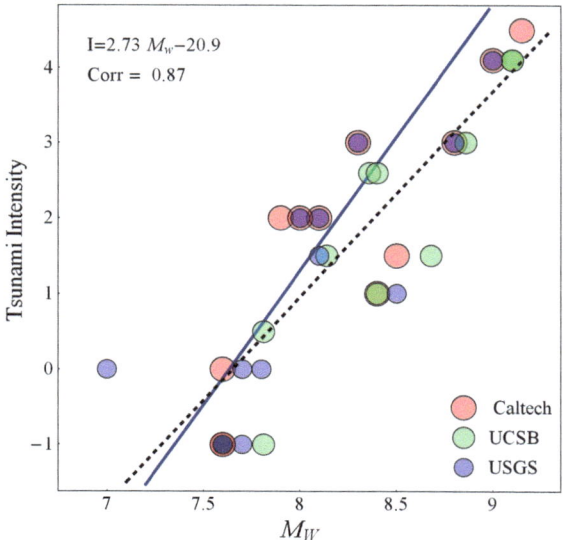

Figure 4
Soloviev–Imamura tsunami intensity versus earthquake moment magnitude. Linear regression is shown by *black dashed line*. Regression equation and correlation coefficient are depicted in the *left upper corner*. *Blue solid line* indicates the theoretical dependence obtained by CHUBAROV and GUSIAKOV (1985)

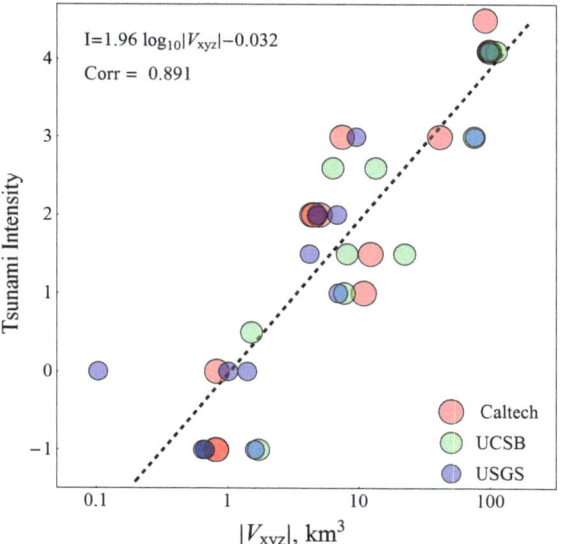

Figure 5
Soloviev–Imamura tsunami intensity versus absolute value of displaced water volume. Linear regression is shown by *black dashed line*. Regression equation and correlation coefficient are depicted in the *left upper corner*

we restrict our analysis to these 36 cases. The black dashed line indicates the regression, which is determined by the equation

$$I = 2.73 M_{\mathrm{W}} - 20.9. \qquad (16)$$

The theoretical dependence (15) obtained by CHUBAROV and GUSIAKOV (1985) is depicted in Fig. 4 by a blue solid line. The newly obtained regression (16) is in acceptable agreement with the theoretical dependence (15).

The dependence shown in Fig. 4 exhibits a large spread of the data. The correlation coefficient amounts to Corr = 0.87. We recall that the reasons for such a large spread have been pointed out by GUSIAKOV (2011) (see Sect. 1). Since tsunami intensity and moment magnitude are poorly correlated, the moment magnitude is not an ideal measure of the tsunamigenic potential of an earthquake.

Let us try to consider the displaced volume and the energy of initial elevation as measures of tsunamigenic potential. More exactly, we consider the decimal logarithms of the energy and absolute value of the volume. These two quantities, as compared with the moment magnitude, have at least two

advantages. First, in the calculation of both quantities, the earthquake mechanism and the mechanism of tsunami generation are taken into account. Second, both quantities represent integrals of motion in the hydrodynamic tsunami problem. So, it is expected that the displaced volume and the energy of initial elevation will exhibit better correlation with the tsunami intensity than moment magnitude.

Tsunami intensity versus absolute value of displaced water volume is plotted in Fig. 5. The regression, as usual, is depicted by a black dashed line. This dependence exhibits a slightly higher correlation (Corr = 0.891) compared with the dependence of tsunami intensity versus moment magnitude shown in Fig. 4. The dependence of tsunami intensity versus potential energy of initial elevation (Fig. 6) exhibits even higher correlation (Corr = 0.944).

These observed higher correlation coefficients clearly demonstrate that the displaced volume and the potential energy of initial elevation indeed represent better measures of the tsunamigenic potential of an earthquake compared with the moment magnitude. Unfortunately, for many weak tsunami events, the Soloviev–Imamura tsunami intensity has not been

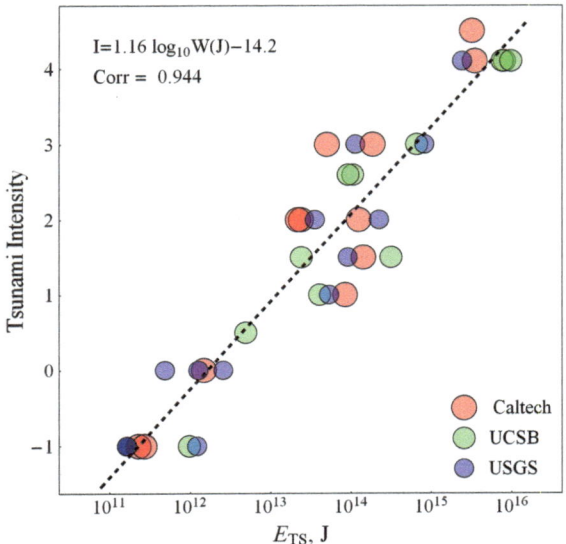

$I = 1.16 \log_{10} W(J) - 14.2$

$Corr = 0.944$

Figure 6

Soloviev–Imamura tsunami intensity versus potential energy of initial elevation. Linear regression is shown by *black dashed line*. Regression equation and correlation coefficient are depicted in the *left upper corner*

estimated yet. If we had estimations of tsunami intensity for all events listed in Table 1, including weak tsunamis, the increase of the correlation coefficient would be even more convincing.

5. Conclusions

We carried out analysis of 44 ocean-bottom earthquakes (1994–2012) for which finite fault models (FFMs) had been provided by Caltech, UCSB or USGS. The vector fields of co-seismic bottom deformation caused by each of these earthquakes were calculated analytically on the basis of FFMs by using the Okada formulae. Realistic co-seismic bottom deformations in consideration of bathymetry (GEBCO) were recalculated into displaced water volume and potential energy of initial elevation (tsunami energy).

It was shown that the absolute value of the relative contribution of the horizontal components of bottom deformation to the displaced volume varies from 0.07 to 55 %, on average amounting to 14 %. In the majority of cases, the horizontal components provide an additional contribution to tsunami and

virtually never diminish the contribution of the vertical component. So, it is important to take into account the horizontal components of bottom deformation in tsunami simulation.

The displaced volume and the energy of initial elevation were examined as functions of the moment magnitude. The regressions (least-squares fits) were derived. The obtained relationships turned out to exhibit good correspondence with existing estimates for the theoretical upper limits.

Tsunami energy calculated on the basis of realistic FFMs in consideration of real bathymetry was compared with the earthquake energy determined from the well-known energy–magnitude relationship by Kanamori. It was shown that tsunami takes from 0.001 to 0.34 % of the earthquake energy, and on average 0.04 %.

The Soloviev–Imamura tsunami intensity was analyzed as a function of the following three quantities: (1) the moment magnitude, (2) the decimal logarithm of the absolute value of displaced volume, and (3) the decimal logarithm of the potential energy of initial elevation. The first dependence (I versus M_W) exhibits rather poor correlation (0.87), whereas the second (I versus $\log_{10}|V_{xyz}|$) and third (I versus $\log_{10}|E_{TS}|$) dependences demonstrate noticeably higher values of correlation coefficient of 0.891 and 0.944, respectively. This gives us grounds to suggest considering the displaced volume and the energy of initial elevation as measures of the tsunamigenic potential of an earthquake.

Acknowledgments

This work was supported by the Russian Foundation for Basic Research, projects 13-05-92100, 13-05-00337, 12-05-31422. We are grateful to USGS, Caltech, and UCSB for slip distribution data. We also thank Dr. V.K. Gusiakov for maintaining the online Historical Tsunami Database for the World Ocean.

REFERENCES

BOLSHAKOVA, A.V., NOSOV, M.A. (2011), *Parameters of Tsunami Source Versus Earthquake Magnitude*, Pure Appl. Geophys. *168*, 2023–2031, doi:10.1007/s00024-011-0285-3.

CHUBAROV, L.B. and GUSIAKOV, V.K. (1985), *Tsunamis and earthquake mechanism in the island arc region*, Sci. Tsunami Hazards. *3*(1), 3–21.

DOTSENKO, S. F., SOLOVIEV, S. L. (1995), *On the role of residual displacements of ocean bottom in tsunami generation by underwater earthquakes*, Oceanology. *35*(1), 20–26.

FRITZ, H.M., HAGER, W.H., MINOR, H.-E. (2003). *Landslide generated impulse waves. 2. Hydrodynamic impact craters*, Experiments in Fluids. *35*(6), 520–532.

FRITZ, H.M., MOHAMMED, F., YOO, J. (2009). *Lituya Bay Landslide Impact Generated Mega-Tsunami 50th Anniversary*. Pure Appl. Geophys. *166*(1–2), 153–175, doi:10.1007/s00024-008-0435-4.

FUJII, Y., SATAKE, K., SAKAI, S., SHINOHARA, M., and KANAZAWA, T. (2011), *Tsunami source of the 2011 off the Pacific coast of Tohoku Earthquake*, Earth Planets Space. *63*, 815–820.

GRILLI, S.T., IOUALALEN, J.M., KIRBY, J.T., WATTS, P., ASAVANT, J., and SHI, F. (2007), *Source Constraints and Model Simulation of the December 26, 2004, Indian Ocean Tsunami*, Journal of Ocean Engineering. *133*(6), 414–428.

GUSIAKOV, V.K. (2011), *Relationship of Tsunami Intensity to Source Earthquake Magnitude as Retrieved from Historical Data*, Pure Appl. Geophys. *168*, 2033–2041.

HATORI, T. (1970), *Vertical crustal deformation and tsunami energy*, Bulletin of the Earthquake Research Institute *4*, 171–188.

IWASAKI, S. (1982), *Experimental study of a tsunami generated by a horizontal motion of a sloping bottom*, Bulletin of the Earthquake Research Institute *57*, 239–262.

KAJIURA, K. (1963), *The leading wave of a tsunami*, Bulletin of the Earthquake Research Institute *41*(3), 535–571.

KAJIURA, K. (1981), *Tsunami energy in relation to parameters of the earthquake fault model*, Bulletin of the Earthquake Research Institute *56*, 415–440.

KANAMORI, H. (1977), *The energy release in great earthquakes*, J. Geophys. Res. *82*, 2981–2987.

KOKETSU, K., YOKOTA, Y., NISHIMURA, N., YAGI, Y., MIYAZAKI S., SATAKE, K., FUJII, Y., MIYAKE, H., SAKAI, S., YAMANAKA, Y, OKADA, T. (2011), *A unified source model for the 2011 Tohoku earthquake*, Earth and Planetary Science Letters *310*(3–4), 480–487.

LAVEROV, N. P., LOBKOVSKY, L. I., LEVIN, B. W. et al. (2009), *The Kuril tsunamis of November 15, 2006, and January 13, 2007: Two trans-pacific events*, Doklady Earth Sciences *426*(1), 658–664.

LEVIN, B.W. and NOSOV, M.A., *Physics of Tsunamis* (Springer 2009).

LØVHOLT, F., KAISER, G., GLIMSDAL, S., SCHEELE, L., HARBITZ, C. B., and PEDERSEN, G. (2012), *Modeling propagation and inundation of the 11 March 2011 Tohoku tsunami*, Nat. Hazards Earth Syst. Sci. *12*, 1017–1028.

MAEDA, T., FURUMURA, T., SAKAI, S., and SHINOHARA, M. (2011), *Significant tsunami observed at ocean-bottom pressure gauges during the 2011 off the Pacific coast of Tohoku Earthquake*, Earth Planets Space Letter *63*, 803–808.

MURTY, T.S., LOOMIS, H.G. (1980), *A new objective tsunami magnitude scale*, Mar. Geod. *4*(3), 267–282.

NOSOV, M. A. (1998), *On the directivity of dispersive tsunami waves excited by piston-type and traveling-wave sea-floor motion*, Volcanol. Seismol. *19*, 837–844.

NOSOV, M. A., KOLESOV, S. V., and LEVIN, B. W. (2011a), *Contribution of Horizontal Deformation of the Seafloor into Tsunami Generation near the Coast of Japan on March 11, 2011*, Doklady Earth Sciences *441*(1), 1537–1542, doi:10.1134/S1028334X11110079.

NOSOV, M. A., MOSHENCEVA, A. V., and LEVIN, B. W. (2011b), *Residual Hydrodynamic Fields near a Tsunami Source*, Doklady Earth Sciences. *438*(2), 853–857, doi:10.1134/S1028334X11060213.

NOSOV, M.A. and KOLESOV, S.V. (2009), *Method of Specification of the Initial Conditions for Numerical Tsunami Modeling*, Moscow University Physics Bulletin. *64*(2), 208–213, doi:10.3103/S0027134909020222.

NOSOV, M.A. and KOLESOV, S.V. (2011), *Optimal initial conditions for simulation of seismotectonic tsunamis*, Pure Appl. Geophys. *168*, 1223–1237, doi:10.1007/s00024-010-0226-6.

NOSOV, M.A. and NURISLAMOVA, G.N. (2012), *The potential and vortex traces of a tsunamigenic earthquake in the ocean*, Moscow University Physics Bulletin. *67*(5), 457–461, doi:10.3103/S0027134912050086.

NOSOV, M.A., MOSHENCEVA, A.V., KOLESOV, S.V. (2013), *Horizontal motions of water in the vicinity of a tsunami source*, Pure Appl. Geophys. *170*(9–10), 1647–1660, doi:10.1007/s00024-012-0605-2.

NOSOV, M.A. SHELKOVNIKOV, N. K. (1997), *The excitation of dispersive tsunami waves by piston and membrane floor motions*, Izvestiya, Atmos. Ocean. Phys. *33*(1), 133–139.

OKADA, Y. (1985), *Surface deformation due to shear and tensile faults in a half-space*, Bulletin of the Seismological Society of America. *75*(4), 1135–1154.

OKAL, E. A. and SYNOLAKIS, C.E. (2003), *A theoretical comparison of tsunamis from dislocations and landslides*, Pure Appl. Geophys. *160*, 2177–2188.

OKAL, E.A. (1988), *Seismic parameters controlling far-field tsunami amplitudes: a review*, Natural Hazards *1*, 67–96.

OKAL, E.A. (2003), *Normal mode energetics for far-field tsunamis generated by dislocations and landslides*, Pure Appl. Geophys. *160*, 2189–2221.

SATAKE, K., FUJII, Y., HARADA, T., NAMEGAYA, Y. (2013). *Time and space distribution of coseismic slip of the 2011 Tohoku earthquake as inferred from tsunami waveform data*. Bulletin of the Seismological Society of America. *103*(2B), 1473–1492.

SOLOVIEV, S. L. (1970), *Recurrence of tsunamis in the Pacific*. In: Adams, W.M., ed. Tsunamis in the Pacific Ocean, Honolulu: East-West Center Press, Honolulu, 149–163.

TANIOKA, Y. and SATAKE, K. (1996), *Tsunami generation by horizontal displacement of ocean bottom*, Geophys. Res. Lett. *23*(8), 861–864.

YAMASHITA, T. and SATO R. (1974), *Generation of tsunami by a fault model*, J. Phys. Earth *22*, 415–440.

(Received March 14, 2013, revised October 22, 2013, accepted October 22, 2013, Published online November 22, 2013)

Reprinted from the journal

Pure Appl. Geophys. 171 (2014), 3527–3538
© 2014 Springer Basel
DOI 10.1007/s00024-014-0791-1

∎ **Pure and Applied Geophysics**

Tsunami Recurrence Function: Structure, Methods of Creation, and Application for Tsunami Hazard Estimates

VICTOR KAISTRENKO[1]

Abstract—This paper considers a theoretical basement for a Poissonian probability model for tsunami run-up heights, with emphasis on the tsunami recurrence function. It is shown that the tsunami recurrence function of a general type contains at least two scale parameters: asymptotic frequency of big tsunamis f related to the considered region and characteristic tsunami height H^* for the considered location in the region. A method for the correct statistical evaluation of the parameters f and H^*, and their variations, using observational data from tsunami catalogues, is created. The paper considers some theoretical and applied problems related to the tsunami recurrence function, an example of a two-parameter tsunami hazard map, and also the problem of probabilistic tsunami hazard estimation.

Key words: Earthquake, Tsunami, Tsunami hazard, run-up heights, distribution, recurrence function, probabilistic model.

1. Introduction

The tsunami recurrence function (RF) is one of the most important characteristics of tsunami activity. Knowledge of it enables tsunami risk to be estimated quantitatively. The tsunami RF $\varphi(h)$ is defined as the average frequency of tsunami occurrence for each selected coastal location x, with maximum run-up height being equal to or more than the "threshold" height h:

$$\varphi(h) \overset{def}{=} \overline{\left(\frac{N(\geq h)}{T}\right)}, \tag{1}$$

where $N(\geq h)$ is the number of tsunamis with maximum height $\geq h$ occurring during the time period T at selected location x. This function is positive and decreasing. Some of the earliest estimates of tsunami RF were created by WIEGEL (1964, 1965), MAGOON and ARNO (1973), WIGEN (1983), and OKADA and TADA (1983), using observational data analysis. The analytical form of this function is unknown, and one of the main problems is to create an adequate analytical approximation for it, especially for big values of tsunami heights h. A similar problem exists for the creation of the recurrence function (Gutenberg-Richter Law) for earthquake magnitude (GUTENBERG and RICHTER (1954), and KAGAN (1999, 2010).

2. Main Properties of the Tsunami Recurrence Function

The tsunami RF is a function of tsunami height threshold h and coastal location x (usually expressed as coordinates). Tsunami RF can also depend on several other parameters of tsunami activity.

The values φ and h are physical parameters having physical dimensions. According to the key theorem of Buckingham, in physical dimensional analysis (SEDOV, 1993), the physical relationship (Eq. 1) should have a dimensionless form. Generally, this means that there are scale parameters f and H^*, representing frequency and height, respectively, for each location x, which make the relationship (Eq. 1) dimensionless:

$$\varphi(x, h) = f(x)\Phi\left(x, \frac{h}{H^*(x)}\right). \tag{2}$$

Special variants when the tsunami RF contains one scale parameter (Φ is power function of h), and many (more than two) scale parameters, will be investigated in parts 4, 5 and 7. Parameter x in Eq. 2 is a symbol of a given location.

---
[1] Institute of Marine Geology and Geophysics, Russian Academy of Sciences, Nauki Str., 1 B, Yuzhno-Sakhalinsk 693022, Russia. E-mail: victor@imgg.ru

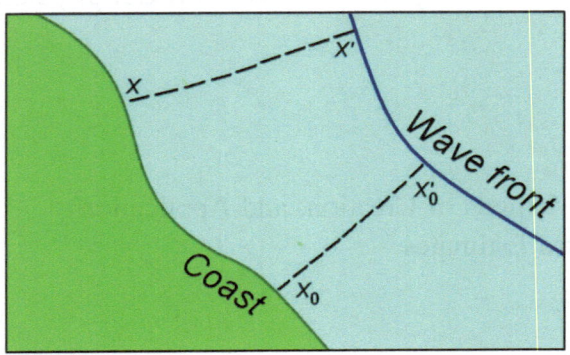

Figure 1
Scheme of the tsunami propagation in a region with a size of a few
tens of kilometers

Physical parameters f and H^* can be studied by examination of the properties of tsunami propagation from the open ocean to the coast (Fig. 1). Big tsunamis are long waves, and their length in the open ocean is several tens of kilometers or more. Therefore, the wave field in the open ocean should be smooth, with no substantial changes over distances of a few tens of kilometers. Tsunami heights in locations x and x_0 near to the coast can be significantly different, but tsunami heights in corresponding locations x' and x_0' in the open ocean should be approximately the same, i.e., x' is the slow parameter.

Each tsunami, prior to its detection at the coastal point x with a maximum height h, is detected at the corresponding point x' in the open ocean with a maximum height h'. We can estimate the average coefficient $K(x', x) = \overline{k(x', x)}$ of tsunami height transformation (amplification) during wave propagation from point x' in the open ocean, to the corresponding coastal point x, and on average, $h = K(x', x) \cdot h'$. Taking into account that maximum wave heights on the shore, given by linear and nonlinear 1-D shallow water models, are exactly equal for non-breaking waves on the slope (PELINOVSKY and MAZOVA, 1992), the "linear" approach with average transformation coefficient K for maximum tsunami heights is used here, for the 2-D domain. Hence, we can make a chain of equations:

$$
\begin{aligned}
\varphi(x, h) = \varphi(x', h') &= f(x') \cdot \Phi\left(x', \frac{h'}{H^*(x')}\right) \\
&= |\text{substitution}: h = K(x', x) \cdot h'| \\
&= f(x') = f \cdot \Phi\left(x', \frac{h}{K(x', x) \cdot H^*(x')}\right) \quad (3)
\end{aligned}
$$

The variable x' in (3) is a "slow" parameter and dependence on it can be neglected for a region with a length of several tens of kilometers. According to (3), regional tsunami RF is a function of the one combined variable:

$$
\varphi(x, h) = f \cdot \Phi\left(\frac{h}{H^*(x)}\right). \quad (4)
$$

Parameter f is the asymptotic frequency of big tsunamis which depends slowly on the location x, and can be considered to be constant for a region of considered size. This has a clear physical explanation, since all strong tsunamis should be detectable within a region of such size.

Parameter H^* is the characteristic tsunami height for selected location x, which is proportional to the average coefficient $K(x)$ of the tsunami height transformation (amplification) from the open ocean to the coastal location x.

$$
H^*(x) = K(x', x) \cdot H^*(x'). \quad (5)
$$

Really, this parameter is very changeable along the coast. Tsunami height transformation function $K(x)$ can be estimated using numerical modeling.

Physically, the tsunami RF can be finite if the maximum possible tsunami height h_{mp} exists. The latter is unknown, but for this case we can formally define the tsunami RF as being equal to zero for $h \geq h_{mp}$.

3. Creation of the Tsunami Recurrence Function using Observational Data

For a given location, there is a random set of tsunami run-up heights collected in catalogues and databases [Tsunami Data and Information; Historical Tsunami Database for the World Ocean (HTDB/WLD)], and these can be sorted $h_1 > h_2 > h_3 > \cdots$ in order of magnitude. A sequence of tsunamis occurring at a given location, with a maximum run-up height exceeding a selected threshold h, can be considered to be a Poissonian flow. The distribution of tsunami inter-event times only differs substantially from a stationary Poisson process for short inter-event times, i.e., for frequent and weak events (GEIST 2008). So, the Poissonian probability P_n of the

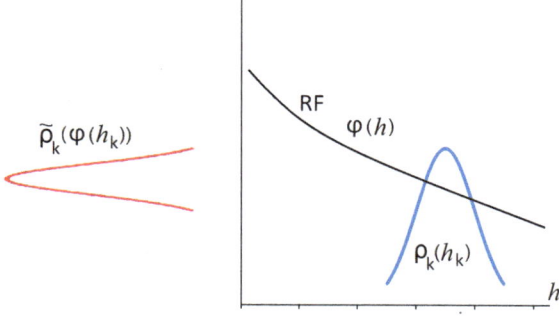

Figure 2

Schematic relationship between two probability density functions ρ_k (h_k) and $\tilde{\rho}_k(\phi_k)$ for the ordered tsunami run-up height and corresponding frequency

occurrence of n tsunamis with a maximum run-up height being equal to or more than the "threshold" height h during time period T at a selected location is given by the formula (FELLER, 1968):

$$P_n(>h) = \frac{[\varphi(h) \cdot T]^n}{n!} \cdot e^{-\varphi(h) \cdot T}, \qquad (6)$$

where $\varphi(h)$ is the tsunami RF. Starting from an ordered set of tsunami heights $h_1 \geq h_2 \geq h_3 \geq \cdots$ we should find corresponding ordinates for tsunami frequencies $\varphi_1 \leq \varphi_2 \leq \varphi_3 \leq \cdots$. To create a linear regression model with a semi-logarithmic scale, these ordinates should be average values $\overline{\ln \phi_k}$. Using the ordinary least squares method (OLSM) for the tsunami RF approximation is not correct (HIMMELBLAU, 1970), because variations of values $\ln \varphi_k$ are not equal (i.e., $D(\ln \varphi_k) \neq D(\ln \varphi_i)$ for $k \neq i$), and they are also correlated. The equation for the linear regression model should be the following:

$$\overline{\ln \varphi_k} = a + b \cdot h_k + e_k, \qquad (7)$$

where e_k are centered casual differences, and the weighted least squares method (WLSM) should be used for estimation of the tsunami activity parameters a and b. The maximum likelihood method (MLM) can be used also for this goal (CHARNES et al., 1976; NAYLOR et al., 2009).

Parameter $a = \ln f$ is the position at which the regression line crosses the ordinate axis, and $b = -1/H^*$ is the slope coefficient of the regression line. The statistical approach used is illustrated in Fig. 2.

Each value of h_k is a random value with a probability density function (PDF) $\rho_k(h_k)$ (colored

blue), which is dependent on an unknown RF $\varphi(h)$. Such PDFs are wide for small values of k (maximum tsunami heights are not stable), and they are narrow and sharp for large values of k. Each tsunami height h is related to a corresponding frequency $\varphi(h)$, and each PDF $\rho_k(h_k)$ is related to a distribution $\widetilde{\rho_k}(\varphi_k)$ (colored red) of corresponding frequencies $\varphi_k = \varphi(h_k)$. The distribution $\widetilde{\rho_k}(\varphi_k)$ is created from two unknown functions, $\rho_k(h_k)$ and $\varphi(h)$, but it is not dependent upon the shape of the unknown tsunami RF $\varphi(h)$, and can be calculated analytically.

Consider the cumulative distribution function $P(k; h)$ for the ordered tsunami height h_k. The situation $\{h_k \leq h\}$ can be realized if the number of tsunamis with height exceeding the threshold height h is less than $(k - 1)$ and the needed probability is the sum:

$$P(k; h) = P(h_k \leq h)$$
$$= \sum_{s=0}^{k-1} P_s(>h) = e^{-\varphi(h) \cdot T} \sum_{s-0}^{k-1} \frac{[\varphi(h) \cdot T]^s}{s!}. \qquad (8)$$

The tsunami RF $\varphi(h)$ is monotone decreasing and $P(k;h)$ can be considered as the cumulative distribution function for the frequency $\varphi_k = \varphi(h_k)$ corresponding to change of variable $h_k \rightarrow \varphi(h_k)$. The final formula for statistical moments is the following:

$$\overline{\varphi(h_k)^m} = \int (\varphi(h_k))^m \cdot d(P(k; h_k)) = M_m(k, T), \qquad (9)$$

where the differential is given by the formula

$$d(P(k; h_k)) = -e^{-\varphi(h_k) \cdot T} \frac{[\varphi(h_k) \cdot T]^{k-1}}{(k-1)!} T \cdot d(\varphi(h_k)). \qquad (10)$$

The independence of the distribution of $\widetilde{\rho_k}(\varphi_k)$ and related statistical moments from the unknown RF $\varphi(h)$ is related to the formulae (6) and (8): $P(k; h)$ is dependent on $\varphi(h)$, but is not directly dependent on tsunami height h. Because of this, the unknown tsunami RF $\varphi(h)$ can be used in (9) as an integrating parameter.

Actually, for creation of the regression line using the WLSM, the average values and variations of frequencies φ_k need to be calculated:

Pure Appl. Geophys.

$$\overline{\varphi_k} = \overline{\varphi(h_k)} = -\int \varphi(h_k) \cdot e^{-\varphi(h_k) \cdot T} \frac{[\varphi(h_k) \cdot T]^{k-1}}{(k-1)!} T$$

$$\cdot \, d(\varphi(h_k)) = |\text{substitution} : y = \varphi(h_k), 0 < y < \infty|$$

$$= \frac{T^k}{(k-1)!} \int_0^\infty y^k e^{-Ty} dy = \frac{k}{T} \qquad (11)$$

All the integrals with differential (10) representing average values and its variations in (11–14) are Laplace transforms (BATEMAN and ERDELYI, 1954; http://eqworld.ipmnet.ru/en/auxiliary/auxinttrans.htm).

$$D(\varphi_k) = \overline{(\varphi_k)^2} - \overline{(\varphi_k)}^2 = \frac{k}{T^2}. \qquad (12)$$

Using a logarithmic scale, the formulae are as follows (KAISTRENKO, 1989):

$$\overline{\ln \varphi(h_k)} = \sum_{s=1}^{k-1} \frac{1}{s} - 0.577\ldots - \ln T, \left\{ \underset{k \to \infty}{\to \ln(k/T)} \right\}, \qquad (13)$$

$$D_k = D(\ln \varphi(h_k)) = \frac{\pi^2}{6} - \sum_{s=1}^{k-1} \frac{1}{s^2} \{ \underset{k \to \infty}{\to 0} \}. \qquad (14)$$

It is interesting to note that variations $D(\ln \varphi_k)$ of $\ln \varphi_k$ are not dependent on observation time T. The standard deviation $\sigma(\ln \varphi_k)$ can be used for estimating relative a priori errors for recurrence frequencies:

$$\sigma(\ln \varphi(h_k)) = \sqrt{\frac{\pi^2}{6} - \sum_{s=1}^{k-1} \frac{1}{s^2}}. \qquad (15)$$

These values are large for the first order numbers $k = 1, 2,\ldots$ and the maximal one $\sigma_1 = \sigma$ (ln φ_1) = 1.28 is related to an event with maximum run-up height h_1. Large values of the firsts variations D_1, $D_2,..$ means that evaluation of the logarithm of empirical frequencies related to maximal events is unstable, and the estimate of the recurrence period corresponding to the maximum observed tsunami height is not correct. The range of permissible values is too large ($T/3.6$—$3.6 \cdot T$), where T is the observation period and $\exp(\sigma_1) \approx 3.6$. Developing technology is the answer to the problem of error estimation for tsunami recurrence periods, as indicated by CAMFIELD (1978).

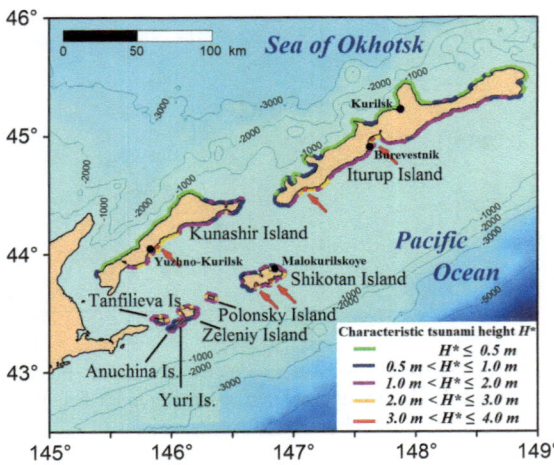

Figure 3
Region of the Southern Kuril Islands with distribution of the characteristic tsunami height H^* (KAISTRENKO et al., 2009). Red arrows denote positions of the coastal parts with the highest level of tsunami hazard

Data analysis shows that an exponential approximation can be used as an asymptotic estimate of a regional empirical tsunami recurrence function for big values of tsunami run-up heights (small tsunamis are not hazardous):

$$\varphi(x, h) = f \cdot e^{-\frac{h}{H^*(x)}}. \qquad (16)$$

In the case of $n = 0$, the formula (6) with (16) for the probability P_0 is:

$$P_0(> h) = e^{-f \cdot T \cdot e^{-\frac{h}{H^*(x)}}}. \qquad (17)$$

The latter formula is in accordance with a double negative exponential law for extreme values (GUMBEL, 1958; GALAMBOS, 1978).

Developing technology was applied to the Southern Kuril region (Fig. 3). An example of a cumulative tsunami RF for Yuzhno-Kurilsk, Kunashir Island, Kurile, is shown in Fig. 4. Generally, variation of tsunami activity parameters f and H^* is mostly accounted by a priori variations $D(\ln \varphi_k)$ of ln φ_k because these coordinates (h_k; ln φ_k), shown in Fig. 3, are located close to the regression line, and the ordinary least squares method gives too small, and too unrealistic, variations for the parameters f and H^*. In reality, such variations are not small.

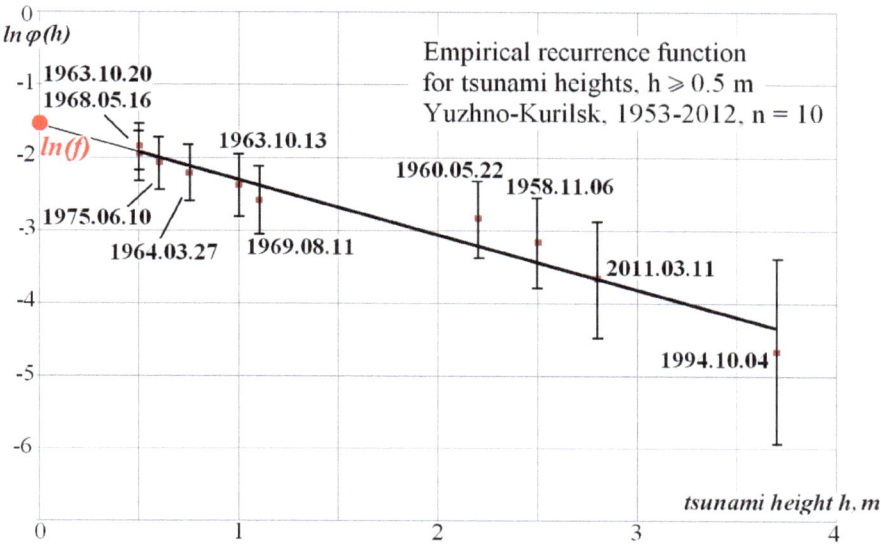

Figure 4

Empirical cumulative tsunami RF for Yuzhno-Kurilsk created using tsunami observational data with height $h \geq 0.5$ m during the time period 1953–2012. All the values $\overline{\ln \phi(h_k)}$ are accompanied by corresponding standard deviations. Asymptotical frequency of big tsunamis for Yuzhno-Kurilsk is $f = 0.16 \pm 0.24$ 1/year and its position is indicated by *red circle* in the crossing point of the ordinate axis by the regression *straight line*. Characteristic tsunami height for Yuzhno-Kurilsk is $H^* = 1.5$ m (the slope coefficient of the regression line) and $1/H^* = 0.7 \pm 0.2$ 1/m

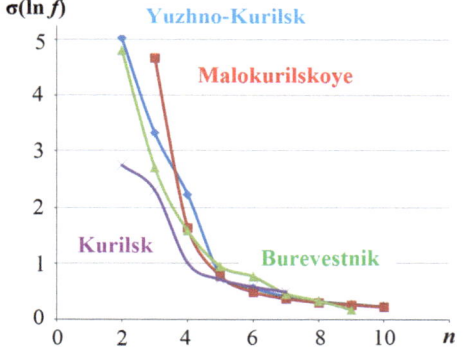

Figure 5

Dependence of standard deviation $\sigma(\ln f)$ of the estimated asymptotic tsunami frequency f from the number n of the used observational tsunami height data

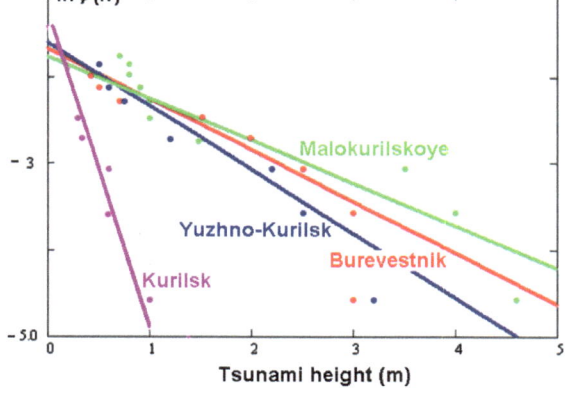

Figure 6

Tsunami RF for several locations in the region of the Southern Kuril Islands (KAISTRENKO, 2011)

The standard deviation $\sigma(\ln f)$ of the asymptotic tsunami frequency of big tsunamis f can be considered as an estimate of its relative error, and it depends very much on the number n of tsunami height observations (Fig. 5). Evaluation of the big tsunami frequency f with the value of $\sigma(\ln f)$ more than 1 (100 %) should be considered as unacceptable. Figure 5 shows that the number of tsunami height observations used for estimating the tsunami RF for selected locations should be not <5.

The asymptotic frequency of big tsunamis f can be considered to be constant for a region with a length of a few tens of kilometers, because of the large wave length of tsunamis in the open ocean. In fact, this might also apply to substantially larger regions, (see Figs. 3, 6).

Table 1

Parameters of tsunami activity f and H*, and their standard deviations, for several locations in the region of the Southern Kuril Islands

Locations	Ln f	f, 1/year	$\sigma(\ln(f))$	$1/H^*$, m^{-1}	$\sigma(1/H^*)$	H^*, m	$\delta H^*/H^*$
Burevestnik	−1.7	0.14	0.3	0.6	0.19	1.6	0.3
Kurilsk	−1.4	0.24	1.4	3.3	2	0.3	0.6
Malokurilskoye	−1.6	0.15	0.2	0.5	0.15	2.0	0.3
Yuzhno-Kurilsk	−1.6	0.16	0.2	0.7	0.2	1.5	0.3

Tsunami RF figures for several locations in the region of the Southern Kuril Islands, and corresponding parameters of tsunami activity, are shown below in Fig. 6 and Table 1, respectively.

The asymptotic tsunami frequencies for all locations considered are approximately the same, in frames of corresponding standard deviations $\sigma(f)$. The size of this region is more than 200 km. The product of $\sigma(1/H^*) \cdot H^*$ can be considered to be a relative error $\delta H^*/H^*$ of characteristic tsunami height H^* related to its standard deviation, and all these values are not small.

4. Tsunami Recurrence Function for Small and Moderate Tsunamis

There are different analytical approximations of tsunami RF for the small and large tsunami heights. WIGEN (1983), OKADA and TADA (1983), and BURROUGHS and TEBBENS (2005) used the power function $\varphi(h) = Ch^\alpha$ as an approximation of tsunami RF, and their analysis of observational data gave values of α in the range of −1.3 to −0.6. Taking into account estimation errors, this parameter should be considered as a uniform parameter, so that $\alpha = -1$. With this value, C has the physical dimension cm/year for a speed. Non-integer values of α do not have a straightforward physical interpretation. All other values around −1 do not result in a physical dimension for the parameter C (KAISTRENKO, 2011). The power law for tsunami heights is created by the propagation of tsunamis in the shelf zone and near the coast, where the seabed has an irregular topography, and, because of this, the power law for tsunami RF is related to log-normal tsunami height distribution along the coast (KAJIURA, 1983; CHOI et al., 2002, 2006).

The power law provides a good analytical approach for estimating RF for small and medium-sized tsunamis, and it was used for the general tsunami data sets because most tsunami heights are related to small and moderate events. It is important to note that the power law in the form Ch^α cannot be valid for large tsunami heights because of energetic restrictions if $\alpha > -2$. Power law uses special functions which contain only one scale parameter C. The parameter C is expressed in cm year^{-1} and has values of several cm year^{-1} evaluated for coastal locations (WIGEN, 1983; OKADA and TADA, 1983; BURROUGHS and TEBBENS, 2005). Tsunami heights in the nearest parts of ocean are approximately ten times less than in the coastal area and corresponding values of parameter C for tsunami RF for the ocean area should be of several mm per year. Such values coincide with the rate of vertical deformation of the Earth's crust in seismically active tsunami source zones located in the ocean, and this appears to be a causal connection.

5. General Problem of Tsunami Activity Parameters: Esteva's Principle and Kagan's Universality Applied to the Tsunami Recurrence Function

There is an essential question: how many parameters are needed, in addition to parameters f and H^*, to adequately describe tsunami activity? Generally, tsunami RF can depend on many other parameters:

$$\varphi = \varphi(h; c_1, \ldots, c_n). \qquad (18)$$

We propose that physical processes causing tsunamis are similar everywhere, and that recurrence functions for different regions should be the same, but with different sets of parameters. This was described by KAGAN (1999, 2010) as "universality" for earthquake RF. Earlier, a similar idea was proposed by ESTEVA (1976).

Consider the big tsunamigenic zone C to be divided into two parts: A and B. Therefore, partial recurrence functions are dependent on sets of parameters $(a_1,...,a_n)$ and $(b_1,...,b_n)$. The full recurrence function for a given location is the sum of the partial ones:

$$\varphi(h; c_1, \ldots, c_n) = \varphi(h; a_1, \ldots, a_n) \\ + \varphi(h; b_1, \ldots, b_n), \qquad (19)$$

and all the parameters c_k are dependent on the parameters $a_1,...,a_n; b_1,...,b_n.$

$$c_k = c_k(a_1, \ldots, a_n; b_1, \ldots, b_n) \qquad (20)$$

Mathematically, the latter equation introduces a commutative semi-group (JACOBSON, 2009):

$$c = a \cdot b \qquad (21)$$

because all the considered sets of parameters are objects of the same type. We can propose that $c(a; b)$ is a smooth function, and identify element (or elements) i as a set of parameters $i = (i_1,...,i_n)$ which is related to a non-tsunamigenic zone, and, because of it, there are $\varphi(h; i) = 0$ and $c(a; i) = a$. A derivative of $c(a; i)$ with respect to a gives the Jacobi matrix $c'_a(a; i) = I$ (BULDYREV and PAVLOV, 1985). The latter formula means that the Jacobi matrix $c'_a(a; b)$ has an inverse one in the vicinity of $b = i$.

Partial differentiations of (19) with respect to a and b give:

$$\varphi'_c(h; c) \cdot c'_a(a; b) = \varphi'_a(h; a) \overset{def}{=} z(h; a) \\ \varphi'_c(h; c) \cdot c'_b(a; b) = \varphi'_b(h; b) \overset{def}{=} z(h; b) \qquad (22)$$

In the vicinity of $b = i$ (22) gives:

$$z(h; b) = z(h; a) \cdot \left\{ \left[c'_a(a; b) \right]^{-1} \cdot c'_b(a; b) \right\}. \qquad (23)$$

Let parameter a be fixed in formula (23). It can be done because $z(h; b)$ is not dependent on a. The last equation means that $z(h, b)$ is the linear combination of functions depending on h only with coefficients depending on b only. So the structure of the recurrence function has a form similar to the function shown in (23):

$$\varphi(h, c) = \sum_{k=1}^{m} f_k(c) \cdot F_k(h). \qquad (24)$$

All the coefficients f_k are frequencies ($m \le n$) which can be considered to be more appropriate parameters than $(c_1,...,c_n)$. The dimensionless structure of (22) can be expressed in more detail, as follows:

$$\varphi(h) = \sum_{k=1}^{m} f_k \Phi_k \left(\frac{h}{H_k^*} \right). \qquad (25)$$

The latter formula can be interpreted as the sum of accounts of several parts of the tsunamigenic zone. According to universality (or Esteva's principle) all the partial functions Φ_k should be the same and the final structure of the recurrence function is as follows:

$$\varphi(h) = \sum_{k=1}^{m} f_k \Phi \left(\frac{h}{H_k^*} \right). \qquad (26)$$

Tsunami RF for the small and moderate tsunami heights can be represented by the power function $\varphi(h) = Ch^{-1}$. In this case each partial tsunami RF is a power function because it is created by the same process in the near coastal area (KAISTRENKO, 2011) and $f_k \Phi \left(h/H_k^* \right) = C_k h^{-1}$.

Obviously, the relative account of each sub-zone in the RF for the small and moderate tsunami heights should be constant being not dependent on the tsunami height threshold h:

$$\psi_{kj}(h) = \frac{f_k \Phi \left(\frac{h}{H_k^*} \right)}{f_j \Phi \left(\frac{h}{H_j^*} \right)} = \frac{C_k}{C_j}. \qquad (27)$$

Generally, a ratio similar to (27) can be used as an indicator of workability of the power law with arbitrary exponent α for the general representation of the tsunami RF.

6. Tsunami Recurrence Function for Large and Catastrophic Tsunamis Applied to the Sanriku Coast, NE Honshu

Let us consider the ratio of the two partial tsunami RF values for two sub-zones, $\Psi_{12}(h) = f_1 \Phi \left(h/H_1^* \right) / f_2 \Phi \left(h/H_2^* \right)$, as being a function of

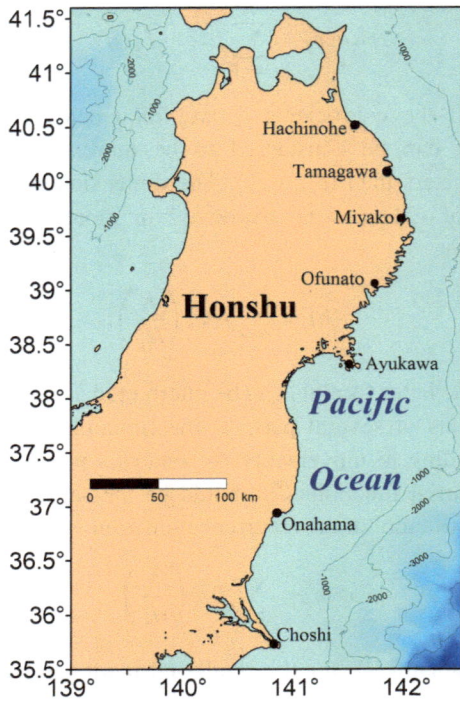

Figure 7
NE coast of the Honshu

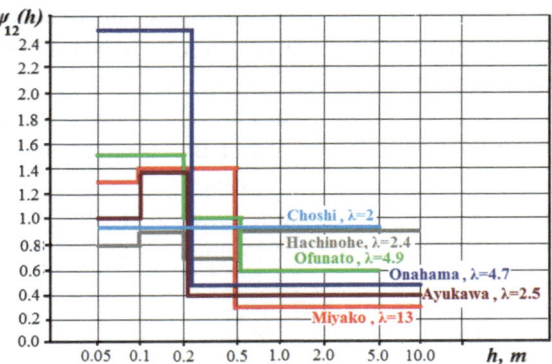

Figure 8
Relative tsunami RF $\Psi_{12}(h)$ for several locations on the Sanriku coast, NE Honshu

tsunami height threshold h. The function $\Psi_{12}(h)$ cannot be a needed indicator for the tsunami RF if $H_1^* = H_2^*$, because in this case the function $\Psi_{12}(h)$ is independent of the form of the function Φ, having the same arguments in the numerator and denominator in (27). To investigate possible deviation of the tsunami RF from the power law, the corresponding characteristic tsunami heights should be essentially different, for example $H_1^* \ll H_2^*$. So, the function $\Psi_{12}(h)$ is dependent very significantly on the ratio $\lambda = (H_1^*/H_2^*) \neq 1$ especially for the large values of tsunami height threshold h.

In the case $\lambda > 1$, the numerator in $\Psi_{12}(h)$ is decreasing faster than the denominator, and if the maximum possible tsunami height h_{mp} exists, then the numerator reaches zero faster than the denominator, and we can set $\Psi_{12}(h)$ to zero for more large tsunami height values (see part 2).

The function $\Psi_{12}(h)$ can be examined using the observed tsunami heights on the Sanriku coast in NE Honshu (Fig. 7), including data related to three catastrophic events in 1896, 1933 and 2011. Let the first sub-zone contain only far

sources and the second one contain only sources being close to the given coastal location. Obviously, there is $H_1^* < H_2^*$ for locations on the Sanriku coast.

The boundary distance $R = 600$ km dividing the Pacific tsunamigenic zone is elected to have an approximately equal number of tsunamis generated in its nearest and distant parts. Of course, this might be different for the different coastal locations. The tsunami data sets from Tsunami Data and Information (http://www.ngdc.noaa.gov/hazard/tsu.shtml) for six locations on the Sanriku coast were used for this analysis. Selection of these locations was governed by the need to have a large number of observational tsunami heights, including data related to the greatest nearest (1896, 1933 and 2011) and furthest (Chilean 1960 tsunami) events for this coast; (tsunami data sets for Onahama and Choshi do not contain tsunami heights for 1933 and 1896, respectively). $\Psi_{12}(h)$ figures and related information are shown in Fig. 8 and Table 2. To avoid instability of ratios, if the number of nearest sources (denominator) is equal to 4 or fewer, we have combined the corresponding ranges to the neighboring ones.

The depth of historical experience residing in Japanese tsunami data for this coast enables one to conclude that all catastrophic tsunamis with run-up heights of more than 20 m were related only to sources located in the nearest parts of the tsunamigenic zone. The maximum historical tsunami height related to a far source (the 1960 Chilean event) was 8.1 m, at Tamagawa, in the Northern part of the

Table 2

Distribution of the number of tsunamis with heights in specified ranges, related to distant and nearest sources for several locations on the Sanriku coast, NE Honshu

Coastal locations and max tsunami heights related to the nearest/ distant sources	Number of tsunamis with heights in the indicated ranges, related to distant/nearest sources, with ratios						
	5–10 cm	11–20 cm	21–50 cm	51–100 cm	1–5 m	5–10 m	>10 m
Hachinohe 9 m/3.8 m = 2.4	10/13 0.8	10/11 0.9	8/11 0.7	4/2 0.9	2/4	0/1	0/0
Miyako 12 m/0.9 m = 13	16/12 1.3	12/4 1.4	5/8	2/2 0.3	0/5	0/1	0/1
Ofunato 24 m/4.9 m = 4.9	8/8 1.5	10/4	5/5 1	½ 0.6	2/3	0/0	0/1
Ayukawa 7.6 m/3.1 m = 2.5	13/13 1	13/9 1.4	12/13 0.9	2/3 0.4	1/4	0/1	0/0
Onahama 8 m/1.7 m = 4.7	14/9 2.5	18/4	3/7 0.5	½	1/0	0/2	0/0
Choshi 3 m/1.5 m = 2	6/7 0.9	4/4 0.9	6/2	0/0	1/2	0/0	0/0

Sanriku coast. Catastrophes on this coast related to far-field tsunamis were not registered during many centuries. It means that such events are extremely rare or absent. Because of this, we can set $\lim_{h \to \infty} \Psi_{12}(h) = 0$ as the related mathematical model.

Let us consider the analytical properties of the limited positive function $\Phi(x)$, satisfying the condition:

$$\lim_{x \to \infty} \frac{\Phi(\lambda x)}{\Phi(x)} = 0, \lambda = \frac{H_2^*}{H_1^*} > 1. \tag{28}$$

Functions of this type should be decreasing, because condition (28) means that the ratio $\Phi(\lambda x)/\Phi(x)$ is less than 1 for large values of x and $\lim_{x \to \infty} \Phi(x) = 0$. Obviously, this ratio decreases faster for larger values of parameter λ. For coarse evaluation of parameter λ, the ratio of maximum observed tsunami heights from nearest to distant sources can be used. Tsunami data for Hachinohe ($\lambda \approx 2.4$) and Choshi ($\lambda \approx 2$) do not demonstrate any decrease in the corresponding value of $\Psi_{12}(h)$ for $h < 5$ m. Function $\Psi_{12}(h)$ for Miyako ($\lambda \approx 13$) and Onahama ($\lambda = 4.7$) demonstrate a very fast decrease when $h > 0.5$ m.

Interestingly, the function $\Phi(x) \cdot x^\alpha$ with arbitrary α satisfies the same condition (28), and consequently has the same properties. This enables one to conclude that tsunami RF for large values of tsunami height decreases faster than each power function. Because of this, a negative exponential function like (16) can provide a good analytical approximation for the right tail of the tsunami RF.

7. Some Additional Remarks Concerning the Tsunami Recurrence Function

We can assume physical limitations on tsunami height, but statistical procedures for evaluation of the maximum possible tsunami height are not stable. This is because such a maximum can substantially exceed an empirical maxima (PISARENKO and RODKIN, 2010).

The general form of the tsunami RF for small and medium tsunami heights contains only one parameter of tsunami activity:

$$\varphi(h) = \sum_{k=1}^{m} C_k \cdot h^{-1} = \left(\sum_{k=1}^{m} C_k \right) \cdot h^{-1} = Ch^{-1}. \tag{29}$$

Decreasing $\Psi_{12}(h) \to 0$ for $\lambda > 1$ means that all the parts of equation (26), except a part (or parts) with the maximum value of the corresponding H^*, can be ignored when considering the biggest tsunami heights. Let the H_1^* in (26) be the biggest one, then because of (28) there is the formula:

$$\varphi(h) = \sum_{k=1}^{m} f_k \Phi\left(\frac{h}{H_k^*}\right)$$

$$= f_1 \Phi\left(\frac{h}{H_1^*}\right) \cdot \left[1 + \sum_{k=2}^{m} \frac{f_k \Phi\left(\frac{h}{H_k^*}\right)}{f_1 \Phi\left(\frac{h}{H_1^*}\right)}\right] \rightarrow f_1 \Phi\left(\frac{h}{H_1^*}\right)$$

$$\tag{30}$$

So, tsunami activity for the biggest tsunami heights is described by two parameters f_1 and H_1^*. The corresponding part (or parts) of the tsunamigenic zone with the largest value of the corresponding H^* should be considered to be the most tsunami hazardous for a given location. The case of the Sanriku coast shows that the most tsunami hazardous zone there is the nearest part of the tsunamigenic zone. For example, the situation for Hawaii looks unusual. Approximately the same historical maximum tsunami heights (Tsunami Data and Information) there were related to at least as far as local sources, i.e., 1946 Alaska tsunami (16.8 m), 1903 Hawaiian tsunami (15.7 m), 1952 Kamchatka tsunami (9.0 m) and Chilean tsunami (9.5 m). So, almost the entire Pacific tsunamigenic zone is very hazardous for Hawaii.

The analysis carried out in part 6, for the Sanriku coast, cannot be realized for most other coastal locations because of historical tsunami data shortfall. The hope for the needed data is related to the paleotsunami investigations (BOURGEOIS, 2009; CHAGUE-GOFF et al., 2011; DAWSON and SHAOZHONG, 2000; MINOURA et al., 2001; NISHIMURA, 2008; RAZJIGAEVA et al., 2013; Tsunami Data and Information). It should be taken into account that the paleotsunami data are specific. Historical tsunami run-up heights contained in tsunami catalogues are heights reached by tsunamis, but paleotsunami data are related to levels which have been exceeded by tsunamis. Developing technology can use both kinds of data for calculating the tsunami RF (KAISTRENKO and KLYACHKO 2003).

It is probable that both asymptotics for small and large tsunami heights can be combined in gamma-distribution $\sim h^{-1} e^{-\beta h}$.

8. Probabilistic Forecasting of Tsunami Run-up Heights and its Practical Use

The probabilistic approach to tsunami hazard estimation, based on historical tsunami data from catalogues, was developed by several authors (GO et al., 1985, PELINOVSKY, 1999; KAISTRENKO and KLYACHKO 2003; KULIKOV and RABINOVICH 2005). For many coastal areas, there are no sufficient and reliable historical tsunami data. In this case, a method called probabilistic tsunami hazard assessment (PTHA), based on probabilistic seismic hazard assessment (PSHA), is used (GEIST and PARSONS, 2006; GONZALEZ et al., 2009).

Starting from physical parameters H^* and f of tsunami activity, received by arbitrary methods, we can estimate tsunami hazard and risk. Estimation of the tsunami hazard R_T related to exceeding a given level of h in location x by any tsunamis during the time period T can be achieved using formula (17):

$$R_T(h) = 1 - e^{-f \cdot T \cdot e^{-\frac{h}{H^*(x)}}}. \tag{31}$$

Having parameters H^* and f for any coastal location, the probable maximum tsunami height h_T with recurrence period T can be calculated as,

$$h_T = H^* \cdot \ln(f \cdot T). \tag{32}$$

An example of the synoptic map with a distribution of H^* for the Southern Kuril Islands is shown in Fig. 3.

For practical purposes, the inverse approach is more useful: the permissible hazard R and the needed time period T should be defined by developed tsunami-codes, and then the "safe" level h_R for a given location can be estimated by the formula:

$$h_R = H^* \cdot [\ln(f \cdot t) - \ln(-\ln(1 - R))]. \tag{33}$$

According to this formula, the "safe" level h_R is the sum of two parts: the first is the permanent hazardous level, which depends on tsunami activity parameters and time period T, and the second is dependent on characteristic tsunami height H^* and permissible hazard R.

9. Conclusions

Investigation of the structure of tsunami RF has shown that it depends on at least two parameters: f and H^*. The first of these is the asymptotic frequency of big tsunamis, which varies very slightly within a region with a length of few hundreds of kilometers and can be considered as being constant for such region. Parameter H^* is the characteristic tsunami height for a selected location x, which should be proportional to the average coefficient $K(x)$ of the tsunami height transformation (amplification) from the open ocean to the coastal location x. Generally, this parameter can be calculated using numerical modeling. Regional tsunami RF depends on only one combined variable h/H^*, where h is the "threshold" height of tsunamis. The method for correct statistical evaluation of the parameters f and H^* with their variations using observational data from tsunami catalogues has been developed.

An investigation of the general problem of tsunami activity parameters has led to the creation of a more detailed form of the tsunami RF. It has been shown that the power law, as a representation of the tsunami RF, is valid only for small and medium tsunami heights. Tsunami RF for large tsunami height values decreases faster than each power function. Because of this, the negative exponential function can provide a good analytical approximation for the right tail of the tsunami RF.

The problem of probabilistic tsunami hazard estimation has also been considered.

Acknowledgments

I would like to thank my anonymous reviewers for their helpful and productive comments. The work was made within the frames of scientific project 12-I-П4-06 the Fat Eastern Division of RAS and was partially supported by the Russian Foundation for Basic Research, project 11-05-01054-a.

References

BATEMAN, H. and ERDELYI, A., *Tables of Integral Transforms. Vol. 1*, (McGraw-Hill Book Co., New York, 1954).

BULDYREV V. S. and PAVLOV B. S. *Linear algebra and functions of several variables* (Leningrad State Univ. Publ, Leningrad, 1985), 496 pp. (in Russian).

BOURGEOIS, J. (2009), Geological effects and records of tsunamis. In *The Sea, Volume 15: Tsunamis* (Robinson, A.R. and Bernard, E.N., eds., Harvard University Press, 2009), p. 53–91.

BURROUGHS, S.M., and TEBBENS, S.F. (2005), *Power-law scaling and probabilistic forecasting of tsunami runup heights*, Pure Appl. Geophys. 162, 331–342.

CAMFIELD, F. (1978), *Insufficient Data Effect on Tsunami Flood Level Predictions—Summary*. In Proc. Int. Tsu. Symp, Seattle, August 17–19, 1978, pp. 247–251.

CHAGUE-GOFF, C., SCHNEIDER, J-L., GOFF, J., DOMINEY-HOWES, D., and STROTZ, L. (2011), *Expanding the proxy toolkit to help identify past events—Lessons from the 2004 Indian Ocean Tsunami and the 2009 South Pacific Tsunami*. Earth-Science Reviews, 107, 107–122.

CHARNES, A., FROME, E. L., YU, P. L. (1976), *The equivalence of generalized least squares and maximum likelihood estimates in the exponential family*, Journal of the American Statistical Association, 71, pp. 169–171.

CHOI, B.H, PELINOVSKY, E., RYABOV, I., and HONG, S.J. (2002), *Distribution functions of tsunami wave heights*, Natural Hazards, 25, 1–21.

CHOI, B.H, HONG, S.J., and PELINOVSKY, E. (2006), *Distribution of runup heights of the December 26, 2004 tsunami in the Indian Ocean*. Geoph. Res. Lett. 33, L13601, doi:10.1029/2006 GL025867.

DAWSON, A., and SHAOZHONG, S. (2000), *Tsunami Deposits*. Pure appl. geophys., 157, 875–897.

ESTEVA, L. (1976), *Seismicity*. In Seismic Risk and Engineering Decisions (eds. C. Lomnitz and E. Rosenbleuth) pp. 179–224, (Elsevier Scientific Publishing Company, Amsterdam, 1976).

FELLER, W. 1968), *An Introduction to Probability Theory and Its Applications*, Vol. 1, 3rd Edition (John Wiley and Sons, Inc., New York, 1968), 528 pp.

GALAMBOS, J. (1978), *The asymptotic theory of extreme order statistics*, (John Wiley and Sons, New-York-Chichester-Brisbane-Toronto, 1978), 302 pp.

GEIST, E., and PARSONS, T. (2006), *Probabilistic Analysis of Tsunami Hazards*, Natural Hazards, 37, 277–314.

GEIST, E. L., and PARSONS T. (2008), *Distribution of tsunami interevent times*, Geophys. Res. Lett., 35, L02612, doi:10.1029/2007GL032690.

GO, CH. N., KAISTRENKO, V. M. and SIMONOV, K. V. (1985), *A Two—Parameter Scheme for Tsunami Hazard Zoning*, Marine Geodesy, 9, N 44, 469–476.

GONZALEZ, F. I., GEIST, E. L., JAFFE, B., KANOGLU, U., MOFJELD, H., SYNOLAKIS, C. E., TITOV, V. V., ARCAS, D., BELLOMO, D., CARLTON, D., HORNING, T., JOHNSON, J., NEWMAN, J., PARSONS, T., PETERS, R., PETERSON, C., PRIEST, G., VENTURATO, A., WEBER, J., WONG, F., and YALCINER, A. (2009), *Probabilistic tsunami hazard assessment at Seaside, Oregon, for near- and far-field seismic sources*, J. Geoph. Res., 114, C11023.

GUMBEL, E.J. (1958), *Statistics of Extremes*, (Columbia University Press. New York, 1958), 375 pp.

GUTENBERG, B., and RICHTER, C.F. (1954), *Seismicity of the Earth and Associated Phenomena*, 2nd ed. (Princeton University Press, Princeton, 1954), 310 pp.

HIMMELBLAU, D. M. (1970), *Process Analysis by Statistical Methods*, (John Wiley and Sons, Inc., New York, 1970), 464 pp.

Historical Tsunami Database for the World Ocean (HTDB/WLD). http://tsun.sscc.ru/htdbpac/.

JACOBSON, N. (2009), *Basic Algebra I* (Dover Publications; 2 edition, June 22, 2009), 528 pp.

KAGAN, Y.Y. (1999), *Universality of the seismic-moment-frequency relation*, Pure Appl. Geophys. *155*, pp. 537–573.

KAGAN, Y.Y. (2010), *Earthquake size distribution: Power-law with exponent?*, Tectonophysics, *490*, Issues 1–2, pp. 103–114.

KAISTRENKO, V. (1989), *Probability model for tsunami run-up.*, In *Proc. International Tsunami Symposium. Novosibirsk, 1989*, pp. 249–253.

KAISTRENKO, V. (2011). *Tsunami Recurrence versus Tsunami Height Distribution along the Coast*, Pure Appl. Geophys. *168*, pp. 2065–2069.

KAISTRENKO, V., PINEGINA T., and KLYACHKO, M. (2003), *Evaluation of tsunami hazard for the Southern Kamchatka coast using historical and paleotsunami data. In Underwater Ground Failures on Tsunami Generation, Modelling, Risk and Mitigation.* (NATO Advanced Research Workshop, May, 23–25, 2001, Istanbul, Turkey, ed. by Yalciner A.C., Pelinovsky E., Synolakis C.E., Okal E.), Kluwer, 2003, p. 225–235.

KAISTRENKO, V., IVASHCHENKO, A., KHRAMUSHIN, V., and ZOLOTUK-HIN, D. (2009), *Tsunami Hazard for the Northern Kuril Islands.* In Atlas of Kuril Islands (eds. Kotlyakov, V. M., Komedchikov, N. N., Fedorova E. Ya.), Publishing Center "DIK", Vladivostok, 2009, p.137 (in Russian).

KAJIURA, K. (1983), *Some statistics related to observed tsunami heights along the coast of Japan*, In *Tsunamis: Their Science and Engineering* (eds. K. Iida and T. Iwasaki) pp.131–145, (Terra Scientific Publishing Co., Tokyo, 1983).

KULIKOV, E. A., RABINOVICH, A. B., and THOMSON R. E. (2005), *Estimation of tsunami risk for the coasts of Peru and Northern Chile.* Nat Hazards, *35*, 185–209.

MAGOON, O. T., and ARNO, A. L. (1973). *Prediction of long term tsunami inundation of Crescent City, California, USA.* In *Tsunami Waves* (Proc. Intern. Tsunami Symposium, XV General Assembly of the International Union of Geodesy and Geophysics, Moscow, August 1971, ed. by S. Soloviev and V. Kaistrenko), (Sakhalin Complex Research Institute, Yuzhno-Sakhalinsk, 1973), pp. 168–182. (in Russian).

MINOURA, K., IMAMURA, F., SUGAWARA, D., KONO. Y., and IWASHITA, T. (2001), The 869 Jogan tsunami deposit and recurrence interval of large-scale tsunami on the Pacific coast of northeast Japan. Journal of Natural Disaster Science, *23, N. 2*, pp. 83–88.

NAYLOR, M., GREENHOUGH, J., MCCLOSKEY, J., BELL, A. F., MAIN, I. G. (2009), Statistical evaluation of characteristic earthquakes in the frequency-magnitude distributions of Sumatra and other subduction zone regions. Geophys. Res. Lett., *36*, doi:10.1029/2009GL040460.

NISHIMURA, Y. (2008), *Volcanism-Induced Tsunamis and Tsunamiites.* In *Tsunamiites* (Ed. by T. Shiki, Y. Tsuji, T. Yamazaki, and K. Minoura), Amsterdam—Boston—Heidelberg—London—New York—Oxford—Paris—San Diego—San Francisco—Singapore—Sydney—Tokyo), pp. 162–184.

OKADA, M. and TADA, M. (1983), *Historical study of tsunamis at Miyako, Japan.* In *Tsunamis: Their Science and Engineering* (eds. K. Iida and T. Iwasaki) pp. 121–130 (Terra Scientific Publishing Co., Tokyo, 1983).

PELINOVSKY, E., and MAZOVA, R. (1992), *Exact analytical solutions of nonlinear problems of tsunami wave run-up on slopes with different profiles.* Natural Hazards 6, 227–249, 1992.

PELINOVSKY, E. (1999), *Preliminary estimates of tsunami danger for the northern part of the Black Sea.* Phys Chem Earth A, *24(2)*,175–178.

PISARENKO, V., and RODKIN, M. (2010), *Heavy-Tailed Distributions in Disaster Analysis.* (Springer Science + Business Media B.V., 2010), 190 pp.

RAZJIGAEVA, N.G., GANZEY L.A., GREBENNIKOVA, T.A., IVANOVA E.D., KHARLAMOV, A.A., KAISTRENKO V.M. and SHISHKIN, A.A. (2013), *Coastal sedimentation associated with the Tohoku Tsunami of 11 March 2011 in South Kuril Islands, NW Pacific Ocean.* Pure Appl. Geophys., *170*, pp. 1081–1102.

SEDOV, L.I. (1993). Similarity and dimentional methods in mechanics, 10th edn. (CRS Press, Boca Raton, 1993), 479 pp.

Tsunami Data and Information http://www.ngdc.noaa.gov/hazard/tsu.shtml.

WIEGEL, R. L. (1964). *Tsunami information in regard to proposed nuclear power plant site, Pacific Gas and Electric Company, at Bodega Head, California.* (Berkeley, Calif.), 23 pp.

WIEGEL, R.L. (1965), *Protection of Crescent City, California, from Tsunami Waves*, Report for Redevelopment Agency, Crescent City. (University of California, Berkley, California, 1965).

WIGEN, S.O. (1983), *Historical study of tsunamis at Tofino, Canada.* In *Tsunamis: Their Science and Engineering* (eds. K. Iida and T. Iwasaki) pp. 105–119, (Terra Scientific Publishing Co., Tokyo, 1983).

(Received September 24, 2013, revised November 28, 2013, accepted February 4, 2014, Published online March 5, 2014)